METHODS IN PLANT PATHOLOGY

WITH SPECIAL REFERENCE TO BREEDING FOR DISEASE RESISTANCE

METHODS IN PLANT PATHOLOGY

WITH SPECIAL REFERENCE TO BREEDING FOR DISEASE RESISTANCE

by

Z. KIRÁLY, Z. KLEMENT
F. SOLYMOSY, J. VÖRÖS

Edited by

Z. KIRÁLY
Research Institute for Plant Protection, Budapest

ELSEVIER SCIENTIFIC PUBLISHING COMPANY
AMSTERDAM LONDON NEW YORK
1974

The distribution of this book is being handled by the following publishers:
for the U.S.A. and Canada

American Elsevier Publishing Company, Inc.
52 Vanderbilt Avenue
New York, New York 10017

for the East European countries, China, North Korea, Cuba, North Vietnam and Mongolia

Akadémiai Kiadó,
The Publishing House of the Hungarian Academy of Sciences,
Budapest

for all remaining areas

Elsevier Scientific Publishing Company
335 Jan van Galenstraat
P.O. Box 330, Amsterdam, The Netherlands

ISBN 0-444-99881-0

First edition 1970
New impression 1974

Joint edition published by Elsevier Scientific Publishing Company, Amsterdam, The Netherlands and Akadémiai Kiadó, The Publishing House of the Hungarian Academy of Sciences, Budapest, Hungary

Printed in Hungary

PREFACE

This book is first of all intended for research workers in plant pathology, who wish to become informed about the basic methods of plant virology, bacteriology, mycology and plant pathological aspects of breeding for disease resistance. We believe, this volume can help professional research workers as well as beginners in this field by providing them with a source of information directly applicable to the daily work in the laboratory and in the experimental field. To assist the research worker we have included schemes and tables on the composition of media, reagents, etc., and keys for the determination of physiologic races of the pathogens, thereby giving the feature of a guide-book. To assist the beginner in the field of plant pathology we have also included a series of simple figures and explanations on laboratory procedures to illustrate the basic important plant pathological methods.

We did not want to go into too much detail in a particular field of plant pathology, however, our aim was to collect in one volume the most important methods of plant virology, phytobacteriology, and plant pathological mycology. We felt that a complete book of this kind was needed by plant pathologists, plant physiologists, as well as plant breeders.

In this book F. Solymosy was responsible for the part on virology, Z. Klement for bacteriology, J. Vörös for mycology and certain typical fungus diseases, and Z. Király for disease resistance and certain typical fungus diseases.

The authors wish to express their gratitude to those who reviewed and commented upon various chapters: O. H. Calvert, Department of Plant Pathology, University of Missouri, I. A. M. Cruickshank, Commonwealth Scientific and Industrial Research Organization, Division of Plant Industry, Canberra, R. Gáborjányi, Research Institute for Plant Protection, Budapest, R. N. Goodman, Department of Plant Pathology, University of Missouri, R. E. Hampton, Department of Plant Pathology, University of Kentucky, B. H. MacNeill, Department of Botany, University of Guelph, Ontario, I. Manninger, Research Institute of Agriculture of the Hungarian Academy of Sciences, Martonvásár, J. S. Niederhauser, The Rockefeller Foundation, International Potato Program, Mexico, D. J. Paxton, Department of Plant Pathology, University of Illinois, G. Ubrizsy, Research Institute for Plant Protection, Budapest.

The Editor

CONTENTS

PART I

VIROLOGY

Introduction 19

CHAPTER I. Morphology and structure of plant viruses 21
1. Purification of plant viruses 21
Purification of TMV 24
Purification of TMV by differential centrifugation 25
Simple technique for purification of TMV 25
2. Morphology of plant viruses 26
Exudation method for the preparation of rod-shaped viruses for electron microscopy 27
Immersion method for the preparation of rod-shaped viruses for electron microscopy 27
Purification of rod-shaped viruses for electron microscopy 27
3. Structure of plant viruses 28
Chemical structure 28
Substructure 35

CHAPTER II. Biological activity of plant viruses 36
1. Ribonucleic acid as the biologically active component of plant viruses 36
Extraction of TMV-RNA from purified TMV preparations "Phenol-procedure" 38
Preparation of RNase sensitive infectious material from clarified sap of TMV infected tobacco leaves 39
Preparation of RNase sensitive infectious material from crude extracts of virus infected plants 40
Isolation of TMV-RNA by the use of bentonite 40
2. Factors influencing the biological activity of plant viruses 40
Physical factors 41
Determination of the dilution end point 42
Determination of the thermal inactivation point 42
Ageing *in vitro* 43
Chemical factors 44
Determination of the mode of action of inhibitors by a dilution method 45

CHAPTER III. Transmission of plant viruses 47
1. Graft transmission 47
Approach graft 47
Budding 47
2. Seed transmission 48
3. Mechanical transmission 49
A brush-extraction method for the mechanical transmission of plant viruses 50
Quick virus inoculation by rubbing with fresh leaf discs 51

Needle transmission of the virus causing rough dwarf disease of maize 52
Mechanical transmission of the virus causing infectious sterility of hops 54
Phosphate effect in mechanical virus transmissions 55
4. Insect transmission 55
Non-persistent viruses 56
Persistent viruses 57
Semi-persistent viruses 58
Transmission of the persistent potato leaf roll virus by aphids 60
Transmission of the non-persistent cucumber mosaic virus by aphids 60
5. Dodder transmission 61
6. Soil transmission 62

CHAPTER IV. Host-virus relations 64
1. Virus in the host 65
Penetration of plant viruses into host cells 65
Virus multiplication 65
Latin-square design 75
Half-leaf method 76
Agar-gel diffusion method 78
Agglutination test 78
Precipitation end-point method 80
Optimal proportions method 81
Translocation of plant viruses in their hosts 81
2. Symptomatology 82
Symptomatology *s. str.* 83
Starch-iodine lesions on leaves of Samsun tobacco plants infected with tobacco mosaic virus 84
Histopathology 85
Demonstration of amorphous inclusion bodies from epidermal cells of *Solanum nudiflorum* infected with a strain of tobacco mosaic virus (tomato aucuba virus) 86
Fixation of crystalline inclusions in the epidermis of tobacco leaves infected with tobacco mosaic virus 87
Pathophysiology 87
3. Environmental factors influencing host–virus relations 92
Physical factors 92
Chemical factors 93
Biological factors 95

CHAPTER V. Genetics of plant viruses 98
Cross protection test 100

CHAPTER VI. Nomenclature and taxonomy of plant viruses 101
1. Virus nomenclature 101
2. Virus classification 102

CHAPTER VII. Control of plant virus diseases 103
APPENDIX. Some representative plant virus diseases 104
Tobacco 104
Mosaic disease 104
Necrosis disease 106
Tomato 107
Mosaic disease 107

Aspermy disease 107
Cucumber 108
Mosaic disease 108
Potato 109
Mosaic disease 109
Leaf roll disease 110
Streak disease 111
5. Bean 112
Southern bean mosaic disease 112
6. Turnip 112
Yellow mosaic disease 112

PART II

BACTERIOLOGY

Introduction 117

CHAPTER I. General characteristics of plant pathogenic bacteria 119

CHAPTER II. The classification of phytopathogenic bacteria 125

CHAPTER III. Types of symptoms 127

CHAPTER IV. Isolation of the pathogen 135
Microscopic examination of the diseased plant 136
Microscopic examination of diseased tissues by means of stained sections (differential-staining) 137
General laboratory procedures 137
Sterilization of equipment 137
Preparation of media 138
Adjusting the pH of the medium 138
Sterilization of media 139
Preparation of pure cultures 140

CHAPTER V. Identification of the pathogen 142
Pathological test for picking the pathogen isolate 142
Cell and colony morphology 143
Hanging drop examination 144
Flagella stain 144
Gram stain 145
Colony-morphology 146
Biochemical characteristics of bacteria 146
Action on carbon compounds 147
Nitrate reduction 148
Gelatin breakdown 149
Ammonia production 149
Production of H_2S 150
Indole production 150
Hydrolysis of starch 150
Action on litmus milk 151
Acetylmethyl carbinol formation 152
Oxidase reaction 152

CHAPTER VI. Inoculation tests to determine the host range 153
Pretreatment of plants for inoculation 156
Preparation of the bacterial suspension and choosing of the inoculation method 156
Spraying of bacterial suspension onto the surface of the plant 156
Use of high pressure inoculation equipment 157
Injection of bacterial suspension into the intercellular spaces with a syringe 159
Rubbing of leaves 160
Pricking 160
Inoculation of cut tissues by bacterial suspension 161
Determination of inoculum cell number 161
Turbidimetric measurements 162
Plate-count technique 165

CHAPTER VII. Maintenance and preservation of bacterial cultures 167
Preservation of cultures by overlaying under mineral oil 167
Lyophilization 167
Collections of phytopathogenic bacteria 168

CHAPTER VIII. Serology 170
Preparation of the antigen for immunization 172
Immunization of the rabbit 172
Agglutination test 174
Precipitation test in agar-gel 176
Adsorption test 178
Rapid serodiagnostic tests to demonstrate bacterial infections from plant materials 179
A practical method for the demonstration of bacterial infection in bean seeds 179
A rapid serodiagnostic test for potato ring rot 180

CHAPTER IX. Bacteriophages 183
Plaque-count technique 187
Use of bacteriophages in plant pathology 189
Isolation of phages from infected plants 189
Isolation of plaques 191
Detection of an unknown pathogen by a known phage 191
Determination of the relationships of bacteria on the basis of phage sensitivity 192

CHAPTER X. The fate of the pathogen in the plant, the plant's defence mechanisms against bacterial infection 193
Preformed factors of resistance 193
Factors of induced resistance 194
Premunity 194
Hypersensitive reaction 195
Primary and secondary factors of pathogenesis 197
Wilt toxins and enzymes 197
Toxins causing necrosis 198
Factors inducing hypertrophy 199
Enzymes causing soft rot 200
Bacterial cell number determination from the living plant 201
The possibilities of plant infiltration 202
Culture filtrate preparation, filtration, centrifugation and washing of bacterial cells 202
Obtaining fluid from the intercellular spaces of the plant 204

CHAPTER XI. Spread of the pathogen in the field 206
Artificially induced epidemics in the field 207

CHAPTER XII. Control of plant pathogenic bacteria 209

APPENDIX. Some examples of bacterial diseases 211
Spot diseases 211
Bacterial blight of tomato and pepper (paprika) (*X. vesicatoria*) 211
Isolation of the pathogen 213
Artificial inoculation 214
Halo blight of beans (*P. phaseolicola*) 214
Symptoms 215
Differentiation of the four bean pathogens 216
Outline for differentiating the bacterial pathogens of beans 216
Die-back of shoots 217
Bacterial blossom blight and bacterial canker (*P. syringae* and *P. mors-prunorum*) 217
Symptoms 218
Isolation of the pathogen 219
Artificial inoculation 220
Soft rots 221
Blackleg and soft rot of potato tuber and other plants (*E. carotovora*) 221
Symptoms 221
The pathogen 222
Isolation of the pathogen 223
Artificial inoculation 224
Wilts 225
Bacterial wilt of beans (*C. flaccumfaciens*) 225
Symptoms 225
Isolation of the pathogen 226
Identification of the pathogen 226
Artificial inoculation 226
Resistance 227
Southern bacterial wilt of *Solanaceae*, other *Dicotyledoneae* and banana (*P. solanacearum*) 228
Symptoms 228
Isolation and identification of the pathogen 228
Biotypes 229
Artificial inoculation 229
Change of virulence and its demonstration 229
Proliferation of tissues 230
Crown gall (*A. tumefaciens*) 230
The pathogen 231
Isolation of the pathogen 232
Artificial inoculation 233

PART III

MYCOLOGY

Introduction 237

CHAPTER I. General characteristics of fungi 239
Vegetative mycelium 239
Reproductive organs 241
Asexual reproduction 242
Sexual reproduction 248

CHAPTER II. Systematic mycology 263
Naming of fungi 263
The phylogeny of fungi and the formation of the phylogenetic system 264

CHAPTER III. Isolation of phytopathogenic fungi 270

CHAPTER IV. Maintenance of fungi 273

CHAPTER V. Microscopic and other laboratory investigation methods 276

CHAPTER VI. Media used for the cultivation of fungi 281
Natural media 283
Synthetic media 285
Semi-synthetic media 287

CHAPTER VII. Apparatus for collecting and sampling fungus spores 288

PART IV

PRINCIPLES FOR BREEDING PLANTS RESISTANT TO PHYTOPATHOGENIC FUNGI

CHAPTER I. Variability and specialization of pathogens 291
Sexual recombination 294
Mutation 295
Adaptation 296
Transformation 298
Transduction 299
Heterokaryosis and parasexuality 300

CHAPTER II. Detection and identification of biotypes and races 302

CHAPTER III. Infection process of fungi 305

CHAPTER IV. Disease resistance of host plant 307
Detection of postinfectionally produced fungitoxic substances: Phytoalexins 309

CHAPTER V. Basic principles of breeding for disease resistance 312
The gene-for-gene theory of host-parasite relations 312
Problems of breeding for disease resistance 313
Multiline hybrid cultivars for stem rust resistance 315
Breeding for disease resistance if pathogens are stable in pathogenicity 316
Adult plant resistance 317

CHAPTER VI. Methods of artificial inoculation 318

CHAPTER VII. Methods of determining disease resistance 320

PART V

TYPICAL FUNGUS DISEASES

CHAPTER I. Plasmodiophoromycetes 323
Clubroot (*Plasmodiophora brassicae* Wor.) 323
Taxonomic position of the fungus 323

Life cycle 323
Symptoms of clubroot 325
Artificial inoculation methods 325
Biological specialization of *Plasmodiophora brassicae* 326
Clubroot resistant cultivars 327
Control of clubroot 327

CHAPTER II. Phycomycetes 328
Potato wart (*Synchytrium endobioticum* [Schilb.] Perc.) 328
Taxonomic position of the fungus 328
Life cycle 328
Symptoms of potato wart 330
Artificial inoculation methods 330
Biological specialization of the pathogen 331
Cultivars resistant to potato wart 332
Late blight of potato and tomato (*Phytophthora infestans* [Mont.] de Bary) 332
Biology and aetiology 333
Physiologic races 334
Cultivation of the fungus 337
Evaluation of demissum resistance based on hypersensitivity 337
Assessment of "field resistance" 338
Laboratory assessment of "field resistance" 338
Field assessment of "field resistance" 339
The peroxidase activity test for assessment of "field resistance" 339
Downy mildew or blue mould of tobacco (*Peronospora tabacina* Adam) 340
Taxonomic position of the fungus 340
Life cycle 340
Symptoms of tobacco *Peronospora* 343
Artificial inoculation methods 343
Biological specialization of the pathogen 347
Cultivars resistant to tobacco *Peronospora* 347
Control of tobacco *Peronospora* 348

CHAPTER III. Ascomycetes 349
Apple scab (*Venturia inaequalis* [Cooke] Wint.) 349
Taxonomic position of the fungus 349
Life cycle 350
Symptoms of apple scab 351
Artificial inoculation method 352
Biological specialization of the pathogen 353
Cultivars resistant to apple scab 353
Control of apple scab 354
Powdery mildew of barley (*Erysiphe graminis* DC. f. sp. *hordei* Em. Marchal) 354
Biology and aetiology 354
Artificial inoculation 355
Cultures in test tubes 356
Detached leaf culture method using benzimidazole 356
Direct microscopic observation of the fungus on the host plant 357
Physiologic races 357
Resistant cultivars 360
Chemical control of powdery mildew 360
Powdery mildew of wheat (*Erysiphe graminis* D. C. f. sp. *tritici* Marchal) 361
Apple mildew (*Podosphaera leucotricha* [Ell. et Ev.] Salm.) 362

Life cycle and aetiology 362
Artificial inoculation method for greenhouse evaluation of fungicides 362
A slide spore-germination method of evaluating fungicides against mildews 364

CHAPTER IV. Basidiomycetes 366

Rust diseases 366
Stem rust of wheat (*Puccinia graminis* Pers. f. sp. *triciti* Erikss. et Henn.) 368
Artificial inoculation of wheat seedlings in the greenhouse 368
Injection of wheat seedlings with a spore suspension 370
Rust nurseries 371
Collection of rust samples and preservation of inoculum 372
Identification of physiologic races of stem rust 372
Infection types (reaction of cultivars) 373
Identification of races in mixtures 375
Physiologic races of *Puccinia graminis* f. sp. *tritici* 375
Subdivision of races of stem rust and the supplemental differential cultivars 385
Identification and detection of races in the greenhouse and in the field (rust trial projects) 385
Resistance to stem rust 386
Sources of resistance to stem rust 386
Physiologic races of *Puccinia recondita* Rob. ex Desm. f. sp. *tritici* Erikss. et Henn. 387
Supplemental differential cultivars of wheat for identification of races of *Puccinia recondita* f. sp. *tritici* 392
Physiologic races of *Puccinia striiformis* West. f. sp. *tritici* Erikss. et Henn. 394
The use of benzimidazole and kinetin for the detached leaf culture of leaf- and stripe-rust 396
Bean rust (*Uromyces phaseoli* [Pers.] Wint.) 397
Artificial inoculation in the greenhouse 397
Artificial inoculation in the field 398
Physiologic races 398
Smut diseases 399
Biology and aetiology 399
Artificial inoculation 401
Wheat bunt (*Tilletia caries* [DC.] Tul. and *Tilletia foetida* [Wallr.] Liro) 403
Physiologic races 403
Controlled environment during infection 405
Testing the effectiveness of fungicides with spore germination-tests 406
Dwarf bunt of wheat (*Tilletia controversa* Kühn) 406
Methods of artificial inoculation 406
Physiologic races 407
Loose smut of barley and wheat (*Ustilago nuda* [Jens.] Rostr. and *U. tritici* [Pers.] Rostr.) 408
Methods of artificial inoculation 409
Resistance to loose smut 410
Quick methods for detecting loose smut mycelium in embryos 410
Detection of loose smut mycelia in seedlings 412
Physiologic races of loose smut of barley 412
Physiologic races of loose smut of wheat 412
Black loose smut (*Ustilago nigra* Tapke) 413
Covered smut of barley (*Ustilago hordei* [Pers.] Lagern.) 414
Methods of artificial inoculation 414
Physiologic races 414
Smuts of oats (*Ustilago avenae* [Pers.] Rostr. and *Ustilago kolleri* Wille) 415
Methods of artificial inoculation 416

Physiologic races 416
Corn smut (*Ustilago zeae* [Beckm.] Ung.) 417
Methods of artificial inoculation 418
Physiologic races 419
Resistance 419
Head smut (*Sphacelotheca reiliana* [Kühn] Clint.) 419
Physiologic races 419
Sources of resistance 420

CHAPTER V. Deuteromycetes (Fungi imperfecti) 421
Anthracnose of cucurbits (*Colletotrichum orbiculare* [Berk. et Mont.] Arx [=*Colletotrichum lagenarium* (Pass.) Ell. et Halst.]) 421
Taxonomic position of the fungus 421
Life cycle 422
Symptoms of anthracnose 422
Artificial inoculation methods 422
Biological specialization of the pathogen 424
The nature of resistance to anthracnose and resistant cultivars 425
Control of anthracnose 426
Verticillium wilt (*Verticillium albo-atrum* Reinke et Berth., *V. dahliae* Kleb. etc.) 427
Taxonomic position of the fungus 427
Life cycle 427
Symptoms of Verticillium wilt and the role of toxins 428
Methods for the differentiation of *Verticillum* species causing wilt 430
Artificial inoculation methods 431
Cultivars resistant to Verticillium wilt 432
Control of Verticillium wilt 432
Rice blast (brusone) disease (*Piricularia oryzae* Cav.) 433
Taxonomic position of the fungus 433
Life cycle 433
Symptoms of rice blast 434
Artificial inoculation methods 434
Biological specialization of *Piricularia oryzae* 436
Rice blast resistant cultivars 437
Fusarium wilt (*Fusarium oxysporum* Schl. emend. Snyder et Hansen) 437
Taxonomic position of the fungus 437
Life cycle 438
Symptoms and the role of toxins 439
Methods used for isolation and artificial inoculations 440
Biological specialization of the pathogen 441
Fusarium wilt resistant cultivars 442
Control of Fusarium wilt 442
Literature 443

Author index 478

Subject index 487

PART I

VIROLOGY

INTRODUCTION

From a microbiological point of view plant viruses are submicroscopic obligate parasites with no metabolism of their own. Chemically they are nucleoproteins.*

On the basis of these properties plant viruses are of both theoretical and practical interest.

Theoretically they are excellent models for studying the relationship between structure and function, due to the relative simplicity of their structure and to the lack of metabolism of their own on the one hand and to their biological activity (infectivity) on the other.

From a practical point of view plant viruses are studied because of their pathogenicity. In order to work out efficient control measures against these agents their properties have to be known. One of the most efficient ways of controlling plant viruses is plant breeding.

The aim of this chapter is to help plant pathologists and breeders become acquainted with the basic methods indispensable in problems connected with plant viruses. Those methods which need special equipment will not be dealt with in detail. However, an attempt will be made, to cover those methods which may be helpful to a plant breeder or to a plant pathologist who is to work with plant viruses and has no special training. In order to provide the reader with a coherent picture of the methods used in plant virology, the individual methods to be described will be compiled in a framework which could be called a basic plant virology. This way of presentation has the advantage of clarifying at once the correlation between any method described and its theoretical background.

In order to get a coherent picture of plant virology we shall deal first with the morphology and structure of plant viruses (Chapter I). Then we shall study the relation between the structure and infectivity of plant viruses (Chapter II). After that we shall answer the question of how plant viruses are transmitted to their hosts (Chapter III) and what happens when plant viruses enter their host (Chapter IV). The intricate interaction between virus particles and host cells cannot be understood without a basic knowledge of the genetics of plant viruses (Chapter V). This leads us to the taxonomy of plant viruses (Chapter VI). The last chapter (Chapter VII) will be a short summary of the control measures against plant virus disease.

* Some viruses, e.g. tobacco rattle virus (Cadman, 1962), strain PM 1 of tobacco mosaic virus (Siegel et al., 1962) seem to exist in the host plant in the form of naked uncoated RNA.

CHAPTER I

MORPHOLOGY AND STRUCTURE OF PLANT VIRUSES

1. PURIFICATION OF PLANT VIRUSES

The aim of virus purification is to separate the virus nucleoprotein from other components of the host cell and to obtain it in a biologically active (infective) state in the form of a concentrated monodisperse suspension.

The success of purification depends partly on choosing the right host plant from which the virus is to be isolated. The host plant should be easy to infect, should not contain inhibitors and dark pigments, should not be infected simultaneously with another virus and should ensure high virus yields. If there is no need for a particular virus strain, the strain to be purified should be genetically stable, should cause characteristic symptoms, should not form complexes with normal host proteins and should reach a relatively high concentration in the host from which it is to be extracted.

The actual purification procedure involves the following steps:

Homogenization of the plant material(1)
↓
Clarification of the sap(2)
↙ ↘
Solution or supernatant fluid — Precipitate or sediment
↓ ↓
Concentrating the virus suspension(3) — Discard
↙ ↘
Solution or supernatant fluid — Precipitate or sediment
↓ ↓
Discard — Crystallization(4)
↓
Checking the homogeneity and (specific) infectivity of the virus preparation(5)

In these steps either physical or chemical methods are used. In most cases both are applied together.

1. For a source of virus, leaves are most often used. *Homogenization* can be carried out in a mortar with a pestle, in a Waring Blendor, or, if large quantities of leaves are used and losses are irrelevant, in an ordinary meat grinder. The homogenate is then pressed through cheese cloth.

Deep-freezing of the leaves before homogenization usually leads to higher virus yields. This procedure can, however, be used only with viruses whose infectivity is not destroyed by deep-freezing the infected plant tissues. Labile viruses can be protected from inactivation by carrying out homogenization in the presence of a buffer of the desired pH or in the presence of reducing agents, such as ascorbic acid or cysteine.

2. *Clarification* of the sap is applied to eliminate most cell components (nuclei, chloroplasts, pigments, proteins) other than the virus nucleoprotein. It is carried out by treating the sap with heat, organic solvents (butanol, chloroform, ether etc.) or simply by filtration or centrifugation. At the end of this process the virus is in the solution.

3. *Concentrating* the virus suspension and elimination of some more contaminating material can be carried out by two alternative methods: a) chemical and b) physical.

a) Chemical procedures include *salting-out* (usually with ammonium sulphate) and *precipitation* (alcohol, acetone, isoelectric point). Any of these steps is followed by centrifugation. Now the virus particles are in the sediment and can be resuspended in an appropriate buffer.

b) Physical procedures have recently been used to an increasing extent because they are more gentle than chemical methods, and therefore can be used for the purification of labile viruses without any appreciable loss of infectivity (cf. Black, 1955). Two basic properties of plant viruses lend themselves to their separation by physical methods: One is their size (and shape) and the other their electric charge. Plant viruses are separated on the basis of their size by ultracentrifugation and molecular sieving techniques and on the basis of their charge by electrophoresis and ion exchange chromatography.

During *ultracentrifugation* the virus particles and other large particulates are pelletted. The pellet is then resuspended. Much of the non-virus protein of the pellet remains aggregated and can be removed by low-speed centrifugation. Repeated low-speed and high-speed centrifugation leads finally to a virus preparation of reasonable purity. Another way of separating virus particles from other high molecular weight contaminants by ultracentrifugation is sucrose *density gradient centrifugation* (Brakke, 1960). In this method the sap containing the virus is layered on top of a linear gradient of sucrose in a centrifuge tube and subjected to high-speed centrifugation. The various components will start to sediment and if enough time and speed is given, they will reach a position in the tube where the density of the sucrose solution is equal to the density of the particles. The particles of a given density will be unable to move farther down in the centrifuge tube. This procedure has been called *equilibrium centrifugation* (Fig. 1). The method in which the centrifuge is stopped before the various components have reached their own density level, but are still separated to an acceptable extent is designated *rate zonal centrifugation*. In both cases the tube is punctured at the bottom with a hypodermic needle and the layer containing the virus particles is drained.

Another useful physical method for the separation of virus particles on the basis of size and shape is *molecular sieving*. For this purpose Sephadex may be

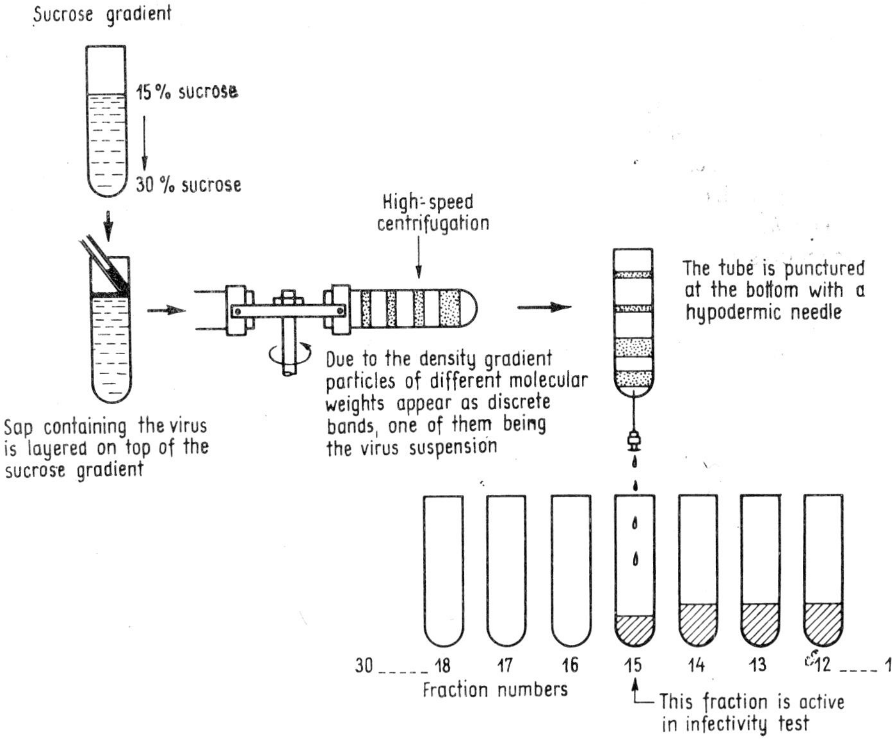

Fig. 1. Purification of plant viruses by density gradient centrifugation

used. Sephadex is a modified dextran of cross-linked macromolecules which gives a three-dimensional network of polysaccharide chains. Sephadex beads swell considerably in water and electrolyte solutions. In a column packed with Sephadex there are actually two routes available for the molecules passing through the column. One is a "free" route between the gel granules and the other a "retarded" route which involves the penetration of the molecules into the pores of the gel granules. The route chosen by the molecules will depend on their size. Molecules larger than the largest pores of the swollen Sephadex cannot enter the gel particles and therefore pass around and between the gel particles. They are thus eluted first. Smaller molecules, however, enter the gel granules to a varying extent depending on their size and shape. Molecules are thus eluted in the order of decreasing molecular size.

Ion exchange chromatography has successfully been applied by Shainoff and Lauffer (1957) for the purification of plant viruses.

Electrophoresis in agar-gel columns (zone electrophoresis) has been worked out by Townsley (1959a) for the purification of TMV. Townsley (1959b) was able to separate TMV from potato virus X by using this method.

4. Some plant viruses can be *crystallized*. Salting out is one of the best methods to get plant viruses in a crystalline state (tomato bushy stunt virus, southern bean mosaic virus, TMV). When tobacco ring spot virus is stored at 4 °C in phosphate buffer, it crystallizes (Steere, 1956). Turnip yellow mosaic virus (Markham and Smith, 1949) as well as tobacco necrosis virus (Bawden and Pirie, 1945) are obtained in a crystalline form at the end of the purification procedure. Different viruses crystallize in distinct forms.

5. The *homogeneity* of the purified virus preparation has to be checked by electrophoresis, ultracentrifugation and electron microscopy. The infectivity per unit volume of the preparation should be compared to that of the original sample. This kind of comparison will show the *specific infectivity* of the preparation.

A detailed procedure for the purification of TMV will be cited below.

*Purification of TMV**

Frozen infected tobacco leaves are ground in a meat grinder. The sap and pulp are collected in a beaker. The sap is then drained through cheese-cloth and as much sap as possible is pressed from the pulp. For each 100 ml of sap 30 ml of 0·3 *M* Na_2HPO_4 is added to the pulp. This is mixed thoroughly, the pulp is ground again and pressed through cheese-cloth. This extract is added to the original sap. The sap is centrifuged at 4000 rev/min for 20 min. To the supernatant $(NH_4)_2SO_4$ is added at the rate of 25 g per 100 ml of sap with stirring to dissolve the salt. The solution is allowed to stand about half an hour. Then the precipitate which contains the virus is centrifuged off. The precipitate is taken up in 3 ml of water per 50 ml of the original sap while stirring for several min. This is followed by centrifugation at 4000 rev/min for 15 min. The supernatants from all the centrifuge tubes are combined. This extraction with water is repeated 4 to 5 times using quantities of water small enough so that the total combined supernatants are only about one-fourth the volume of the original sap. The remaining precipitate is discarded. To the combined supernatant NaOH is slowly added, with vigorous stirring, to bring the pH to 6·5. Again $(NH_4)_2SO_4$ is added at the rate of 25 g per 100 ml. The precipitate (virus) is centrifuged and the supernatant discarded. The virus is redissolved by 4 to 5 extractions with small amounts of water as before. To the combined supernatant 0·1 per cent HCl is added to bring the pH to 3·3. This is allowed to stand for about half an hour. Then the precipitated virus is centrifuged off. The precipitate is suspended in about 10 ml of water and dialyzed for 1 to 2 days against running tap water. As the acid diffuses from the dialysis bag the pH will rise and the virus will dissolve again. Some plant material will remain insoluble and should be removed by centrifugation. The supernatant may be kept in a refrigerator and used as purified tobacco mosaic virus.

For further details on the purification of plant viruses the excellent review of Steere (1959) should be consulted.

* Bawden and Pirie (1943).

*Purification of TMV by differential centrifugation**

About 3 weeks after inoculation with TMV, leaves of a systemic tobacco host are harvested and frozen at —30 to —60 °C. Frozen leaves are easier to grind up into a pulp. 100 g of frozen leaves are powdered in a mortar and 150 ml of 0·1 *M* sodium phosphate buffer, pH 7 is added which keeps the plant acids neutralized. Then the extract is pressed through several layers of cheese-cloth.

The juice is then centrifuged at 8000 rev/min for 20 min. This speed is higher than that obtainable with an ordinary laboratory centrifuge. Debris and chloroplasts will be in the pellet and the supernatant fluid will contain the virus. The supernatant is then centrifuged in an ultracentrifuge at 22,500 rev/min, in other words at about 60,000 × g, for 1 hr. This results in the sedimentation of viruses in the form of gels at the bottom of the tube (pellet). The supernatant fluid, which contains sugars, soluble proteins, free chlorophyll and salts, will be discarded. The pellet is redispersed in 0·1 *M* phosphate buffer, pH 7 overnight. This procedure is carried out at 2–10 °C. Our TMV solution usually looks green, and opalescent. Therefore, the same procedure is repeated several times: low speed centrifugation (removing big debris) and high speed centrifugation (sedimenting the virus and removing small-molecular materials). This is called differential centrifugation. When the pellet is colourless and almost water-clear it is probably composed entirely of TMV particles. 100–200 mg of virus/100 g of leaf can be expected. Finally, the pellet is taken up in water. A dilute solution of TMV (10–100 μg/ml) can be used as inoculum to tobacco leaves. Absorbancy of 100 μg/ml at 260 mμ is 0·27 O.D. units.

*Simple technique for purification of TMV***

This procedure is carried out at room temperature. Homogenize frozen, infected leaves of a systemic tobacco host for 5 min in 0·5 *M* Na_2HPO_4–KH_2PO_4 buffer at pH 7·2 containing 1 per cent 2-mercaptoethanol. 1 g of tobacco leaf is homogenized in 1 ml of buffer. Strain homogenate through cheese-cloth and add 8 ml of *n*-butanol/100 ml of extract. Stir extract during this mixing procedure. After the chloroplasts coagulate, stir extract for an additional 15 min and centrifuge at 10,000 g for 30 min. While stirring, add 4·0 g of polyethylene glycol (M.W. 6000) per 100 ml to the supernatant and centrifuge at 10,000 g for 15 min. The pellet is resuspended in 20 ml of 0·01 *M* PO_4 buffer, pH 7·2, per 100 ml of initial extract. Finally clarify at 10,000 g for 15 min. The supernatant will consist of a suspension of relatively pure virus.

Further purification may be obtained by a second precipitation with polyethylene glycol. Add 0·4 g of NaCl and 0·4 g of polyethylene glycol, while stirring, for each 10 ml of virus suspension. After the NaCl and polyethylene glycol dis-

* Cf. Fraenkel-Conrat (1966).

** Gooding and Hebert (1967).

solve, centrifuge at 10,000 g for 15 min. The pellet is resuspended in 2 ml of 0·01 *M* PO_4 buffer for each 100 ml of initial extract. Centrifuge at 10,000 g for 5 min. The supernatant will be a solution of TMV and will appear milky white. Approximately 100 mg of TMV is recovered per 100 g leaf.

2. MORPHOLOGY OF PLANT VIRUSES

With respect to shape, plant viruses can be divided into two main groups: Anisometric viruses with elongated particles (rod-like or thread-like) and isometric viruses with spherical particles. Size and shape are constant characteristics of a virus species (cf. Protzenko and Legunkova, 1960).

The length of anisometric viruses varies between the two extreme values of 55 mμ for alfalfa mosaic virus and 1250 mμ for sugar beet yellow virus. Most anisometric viruses, however, have a length somewhere between 300 mμ (tobacco mosaic virus) and 750 mμ (pea mosaic virus) and are about 10 to 20 mμ wide. The diameter of isometric plant viruses is around 30 mμ (cf. Paul, 1961). Potato yellow dwarf virus has exceptionally large ellipsoidal particles, 200 mμ long and 50 mμ wide. At present the morphology of more than fifty plant viruses is known (cf. Paul, 1961). For more details concerning the morphology of plant viruses works by Protsenko and Legunkova (1960) and Paul (1961) should be consulted.

Two parameters have to be known for the determination of the morphology of plant viruses: size and shape. These data can be obtained by the use of any of the following three methods: a) Filtration, b) Determination of molecular constants by physico-chemical methods and c) Electron microscopy. All these methods require the use of special equipment. Therefore, they will not be dealt with in detail. However, some methods yielding virus preparations which can be used directly for electron microscopy will be described.

a) *Filtration*. The principle of this method is that the purified virus preparation is filtered through a series of filters with gradually decreasing known pore sizes, and after each filtration, infectivity tests are made. After passing through a filter of a certain pore size the filtrate will lose its infectivity. The size of the virus particles can be calculated from the pore size and from the rate of decrease of infectivity.

b) *Molecular constants* (molecular weight, length, and diameter of the particles) can be calculated from data obtained by sedimentation, diffusion and light scattering measurements.

c) *Electron microscopy*. Due to their small size, plant viruses cannot be distinguished with an ordinary microscope. This is connected with the low resolution power of the light microscope. The resolution power, however, can be increased by decreasing the wave length. Since beams of electrons have associated with them a wave length about 100 thousandth that of visible light, a much higher resolution can be achieved with the aid of an electron microscope than with a light microscope. The resolving power of an electron microscope can be increased up to 20 Å and a 200,000 fold magnification can be achieved. The most important point in the preparation of plant viruses for electron microscopy is to avoid altering virus structure during the stages of preparative work. On the other hand, such preparations should contain a minimum amount of normal cellular particles which could be mistaken for virus particles under the electron microscope. With spherical viruses these two aspects are difficult to reconcile because ribosomes might appear similar to virus particles. Therefore, in this case a preparation of a higher purity should be aimed at than with rod-shaped viruses. In order to keep the structure of spherical viruses as little affected as possible, they have to be purified by using very gentle methods (e.g. density gradient centrifugation). With rod-shaped viruses, however, the preparations may contain more contaminating particles. The three methods to be described below can be used for the preparation of rod-shaped viruses for electron microscopy. Of course, with all these methods, sap from healthy plants should also be subjected to the same treatment in order to ensure a reliable control.

*Exudation method for the preparation of rod-shaped viruses for electron microscopy**

A cork borer is used to make a hole in the middle of a rubber stopper. A part of stem or the petiole of the virus infected plant is pushed through the hole. The stopper with the cut stem or leaf is placed on a water pressure pump. On opening the tap, drops of liquid (exudate) will appear at the cut ends of the stem or leaf. These drops can be investigated directly in the electron microscope.

*Immersion method for the preparation of rod-shaped viruses for electron microscopy***

This method is particularly successful with virus infected leaves and dark sprouts of potato. It is carried out in the following way: the grid coated with a layer of film, is wetted with a drop of distilled water. From the plant to be investigated a small part is cut out with a sterile razor blade. The cut surface is dipped for a short time in the drop of distilled water. The preparation is dried in the air, shadowcast, and directly examined under the electron microscope.

*Purification of rod-shaped viruses for electron microscopy****

Leaves exhibiting severe symptoms are used. The main veins are cut out and discarded. The leaf blades are slightly dusted with Na_2HPO_4, homogenized in a mortar and pressed through cheese-cloth. The expressed sap is kept frozen for

* Johnson (1951).
** Brandes (1957).
*** Bawden and Nixon (1951).

several hours at —10 °C and thawed. Thereafter the homogenate is centrifuged at 7000 rev/min for 10 min. The supernatant can be examined in the electron microscope for the presence of virus particles.

Electron micrographs have to be evaluated properly. This means that the size and shape of virus particles as established from the electron microscopic picture should reflect the actual size and shape of virus particles as they occur in the plant. To this end a statistical analysis of the electron micrographs is necessary. This will eliminate the possibility of regarding aggregates and/or broken virus particles as being characteristic representatives of the virus species in question. For the estimation of the size of rod-shaped viruses the term "normal length" has been introduced. A detailed method for the estimation of normal length in the case of potato virus S is given by Wetter and Brandes (1956).

3. STRUCTURE OF PLANT VIRUSES

CHEMICAL STRUCTURE

Plant viruses are nucleoproteins, i.e. they consist of protein and nucleic acid.

Proteins are known to consist of a number of repeating units, the L-amino acids, which are joined one to the other by a peptide bond. The general formula of amino acids is

$$\begin{array}{c} \quad\;\, R \\ \quad\;\, | \\ NH_2—CH—COOH \end{array}$$

where R stands for a certain organic radical. There are 20 amino acids which take part in the building up of proteins. They are the following: alanine (Ala), arginine (Arg), aspartic acid (Asp), asparagine (Asp-NH_2), cysteine (Cys), glutamic acid (Glu), glutamine (Glu-NH_2), glycine (Gly), histidine (His), isoleucine (Ileu), leucine (Leu), lysine (Lys), methionine (Met), phenylalanine (Phe), proline (Pro), serine (Ser), threonine (Thr), tryptophan (Try), tyrosine (Tyr), valine (Val). In the formation of peptide bonds the —COOH group of one amino acid is linked to the —NH_2 group of the other amino acid with loss of one molecule of water.

$$H_2N—\overset{\overset{R}{|}}{C}H—COOH + H—NH—\overset{\overset{R}{|}}{C}H—COOH \rightarrow$$

$$\rightarrow H_2N—\overset{\overset{R}{|}}{C}H—CO—NH—\overset{\overset{R}{|}}{C}H—COOH$$

A large number of amino acids joined together are polypeptides. It can easily by realized that irrespective of the number of amino acids linked together to form a polypeptide chain the COOH group of the amino acid occurs at one end of the chain and at the other end of the same chain the NH_2 group of the amino acid is free. The former is called the C-terminal and the later the N-terminal amino acid of the protein. Amino acids occur in proteins in different proportions and sequences. Thus a practically unlimited range of variations is obtained with respect to the structure of proteins. The amino acid sequence in a certain protein is strictly defined and this is the most important background of the specificity of proteins.

The protein moiety of plant viruses consists of polypeptide chains with defined amino acid sequences. These polypeptide chains form morphological units and are called subunits. In the formation of the protein part of tobacco mosaic virus 2130 subunits are known to take part (Franklin et al., 1959). Each subunit consists of 158 amino acid molecules. Their sequence has been determined by Anderer et al. (1960) and independently by Tsugita et al. (1960).

Ribonucleic acids consist of three kinds of molecules: heterocyclic base, pentose and phosphoric acid. The backbone of the ribonucleic acid macromolecule consists of a long chain of alternating pentose and phosphate residues. The phosphate residue is attached to the 3′ C atom of one sugar residue and to the 5′ C atom of the other sugar molecule. Attached to the 1′ C atom of the sugar molecule is the third constituent of nucleic acid, a heterocyclic base. There are four kinds of bases in ribonucleic acids. These are the purine bases adenine (A) and guanine (G) and the pyrimidine bases cytosine (C) and uracil (U). As far as the specificity of nucleic acid macromolecules is concerned, these bases play a role analogous to that of amino acids in proteins. According to our knowledge, their sequence is strictly determined for each nucleic acid molecule.

On the basis of the above, a nucleic acid chain is built up in the following way:

```
          BASE
            |
      SUGAR
        |
        P       BASE
        |         |
        SUGAR
            |
            P      BASE
            |        |
            SUGAR
```

where P = phosphate residue.

By using appropriate enzymes and chemicals, ribonucleic acids can be degraded to their structural components, which are the following:

Nucleoside = base + sugar (adenosine, guanosine, cytidine and uridine)

Nucleotide = base + sugar + phosphate residue (adenylic acid, guanylic acid, cytidylic acid and uridylic acid).

The amino acid and nucleotide composition of individual plant virus species is different. Individual plant viruses also differ in their nucleic acid content (cf.

Markham, 1959). The RNA content of tobacco mosaic virus is 5·1 per cent (Knight and Woody, 1958), of potato virus X 5·0 per cent (Bawden and Pirie, 1938), of alfalfa mosaic virus 19 per cent (Frisch-Niggemeyer and Steere, 1961), of tobacco necrosis virus about 18 per cent, of tobacco ringspot virus 35 per cent (cf. Markham, 1959), of turnip yellow mosaic virus 37 per cent (Markham, 1959).

As far as the primary structure (nucleotide sequence) of plant virus nucleic acids is concerned very little has been achieved up to now as compared to the structural analysis of plant virus proteins. The reason for this is no doubt due to difficulties in methods.

Both proteins and nucleic acids are macromolecules consisting of building blocks, amino acids and nucleotides, respectively. This means that the fundamental principles involved in their structural analysis are similar. The corresponding methods consist of either breaking up the macromolecules into fractions (partial sequences) or chopping off the building blocks from one end of the macromolecule one by one (stepwise degradation) and identifying them. In most cases the two methods are used in combination.

There are, however, two basic differences in the structure of proteins and nucleic acids of plant viruses, and these are responsible for the peculiar difficulties encountered in the elucidation of the structure of nucleic acids:

(I) Whereas there are 20 different amino acids in proteins, in plant virus nucleic acids only the four "classical" nucleotides have been found so far. It is not difficult to visualize that the less variability there is in the building blocks the less choice in specific enzymes and the higher the probability of getting fractions of the same sequence.

(II) The proteins of plant viruses studied so far consist of subunits, each with a comparatively limited number of amino acid residues (e.g. 158 with TMV), whereas the RNA moiety of TMV particles and probably of other plant viruses too is a single-stranded macromolecule without any apparent repetition in its structure (about 6000 nucleotides in TMV-RNA). Obviously it is much easier to elucidate the sequence of 158 building blocks than that of 6000.

In spite of these drawbacks structural studies on plant virus nucleic acids were begun and are being continued on the basis of the above-mentioned two principles: partial sequence studies and stepwise degradation. Of course, the methods used so far do not permit the elucidation of the complete nucleotide sequence of plant viral nucleic acid but may be useful in detecting some rough differences between viruses and virus strains with respect to the structure of their nucleic acids.

Partial sequences in plant virus nucleic acids. Although the principle involved in partial sequence studies is the same for both proteins and nucleic acids, the amount of information one can get is quite different with these two types of molecules. With subunits of plant virus proteins several specific enzymes can be used independently and by analysing individual oligopeptides obtained with enzymes of varying specificities the overlapping portions can be identified and the complete sequence established. With plant virus RNAs this approach does not work, except for *t*-RNA (Holley et al., 1965) which consists of about 70 to 80 nucleotide residues, due to the very large number of building blocks in a single molecule and to the limited variability in the building blocks. All one can do is to split specifically some of the internucleotide bonds and to get in this way a quantitative liberation of oligonucleotides of different length and different composition. By a reasonable separation of the resulting oligonucleotides and by establishing their amounts, the relative frequency in which certain sequences occur in ribonucleic acids can be established. To achieve this two requirements have to be fulfilled: to get reproducible fractions, and to get a reproducible separation of as many of the fractions obtained as possible.

Getting reproducible fractions of plant virus RNA is the less difficult problem of the two. It can be achieved by using highly specific enzymes. The two most widely used nucleolytic enzymes are RNase and ribonuclease T_1 (Sato and Egami, 1957). RNase is known to hydrolyze uridine-3′ phosphoryl and cytidine 3′-phosphoryl bonds in ribonucleic acids giving rise to cytidylic acid, uridylic acid and fragments of varying lengths terminated by either cytidylic acid or uridylic acid. Ribonuclease T_1 specifically cleaves the secondary phosphate ester bonds of guanosine 3′ phosphate, giving rise to guanylic acid and several oligonucleotides of different lengths terminated by guanylic acid.

For the *reproducible separation* of the fractions obtained, paper chromatography combined with electrophoresis and column chromatography are the methods most widely used.

As early as 1952 Markham and Smith by using *paper chromatography* followed by the *electrophoretic separation* of the eluates of the chromatographic bands were able to detect some of the smaller products of ribonuclease digestion of TYMV-RNA. Rushizky and Knight (1960a, b) have developed a mapping procedure similar in its lay-out to finger printing as used with peptides. After enzymic digestion the fragments were separated first by paper electrophoresis in one direction and then with the paper turned 90° from the direction used in electrophoresis by paper chromatography in the other. In this way RNase digests of TMV-RNA were resolved into eighteen spots all of which were identified. The advantage of this method lies in the fact that comparatively small amounts of RNA are needed. The disadvantage, however, is, as Markham (1963) has put it, that "it is almost impossible to correct for paper blank variations". At any rate, the method proved to be quantitative enough for the demonstration of the preferential specificity of RNase (Rushizky et al., 1961), and of the striking differences in partial nucleotide sequences between strain HR, and two other strains (M and common strain) of TMV upon digestion with either RNase (Rushizky and Knight, 1960b) or ribonuclease T_1 (Rushizky and Sober, 1962). This method was used also for elucidating the mechanism of action of micrococcal nuclease employing TMV-RNA as a substrate (Rushizky et al., 1962).

The other most promising method for the separation of oligonucleotide fragments obtained by enzyme digestion is *ion exchange chromatography*. The advantage of such methods is that it is much easier to estimate the amounts of the individual fractions than with methods involving the use of chromatographic papers. The disadvantage, however, is that fairly large amounts of starting material are needed.

As early as 1953 Volkin and Cohn succeeded in separating products of ribonuclease action by ion exchange column chromatography under acidic conditions using Dowex-1 and stepwise increments of NaCl for elution. By this procedure a fairly large number of fractions could be separated, there was, however, considerable overlapping (e.g. ApUp came off together with ApGpCp in a single peak) making quantitative estimations difficult. In addition, not all the possible sequences (up to the pentanucleotide level) could be detected because of a considderable overlapping of the peaks. Staehelin's procedure (1961a, b, c) in which DEAE-cellulose is used as a resin and elution is carried out under slightly alkaline conditions with increasing concentrations of ammonium bicarbonate offers some advantages over the method mentioned above in that polyvalent anions are not bound so tightly by DEAE-cellulose than by Dowex-1 and therefore less time is needed for the elution procedure. Even with this method, however, there is overlapping. In cases where two oligonucleotides were not completely separated additional techniques such as *rechromatography* at a lower pH and *paper electrophoresis* had to be used to obtain the exact amounts of the individual components. Symons et al. (1963) used the above method to analyze six strains of TYMV and one each of wild cucumber mosaic virus and turnip crinkle virus for the products of pancreatic ribonuclease digestion of their RNA. They reported the amounts of the mono-, and dinucleotides, most of the trinucleotides and of one of the tetranucleotides in the individual virus strains.

It is easy to realize that the higher the oligonucleotides that can be isolated and identified the more information can be obtained on the structural characteristics of an RNA molecule. Solymosy et al. (1965) set out to find a method which would permit the isolation of oligomers larger than trinucleotides. In their work the ribonucleic acid of TYMV was digested with RNase and the resulting oligonucleotides were separated by combining two column chromatographic methods. In the first column (DEAE-cellulose) elution by neutral salt solutions containing 7 *M* urea to suppress secondary binding forces between the oligonucleotides and the DEAE-cellulose (Tomlinson and Tener, 1963a, b) led to the separation of the mixture into fractions containing oligomers of equal length. These individual fractions were then subfractionated under acidic conditions on Dowex-1 resins: a separation which depends not only on the charge carried by each oligomer but also on hydrophobic interaction between the resin and the base residues (Volkin and Cohn, 1953). An extensive fractionation of the digest thus resulted. The fractions were then characterized with respect to both their base composition and the portion of the total represented by them.

End group determination and stepwise degradation with plant virus nucleic acids. Before discussing this topic one has to be reminded again of the formidable difficulties involved in such experiments. The aim of such experiments is to determine the sequence of at least some of the terminal nucleotides each of which represents one out of about 6000 to 6500 nucleotides. For this type of work mostly labelled (P^{32} or C^{14}) viral RNA is used. In many cases only the determination of the terminal nucleoside or nucleotide at either end of the chain is attempted. The following methods have been used alone or in combination:

Enzymatic methods

a) Removal of terminal phosphate groups with alkaline phosphomonoesterase.

b) Digestion with snake venom phosphodiesterase. Snake venom phosphodiesterase is an exonuclease which attacks poly- and oligonucleotides from the 3′-hydroxyl end in a stepwise fashion giving rise to nucleoside 5′-phosphate units. Upon exhaustive digestion of RNA with this enzyme one diphosphate will be released from the 3′-end of the molecule if it is phosphorylated and one nucleoside from the 5′-end if it is not phosphorylated.

c) Digestion with spleen phosphodiesterase. The action of this enzyme on oligo- and polynucleotides is complementary to that of venom phosphodiesterase. Thus the mode of action is stepwise from the end bearing the 5′-hydroxyl group and results in the successive release of nucleoside 3′-phosphate units. Upon exhaustive digestion of RNA with this enzyme the same end groups can be determined as by alkaline degradation.

d) Another approach to end group analysis of nucleic acids is that proposed by Tomlinson and Tener (1963b): The nucleic acid is degraded with an appropriate nuclease such that the polynucleotide containing the original terminal nucleotide residue has either zero or two terminal phosphate groups, whereas the rest of the polynucleotides, which arise from inside the nucleic acid chain, contain only one terminal phosphate residue. The mixed polymers are separated according to their charge on DEAE-cellulose at pH 7·5 with 7 *M* urea and an increasing concentration of sodium chloride in the eluting solution. Each fraction is isolated and treated with phosphomonoesterase. The polymer derived from the end now differs by two charges from the other polymers in the fraction and can be separated from them by rechromatography in the same system.

Chemical methods

a) Alkaline degradation. Alkali is known to split the phosphate ester bond to the 5′-OH group of the ribose of RNA and to give rise to a mixture of 2′ and 3′-nucleoside phosphates from the middle part of the chain and to one nucleoside diphosphate from the 5′-end of the molecule if that is phosphorylated and to one nucleoside from the 3′-end if that is unphosphorylated.

b) Periodate oxidation followed by cleavage with an amine. Theoretically this method is suitable for the stepwise degradation of an oligonucleotide chain from the 3′-end. If the 3′-end is phosphorylated the terminal 3′-phosphate is first removed by phosphomonoesterase. After this the oligonucleotide chain is treated with periodate which oxidizes the cis-hydroxyls on carbon atoms 2 and 3 of the terminal ribose. Upon subsequent treatment with an amine (cyclohexylamine and lysine have been most successfully used by Yu and Zamečnik, 1960, and Neu and Heppel, 1964, respectively) an unstable amine complex (Shiff-base) is formed which is cleaved to produce the free base of the first nucleotide residue and the exposed 3′-phosphate of the second. Repetition of these steps results in the stepwise liberation of bases from the 3′-end. The mechanism of the reaction is supposed to be as follows (Yu and Zamečnik, 1960) (Fig. 2):

It was even difficult to determine whether or not the 5′-end or the 3′-end nucleoside of TMV-RNA was phosphorylated. In an earlier paper Gordon et al. (1960) reported that under the influence of phosphomonoesterase TMV-RNA releases inorganic phosphate equivalent to 3000 to 5000 nucleotides. This was interpreted as signifying that each intact RNA molecule of about 6500 nucleotides carries, at most, 1 terminal phosphate, and that additional phosphate ends are

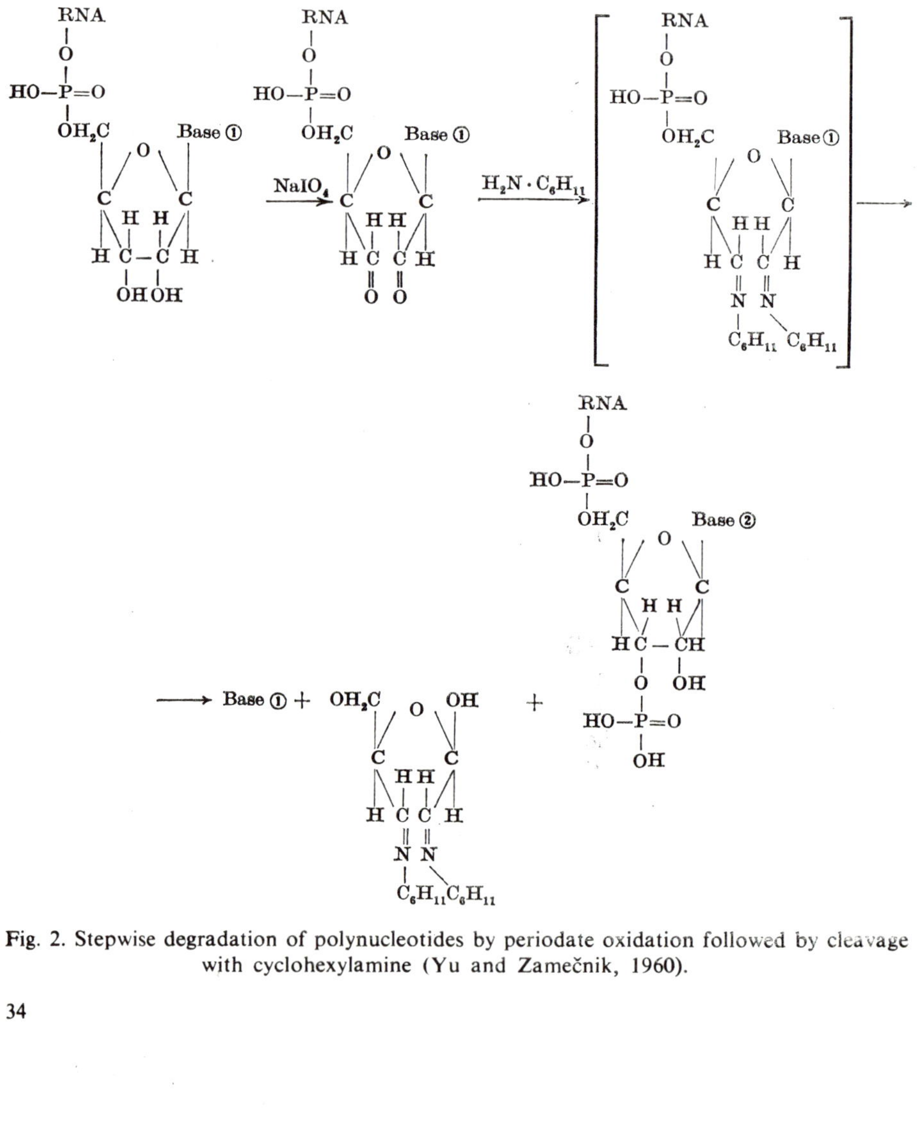

Fig. 2. Stepwise degradation of polynucleotides by periodate oxidation followed by cleavage with cyclohexylamine (Yu and Zamečnik, 1960).

due to chain breakage. It was later found (Fraenkel-Conrat and Singer, 1962) by employing specifically purified RNA preparations free from nucleases, that the phosphate released from TMV-RNA upon digestion with *E. coli*-phosphomonoesterase was much less than mole/mole RNA. This means that the 3′-end of TMV-RNA was not phosphorylated. This conclusion was supported by the finding that degradation with snake venom phosphodiesterase under optimum conditions yielded less than one mole of nucleoside 2′(3′)5′-diphosphates per mole RNA. Alkaline degradation of TMV-RNA yielded no nucleoside 2′(3′)5′-diphosphates indicating the absence of 5′-phosphorylated chain ends. On the basis of these data Fraenkel-Conrat and Singer (1962) concluded that neither the terminal 5′-position nor the terminal 2′(3′)-position of the TMV-RNA chain carried a phosphate group. Using C^{14}-labelled TMV-RNA it was further shown that upon alkaline hydrolysis the nucleoside liberated was adenosine (Sugiyama and Fraenkel-Conrat, 1961) suggesting that the 3′-end of TMV-RNA is an adenosine residue. Exhaustive digestion of C^{14}-labelled TMV-RNA by snake venom phosphodiesterase usually yielded about 0·5 moles of guanosine, cytidine and uridine and 1·5 moles of adenosine (Sugiyama and Fraenkel-Conrat, 1963). The authors have interpreted these findings as suggesting that adenosine represents the 5′-end of the intact RNA chain.

Recently Mandeles (1967) has shown in a very elegant way that the 3′-end base sequence of tobacco mosaic virus ribonucleic acid is CpCpCpA.

SUBSTRUCTURE

The term substructure includes the spatial arrangement of submicroscopic particles. In the case of plant viruses, substructure means the steric configuration of the protein and nucleic acid moieties of the virus particles.

In plant viruses studied so far the RNA component, in the form of a single strand, is embedded in the protein coat. The protein component as mentioned above consists of subunits probably of identical amino acid sequence. The substructure of tobacco mosaic virus has been established by Franklin et al. (1959) by X-ray diffraction. In their model the tobacco mosaic virus rod is of a helical structure. There is a central hole about 40 Å in diameter along the axis of the helix. The maximum diameter of the TMV particle is 180 Å. The protein subunits are arranged about the viral particle axis in a helical array. The structure is repeated after 69 Å in the axial direction, and the repeat contains 49 subunits distributed over three turns of the helix of 23 Å pitch. The nucleic acid strand is located at a 40 Å radius and follows the helical arrangement of protein subunits in which it is embedded.

The existence of protein subunits in virus particles has been proved with the following viruses in addition to tobacco mosaic virus: southern bean mosaic virus, turnip yellow mosaic virus (Nixon and Gibbs, 1960), wild cucumber mosaic virus (Yamazaki and Kaesberg, 1961), potato virus X (Reichmann, 1960), tomato bushy stunt virus and wound tumour virus (Bils and Hall, 1962).

CHAPTER II

BIOLOGICAL ACTIVITY OF PLANT VIRUSES

Plant viruses consist of two components: RNA and protein. The question arises whether both components are needed for the initiation of the infection process, or whether, only one of them suffices. If the ability of initiating the infection process is termed biological activity the question is whether the RNA or the protein or both are responsible for the biological activity of plant viruses. This will be answered in point 1 of this chapter. The other question which is connected with the biological activity of plant viruses is how far external conditions influence the biological activity of plant viruses under *in vitro* conditions. This will be dealt with in point 2 of this chapter.

1. RIBONUCLEIC ACID AS THE BIOLOGICALLY ACTIVE COMPONENT OF PLANT VIRUSES

Early experiments (Schramm and Müller, 1940; Miller and Stanley, 1941 have pointed out that the protein component of TMV does not necessarily play an important role in the infection process, because a considerable amount of the amino-, phenol-, and indole-groups of TMV protein can be substituted (e.g. acylated) without a considerable loss of infectivity. After Hershey and Chase (1952) found that in bacteriophages only the nucleic acid part of the phage enters the bacterial cell and initiates the infection, the attention of investigators was directed to the specific role of the RNA moiety in plant viruses (Jeener, 1956). Finally Gierer and Schramm (1956) demonstrated that the RNA component of TMV can be isolated by phenol extraction and that this RNA preparation containing less than 0·02 per cent protein is in itself infectious. The infectivity of such RNA preparations, however, is only about 0·2 per cent that of the intact virus (on the basis of RNA content). This discovery led to the general assumption that the biological activity of tobacco mosaic virus is exclusively inherent in its RNA component.

Two important points, however, should not be neglected. One of them is that infectious RNA preparations do contain protein contamination even if in negligible amounts. The other difficulty lies in the fact that the infectivity of TMV-RNA is very small compared to that of intact TMV. As to the nature of protein contamination, there are two possibilities:

a) Protein contamination is evenly distributed in the TMV-RNA population, i.e. each RNA molecule contains an extremely small amount of protein. According to Commoner (1959) a protein contamination of 0·2 per cent is enough to account for one protein subunit being attached to each infectious RNA macromolecule.

b) Protein contamination is unevenly distributed in the TMV-RNA population. Some RNA molecules have a considerable amount of protein attached to them and others have none. It may be supposed that only those RNA molecules are infectious which are combined with a large amount of protein.

Some experimental results seem to support this latter hypothesis. Takahashi et al. (1958) separated a TMV-RNA preparation into 24 fractions by electrochromatography. Each fraction was simultaneously tested for RNA concentration and infectivity. Only in two cases did they find that an increased infectivity was associated with higher RNA concentration, in 8 cases the reverse was true. The most infectious fractions were stained most strongly with bromophenol blue (a protein reagent) and migrated very slowly towards the cathode (similar to intact TMV). Using infectious RNA prepared from ^{35}S labelled TMV, he found, that among the RNA fractions obtained by electrochromatography, the ones with the highest counts (containing most protein) were most infectious.

The low infectivity of TMV-RNA compared to intact TMV may be due to the fact that the RNA molecules are degraded during the extraction procedure or in the course of coming in contact with the living plant cell. The latter problem has been intensively investigated by Pirie (1959) without reaching any definite conlusion. The possibility of the degradation of TMV-RNA during extraction has been strongly supported by the experimental results of Fraenkel-Conrat et al. (1961) and Singer and Fraenkel-Conrat (1961). They have shown that if RNA is prepared from TMV in the presence of bentonite (an absorbant of RNase) the RNA preparation is ten times more infectious than TMV-RNA isolated in the absence of bentonite.

Summing up, the general assumption that TMV-RNA is in itself infectious has not yet been unequivocally proved, mostly due to difficulties in methods. It is namely, extremely difficult to exclude the possibility of the presence of minute amounts of intact TMV in TMV-RNA preparations. From a theoretical point of view, however, the presence of minute amounts of protein contamination in TMV-RNA preparations is irrelevant. The fact that chemical modification of TMV-RNA molecules (Mundry and Gierer, 1958) leads to mutation, that this mutation is reflected in the primary structure of the virus protein (Wittmann, 1961), and that RNase sensitive infectious material obtained from several other viruses (Kaper and Steere, 1959a, b; Diener and Weaver, 1959; Harrison and Nixon, 1959; Kassanis, 1960; Rushizky and Knight, 1959; Reichmann and Stace-Smith, 1959) is infectious, strongly supports the view that the RNA moiety of plant viruses is potentially infectious. The specific infectivity of TMV-RNA is highly increased when it is reconstituted *in vitro* with TMV protein (Fraenkel-Conrat and Williams, 1955; Lippincott and Commoner, 1956). The fact that some viruses (Cadman, 1962; Siegel et al. 1962) occur in the plant in the form of naked

RNA also points to the exclusive role of plant virus RNA in biological activity.

RNase sensitive infectious material can be prepared from both purified virus preparations and crude plant extracts. In the latter case the RNA preparation will be, of course, contaminated with RNA originating from normal host constituents. Therefore, viral RNA extracted directly from virus infected host tissues cannot be used for studies on the structure of viral RNA. For such purposes viral RNA has to be prepared from purified virus preparations.

The following methods can be used for the extraction of RNA from purified viruses:

a) Extraction with phenol (Gierer and Schramm, 1956)
b) Denaturation with guanidine (Reichmann and Stace-Smith, 1959)
c) Use of detergent (Fraenkel-Conrat and Williams, 1955)
d) Heat denaturation (Lippincott, 1961).

With crude plant sap, only modifications of the phenol method have been used up to the present.

Three methods will be described below for the extraction of viral RNA: One for purified virus preparations, another for clarified plant sap, and the third one for crude extracts.

*Extraction of TMV-RNA from purified TMV preparations**
"Phenol-procedure"

A 10 per cent solution of purified crystalline TMV in 0·02 *M* phosphate buffer, pH 7·3 is shaken vigorously for 8 min at 4 °C with an equal amount of water saturated phenol. The concentration of the phenol is approximately 80 per cent (8 g of phenol, preferably redistilled *in vacuo*, plus 2 ml of water). The phases are separated from each other by centrifugation for 1 min at 12,000 rev/min. The upper layer (aqueous phase) containing the RNA is pipetted off and shaken twice more for two minutes with phenol. Thereafter phenol is removed from the aqueous phase by shaking several times with peroxide-free ether. Finally, the ether is removed by bubbling nitrogen gas through the RNA-containing aqueous phase. The whole procedure is carried out in the cold at 4 °C. It has been found, however, that equally infectious RNA preparations can be obtained if shaking with phenol is carried out at room temperature instead of in the cold. Recently, it has been shown that the deproteinizing effect of phenol can be increased by adding *m*-cresol (10 per cent) to the phenol solution. Furthermore, 8-hydroxy-quinoline (0·1 per cent) is useful in inhibiting oxidation of the phenol. Therefore, it is recommended to apply a freshly distilled water saturated phenol solution containing 10 per cent *m*-cresol and 0·1 per cent 8-hydroxy-quinoline (Fig. 3).

* Gierer and Schramm (1956).

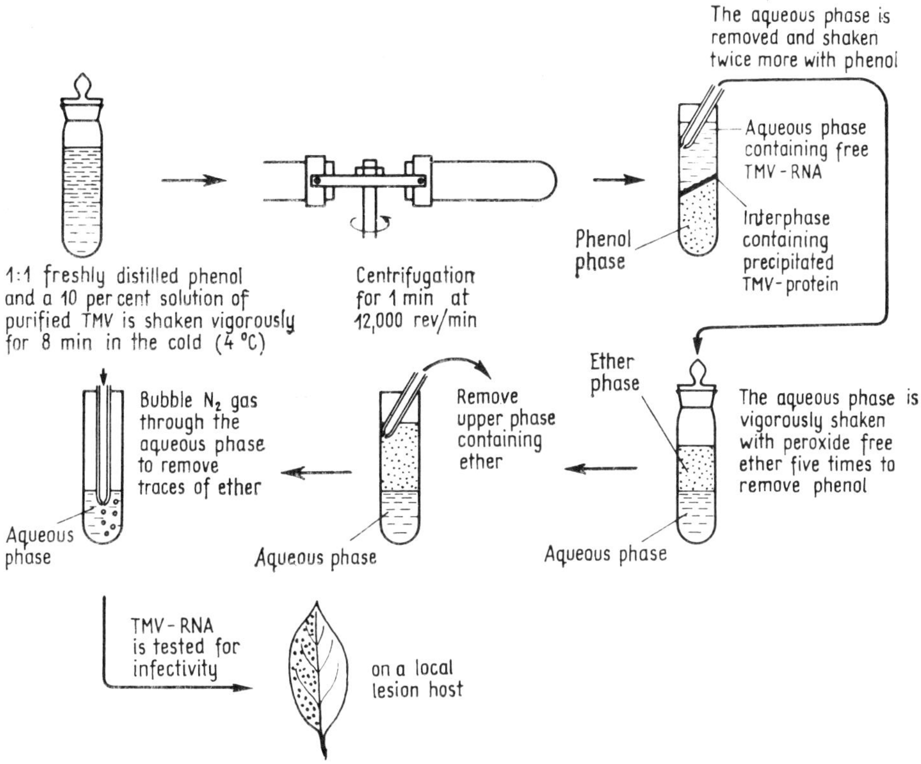

Fig. 3. The "phenol procedure" for extraction of TMV-RNA from purified TMV preparations

*Preparation of RNase sensitive infectious material from clarified sap of TMV infected tobacco leaves**

The leaves to be used are infected with a very concentrated inoculum which causes coalesceing of local lesions. The leaves with well developed symptoms are ground in a meat-grinder and the brei is pressed through cheese-cloth. The remaining cake is mixed with 0·06 *M* phosphate buffer, pH 7·0, of a volume equal to that of the original sap and pressed through cheese-cloth. The two filtrates are combined and centrifuged for 5 min at 10,000 rev/min. To two millilitres of the supernatant an equal volume of water saturated phenol is added, and the mixture is shaken vigorously for 1 min. The two phases are separated from each other by centrifugation at 10,000 rev/min for 4 min. Traces of phenol are removed from the nucleic acid-containing water phase by shaking with 40 ml of ether. The whole procedure is carried out at 4 °C. A test for the absence of intact virus particles in the RNA solution consists of infectivity tests with the RNA solution to which

* Kassanis (1960).

ribonuclease at a concentration of 5 μg/l has been added and which has been incubated with the enzyme for 20 min at 20 °C.

*Preparation of RNase sensitive infectious material from crude extracts of virus infected plants**

The plant tissue is homogenized for 5 min in a mortar with 4 times its volume of a solution consisting of a 1 : 1 mixture of water saturated phenol and 1 per cent tetrasodium pyrophosphate. The solutions and glassware are chilled before use and the whole procedure is carried out in a cold room at 5 °C. The homogenate is centrifuged for 5 min at 6000 rev/min. The nucleic acid-containing aqueous phase is pipetted off and shaken twice with an equal amount of water saturated phenol. Phenol from the aqueous phase is removed by three shakings with peroxide-free ether. Finally the nucleic acids are precipitated with two volumes of ice cold 96 per cent ethanol. The precipitate is sedimented by centrifugation at 6500 rev/min for 10 min and the nucleic acid-containing pellet is resuspended in cold 1 per cent K_2HPO_4 solution.

Isolation of TMV-RNA by the use of bentonite

Bentonite, an acidic clay, is able to inactivate ribonuclease, and thereby, greatly increase the relative infectivity of TMV-RNA without affecting infectivity of the intact TMV (Singer and Fraenkel-Conrat, 1961).

5 mg bentonite is added to 10 mg of TMV and carried through the phenol-procedure described above by Schlegel (1960). Bentonite is prepared as follows. Two g of bentonite is suspended in 40 ml of water and centrifuged at 2500 rev/min for 15 min. The supernatant is centrifuged at 8500 rev/min for 20 min. The second supernatant is discarded and the pellet is resuspended and held for two days in 0·1 *M* EDTA, pH 7 at room temperature. Then it is sedimented again at 8500 rev/min. The next centrifugation is made from 0·01 *M* sodium acetate, pH 6, and the bentonite is taken up in the same medium to a concentration of 2—6 per cent, as determined by dry weight.

After the bentonite is added to TMV, the aqueous phase will be turbid. Most of the bentonite is removed by ultracentrifugation prior to the first alcohol precipitation (at 40,000 rev/min). The rest of the bentonite will be removed by the final ultracentrifugation.

2. FACTORS INFLUENCING THE BIOLOGICAL ACTIVITY OF PLANT VIRUSES

Plant viruses as it is known, are nucleoprotein particles with a well defined substructure. The integrity of the chemical components (first of all that of RNA) and of the substructure is necessary for the biological activity. Some factors which

* Schlegel (1960).

act *in vitro* on a virus preparation may cause considerable changes both in the chemical composition and substructure of virus particles, leading thereby to the loss of the biological activity of plant viruses. The loss of the biological activity of plant viruses is called inactivation *sensu lato*. Plant viruses are inactivated by physical factors (dilution, temperature, ageing, pressure, ultrasonic treatment, irradiation) and chemical factors (inorganic and organic compounds, inhibitors isolated from microorganisms, higher plants and animals).

Among the above factors the effect of dilution, temperature and ageing are of diagnostic value and the mode of action of inhibitors is of theoretical interest. Therefore, kinds of methods concerned with the determination of the effect of these factors will be dealt with in detail.

PHYSICAL FACTORS

Crude sap from virus infected plants is usually used for the determination of the effect of dilution, temperature and ageing on the biological activity of plant viruses. This involves a high source of error with respect to the reliability of the results obtained. The results will vary greatly depending on the physiological state of the plant from which the inoculum is taken, as well as on the physiological state of the plant to be infected with the treated inoculum. The pH, the oxidizing ability, and the inhibitor content of various plants is different and so is the virus concentration in different plants or in different organs of the same plants. Environmental conditions (illumination, temperature, nutrition) may greatly influence the susceptibility of the test plants which are infected with the treated inoculum. These variations can be minimized to a certain extent by setting up an international standard of methods used (Bos et al., 1960). In the following, efforts to make such standardization will be taken into consideration when describing the methods to be used for the determination of the effect of dilution, temperature and ageing on the biological activity of plant viruses.

General considerations:

1. Both the plants from which the inoculum is taken and those which are to be infected are kept at 20–25 °C under controlled illumination.
2. The inoculum should be taken from young plants infected for about three weeks and exhibiting conspicuous symptoms.
3. The indicator plants should react, whenever possible, with local lesions in order to ensure statistical evaluation of the results. If there is no host plant available which would react with the formation of local lesions to the virus under investigation, young plants in the three to four leaf stage should be infected. The plants should be uniform.
4. The expressed sap should be diluted 1 : 1 with 0·01 *M* phosphate buffer pH 7·0 prior to the treatments.

Determination of the dilution end point

Expressed sap diluted 1 : 1 with 0·01 *M* phosphate buffer, pH 7·0 serves as the starting solution. Seven test tubes are prepared with 9 ml of distilled water in each. From the starting solution 1 ml is added to the first test tube. This will be the 1 : 10 dilution. The pipette is rinsed with this diluted sap three times by repeated sucking and blowing of the sap and 1 ml of this diluted sap is added to the next test tubes. This will be the 1 : 100 dilution. This procedure is continued up to a final dilution of 1 : 1,000,000. With each dilution the same number of uniform test plants are infected, starting with the highest dilution. In the case of higher dilutions, the incubation time is longer in host plants reacting with systemic symptoms. Therefore, the evaluation should take place later than with the plants infected with the most concentrated solution. The dilution end point of a given virus is the highest dilution which still results in the formation of symptoms.

Determination of the thermal inactivation point

Whenever possible the thermal inactivation point should be determined by using a Höppler ultrathermostat where the temperature of the water bath can be adjusted very accurately (with a variation of only some hundredth of a degree). If no ultrathermostat is available a water bath of a large volume (1 or 2 l), regulated with the microflame of a Bunsen lamp can be used. The treatments are made in very thin-walled glass tubes (wall thickness 0·7 mm, outer diameter 10 mm, length 10 cm). 1 or 2 ml of the starting solution (crude extract diluted 1 : 1 with 0·01 *M* phosphate buffer, pH 7·0) is placed in each tube. The pipetting of the sap into the tubes should be carried out very cautiously in order to avoid contact of the sap with the inner wall of the test tubes. In such a case the drops adhering to the wall of the test tube above the level of the sap will escape heat treatment and this will interfere with the results. The heat treatments are carried out for 10 min at each temperature starting from 50 °C to 95 °C with 5 °C intervals. The test tubes have to be placed in the water bath in such a way that the water level should be about three cm above the level of the sap in the test tube. After heat treatment the sap is cooled in running tap water and used for the inoculation of an equal number of test plants of the same age for each treatment. In the evaluation it should be taken into consideration that inoculation with sap treated at higher temperatures will result in symptom development later than if inocula treated at lower temperature are used for infection. After this experiment the thermal inactivation point can be determined more accurately by treating inocula at temperatures near the thermal inactivation point found at 2 °C intervals. If the test plants employed react with local lesions, the thermal inactivation coefficient of the virus investigated can be calculated using the following equation:

$$Q_t = \frac{C_0 - C_T}{C_0 - C_{T-t}}$$

here Q_t = thermal inactivation coefficient related to a temperature difference, t

C_0 = original virus concentration (= number of local lesions induced by the sap which has not been heat-treated)

C_T = virus concentration after heat-treatment at T

C_{T-t} = virus concentration after heat-treatment at T–t

Results are generally related to a temperature difference of 10 °C (t = 10 °C). In this case the thermal inactivation coefficient (Q_{10}) is

$$Q_{10} = \frac{C_0 - C_T}{C_0 - C_{T-10}}$$

Reliable results are obtained if T is chosen such as to be somewhat below the thermal inactivation point.

Ageing in vitro

The expressed sap diluted 1 : 1 with phosphate buffer is stored at room temperature (20–22 °C) and at given intervals applied as inoculum to the same number of test plants. As a preliminary experiment it is advisable to take the samples 1, 2, 4, 8, 16, 32 and 64 days after the preparation of inoculum. The results obtained in this preliminary experiment will give an indication of whether the next experiment should be carried out with time intervals shorter than 1 day or longer than 2 months. Of course, in order to preserve the sap for such long periods fungicides and bactericides should be added to the inoculum. However, one has to be very careful in the application of such substances because they may influence the biological activity of plant viruses. Their use should be preceded by careful tests on their possible interfering effect.

When it is possible to use test plants reacting with local lesions the half-life time of viruses can be determined as described by Yarwood and Sylvester (1959). Half-life time is that period of time during which a homogeneous population loses half of its activity:

$$t_{1/2} = \frac{T \log 2}{\log P_0 - \log P_1}$$

where $t_{1/2}$ = half-life time

P_0 = the activity of the original population

P_1 = the activity of the population after time T

T = duration of the treatment

The use of this equation for the determination of the half-life time of a given virus will be demonstrated with an example based on the results of Bennett (1944). Bennett, in studying dodder latent mosaic virus has found that samples taken from the inoculum 0, 2, 6, 12, 24, 48 and 72 hr after its preparation caused

93, 87, 93, 41, 5, 2, and 0 lesions, respectively. The *in vitro* stability of the virus falls between 48 and 72 hr. The half-life time is

$$\frac{T \log 2}{\log P_0 - \log P_1} = \frac{48 \times 0{\cdot}301}{\log 93 - \log 2} = \frac{14{\cdot}4}{1{\cdot}969 - 0{\cdot}301} = 8{\cdot}6 \text{ hr}$$

CHEMICAL FACTORS

Physical factors are known to act on the virus itself in the inoculum. However, if a chemical is added to a virus preparation it will enter the host cell together with the virus upon inoculation. If the infectivity of such treated preparations is affected it is uncertain whether the chemical had an effect on the virus itself or on the susceptibility of the host cell. There are, however, certain criteria which make it possible to decide whether the chemical acted on the virus or on the host cell (Slagle et al., 1952; Gupta and Price, 1950; Bawden and Freeman, 1952; Gendron and Kassanis, 1954; Crowley, 1955). It should be stressed, however, that none of the criteria listed below is decisive in itself, they have to be evaluated together.

Differential criteria for the elucidation of the mode of action of inhibitors

The inhibitor acts on the virus *in vitro*	The inhibitor acts on the metabolism of the host celi
1. The effect of the inhibitor on the virus preparation increases as a function of time after mixing the two.	1. The inhibitor acts on the virus preparation immediately upon its addition and the degree of inhibition does not increase with time.
2. The inhibitor attached to the virus particles can not be separated from it by ultracentrifugation.	2. Fully infectious virus particles can be obtained by ultracentrifugation from the virus-inhibitor mixture with reduced or no infectivity.
3. The non-infectious virus preparation obtained by treatment with the inhibitor does not regain its infectivity upon dilution.	3. The non-infectious virus preparation obtained by treatment with the inhibitor regains its infectivity upon dilution.
4. The degree of inhibition does not depend on the host species used.	4. The degree of inhibition varies with different host species.
5. The inhibitor acts only if mixed with the virus preparation prior to inoculation.	5. The inhibitor acts also if applied to the host to be infected independently from the inoculum (e.g. the lower surface of the leaf is treated with the inhibitor and the upper leaf surface is inoculated with the virus; the plant is treated first with the inhibitor and infected several hours or days later).

*Determination of the mode of action of inhibitors by a dilution method**

The method is based on the following principle: When a test plant is inoculated with a virus which causes local lesions the number of local lesions is in a certain concentration range linearly proportional to the virus concentration. Above a certain concentration, however, the curve representing the number of local lesions as a function of virus concentration tends to flatten out. This is due to the fact, that with a high concentration of infectious virus particles the number of susceptible areas becomes the limiting factor in the formation of lesions. This means that at lower virus concentration it is the number of virus particles which is the limiting factor in the formation of local lesions whereas at high virus concentrations it is the number of susceptible sites which limits the number of local lesions. This correlation is graphically represented in Fig. 4.

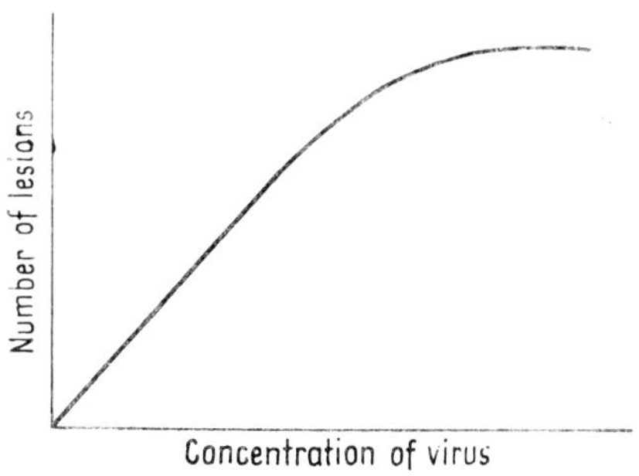

Fig. 4. Increase in the number of necrotic local lesions as a function of virus concentration in the absence of inhibitor (Caldwell, 1936)

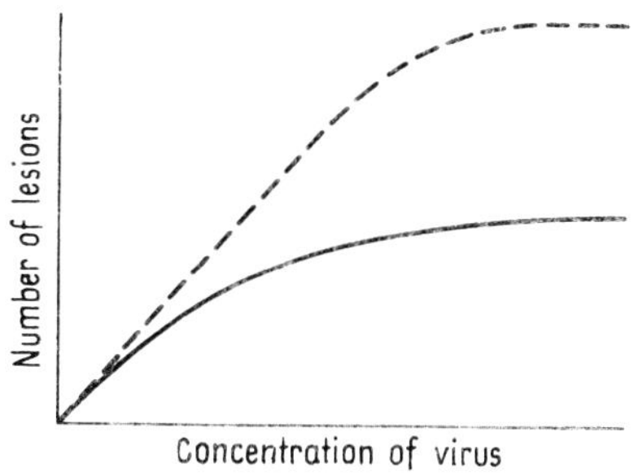

Fig. 5. Increase in the number of necrotic local lesions as a function of virus concentration in the presence of an inhibitor which decreases host susceptibility without affecting virus particles *in vitro* (Caldwell, 1936)

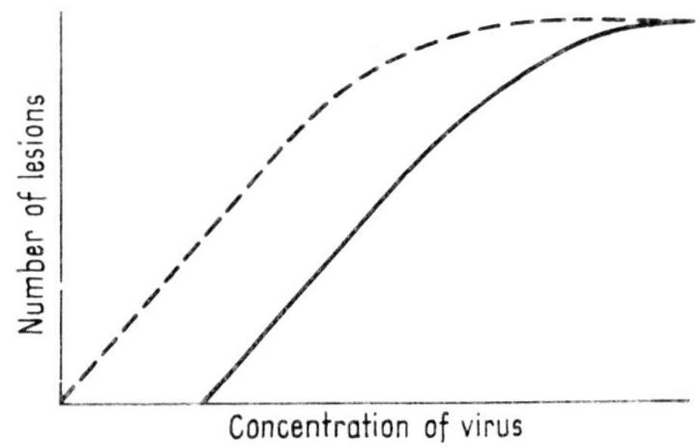

Fig. 6. Increase in the number of necrotic local lesions as a function of virus concentration in the presence of an inhibitor which acts on the virus particles *in vitro* without influencing host susceptibility (Caldwell, 1936)

* Caldwell (1936).

If equal amounts of an inhibitor which decreases host susceptibility without affecting the virus *in vitro* are added to different concentrations of the inoculum we have the following situation: at low virus concentrations, when, as already mentioned, the virus concentration is the limiting factor, the slope of the curve coincides with that of the control curve. Since the number of susceptible sites will decrease under the effect of the inhibitor the limiting role of susceptible sites will be manifest at higher virus concentrations earlier than in the case of the control. Consequently the curve will start to flatten out at lower virus concentrations than with the control. This is represented in Fig. 5.

On the other hand, if equal amounts of an inhibitor which act on the virus *in vitro* without influencing the susceptibility of host cells are added to different concentrations of the inoculum the situation is the following: let us suppose that equal amounts of the inhibitor inactivate with all virus concentrations, the same number of infective virus particles. In an ideal case the inhibitor will inactivate all virus particles present at a very high dilution. This means that if very dilute inocula are treated with the inhibitor no local lesions will be formed, i.e. the curve representing the number of lesions as a function of virus concentration will start later than the same type of curve for the untreated inoculum. As the virus concentration in the inoculum is increased, more and more active virus particles will be present in the inoculum, i.e. an increasing number of local lesions will be obtained with increasing virus concentrations. Finally when the virus concentration will be so high that the number of inactivated virus particles related to the number of active infection units can be disregarded, the limiting function of the susceptible sites only will operate, i.e. the curve will reach the level of the control curve. This is shown in Fig. 6.

CHAPTER III

TRANSMISSION OF PLANT VIRUSES

The ways in which a given virus can be transmitted is characteristic of the virus species and is of diagnostic value. Plant viruses can be transmitted in the following ways:

1. GRAFT TRANSMISSION

Theoretically any virus can be transmitted by grafting. The only limiting factor is tissue compatibility of stock and scion. Two methods for graft transmission of plant viruses will be described.

Approach graft

A downward slit 1–2 1/2 cm long is made in the stem of one plant extending about halfway into the centre of the stem. The epidermis is sliced from the outside of the flap of tissue thus formed. In the stem of another plant a similar, but upward slit is cut, and the epidermis is removed from it. The flap of one stem is then fitted into the slit of the other and the union is wrapped firmly with grafting rubber. After the cut surfaces have healed together, the stem of one plant may be cut off below the graft and the stem of the other cut above the graft, leaving the top of one plant to grow on the roots of the other.

Budding

This method is used to transmit virus diseases of woody plants. A T-shaped cut is made in the bark of the stock and the two flaps of bark are lifted up. A dormant diseased bud is sliced from a twig so that the cut surface beneath the bud is wider than the bud, but smaller than the T-shaped opening in the bark of the stock. The bud is inserted downwards into this opening and bound in place with a budding-rubber. The cut surfaces are covered with the rubber, but the bud itself is left exposed.

2. SEED TRANSMISSION

Transmission of plant viruses by the seed is a relatively rare phenomenon. Out of more than 300 plant viruses described so far only about 45 are transmitted through the seed (cf. Crowley, 1957). Basically there are two possibilities of seed transmission: 1. Virus particles are attached to the surface of the seeds and during germination they infect the seedlings through the little wounds caused by small soil particles (surface seed transmission). 2. Virus particles enter the reproductive organs in some phase of the generative process. Consequently the virus is already present in the cells of the developing seedlings (internal seed transmission).

For the existence of surface seed transmission a peculiar structure of the fruit as well as a high degree of stability of the virus are necessary. These prerequisites of surface seed transmission are given for instance in the host virus combinations tomato–tobacco mosaic virus and red pepper–tobacco mosaic virus. In the former case the seeds are embedded in a gelatinous material in the fruits. This virus-containing substance in a dried state will adhere to the seeds and infect them during germination (Taylor et al., 1961). In the case of tobacco plants such a mechanism does not exist, consequently tobacco mosaic virus is not transmitted by tobacco seeds (Taylor, 1962).

With red pepper a mechanism similar to that existing with tomatoes is responsible for the transmission of tobacco mosaic virus by red pepper seeds (Szirmai, 1950). Miličić and Udjbinać (1960) have shown that tissues of red pepper contained a large amount of TMV crystals. Surface seed transmission can be reduced by sterilizing the surface of seeds originating from virus infected plants (Szirmai, 1950; Milinkó, 1956; Canova, 1957).

The presence or absence of internal seed transmission largely depends on the host virus combination (Crowley, 1957). Tobacco ringspot virus, for instance, is transmitted by bean seeds but not by tobacco seeds. Bean as a host plant transmits through the seed bean mosaic virus but not bean yellow mosaic virus and southern bean mosaic virus. To explain the failure of seed transmission in so many host virus combinations several hypotheses have been advanced. None of them, however, is satisfactory. Since in none of the cases where there is no seed transmission could virus be demonstrated from the embryo (Crowley, 1957) the main factor responsible for the existence or absence of seed transmission must be sought in some peculiar conditions created by haploid cells or the early development of the embryo. There are two hypotheses to explain the failure of seed transmission:

1. Viruses which are not transmitted by the seed are unable to infect haploid cells. This hypothesis is supported by the observation that bean yellow mosaic virus which is not transmitted by the seed cannot be demonstrated from the pollen either, whereas bean mosaic virus which is seed transmitted occurs in the pollen of virus infected bean plants (Medina and Grogan, 1961).

2. Plant viruses which are not seed transmitted are unable to multiply in cells with intensive protein synthesis. It is assumed that in actively synthesizing cells

both free nucleotides and free amino acids are used for the synthesis of normal host proteins and in such cases viral RNA cannot overcome the cell's own genetic information for the synthesis of normal proteins (Caldwell, 1962). This hypothesis is supported by earlier findings indicating that plant viruses are unable to multiply in meristematic tissues characterized by a very active normal protein synthesis.

For testing seed transmission of plant viruses there is no need for special methodical instructions. The most important point is that the seeds of the virus infected plant should be sown under "virus free conditions". The seedlings developing from such seeds should be tested individually by symptomatological, serological, and infection tests. The term "virus free conditions" means the exclusion of any possibility of soil and insect transmission. The seeds to be tested are to be sown in sterile soil and in order to avoid insect transmission the seedlings and developing plants are kept in insect proof cages and sprayed at regular intervals with insecticides. The most reliable results are obtained if 300–500 seeds from virus infected plants are simultaneously tested with the application of an identical number of seeds originating from healthy plants.

3. MECHANICAL TRANSMISSION

The principle involved in mechanical transmission consists of introducing infective sap into living plant cells through wounds. Wounds must be sufficient to allow virus entry but not great enough to cause death of the cells.

For mechanical transmission the sap is first expressed from the infected plant which serves as the virus source. Virus yields are usually higher when the inoculum is obtained by grinding infected leaf tissues in a mortar with pestle than when the sap is expressed with a press commonly used in refractometry. From small amounts of leaf, inocula are most conveniently prepared by using a ground glass homogenizer as described by Takahashi (1951). The homogenizer consists of two parts, the mortar and the pestle. On the grinding surface of the pestle there is a continuous groove which extends from the shoulder around the base up to the shoulder on the opposite side. Two such grooves cut in each side of the pestle and crossing on its base serve to supply liquid to the grinding surface and in addition act as gross cutting edges. To prevent the compacting of unhomogenized tissue at the bottom of the homogenizer it is desirable that the base of the pestle be rounded and fit snugly into the rounded bottom of the mortar cavity. The bulge on the mortar serves as a well into which the homogenate may be collected.

The prepared inoculum has to be introduced into the plant cells. This may be done in two ways: either by injection or by rubbing. Presently injection is used only for the solution of some special problems (e.g. introducing the inoculum directly into the phloem). For inoculation several methods are employed.

A pad of cheese-cloth with a surface of 1–2 cm^2 inbibes a sufficient amount of inoculum to infect several leaves. The advantage of using a cheese-cloth pad is that repeated dippings are avoided. The disadvantage of the method is that the hands get contaminated with the virus. Therefore, when different viruses are transmitted in one experiment the hands have to be washed carefully between infections. The advantage of using just the forefinger for inoculation is that it is easy to feel the pressure and thereby avoid damaging the leaves. The disadvantage of this technique is the same as that of the foregoing one. A glass spatula is also a convenient tool for mechanical transmission. The advantage of using a glass spatula is that the hands do not become contaminated with the inoculum. However, the leaves can be damaged much more easily than with any of the former techniques. To overcome some of the disadvantages of the methods mentioned above, an inexpensive, disposable, readily available tool, the pipe cleaner, has been adapted by Kreitlow (1961) to inoculate plants artificially with mechanically transmitted viruses. "Pipe cleaners can be bent into any desired shape. It has been most satisfactory to bend a pipe cleaner double, preferably around a pencil tip or similar small object to form a loop of the desired size. The ends of the pipe cleaner then are twisted to form a handle. The doubled pipe cleaner possesses sufficient stiffness to perform most rubbing operations. If the pipe cleaner loop is placed in the virus inoculum, enough inoculum usually is soaked up or held in the loop to inoculate a bean or a tobacco leaf adequately without repeated dipping into the inoculum. Carborundum powder or similar abrasives may be dusted on the leaf or suspended in the inoculum prior to rubbing. If the pipe cleaner is held by the twisted ends and the leaf is supported on a pot label or by a plastic bag covering the hand, the fingers are not contaminated with virus inoculum and the pipe cleaner can be discarded after use."

After mechanical inoculation the leaves can be rinsed immediately with distilled water. According to Holmes (1929) rinsing of the inoculated leaves results in an increased number of local lesions as compared to non-rinsed controls. This advantageous effect of rinsing may be explained by the fact that some plant cells being extensively damaged during rubbing regain their turgor more easily upon rinsing and are thus able to support virus multiplication.

After inoculation, the plants should not be exposed to direct sunshine because they may wilt rapidly. If some cloudy days are directly followed by a sunny day and the inoculation has to be carried out on such a day; it is advisable to water the soil in the pots before inoculation and to cover the plants which had been inoculated with wet newspaper.

To avoid the tedious procedure of preparing inocula and rubbing, Yarwood (1957) has developed a simple brush-extraction method for the transmission of viruses.

*A brush-extraction method for the mechanical transmission of plant viruses**

A stiff poster brush is dipped in water and stroked over the Carborundum-dusted surface of the donor host. The same brush is then immediately stroked over the Carborundum-dusted surface of the suscept. This combined operation of extrac-

* Yarwood (1957).

tion and inoculation (not including, of course, dusting of the leaves with Carborundum and washing of the brush) requires about 5 sec. By contrast, the standard method of grinding the infected leaf tissue in a mortar and rubbing the homogenate over a Carborundum-dusted leaf surface requires about 20 sec. The success of this brush-extraction method depends on the degree of stiffness of the brush used and on the pressure used in stroking the leaves.

The success of mechanical transmission depends on three factors (Yarwood, 1957): Properties of the virus; properties of the infected host yielding the inoculum (donor plant); properties of the healthy host plant (acceptor plant or suscept) to be infected.

Each of these factors can be influenced by appropriate methods and the success of mechanical transmission can be increased in this way.

One major property of plant viruses which are easily transmitted mechanically is that they are more resistant to unfavourable environmental conditions (ageing of the inoculum, temperature, dilution), than those viruses which are mechanically transmitted with difficulty or not at all (stable viruses as compared to labile viruses). The following method yields successful mechanical infections with some labile viruses.

*Quick virus inoculation by rubbing with fresh leaf discs**

One or more 11 mm diameter discs (four is about the optimal number, size is not important) are cut from leaves of the donor source with a cork-borer, stacked one on top of another, and held by means of cover-glass forceps with the cut edges protruding, these are trimmed to a straight edge with a scalpel, and rubbed over the Carborundum-dusted suscept surface. In this method the donor tissue is handled only with cork-borer, forceps, and scalpel, each of which can be flame sterilized. The usefulness of this method is well demonstrated by the fact that tomato spotted wilt virus which, due to its high degree of lability, could not be transmitted from tobacco to bean by the juice-carborundum method was easily transmitted from tobacco to bean when this quick method was applied.

Another major property of viruses which are easily transmitted mechanically is that they occur in the epidermis and parenchymatic tissues, i.e. they belong to viruses causing mosaic symptoms. Their insect vectors are aphids not leafhoppers. Viruses causing symptoms of yellowing seem to be localized in the phloem of their hosts, and are mechanically transmitted only with difficulty. One way of mechanically transmitting such viruses consists of injecting the expressed sap of the donor into the phloem of the acceptor plants. This method proved to be successful in transmitting sugar-beet curly top virus, clover wound tumour virus and a virus causing rough dwarf disease in maize (Harpaz, 1959). This latter method is described below:

* Yarwood (1953).

*Needle transmission of the virus causing rough dwarf disease of maize**

"Hybrid maize plants showing severe dwarf symptoms were ground in a meat mincer, the sap squeezed through cheese-cloth and then centrifuged for 5 min at 3000 rev/min. The supernatant fluid was injected by means of a 1 ml tuberculin syringe into the stalks of 3-week-old hybrid maize seedlings grown under insect-proof conditions. The dosage was about 0·2 ml per seedling divided into 5 punctures at different sites. A control series was injected with healthy sap in the same manner. Three out of twelve plants injected with diseased sap, in two different series, developed both stem and leaf symptoms (including the rare symptom of split blade) within two months. As this was done in winter, without artificial illumination, it is believed that during summer the development of symptoms should be faster. Infectivity of the sap, when frozen, was retained for at least 24 hr. Since the virus is not transmitted by seed, the reliability of the test is unquestionable."

The properties of the donor host influence the sucess of mechanical transmissions in two ways: on the one hand different host plants, individuals of different ages of the same host plant or different parts of a plant may contain the virus in different concentrations, on the other hand their inhibitor content may vary to a considerable extent. Therefore, when mechanical transmissions are made the inoculum should be taken from such a donor plant or from part of such a given donor where the virus concentration is highest and a possible inhibitor content is minimal.

Usually the virus concentration is higher in young plants than in older ones (Bercks, 1954). Potato virus Y, for instance, can no longer be detected in potatoes in the ripening period (Bartels, 1958) and the concentration of potato virus X also decreases (Bartels, 1955). Generally speaking virus concentration in a plant reaches its maximum about 3–4 weeks after infection and then it gradually decreases (Bartels, 1954). The dependence of virus concentration on the age of the donor plant has a particular influence on the success of mechanical transmission of labile viruses. As to the role of different plant organs it has been shown that less tobacco mosaic virus can be extracted from tobacco roots than from tobacco leaves.

Inhibitor effects can be eliminated or reduced in a number of ways. In this context the term "inhibitor effects" will be used in its broadest sense. Inhibitor effect is meant to mean not only the effect of some well defined inhibitor but also the deleterious effect of some physical or chemical properties (e.g. pH, oxidizing ability) of the expressed sap.

The easiest way of eliminating the inhibitor effect is to take the inoculum from host species which do not contain inhibitors. There are two possibilities of transmitting the virus from an inhibitor-containing infection source to a host plant which does not contain inhibitors. One possibility consists of choosing a method

* Harpaz (1959).

of transmission other than mechanical transmission where there is no inhibitor effect. An example illustrates this procedure. Since *Abutilon*-mosaic virus cannot be transmitted mechanically directly from *Abutilon striatum* var. *thompsonii*, Costa and Carvalho (1960) transmitted the virus first from *Abutilon* to *Sida micrantha* by grafting and used this latter host which does not contain inhibitor, as a donor plant for further mechanical transmissions. From the inhibitor-containing *Dahlia* cucumber mosaic virus can easily be transmitted by aphids to tobacco which does not contain an inhibitor against cucumber mosaic virus. Inhibitors usually interfere only with the success of mechanical transmissions.

The next method of eliminating inhibitor effects is based on the fact that inhibitors have a reduced effect on the susceptibility of host species which are taxonomically related to the inhibitor-containing plant. It is known for example that carnation mosaic virus cannot be transmitted mechanically from carnation plants to hosts which are taxonomically remote from carnation due to the inhibitor content of *Dianthus caryophyllus*. *Stellaria media* (also the *Caryophyllaceae* family), however, is not sensitive to the inhibitor contained in the sap of carnation. *Stellaria media* is free of inhibitor. Schmelzer (1957) transmitted therefore carnation mosaic virus first to *Stellaria media* by mechanical inoculation and then used these infected *Stellaria* plants as a source of inoculum for further mechanical transmissions to host species which would not have become infected with carnation virus if the inoculum had been taken directly from carnation plants.

When the inhibitor effect results from an unfavourable hydrogen ion concentration of the expressed sap the success of mechanical transmissions can be increased by adjusting the pH of the sap to a value which does not inactivate the virus. For instance the mosaic of rose can be mechanically transmitted from *Vigna sinensis* only if the pH of the sap is adjusted to 7·0 with phosphate buffer (Fulton, 1952). Prunus virus B is most infectious at pH 8·7, prunus viruses A, E and G at pH 8·0.

The effect of inhibitors influencing host susceptibility can be neutralized to a certain extent by diluting the expressed sap (cf. Bawden, 1955). The principle underlying this technique is that upon dilution the inhibitor will be diluted to such an extent as to lose its activity whereas the virus still remains infectious at the same dilution. Szirmai (1941) succeeded in eliminating the effect of an inhibitor present in red pepper plant extracts on cucumber mosaic virus by diluting the expressed sap 1 : 10.

When working with viruses which are susceptible to oxidation due to some oxidative processes in the expressed sap the success of mechanical transmission can be increased by the administration of either reducing agents or of chelating agents inhibiting the activity of the copper requiring enzyme, polyphenoloxidase.

The most commonly used reducing agents are sodium sulphite, cysteine and ascorbic acid. It is advisable to homogenize the plant tissues in the presence of one of the above reducing agents. Yarwood (1955) used a 0·5 per cent sodium sulphite (Na_2SO_3) solution to promote the success of transmitting apple mosaic virus. Cysteine is to be used at a 0·01 *M* concentration in phosphate buffer, pH 7·0. The cysteine solution should be prepared freshly before use, because cysteine is known to undergo rapid oxidation when in solution.

To inhibit polyphenoloxidase activity Hampton and Fulton (1961) homogenized cucumber cotyledons infected with sour-cherry necrotic ringspot virus in 0·01 *M* phosphate buffer pH 8·0 containing 0·01 *M* Na-diethyldithiocarbamate (DIECA). The inoculum prepared in this way produced 160 local lesions on leaves of *Momordica balsamina* in contrast to one local lesion induced by inocula prepared in the conventional way.

Some plants (strawberry, hop, *Prunus* species) contain fairly large amounts of tannins (Bawden and Kleczkowski, 1945) which have a virus inactivating effect due to their protein precipitating ability. When the tannins are precipitated with nicotine sulphate the inhibitory effect of the sap disappears (Thresh, 1956).

The virus causing the so-called infectious sterility of hops can be mechanically transmitted only if tannins are precipitated, oxidation processes reduced and the inoculum clarified. A method of doing this is described below.

*Mechanical transmission of the virus causing infectious sterility of hops**

Two grams of the leaf tissue without the midribs are homogenized in 2 ml of a solution prepared in the following way: 2 ml of nicotine and 2 grams of ascorbic acid are dissolved in 100 ml of 0·1 *M* Na_2HPO_4 and the solution is neutralized with NaOH. The homogenate is kept at —15 °C overnight. After thawing the homogenate is centrifuged at 3000 g for 1 hr and the supernatant is salted out with ammonium sulphate (0·6 per cent saturation). The precipitate is pelletted at 3000 g for 30 min, washed with saturated ammonium sulphate solution and dialyzed for 3 days against running tap water. The dark coloured precipitate formed during dialysis is removed by centrifugation and the light brown supernatant, which can be stored at —15 °C without any appreciable loss of infectivity, is used as inoculum.

The susceptibility of the acceptor host can be influenced by either physical or chemical means.

a) Physical methods. One of the most widely employed methods of increasing host susceptibility consists in the use of abrasives. Rawlins and Tompkins (1934) first pointed out that if the leaves are dusted with carborundum prior to inoculation the number of local lesions will be higher than with the control. The mode of action of carborundum and probably other abrasives (e.g. celite) consists of increasing the number of entry points for the virus owing to wounding of the leaf surface (Beraha et al., 1955). Either the leaves of the acceptor host are dusted with abrasive prior to inoculation or the abrasive is added directly to the inoculum to be used for infection.

The susceptibility of host plants can be influenced also by changing the illumination and/or temperature under which the plants are grown. Placing the plants

* Čech et al. (1957).

to be infected in the dark for about 24 hr prior to inoculation usually results in increased susceptibility. *Vigna sinensis* gives a reliable reaction to infection with cucumber mosaic virus only after a preliminary dark treatment (Sill and Walker, 1952).

b) Chemical methods. The most widely used chemical in increasing susceptibility to virus infection is phosphate buffer.

*Phosphate effect in mechanical virus transmissions**

A piece of TMV infected tobacco leaf of a surface of about 1 cm^2 is homogenized with 2 drops of water and a dilution series is made from the homogenate. Each diluted sample is divided into four parts and used to infect Pinto bean leaves in the following combination: control, 1 per cent K_2HPO_4 added, carborundum added, carborundum + 1 per cent K_2HPO_4 added. When both carborundum and 1 per cent K_2HPO_4 were used, the number of lesions was higher by a factor of 2800 compared to the control. With tobacco as acceptor plant the effect was not so dramatic. Phosphate effect was found also with clover mosaic virus, alfalfa mosaic virus and tobacco necrosis virus. Phosphate acts most probably by increasing the susceptibility of the acceptor host. In practice the phosphate-effect is ensured by homogenizing the virus infected leaf tissues in a 1 per cent solution of K_2HPO_4 or Na_2HPO_4.

4. INSECT TRANSMISSION

From an epidemiological point of view insects play the most important role in the transmission of plant viruses. Virus transmitting insects are called *vectors*. For an insect to be a vector it is necessary that its mouth parts be such as to pick up and eject viruses. The most suitable for this purpose are the mouth parts of *Hemiptera*. Most vectors (aphids and leaf-hoppers) belong to *Hemiptera*. In the following our attention will be devoted exclusively to aphids and leaf-hoppers.

In order to be able to give a general survey on the mechanism of insect transmission some basic terms have to be explained. Feeding on the infected plant is called *acquisition feeding*. The time period elapsing from the beginning of the acquisition feeding to the time when the insect is able to transmit the virus is designated as *incubation (latency, celation) period*. The feeding of the vector which has already acquired the virus on the healthy plant to be infected, is called *infection feeding*. The time period during which the vector carrying the virus remains infectious is specified by the term *persistence*.

On the basis of the mode of transmission of plant viruses by vectors 3 groups of plant viruses are to be distinguished: non-persistent viruses, persistent viruses (Watson and Roberts, 1939) and semi-persistent viruses (Sylvester, 1956).

* Yarwood (1952).

Non-persistent viruses (e.g. cucumber mosaic virus, potato virus Y) usually cause mosaic symptoms. They are easily transmitted mechanically. They occur in the highest concentration in the epidermis and parenchymatic tissues of their hosts. Their insect vectors are aphids. In contrast to persistent viruses they are little specialized with regard to their vectors. Onion yellow dwarf virus, a non-persistent virus, has for instance more than 60 aphid vectors (cf. Köhler and Klinkowski, 1954). As far as the mechanism of transmission is concerned, non-persistent plant viruses are characterized by the following properties (cf. Heinze, 1957):

1. A short acquisition time (from 30 sec to 2 min) is enough for the aphid to acquire the virus. The stylets of the aphid do not penetrate deep into the host tissues during acquisition (Bradley, 1956). With potato virus Y, the longer the acquisition period the less is the percentage of transmission (Bradley, 1954). This may be explained by the fact that non-persistent viruses causing mosaic symptoms occur at the highest concentration in the uppermost cell layers of the leaf (Bawden et al., 1954). If the acquisition period is prolonged, the stylets of the aphids will penetrate to deeper cell layers and will thus have access to decreasing amounts of virus.

2. Preacquisition fasting enhances transmission (Hamlyn, 1953). The phenomenon may have three possible explanations. One of them is that during fasting no saliva is excreted (Watson, 1938) and production of saliva does not start at the beginning of the acquisition period either. The possible inhibitory effect of saliva decreases during fasting. Another factor which may contribute to the increase in the percentage of infection upon preacquisition fasting is that aphids which have been subjected to fasting will immediately start to feed and will take up the sap after some probings right from the upper cell layers, i.e. the epidermis (Watson and Nixon, 1953). As already mentioned, virus concentration is higher here than in lower cell layers, and, as shown by Simons (1958) there is a direct correlation between virus concentration in the plant and percentage of aphid transmission, at least in the case of the three non-persistent viruses studied. A third factor which may contribute to an increase in the percentage of transmission upon fasting is that aphids which have been subjected to fasting and a very short acquisition feeding will start to feed immediately when placed on the healthy plants. Consequently the same mechanism may be involved in infection feeding as in acquisition feeding.

3. No latent period is required for a successful transmission. This points to the fact that in the case of non-persistent viruses only a "superficial" relation exists between the vector and the virus. The nature of this "superficial" relation will be elucidated below.

4. The virus is retained in the vector for very short periods. The virus is retained in the aphids for a longer period when they are starved after the acquisition feeding than when serially transferred from one plant to the other (Hamlyn, 1953).

5. Moulding of a vector carrying a non-persistent virus results in the loss of its ability to transmit the virus.

These two latter criteria suggest that for the transmission of non-persistent viruses it is not necessary for the virus to get into the blood stream of the vector for transmission to occur. It seems sufficient that the virus be bound in some way to the exuvium. The most acceptable hypothesis compatible with the criteria mentioned above is that non-persistent viruses are simply attached to the tip of the stylets of the vectors during the transmission process. This assumption seems to be supported by the findings of Bradley and Ganong showing that treating the tips of the stylets of aphids carrying non-persistent viruses with formaline (Bradley and Ganong, 1955b) or UV-irradiation (Bradley and Ganong, 1955a) results in the loss of their ability to act as vectors.

Despite the fact that a great number of experiments were carried out to elucidate the mechanism of vector transmission of non-persistent viruses (cf. Heinze, 1957) several phenomena have still remained unanswered. No explanation can be given e.g. of why tobacco mosaic virus and potato virus X, which are known to be easily transmitted mechanically, cannot be transmitted by aphids. On the other hand if aphids are allowed to feed on tobacco leaves, surface-treated with tobacco mosaic virus suspension and are transferred to *Nicotiana glutinosa*, typical local lesions will appear on this plant (Teakle and Sylvester, 1962). It is still an open question why aphids are not able to carry non-persistent viruses if their mouth parts are dipped into a virus containing solution or if the aphids are allowed to suck from a virus suspension through a membrane (Heinze, 1959).

PERSISTENT VIRUSES

Persistent viruses usually cause yellow type symptoms. In most cases they are restricted to the phloem of the host plant. They are either not transmitted at all or only with difficulty by mechanical means. Their insect vectors are mostly leaf-hoppers (e.g. aster yellows virus, stolbur virus) and seldom aphids (potato leaf roll virus). The former are highly specialized to their vectors whereas the latter are less specialized (Heinze, 1959). The mechanism of transmission of persistent viruses is characterized by the following features:

1. The vector will acquire the virus only after an acquisition feeding of comparatively long duration (minimum half hour). This is explained by the fact that, since yellow type persistent viruses are localized in the phloem of the plant (Rawlins et al., 1956) the stylets have to reach the phloem to pick up the virus. The existence of such a mechanism is supported by the finding that the percentage of transmission is in direct correlation with the duration of the acquisition and infection feedings.

2. Preacquisition fasting has little effect on the percentage of transmission.

3. For successful transmission a latent period is required after acquisition feeding. During this incubation period the vector is unable to transmit the virus.

With persistent viruses apparently a physiological relation exists between the virus and its vector. This assumption is supported by some experimental results indicating that if vectors of persistent viruses are allowed to pick up the virus through a membrane or if the virus is injected into the body of the vectors they will transmit the virus (Harrison, 1958; cf. Maramorosch, 1955). During the incubation period some persistent viruses pass through the gut wall into the blood stream and thence into the salivary glands of the vector from which they are ejected (cf. Bawden, 1950).

Fukushi (1933) demonstrated transovarial passage of rice stunt virus. Fukushi (1940), Black (1941), and Maramarosch (1952) have established that some persistent viruses multiply in leafhoppers. Some leafhopper-transmitted viruses, however, such as sugar beet curly top virus, do not multiply in their vectors. An interesting observation pointing to the "physiological relation" between a persistent virus and its vector has been made by Sukhov and Razvyazkina (1955). They have shown that the virus causing oak dwarf disease induces the same type of inclusion bodies in both its vector, *Liburnia striatella* and its host plant. In some cases, even a pathological effect of the virus on its vector could be demonstrated. This pathogenic effect may appear as cytological changes in the fat bodies of the vector (Littau and Maramarosch, 1960) or in the premature death of the leafhopper carrying the virus (Jensen, 1959; Watson and Sinha, 1959).

SEMI-PERSISTENT VIRUSES

With respect to their properties semi-persistent viruses represent a transitional group between non-persistent and persistent viruses. Sylvester (1956) established the group of semi-persistent viruses through a detailed study of sugar beet yellows virus. This virus is characterized by the following properties:

1. Preacquisition fasting of the aphids does not enhance transmission (in contrast to non-persistent viruses).

2. No latent period is required for transmission (in contrast to persistent viruses) in spite of the fact that relatively long acquisition feeding (several hours) is necessary for the vector to be able to transmit the virus (in contrast to non-persistent viruses).

3. The virus is retained in the vector for a maximum of 2–3 days in contrast to persistent viruses where the vector remains infective for months and sometimes even for the whole duration of its life time.

The mechanism of transmission of the semi-persistent sugar beet yellows virus is unknown. Treating the tips of the stylets with formalin did not result in the loss of infectivity of aphids carrying sugarbeet yellows virus.

Since the mechanism of insect transmission of plant viruses is far from being elucidated, the above way of grouping is somewhat artificial.

The success of insect transmission depends on several factors.

1. The species or clone of the vector. The non-persistent southern cucumber

mosaic virus is transmitted by different clones of *Aphis gossypii* with varying degrees of efficiency (Simons, 1959). Vector specificity of persistent viruses is a well known phenomenon.

2. The host species and the actual physiological state of the host. This factor influences the efficiency of insect transmission through its action on the virus concentration in the host plant. Bradley (1962) has shown for instance that the uneven distribution of potato virus Y in tobacco leaves strongly influences the percentage of aphid transmission.

3. Properties of the virus. A strain of cucumber mosaic virus isolated from spinach has been maintained by Badami (1958) for nine years during which it was transmitted only mechanically. After these nine years of serial mechanical transfers the virus which was originally transmitted by aphids at a high rate, completely lost its ability to be transmitted by *Myzus persicae*. Potato yellow dwarf virus could not be transmitted by its leafhopper vector *Aceratogallia sanguinolenta*, when it was transmitted for several years by mechanical means. Some viruses can be transmitted by aphids only from plants which in addition to the virus in question are infected by another virus. For instance potato aucuba mosaic virus is transmitted by *Myzus persicae* only from potato plants simultaneously infected with potato virus Y or potato virus A (Kassanis, 1961).

4. Environmental factors acting on the vector. Brčak (1959) has shown that sugar beet mosaic virus was most efficiently transmitted by aphids which were fasting before acquisition feeding at a relative air humidity of 34 per cent. An air humidity higher or lower than the above had a deleterious effect on the percentage of transmission.

For more details concerning insect transmission we refer to the papers of Black (1953, 1958, 1959), Maramarosch (1955, 1958, 1960), Limasset (1956), Heinze (1957), Smith (1958a), Jensen (1959), Schmidt (1960), Schmutterer (1960), and Watson (1960a).

The two most important groups of vectors are, as described above, aphids and leafhoppers. The techniques involved in culturing these insects are different with these two groups. Therefore the methods of maintaining stock colonies and carrying out transmissions with aphids and leafhoppers will be dealt with separately.

Transmission of plant viruses by aphids. In general every plant virus laboratory has a stock colony of aphids. The aphids (*Myzus persicae* Sulz. and *Doralis fabae* are the most commonly used species) have to be kept under strictly controlled virus free conditions. The most convenient way of preventing the plant (e.g. sugar beet) on which the aphids are cultured from becoming virus infected is to keep the plants in cages. Many types of cages have been used for aphids. Cloth or screen of plastic or wire can equally be used. The pot with the plant on which the aphids are kept is placed on another empty pot turned upside down, the pot below being in a tray half-filled with water.

The actual experimental procedure using aphids as virus vectors requires a method of transferring aphids from one plant to another and a method for removing aphids from plants at the end of the experiment.

The most important general rule which has to be observed during the transfer of aphids from the donor plant to the acceptor host is to avoid breaking off the stylets of the aphids. Transfers of aphids can be made conveniently by picking up the insects with a small slightly moistened, camel's hair brush. A feeding aphid can often be induced to withdraw its stylets from the leaf by a gentle touch with the brush. This method is less likely to injure aphids than attempting to brush them from the leaves. In picking aphids from virus-infected leaves, the tip of the brush may become contaminated with the virus. Thus, a virus can easily be transmitted with the brush. One method of avoiding this is to place a small piece of clean paper on the healthy plant and place the aphids on this. They will rapidly move to the leaves. A large number of aphids can be transferred rapidly by placing the leaf or part of a leaf on which they are feeding on a piece of paper laid on the healthy plant. As the detached leaf wilts and dries, the aphids will move from it to the healthy plant. At the termination of an experiment aphids may be killed by fumigating or spraying with an insecticide. The fumigating chamber should be in a well aerated area well away from the stock culture of aphids.

Transmission of the persistent potato leaf roll virus by aphids.

Potato leaves infected with potato leaf roll virus are placed in Petri dishes and 10–15 *Myzus persicae* aphids are placed on the leaves from the virus free stock culture. After an acquisition feeding of three days the aphids will certainly be viruliferous. After this period of time they are transferred in groups of two to *Physalis floridana* plants in the 2 leaf stage. After an infection feeding of three days the *Physalis* plants are fumigated with an insecticide and transplanted in wooden boxes. As a control the same experiment is carried out with virus free aphids to check the effect on growth due to the feeding of the aphids. As another control untreated *Physalis floridana* plants are also planted in separate boxes. To ensure a rapid development of characteristic symptoms the plants are illuminated from dawn to 7 o'clock in the morning with tungsten lamps of 250 W placed at a distance of 60 cm. Leaf roll symptoms will appear on *Physalis floridana* in 8–12 or in the case of artificial illumination in about 5–7 days. The leaf blades will be reduced in size and the internodes will shorten. The plants on which non-viruliferous aphids were fed will show growth reduction.

Transmission of the non-persistent cucumber mosaic virus by aphids

Ten aphids (*Myzus persicae*) are removed from the healthy stock plant and placed on tobacco infected with cucumber mosaic virus. After an acquisition feeding period of 10 min they are transferred to a healthy plant. They are allowed to feed for 10 min and killed by spraying or fumigating with insecticide.

Transmission of plant viruses by leafhoppers. Techniques of handling leafhoppers are somewhat different from those used with aphids primarily because leafhoppers move rapidly and may fly for considerable distances. More precaution must be taken against accidental escape because once a few insects have escaped in the greenhouse it is nearly impossible to recapture them.

Leafhoppers reproduce by means of eggs deposited in plant stems. It should be noted that such eggs may not be killed by treatments effective against adults. For making transfers of leafhoppers, aspirators are commonly used. A convenient method of collecting leafhoppers from large cages is to have, in one wall of the cage, a cloth sleeve with a cotton glove sewed on the end. The aspirator is inserted into the cage through a slit in the palm of the glove and the arm placed into the cage through the sleeve. Insects disturbed on plants when they are feeding by gently shaking will be attracted to the light shining into one side of the cage and can be collected rapidly. The same type of cage can be used to feed insects on experimental plants.

Cages consisting of cylinders of plastic or screen with "skirts" of cloth at both ends can be used to tie over single pots, or single branches of larger plants. Insects are transferred in and out of such cages through a small hole in the cylinder which is kept plugged with cotton between transfers. A convenient method of handling small numbers of leafhoppers is by means of a "pillbox"-type cage clamped against the surface of the leaf. The side of the cage applied to the leaf is made of thin material such as bolting cloth or nylon hose, through which the insects feed on the leaf. Insects are introduced into the cage by turning the outer side wall until the hole in it is lined up with the hole in the inner wall. In making transfers from one plant to another the entire cage is transferred.

5. DODDER TRANSMISSION

It was Bennett (1940) who first reported the successful transmission of sugar beet curly top virus by *Cuscuta subinclusa*. His method consisted of parasitizing the diseased plants with dodder and then transferring stem parts of the dodder to healthy plants. After some days the dodder stems were removed from the healthy plants. The plants showed signs of disease after the incubation period. Later it was shown by Bennett (1944) that tracheal parts and the phloem of both dodder and sugar beet plants were in an anatomical contact when the transmission occurred.

Different viruses are transmitted by *Cuscuta* species with various degrees of efficiency (Schmelzer, 1956a). This may be connected with different amounts of inhibitors present in individual Cuscuta species (Schmelzer, 1956b).

When a stem of dodder (*Cuscuta campestris* and *Cuscuta subincluse* are most commonly used) is allowed to establish itself on a virus-infected plant and then is allowed to parasitize a healthy plant it will transmit most viruses. In some cases the virus multiplies in the dodder whereas in others the dodder serves merely as a channel through which the virus moves. Since dodder will parasitize many plants in different families the method of dodder transmission has been valuable in transmitting viruses between plants which cannot be grafted. By this technique it has been possible to extend host ranges of viruses and to get a virus out of one host where it cannot be studied into another where it can be studied.

A culture of dodder may be started from seed or by transferring stem tips. Dodder seed may be sown with the seed of some host on which the dodder will establish itself after the seeds germinate. Another method is to germinate dodder seeds in a Petri dish. After seedlings are 2–5 cm long the lower end is placed in a small vial containing water, which is tied to the stem of the host. The dodder seedling should also be tied to the host to support it until haustoria are formed.

Virus transmission may be aided by darkening the healthy plant. This procedure presumably induces a flow of food material and virus from the diseased plant. Another way of promoting the efficiency of virus transmission by dodder consists of having the interconnecting dodder strands as short as possible. This is done by placing diseased and healthy stems together and allowing the dodder to twine around both.

6. SOIL TRANSMISSION

For soil transmission to be successful it is necessary that the virus should not lose its biological activity in the soil for prolonged periods and should be able to get in contact with the healthy plant and to infect it. There are several mechanisms of soil transmission:

a) In the simplest case the virus particles are so stable that they retain their biological activity in the soil without the help of any other factors (e.g. adsorption to soil fractions, organisms). In this case soil transmission can be effected by adding virus containing sap to the soil. The healthy plants are infected either through minute wounds on the roots or through wounds of the lower leaves which are in contact with the soil. TMV and tobacco necrosis virus are known to be soil transmitted in this way.

b) Viruses of the wheat mosaic group (Harrison, 1960) are so labile in expressed sap that their transmission through the soil cannot be explained by the mechanism mentioned above. When the sap of wheat plants infected with wheat mosaic virus is added to the soil no soil transmission will occur. Therefore, it has to be assumed that the virus particles are stabilized by some factor in the soil to make them soil transmissible. Miyamoto (1959a; 1959b) has shown that barley yellow mosaic virus and wheat yellow mosaic virus are adsorbed to soil colloids less than 2 μ in diameter.

c) Ringspot viruses (Harrison, 1960; tomato black ring virus, raspberry ringspot mosaic virus, Arabis mosaic virus, cherry rosette virus) are also too labile to survive in the soil. In contrast to the viruses of the wheat mosaic group in this particular case, some organism seems to be responsible for soil transmission. Cadman and Harrison (1960) have demonstrated that soil samples which had been soaked with a suspension of potato black ring virus did not transmit the virus and treatment of contaminated soil with pentachloronitrobenzene or tetramethyl-thiuram-disulphide, known to be without effect on the virus itself, abolished virus transmission. With grapevine fanleaf virus Hewitt (1956) has provided conclusive evidence for the fact that the soil inhabiting nematode, *Xyphinema index* is responsible for the soil transmission of the virus. Tobacco necrosis virus has been shown to be transmitted by the zoospores of *Olpidium brassicae* (Teakle, 1960).

For more details on the soil transmission of plant viruses we refer to Harrison's review (Harrison, 1960).

CHAPTER IV

HOST-VIRUS RELATIONS

Whenever a plant virus interacts with a plant, one of the following types of host responses will occur:

1. The host plant becomes infected and reacts with visible symptoms. The virus multiplies in the host (susceptibility).

2. The host plant becomes infected but fails to show symptoms despite the fact that the virus multiplies in it (tolerance). This type of infection is called latent infection. The cause of latency is not yet known. Bennet (1956) suggested that tolerance is due to a decrease in virus concentration in the host. In this respect tolerance may be regarded as a form of normergic defense rection (see below).

3. The host plant becomes infected, reacts with symptoms, the virus multiplies in it, at the same time, however, the plant sets in work its defence reaction (resistance).

Resistance can be either normergic (plasmatic defence reaction) or hyperergic (hypersensitive). When the virus becomes localized in certain parts of the host plant (e.g. TMV in inoculated leaves of sugar beet), or the infected host recovers from the disease (e.g. tobacco plants infected with tobacco ringspot virus, Benda and Naylor, 1954), resistance is normergic. Hyperergic defence reactions may occur in either of the following ways:

a) The host eliminates the virus by abscission of the infected leaves (e.g. certain cultivars of red pepper infected with TMV) before the virus particles can move into the phloem (amputative hypersensitivity). Since decrease in auxin content is an important factor in leaf abscission and since the auxin content of virus infected plants usually decreases, it can be assumed that a sudden decrease in auxin content plays an important role in amputative hypersensitivity. This suggestion has been supported by Dyson and Chessin (1961). They have shown that spraying of plants which normally react to virus infection with leaf abscission with 1000 ppm naphthalene-acetic acid prevents leaf abscission. It has to be mentioned, however, that indoleacetic acid (natural auxin) had no such effect.

b) Infected groups of cells of the host die (e.g. TMV-*Nicotiana glutinosa*) preventing thereby the further spread of the virus (necrogenic hypersensitivity).

To get a coherent picture of host-virus relationship we first have to deal with the fate of the virus in the host and then with the ways the host reacts to virus infection. Finally external factors which have an influence on host-virus relations will be discussed.

1. VIRUS IN THE HOST

The first step in the development of the host-virus relationship consists in the penetration of virus particles into host cells (Penetration of plant viruses). After entering the host cell, the virus will multiply (Virus multiplication). At the same time the infectious entities will invade the entire host plant or a part of it (Translocation).

PENETRATION OF PLANT VIRUSES INTO HOST CELLS

It has been known since Caldwell's (1932) classical experiment that plant viruses can enter living cells only through wounds. Caldwell (1932) infiltrated TMV solutions into tobacco leaves but no infection occurred. From a microbiological point of view plant viruses are obligate wound parasites.

VIRUS MULTIPLICATION

Virus multiplication (virus biosynthesis) can be studied from different standpoints. First the amount of infectious virus particles can be titrated by biological methods during virus multiplication. Secondly different phases of virus multiplication can be approached by metabolic studies. Thirdly, *in situ* electronmicroscopic investigations may be used to reveal structural aspects of virus multiplication in plant cells.

It has been shown that the RNA moiety of TMV (Gierer and Schramm, 1956) and of several other plant viruses (Kaper and Steere, 1959a,); Diener and Weaver, 1959; Harrison and Nixon, 1959; Kassanis, 1960; Rushizky and Knight, 1959; Reichmann and Stace-Smith, 1959) is capable of initiating infection on its own and is able to trigger the synthesis of complete virus particles. This means that the whole genetic information of the complete virus particle is contained in its RNA. If this is so, plant virus nucleic acids have to fulfil the following requirements: 1. once in the host cells they have to separate from their protein coat; 2. they have to be capable of self-reproduction; 3. they have to direct the synthesis of their coat protein.

Separation of the viral nucleic acid from the protein coat. The following observations if taken together point to the fact that TMV-RNA actually separates from the protein at a very early stage of infection:

(i) Infiltration of tobacco leaves with RNase prior to or within 2 hr after inoculation with TMV leads to the abortion of the infection (Hamers-Casterman and Jeener, 1957). Since intact TMV unlike TMV-RNA is not sensitive to the action of RNase the above finding suggests that for about 2 hr after inoculation TMV-RNA occurs in a free state in host cells.

(ii) If *Nicotiana glutinosa* leaves are inoculated with TMV and TMV-RNA, respectively, and UV-irradiated at different intervals after inoculation, the UV-resistance of both TMV and TMV-RNA starts to increase after a certain period of time as judged by the number of local lesions which later appear (Siegel et al., 1957). The length of the lag period preceding the increase in resistance, however, is different with TMV and TMV-RNA. With TMV it lasts for about 2 1/2 and 5 hr, respectively, depending on the virus strain used, whereas with TMV-RNA it is much shorter than that or is practically non-existent. These results have been interpreted in the following way: Since the viral RNA is much more sensitive to UV-irradiation than the intact virus, the UV-sensitive phase observed suggests that during this lag period viral nucleic acid exists in a free state in the host cell. The increase in UV-resistance, on the other hand, shows that the nucleic acid macromolecules have been bound by some receptor in the host cell, losing thereby their sensitivity to UV-irradiation. The fact, that infection with intact TMV results in a rather long UV-sensitive period indicates that nucleic acid separates *in vivo* from the protein moiety of the virus particle and occurs in a free state in the cells. TMV-RNA, on the other hand, if introduced into host cells, will be bound immediately by the cell-receptor(s). The same results are obtained with the TNV-bean virus-host combination (Kassanis, 1960) as well.

(iii) Maximum number of local lesions (Fraenkel-Conrat et al., 1958; Kassanis, 1960) with both TMV and TNV, infective TMV-RNA in a free state (Schramm and Engler, 1958) and infective intact TNV (Kassanis, 1960) appear in the infected tissues earlier upon inoculation with the viral RNA than with the intact virus. These time differences presumably correspond to the time periods needed by the viral RNA to separate from its protein coat.

The mechanism of the *in vivo* separation of the viral nucleic acid from the protein moiety of the virus particles is not known. The results of Gordon and Smith 1960, 1961), however, suggest that the enzymatic machinery of the host cells may be involved. The above authors found that at low light intensities *Rhoeo discolor* is immune from infection with intact TMV, but can be infected with TMV-RNA. This indicates that at least at low light intensities cells of *Rhoeo discolor*, though able to support virus multiplication once it has been started, are unable to cooperate in the initiation of the infection process, most probably in the separation of TMV-RNA from the viral protein.

Self-reproduction of plant viral RNA. Once in the host cell in a free state, plant virus RNA has to induce the synthesis of complete virus (nucleoprotein) particles (i.e. of both RNA and protein).

The problems involved in this process are very similar to those of normal protein biosynthesis. There are, however, some basic differences. In the synthesis of normal proteins (e.g. enzymes) the message coded in nuclear DNA is transferred by the *m*-RNA to the cytoplasmic ribosomes and read by *t*-RNA to which the specific amino acids are attached:

DNA → RNA → Protein
Nucleus Cytoplasm Ribosomes

The characteristics of this process at the nucleic acid level are as follows:

1. The original information is contained in the double stranded nuclear DNA which has a double function: on the one hand it reproduces itself according to the Watson-Crick model and on the other hand it serves as a template for *m*-RNA (DNA-dependent RNA synthesis). Nuclear DNA does not leave the nucleus and does not become a structural component of the product of the synthesis (enzyme).

2. The short-lived *m*-RNA serves only to transmit information and does not get incorporated into the product either.

In the case of plant virus biosynthesis we have the following situation:

The original information is contained in the single stranded viral RNA which is introduced into the cytoplasm. This single stranded RNA has to play the role of the double stranded DNA: on the one hand it has to reproduce itself and on the other it has either to serve as a template for some sort of *m*-RNA or to act as an *m*-RNA itself.

The question arises as to how viral RNA can cope with these requirements. Most of the pertinent experiments have been performed with TMV-RNA. These will be summarized below.

As mentioned earlier the results of Siegel et al. (1957) suggest that after TMV-RNA has separated from its protein coat, it is adsorbed by some "receptor" in the host cell. It seems very likely that this "receptor" is the cell nucleus. Examination of nuclei isolated from tobacco leaves infected with P^{32} labelled TMV-revealed the presence of a radioactive RNA fraction, which was nonsedimentable at 105,000 g and digestible with RNase (Reddi, 1966). Its nucleotide composition was similar to that of TMV-RNA. These results seem to indicate that the radioactive RNA present in the nuclear fraction was parental TMV-RNA. As a next step of TMV biosynthesis this parental RNA is expected to multiply. The earliest report suggesting that shortly after infection with TMV there is an accumulation of RNA in the nuclei of infected cells comes from Zech and co-workers (Zech, 1954; Zech and Vogt-Köhne, 1955). By the use of UV microscopy, they have shown that in TMV-infected tobacco leaf hair cells, the absorbancy of the nucleus at 260 mμ increases shortly after the cell has been infected. The hypothesis that the initial site of viral RNA synthesis may be the nucleus and/or nucleolus has been substantiated by later studies. Hirai and Wildman (1963) have shown by cytochemical methods involving differential staining of RNA and DNA that it is indeed RNA which accumulates in the nuclei of TMV-infected cells. Bald (1964) came to the same conclusion by using phase contrast microscopy and digestion with RNase. A further suggestion that cell nuclei might be the initial site of TMV-RNA synthesis was given by Yasuda and Hirai (1964). They examined the incorporation of H^3-uracil into TMV infected tobacco leaf epidermis by means of microautoradiography and have found that the rate of incorporation into the nuclei was much higher with TMV-infected cells than with healthy ones.

All these results clearly show that shortly after infection with TMV there is an accumulation (Zech, 1954; Zech and Vogt-Köhne, 1955; Hirai and Wildman,

1963; Bald, 1964) or rather synthesis (Yasuda and Hirai, 1964) of RNA in the nucleus. Two important questions, however, still remain unanswered, namely:

1. What are the steps involved in RNA-synthesis in the nucleus, and
2. Is it really viral RNA which is synthesized.

As to the first question there are two possibilities: viral RNA is synthesized either via the "normal" pathway, i.e. there is a DNA-dependent RNA-synthesis, or by some other mechanism in which the participation of DNA is excluded. The results of Sänger and Knight (1963) point to the second possibility. They have demonstrated that actinomycin D, a potent inhibitor of DNA-dependent RNA synthesis, inhibits normal nucleic acid metabolism in plant cells, but not TMV-RNA synthesis. Some further data also show that the participation of host nuclear DNA is not needed for TMV-RNA synthesis and throw some more light on the mechanism of this type of RNA-directed RNA synthesis, which consists in the formation of double-stranded RNA.

Shipp and Haselkorn (1964) were the first to report the occurrence of the so-called replicative form of TMV-RNA. In the nucleic acid fraction of TMV-infected plants, they have found a component which when mixed with P^{32}-labelled TMV-RNA made the latter resistant to the action of RNase probably by hybridization. The product thus formed was resistant also to DNase, could be heat-denatured and reannealed with P^{32}-labelled TMV-RNA and on melting exhibited a sharp thermal transition curve of RNase susceptibility. These data have been confirmed and the double-stranded RNA has been further characterized by Burdon et al. (1964). Their results on the GC content, melting point, buoyant density and sedimentation constant of this particular partially purified nucleic acid fraction of infected leaves, fully support the idea that it is a double-stranded, TMV-specific RNA.

As to the second question, i.e. whether it is a virus-specific RNA which is synthesized in the nuclei of virus-infected cells, Smith and Schlegel's (1965) paper seems to give an unequivocal answer. In order to distinguish between host cell RNA synthesis and virus RNA synthesis the above authors have made use of the results obtained by Sänger and Knight (1963). They found that actinomycin D inhibited normal nucleic acid metabolism in plant cells without affecting the synthesis of TMV-RNA. Using autoradiographic techniques Smith and Schlegel (1965) demonstrated that tritiated uridine was selectively incorporated into the nucleus and nucleolus of cells of actinomycin D-treated tobacco leaf discs infected with TMV and *Vicia faba* root tips infected with clover yellow mosaic virus. Healthy tobacco and *V. faba* cells treated with actinomycin D did not show such an incorporation. With long periods of incorporation, the label appeared in virus inclusion bodies in TMV-infected cells. Especially with virus-infected and actinomycin D-treated *Vicia faba* roots it was clearly visible that the label was incorporated primarily in the nucleolus. Neither RNase nor DNase nor a combination of the two removed this label entirely. This phenomenon is compatible with the hypothesis that it was a double stranded RNA which had incorporated the labelled uridine. On the basis of the above results the authors suggested that virus-directed

RNA synthesis takes place in the nucleolus and that at least some of this RNA is virus RNA.

The role in plant virus multiplication of the supposedly nucleolar double stranded virus specific RNA has not been established yet, but on the basis of our present knowledge on nucleic acid self-duplication the following asymmetric, semi-conservative mechanism can be hypothesized:

After the penetration of the complete virus particle the single stranded RNA (plus strand) separates from the protein coat and is absorbed by the cell nucleolus. Here, serving as a template, it induces the synthesis of a complementary minus strand. The plus and minus strands combine to give the double stranded replicative form. After that the minus strand serves as a template for the formation of more plus strands.

In the following, the plus strands which have been formed have to leave the nucleolus and nucleus, get into the cytoplasm and induce the synthesis of specific virus protein.

The fact that the virus-specific RNA actually diffuses from the nucleus into the cytoplasm has been shown (Zech, 1954; Zech and Vogt-Köhne, 1955; Hirai and Wildman, 1963; Bald, 1964). On the other hand, as early as 1960 (Engler and Schramm, 1960) the accumulation of infectious RNA in a free state was established in the cells of TMV-infected tobacco leaves and later confirmed by Diener (1963). The amount of infectious RNA has been shown to increase for about 40 hr after inoculation (Engler and Schramm, 1960). After about 40 hr a sudden decrease sets in. In the authors' opinion this is the time when the free viral RNA assembles in virus nucleoprotein particles.

The experimental results mentioned so far seem to give a good coherent picture of the early events of TMV biosynthesis at the nucleic acid level. One has to bear in mind, however, that this applies only to TMV and even here the results are sometimes questionable. Ralph and Clark (1966) were unable to demonstrate the presence of double stranded TMV-RNA in the cell nuclei. Instead they suggest on the basis of sedimentation studies that the viral duplex RNA may be associated with mitochondria. TYMV duplex RNA has been shown to replicate viral RNA by an asymmetric semiconservative mechanism (Ralph et al., 1965) and to appear in TYMV-infected Chinese cabbage leaves 3 days after inoculation, i.e. during the latent period preceding the appearance of complete virus particles (Ralph and Clark, 1966). These findings strongly support the view that TYMV duplex RNA is a replicative form, as also postulated for TMV. It has also been shown (Ralph and Clark, 1966), however, that in leaves systemically infected with TYMV, P^{32}-labelled double stranded RNA is associated with the chloroplast fraction and not with the nuclei as is the case with the duplex RNA of TMV.

There are in addition some aspects concerning the role of viral RNA in the infection process, of which extremely little is known. Two of them deserve special attention.

First, as every process in a living organism, the biosynthesis of plant viral RNA has to be catalyzed by specific enzymes. The genetic code for these enzymes has to be contained in the single stranded plant virus RNA, since it is able to initiate the entire infection process alone. The size and structure of TMV-RNA are such as to allow for this, since the protein subunits of TMV consist of 158 amino acid residues (Anderer et al., 1960; Tsugita et al., 1960). Assuming a triplet code, 474 nucleotides out of the 6500 nucleotides present in TMV-RNA are enough to code for the virus protein. The rest would suffice to code for a number of enzymes. Oddly enough, no virus-specific enzyme has been found so far in virus-infected plants, except for a synthetase in TYMV-infected Chinese cabbage (Astier-Manifacier and Cornuet, 1965). These authors claim that the enzyme is able to synthesize single stranded polynucleotides *in vitro* which are identical with TYMV-RNA. These results have to be confirmed. One of the reasons for the failure to find virus specific enzymes in infected cells may be the fact that they occur in such small quantities as to escape detection by present methods.

The other problem for which no definite answer can be given yet is the origin of the building blocks of the newly synthesized virus-specific nucleic acids. A logical initial step for the elucidation of this problem is to fractionate nucleic acids in both infected and healthy leaves and to see whether there is any remarkable shift in the amounts of the individual fractions. Kubo et al. (1965) using MAK column chromatography have carried out such experiments with leaves harvested 2 days after inoculation. The overall chromatographic pattern of nucleic acids (*s*-RNA, DNA, 18 S and 28 S *r*-RNA) from healthy and TMV infected tissues except, of course, the occurrence of TMV-RNA in the latter was practically the same. Reddi (1963a), focusing his attention primarily on *r*-RNA has shown that upon infection of tobacco leaves with TMV, *r*-RNA is rapidly degraded and this degradation is accompanied by a concomitant increase in TMV-RNA (Reddi, 1963a). From this and from the results of labelling experiments using tritiated uridine he has concluded that the breakdown products of ribosomal RNA are actually utilized in the synthesis of TMV-RNA. Further studies with P^{32} labelling have suggested (Reddi, 1963b) that the breakdown products which are used in TMV-RNA synthesis are ribonucleosides. Kubo (1966) using P^{32} labelling at various time intervals after inoculation and separating nucleic acid fractions by means of MAK column chromatography, demonstrated that before the highest relative rate of TMV-RNA synthesis (48 hr after inoculation) was reached an enhanced synthesis of host-RNA had taken place as judged by the amount of P^{32} incorporated into nonviral RNA. This finding agrees with Röttger's data (1965) of a dramatic increase in RNA which had a base composition similar to normal leaf RNA in TMV-infected tobacco leaves at early stages of infection.

The same problem, i.e. the origin of the building blocks for viral RNA synthesis has been recently reinvestigated by Babos (1966) using sucrose density gradient centrifugation for the separation of ribosomes and virus particles. Using tobacco leaves which were growing by cell expansion only, (unlike the young leaves which were growing by both cell division and cell expansion used by Reddi [1963a, b]) he found that the decrease of ribosome content was independent of the amount of virus synthesized in the infected leaves. The rate of incorporation of P^{32} into ribosomes was the same with both virus-infected and healthy leaves. In Babos' opinion the discrepancy between his results and those of Reddi (1963a, b) may be due to the fact that "the source of materials for the synthesis of virus RNA is variable and depends on the physiological state of the cells in which the virus multiplies".

Shigematsu et al. (1966), however, did not find any reduction in the amount of ribosomal RNA two days after infecting young tobacco leaves with TMV. The yields of *s*-RNA and 17 S RNA even showed a four and twofold increase, respectively, upon infection. On the basis of the experimental results cited above it is very hard to come to any definite conclusion about the origin of the building blocks of TMV-RNA. We have no clear picture about the shifts in the pathways of nucleic acid metabolism due to TMV-infection.

Furthermore, it has to be stressed that most of the information we have is gathered from studies on TMV. The infection process may well be quite different with other plant viruses. It has been reported, e.g., that the RNA moieties of both wound tumour virus (Black and Markham, 1963) and rice dwarf virus (Miura et al., 1966) are double-stranded. Consequently, multiplication of these viruses will involve steps different from those of TMV biosynthesis.

Synthesis of virus protein. In the preceding chapter we have seen that the experimental evidence strongly favours the view that plant virus RNA starts to multiply alone and that the multiplication goes through a step which involves the formation of a so-called replicative, double stranded form. We have also seen that before the bulk of infective intact virus particles can be detected, protein-free infectious RNA accumulates in the infected cells in large amounts. Another line of evidence points to the fact that the protein of plant viruses also accumulates at a certain stage of the infection process and in this stage it is not associated with RNA.

As early as 1952 Takahashi and Ishii found that in TMV-infected tobacco leaves there is an accumulation of a protein which is missing in healthy tobacco leaves. This has been called X-protein (by others it is also referred to as abnormal protein, soluble antigen). Later they found (Takahashi and Ishii, 1953) that the X-protein consists of RNA-free spherical particles 15 mμ in diameter; moreover it has an antigenic structure similar to that of intact TMV and polymerizes *in vitro* to form rod shaped particles which resemble intact TMV particles. The amino acid composition of the X-protein and of the protein moiety of TMV is the same (Newmark and Frazer, 1956). Electrophoretic mobility of the X-protein is characteristic of the TMV-strain which produces it (Takahashi, 1955) indicating that the X-protein is strictly virus-specific. This is shown also by the fact that the host does not seem to have any influence on the serological properties of the X-protein (Takahashi and Ishii, 1953). Van Rysselberge and Jeener (1957), by showing that the abnormal protein incorporates C^{14} provided in the form of $C^{14}O_2$ at a much higher rate than the virus, came to the conclusion that it is actually the immediate precursor of the protein part of the virus. The specific radioactivity of the virus and that of the X-protein supported this hypothesis.

On the basis of the results mentioned above it is not difficult to visualize TMV biosynthesis as a process in which viral RNA and protein first accumulate separately and are then assembled as a final step of TMV multiplication. This hypothesis is supported by three independent lines of evidence.

1. Takahashi (1959) has demonstrated that anomalous noninfectious protein isolated from diseased tobacco plants and nucleic acid isolated from TMV could be constituted *in vitro* to form highly infectious TMV particles. There is no reason to suppose that TMV could not assemble *in vivo* by the same mechanism.

2. As early as 1949 Markham and Smith demonstrated that purified preparations of turnip yellow mosaic virus contained two types of particles. One was a nucleoprotein containing RNA packed in a roughly spherical shell of protein, and the other was an apparently identical protein containing no RNA. In the ultracentrifuge the infectious nucleoprotein sedimented at a higher rate (bottom component) than the non-infectious protein (top component). This shows that the synthesis of virus particles does not necessarily imply the simultaneous presence of proportionate amounts of both RNA and protein but may involve a mechanism in which viral protein is produced in excess. It has to be mentioned that neither with TMV nor with TYMV were pools of detectable quantities of X-protein (Commoner et al., 1953; Jeener, 1954) and top component, respectively, found before the appearance of the virus. This, however, does not mean that they cannot be in a way precursors of the virus. It may just signify that at the time of their formation in excess there was not enough viral RNA present to combine with them. Subunits of the protein of TYMV are able to form virus-like particles *in vivo* giving rise to the "top component" whereas subunits of TMV protein for some reason are unable to do so and occur in the infected plant as non-aggregated X-protein.

Later Matthews (1960) using density gradient centrifugation, demonstrated the presence of altogether five virus-like nucleoprotein fractions (in addition to the top component containing no RNA) with approximate RNA contents of 6 per cent, 10 per cent, 20 per cent, 35 per cent, and 36 per cent. Only the fraction containing the highest amount (36 per cent) of RNA was infectious. When virus-infected plants were labelled with P^{32} for short periods the specific radioactivity of RNA in the individual fractions decreased with increasing RNA contents (Matthews et al., 1963). This suggests that the fractions may represent steps in the assembly of the complete virus nucleoprotein.

Although it is difficult to reconcile this interpretation with our present concepts on virus multiplication, the results nevertheless suggest that the synthesis of nucleic acid and protein are probably controlled separately and do not always occur in sequence. A tentative explanation of the origin of these fractions would be to suppose that owing to some disturbance in RNA biosynthesis some RNA threads not yet fully formed are detached from their site of synthesis and coated by the excess protein. This interpretation, however, would not explain the results of the labelling experiments mentioned above. As Bollard and Matthews have recently put it (1966): "So far, there is no conclusive evidence as to the role of any of these anomalous virus proteins in the infected cell."

Macromolecules differing in their RNA content were also found in plants infected with squash mosaic virus (Mazzone et al., 1962), cherry yellow virus (Willison et al., 1961), necrotic ring spot virus (Willison et al., 1961) and wild cucumber mosaic virus (Yamazaki and Kaesberg, 1961).

3. The evidence to be outlined below shows that in some cases plant viral RNA is able to multiply in the infected cell without assembling or inducing the formation of its protein coat.

Cadman (1962) found that the NM form of tobacco rattle virus, although causing severe symptoms, exists in the form of free nucleic acid strands probably attached to cell nuclei. Babos and Kassanis (1962) have described unstable variants of tobacco necrosis virus. These also possibly exist in the cell as uncoated nucleic acid. Siegel et al. (1962) reported the isolation of two defective nitrite mutants of TMV, one of which (PM_1) failed to induce the synthesis of viral protein whereas plants infected with the other one (PM_2) synthesized a virus-like protein. This however, failed to aggregate *in vivo* with the viral nucleic acid to form complete virus particles.

As to the site of virus protein synthesis there is much controversy. Schramm and Röttger (1959) using fluorescent antibodies for *in situ* investigations, showed that TMV-protein in infected tobacco plants appeared first in a zone around the nucleus and then spread all over the cytoplasm. In this later stage of infection most virus protein was found in the microsomal fraction. No virus protein could be demonstrated from either the nuclei or the chloroplasts. From these observations Schramm and Röttger (1959) concluded that virus protein is synthesized in the microsomes. Reddi (1964) on the other hand claims to have been able to isolate complete virus particles from the nuclei of infected tobacco cells. He concludes that the site of virus synthesis (probably that of viral protein synthesis, too) is the nucleus. Reddi (1964) found the chloroplast fraction of TMV-infected tobacco leaves to be free from TMV. In contrast to this Zaitlin and Boardman (1958) showed that small amounts of TMV could be released from the chloroplast fraction of TMV-infected tobacco leaves by extraction with an appropriate buffer. Labelling experiments with C^{14}-aspartic acid suggested that TMV could be synthesized within the chloroplasts and the virus might move from its site of synthesis into the cytoplasm (Boardman and Zaitlin, 1958). This suggestion was later substantiated by Matsushita (1965). He found that most of the infectivity was observed in the chloroplast fraction 40 hr after inoculation. Thereafter, infectivity in the chloroplast fraction decreased and that of the cytoplasm increased with time.

As indicated above, nuclei, microsomes and chloroplasts, have all been suggested to be the organelles in which TMV particles or their protein moieties are synthesized. The reason for this discrepancy may be the difficulty in distinguishing between the site of viral protein synthesis and the site of virus assembly.

Recent reports, however, strongly favour the view that viral protein is synthesized within the chloroplasts. It has been shown by Spencer and Wildman (1964) that cell-free extracts of tobacco leaves were capable of incorporating amino acids into protein and that more than 80 per cent of this activity was associated with particles which sedimented at 1000 g. The 1000 g pellet consisted mainly of nuclei and chloroplasts but most of the activity was associated with this latter fraction. Later it was found (Boardman et al., 1965) that the ability of chloroplasts to incorporate amino acids was due to ribosomes. These ribosomes have a sedimentation constant of 70S and their physical properties are different from the 80S cytoplasmic ribosomes (Boardman et al., 1966). 70S ribosomes are 10 to 20 times more active in protein synthesis (*in vitro*) than 80S ribosomes (Boardman et al., 1966). Since in addition to the 70S ribosomes the chloroplasts seem to contain all the materials including *t*-RNA and/or activating enzymes necessary for protein synthesis (Francki et al., 1965) they may well be the site of active viral protein synthesis.

In extracts of Chinese cabbage, two species of ribosomes have been found (Clark et al., 1964): the 68S ribosomes appear to be confined to the chloroplasts and the 83S ribosomes to the cytoplasm. It has been shown recently (Reid and Matthews, 1966) that TYMV infection has little effect on the concentrations of 83S ribosomes and fraction II protein (both cytoplasmic) but the concentration of fraction I protein (which is in the chloroplasts) is reduced and 68S (chloroplast) ribosomes are eliminated in TYMV-infected tissues. Cytologically, TYMV causes marked abnormalities in the ultrastructure of chloroplasts of infected cells without affecting any other cell structures (Chalcroft and Matthews, 1966).

As regards methodology, the estimation of virus concentration involves a very important problem. Before determining virus concentrations in a purified virus preparation or a virus infected leaf one has to decide whether the relative amount of biologically active infectious units or the absolute amount of viral nucleoprotein as a structurally and chemically defined entity is to be estimated. The two are definitely not the same. For instance, if a virus suspension is treated with an inhibitor, the infectivity of this virus preparation may decrease or even drop to zero despite the fact that the absolute amount of chemically definable virus nucleoprotein remains unchanged.

For the estimation of the amount of infectious units a) biological methods (local lesion test) are used whereas for the quantitative estimations of viral nucleoprotein b) serological and c) optical methods are employed. Biological and serological methods can be used for both crude extracts and purified preparations whereas optical methods will give unequivocal results only if used for highly purified virus preparations.

Estimation of virus concentration by biological methods. Certain viruses and virus strains produce necrotic local lesions on given host plants. As early as 1929 Holmes observed that the number of local lesions produced on *Nicotiana glutinosa* leaves was inversely proportional to the dilution of tobacco mosaic virus inocula. From this he concluded that the local lesion method could be used for the determination of relative virus concentration. Since then, several attempts have been made to correlate mathematically the number of local lesions produced by a given inoculum and the virus concentration in it. The basic problem was whether the plots obtained by representing the number of local lesions as a function of dilution are compatible with the hypothesis that one necrotic local lesion is induced by a single virus particle. In some cases such a correlation could be established. In some others, however, various correction factors had to be applied (cf. Roberts, 1964). The calculation of these correction factors was based on either of the following two hypotheses. According to one hypothesis, deviations from a theoretical dilution curve are due to the fact that if there are virus aggregates in the inoculum the number of local lesions will be determined by the number of these aggregates rather than by the number of individual virus particles. According to the other hypothesis the cause of the incongruity should be sought in the fact that different parts of a leaf are not equally susceptible to virus infection and, therefore, the minimum number of virus particles necessary for the initiation of the infection process is different with different leaf areas.

It is clear that absolute virus concentrations cannot be determined by the use of biological methods. In everyday plant virological practice, however, in most cases this is not important. The question usually involves changes that are induced in the relative infectivity of a given virus inoculum upon different treatments. The lesion numbers which appear after infection using inocula treated in different ways are expressed in percentage values with respect to the control. In such cases the most important point is that the differences in the numbers of local lesions induced by inocula subjected to different treatment should be statistically valid, i.e. they should depend only on the treatments and not on the physiological state of the plants, leaf position, etc. Reliable methods are available to exclude such disturbing factors. Some of them will be enumerated below.

*Latin-square design**

This method can be used when the number of the leaves to be infected is equal to the number of the inocula to be tested. For instance the relative infectivity of 6 inocula in 6 plants, each having 6 leaves can be assayed using the following design (each letter represents 1 inoculum):

Leaf position No.	Plant No.					
	1	2	3	4	5	6
1	A	D	B	E	C	F
2	E	B	D	A	F	C
3	D	A	F	C	E	B
4	C	A	E	F	B	D
5	B	F	C	D	A	E
6	F	C	E	B	D	A

* According to Youden and Beale (1943).

Half-leaf method

This method is very reliable, because differences in the physiological state between two halves of the same leaf are negligible. When the number of inocula, the number of plants, and the number of leaf positions are the same, the following design can be used:

Leaf position No.	Plant No. and half leaf					
	1		2		3	
	R	L	R	L	R	L
1	A	B	B	C	C	A
2	C	A	A	B	B	C
3	B	C	C	A	A	B

If leaf position is the limiting factor, the following arrangement can be suggested (in the Table shown below there are four inocula to be tested, seven test plants, but only three leaf positions):

Leaf position No.	Plant No. and half leaf													
	1		2		3		4		5		6		7	
	R	L	R	L	R	L	R	L	R	L	R	L	R	L
1	A	B	C	B	D	C	C	A	D	B	A	D	B	A
2	B	C	A	C	D	A	B	D	B	A	B	A	C	D
3	C	D	D	A	D	B	B	A	C	B	A	B	C	A

Estimation of virus concentration by serological methods. Before describing the estimation of virus concentration by serological methods some basic principles and common techniques have to be described.

When a macromolecular substance which is antigenic in nature (antigen) is introduced parenterally (not via the digestive tract) into animal tissues, antibodies will be produced. Antibodies are proteins in nature and occur in the γ-globulin fraction of the serum of the animal which has been immunized. A serum containing antibodies is known as an antiserum. Antibodies have the property of combining *in vitro* with the antigen which has induced their production. This *in vitro* reaction is called serological reaction. The most important features of serological reactions are their great specificity and high sensitivity. The former property of serological reactions is used in plant virology for the identification of plant viruses and the latter property for the estimation of virus concentration.

From a methodological point of view we shall deal first with the techniques of immunization and then with different types of serological tests.

1. *Immunization.* Rabbits are the most convenient experimental animals in serological work. The animals can be immunized with either purified virus preparations

or clarified or crude expressed sap. In the latter case, however, one has to take into account that the antiserum will contain antibodies against both the virus and normal plant proteins. There are two ways of rendering antisera obtained after immunization with clarified or crude sap of virus-infected plants virus specific (cf. Bawden, 1950). One method consists of absorbing the antiserum with healthy plant sap. This is done by mixing the antiserum containing antibodies for both the virus in question and healthy plant constituents, with the sap of healthy plants (of course the same species which was used for immunization) in a proportion of 1 : 4. The mixture is incubated until the precipitation is completed (reaction of antibodies to normal host proteins with the host proteins present in the healthy sap), centrifuged and the supernatant, which is now free from antibodies against normal plant constituents, used for carrying out the serological reaction. Another method for eliminating host specific antibodies consists of using inocula for immunization which have been absorbed with antiserum obtained from animals immunized with sap from healthy plants.

Intravenous injection, intramuscular injection or intraperitoneal injection is used for immunization. Usually several (6–8) injections are carried out at 2–3 day intervals. Bleeding of the rabbits should take place about 7–10 days after the last injection. This time period is necessary for the antibodies to accumulate in sufficient amounts. The blood is allowed to stand for some hours at 37 °C, during which time fibrin is precipitated. The blood is then centrifuged and phenol (0·5 per cent final concentration) or merthiolate (0·01 per cent final concentration) is added to the serum to prevent bacterial growth during storage. Another convenient method of storing antisera consists of absorbing them on filter paper which is then dried and used as a source of serum whenever needed (Stapp and Marcus, 1944).

According to the method of preparation and use sera can be divided into the following groups:

I. Normal serum: serum of untreated animals (used as a control to check incidental non-specific serological reaction)

II. Antiserum: serum of animals treated with antigen
   a) homologous antiserum is an antiserum which is reacted with the same antigen which had been used to obtain the antiserum in question
   b) heterologous antiserum is an antiserum which is reacted with an antigen other than the one used to obtain the antiserum in question

2. *Types of serological test.* The term precipitation reaction denotes a serological reaction in which the antigen is of a molecular size range. Since plant viruses are actually macromolecules, the serological reaction between a purified virus preparation and its antiserum may be termed precipitation. This is usually carried out in small test tubes. A modified and very useful variation of this method is the agar-gel diffusion method which was first used with plant viruses by van Slogteren (1955a).

Agar-gel diffusion method

One gram of Agar-noble (Difco) is placed in 100 ml of physiological saline solution in an Erlenmeyer flask and allowed to swell. After 16–24 hr swelling the saline solution is replaced by a new one and 50 mg NaN_3 are added to it. The agar is boiled in a water bath for 5–7 min. As soon as the agar has completely dissolved it is pipetted in 12 ml portions into Petri dishes and allowed to spread to form a uniform layer in the Petri dish. The Petri dishes have to be placed in a horizontal position. After the agar has solidified, 7 holes, each 6 mm in diameter, are cut into the agar plate, one in the middle and six around it at a distance of five mm. The leaf to be tested is homogenized in a mortar and the brei pressed through cheese-cloth. The sap is used to prepare a dilution series (concentrated, 1 : 1, 1 : 2, 1 : 4, 1 : 8, 1 : 16). The hole in the middle of the agar plate is filled with the antiserum and the holes around it are filled with the individual dilutions of the sap. As a control the following combinations should be set up: 1. Antiserum with the same dilutions of sap from healthy plants, 2. Normal serum with a dilution series of sap from healthy plants and 3. Normal serum with a dilution series of sap from infected plants. The Petri dishes are placed in an incubator with a high air humidity at 25 °C. The reaction is evaluated after 24 and 48 hr. In the case of a positive reaction a narrow white band is visible between the antiserum and the antigen.

The term agglutination reaction denotes a serological reaction in which the antigen is of a cellular size range. If virus containing crude sap reacts with its antiserum, corpuscular elements (e.g. chloroplasts) of the sap will precipitate. By analogy to the serological reaction with bacterial cells, precipitation carried out with crude sap is called agglutination.

Agglutination test

Sap from leaves of diseased and healthy plants is obtained with a press commonly used in refractometry. 5 drops of antiserum are placed on two microscopic slides. On two additional slides 5 drops of normal serum and on two other slides 5 drops of physiological saline solution are placed. One or two drops of the sap obtained from healthy plants are added to one of the similarly treated slides and to the remaining ones one or two drops of the sap obtained from diseased plants. After thorough mixing the slides are incubated for about 10 min at room temperature and the reaction is evaluated. In case of a positive reaction, agglutination occurs only with the combination of antiserum and sap from the diseased plant. This method was introduced into plant virology by Chester (1937).

For micro amounts of sap the micromethod developed by Lovrekovich (1959) can be used. The test is carried out as follows: A glass capillary tube, 60 to 70 mm long and about 0·5 mm in diameter is filled with immune serum diluted with physiological saline solution by dipping one end of it into the serum. The serum is preferably diluted 1 : 10. However, when necessary, higher dilutions can be used. The ends of the capillary tube filled with the serum are then sealed with a microflame of a Bunsen burner. Two tests can be made with a single capillary tube.

Sap is expressed from virus infected leaves by means of a press commonly used in refractometry. If virus is to be detected in smaller leaf parts, it is more useful to press leaf parts between two glass slides of quadratic form (20 × 20 mm, about 1 mm in thickness) until a drop of expressed sap appears. The advantage of this method is that, whereas the press must be sterilized after each sample, there is no need for doing so with slides, since a new pair of glass slides can be used for each sample.

Once the expressed sap is ready, the capillary tube is broken in the middle, the open end is dipped into the sap and rotated several times in it. During this procedure a green disc consisting of chloroplasts and cell fragments is formed on the surface of the serum which is in the capillary tube. The capillary tube is then laid horizontally for 3 to 4 min in order to enhance the reaction. After this it is held perpendicularly with the open end upright; working with more capillary tubes it is most convenient to stick them with their sealed ends into a plastilline layer on a microscope slide.

With a positive reaction the particles which form the green disc are clumped together and remain on the surface of the serum or, especially with a capillary diameter higher than 0·5 mm, form a green bulk and slowly sink into the serum.

In case of a negative reaction the particles of the green disc begin to sink into the serum and gradually disperse as soon as the tubes are held perpendicularly. The particles, mainly those in the sap of older leaves, can clump non-specifically. This clumping is a spontaneous reaction or is induced by the effect of the serum. To reveal such possible non-specific reactions, the test must also be carried out with capillary tubes containing normal sera. The reaction can be seen with the naked eye, however, to make the test easier it is better to observe the tubes with a magnifying glass using a white background.

The advantage of the method described above is that it can be readily carried out, since it does not require anything else than capillary tubes prepared in advance and a press or glass slides. Furthermore, by this method the presence of the virus can be tested with as little as 0·0006 ml plant sap.

This method was originally described for potato virus X. There is, however, no reason to suppose that it would not work with other viruses as well.

Complement fixation. To understand the mechanism of complement fixation some basic principles have to be explained first. Any fresh serum (be it normal serum or antiserum) contains non-specific components known as complement. Whenever an antigen and its antibody react, the complement, an aspecific component necessary for the occurrence of the reaction is used up or fixed. On the other hand, it is known that when rabbits are repeatedly immunized with red blood cells of sheep, antibodies are formed which will cause the lysis of these red blood cells. For the hemolysis to occur the presence of the complement is needed. If the anti-sheep-cell serum is freed of its complement by heating to 56 °C for 30 min, no hemolysis will take place. However, if complement prepared from fresh guinea-pig serum is added to the reaction mixture the anti-sheep-cell serum freed of its complement (hemolytic amboceptor) is activated and hemolysis will occur. These two properties of the complement (i.e. that the complement is fixed during any antigen-antibody reaction and that it activates the hemolytic amboceptor) are used for increasing the sensitivity of serological reactions in the following way:

First the virus antigen (A) is combined with the antiserum (E) and the complement (C). The mixture is incubated at 37 °C for 1 hr. After that the indicator system, consisting of sheep red blood cells (V) and the hemolytic amboceptor (H), is added to the mixture. This is followed by incubation at 37 °C. If the serological reaction is positive, the complement will be used up in the reaction between the antigen and the antibody during the first incubation and consequently the red blood cells will not lyze. A negative reaction on the other hand is indicated by the occurrence of hemolysis. In this case the complement is not used up in the reaction between the antigen and the antibody, hence, it is able to activate the hemolytic amboceptor. What has been said above can be summarized schematically as follows:

$A + E + C \,||\, V + H$ no hemolysis, the reaction is positive

$A + E \,||\, C + V + H$ hemolysis, the reaction is negative

This serological test has so far been little used in plant virus work. According to Matthews (1957) this reaction has an advantage with spherical viruses because its use permits the detection of very small amounts of virus.

Anaphylaxis. In this case the reaction between the antigen and antibody is detected by the contraction due to anaphylactic shock of the horns of the uterus removed form virgin female guinea-pigs which had previously been sensitized with the antigen to be tested. The horns of the uterus are placed in a Ringer solution into which the antigen is then introduced. The disadvantage of the method is that it is very time consuming, and that the reaction can be carried out only once with each material. Its advantage lies in its very high sensitivity. It is not widely used in plant virology.

For the determination of relative and absolute virus concentrations the following methods can be used.

*Precipitation end-point method**

"Serial twofold dilutions of each virus sample to be compared are made in 0·14 *M* saline. An equal volume of the antiserum at a constant dilution is added to each tube. The contents are mixed by shaking and incubated in a water bath at 37–50 °C for 1–2 hr and overnight at room temperature. The highest dilution

* Matthews (1957).

of each virus sample to give a visible precipitate is noted. If one sample had an end-point of 1/128 and another 1/32, then the first sample would be taken to contain four times as many viruses as the second. If we knew from tests on a purified virus preparation the weight of virus which gives the smallest visible precipitate, then these results could be converted to weight of virus per ml of solution."

*Optimal proportions method**

"Serial twofold dilutions are made of the virus preparations to be compared and antiserum at a suitable constant dilution is added to each tube. The antiserum is added as quickly as possible to the tubes containing the virus dilutions, beginning with the most dilute. The tubes are shaken immediately and placed in the water bath. The first tube to show visible precipitation is noted. If with one virus preparation the fastest precipitation occurred in the tube containing virus at a dilution of 1/4 and in another at 1/16, then the second virus preparation would be taken to contain four times as many viruses as the first. If the optimal proportions have been determined for a purified virus preparation of known concentration, then these results can be converted to approximate weights of virus per ml."

For more detailed methodological instructions we refer to the excellent work of Matthews (1957).

Estimation of virus concentration by the optical method. This method is based on the fact that both proteins and nucleic acids absorb ultraviolet light with a well defined absorption maximum (280 mμ for protein and 260 mμ for nucleic acid). The amount of UV light absorbed is directly proportional to the protein and nucleic acid concentration, respectively, and can be measured in a spectrophotometer. If the degree of absorption (optical density) of a purified virus preparation is plotted against the wavelengths, an absorption spectrum (absorption curve) is obtained. Since plant viruses are nucleoproteins they have a characteristic absorption spectrum in ultraviolet light. With plant viruses the maximum of the spectrum is at 260 mμ. The virus concentration can be calculated exactly from the optical density at 260 mμ of a purified virus preparation and from the degree of the dilution of the sample. The absorption coefficient differs with different plant viruses. With tobacco mosaic virus it is 2·7 which means that at 260 mμ tobacco mosaic virus at a concentration of 1 mg/ml will give a reading of 2·7 if measured in a cell of 1 cm light path.

TRANSLOCATION OF PLANT VIRUSES IN THEIR HOSTS

Virus particles will spread from the primary infected cells during or after virus multiplication to neighbouring cells and in most cases will invade the whole plant. Thus, translocation consists of two phases: translocation in the parenchyma

* Matthews (1957).

and translocation in conductive tissues (phloem or xylem). If both phases of translocation take place, the infection is systemic, but if only the first (parenchymatic) phase occurs the infection is local.

Translocation in the parenchyma. As early as 1936 Sheffield came to the conclusion that the movement of virus particles from cell to cell occurs through the plasmodesmata. This assumption has not been disproved since (Bennett, 1956). The question is, however, in which form the virus spreads from cell to cell. Shalla (1959) has shown in *in situ* electronmicroscopic studies that the size of the plasmodesmata in tobacco leaves is not large enough to let intact TMV rods pass through them. Based on their kinetic studies of the growth rate of local lesions Rappaport and Wildman (1957) have suggested that TMV in *Nicotiana glutinosa* spreads in the form of naked nucleic acid. This assumption seems to be supported by Shalla's (1959) finding that in non-necrotic degenerating tissues which surround virus induced necrotic local lesions, no typical rod-shaped TMV particles are present.

According to Bennett's (1956) estimates the distance moved by viruses in the parenchyma is at the most 2 mm per day. On the basis of circumstantial measurements Rappaport and Wildman (1957) have calculated that TMV in *Nicotiana glutinosa* moves in a radial direction over a distance of about three to six cell layers per day.

Translocation in the conductive tissues. The virus particles which spread initially from cell to cell will reach the leaf veins in 2 to 3 days after infection. In the case of systemic infection they will invade the whole plant through the conductive tissues.

Most viruses move in the phloem. Pierces' disease virus of grape, however, seems to occur in and move through tracheal elements of the xylem. It has been generally assumed that virus movement in the phloem is related to movement of food materials (Bennett, 1956). The rate of movement in the phloem is much higher than in the parenchyma and is different with various viruses. According to Köhler (1956) potato virus Y moves over a distance of 0·7 cm per hr and potato virus X at a rate of 0·2 cm per hr in tobacco stems. The rate of movement in tomato stems has been found to be 8 cm per hr for both potato X virus and TMV (Capoor, 1949).

Further details on the translocation of plant viruses are to be found in Bennet's (1956) review.

2. SYMPTOMATOLOGY

We have seen that development of host-virus relations is reflected in virus multiplication on the one hand, and in symptom development on the other. Part of the symptoms are visible with the naked eye (Symptomatology *s. str.)*, the plant, however, reacts to virus infection also with symptoms which are discernible only by microscopic (Histopathology) or plant physiological (Pathophysiology) investigations.

Virus induced symptoms can be divided into two groups: primary and secondary symptoms (cf. Smith, 1951).

Primary symptoms are produced upon mechanical inoculation on the inoculated leaves themselves. Primary symptoms appear in the form of chlorotic or necrotic spots, or rings. For instance primary symptoms caused by cucumber mosaic virus in sugar beet consist of chlorotic spots and in *Amaranthus caudatus* of local lesions.

Secondary symptoms appear on non-inoculated parts of the plant. They may be grouped in the following way (Smith, 1951): 1. mosaic mottling, 2. chlorosis, 3. ringspot, 4. necroses, 5. distortions and enations, 6. colour changes.

1. Mosaic mottling appears on the leaves (e.g. TMV in tobacco). It is often preceded by vein clearing. A form of mosaic pattern consists of vein banding. This is characterized by two dark green bands running along the veins on both sides; the rest of the leaf blade is normal green or light green (e.g. certain strains of cucumber mosaic virus in tobacco). Mosaic mottling usually consists of dark and light green spots. With some host virus combinations, however, the light coloured spots are bright yellow (e.g. aucuba strain of TMV in tobacco).

2. Chlorosis consists of a uniform yellowing of the leaves (e.g. sugar beet yellows virus in sugar beet).

3. Ringspot. The rings can be simple or concentric (tobacco ringspot virus in tobacco).

4. Necroses. Areas including a group of disintegrated cells may appear in the form of streaks on the stem (stem necrosis, e.g. TMV in certain varieties of red pepper) and on the leaves (e.g. TMV in *Nicotiana glutinosa*). Sometimes whole parts of plants (e.g. apical necrosis) or entire plants may die from necrosis.

5. Different types of distortion can be distinguished. In the most severe case the appearance of the entire plant may change (e.g. bushy appearance, or dwarfing). These changes in the appearance of the plants may be due either to a shortening of the internodes (e.g. cucumber mosaic virus in cucumber) or to an enhanced production of lateral shoots (e.g. tomato bushy stunt disease). Deformations are sometimes restricted to the leaves (e.g. potato leaf roll disease). Enations are special forms of leaf distortion. They are actually leafy outgrowths of variable size (e.g. enations on the undersides of pea leaflets and stipules on pea leaves infected with pea enation mosaic virus). Conspicuous enations occur in aspermy disease in *Nicotiana glutinosa* L., generally as leafy outgrowths from the lower surface of the midrib, producing a structure resembling a small new leaf. In some host-virus combinations (e.g. red pepper-cucumber mosaic virus) flowers and fruits may also be distorted.

6. Colour changes are most conspicuous on flowers (e.g. tulips infected with tulip mosaic virus or petunia infected with tomato spotted wilt virus).

As far as symptomatology *s. str.* is concerned there is no need for special methods. What should be mentioned in connection with plant virus symptomatology is the use of special abbreviations of symptoms produced by plant viruses. Abbreviations are mainly used if symptoms produced by several viruses or virus strains are to be compared in a table. The international abbreviations for the most common symptoms are the following:

0 = No virus multiplication and no symptoms produced
1 = The virus multiplies only locally and no symptoms are produced
2 = The virus spreads in the plant systemically but no symptoms are produced
3 = The virus multiplies locally and causes local symptoms
4 = The virus spreads systemically in the plant and causes systemic symptoms
Ab = Leaf abscission
Bl = Black
Br = Brown
C = Chlorosis or chlorotic
Co = Concentric pattern
Col = Collapse of the plant
D = Death of the plant
Dif = Diffuse
Dis = Distortion or distorted
E = Etch symptoms
En = Enations
Ex = Extreme
Fcd = Colour of flower deepens
Gr = Green or greenish
K = Often escaping infection
L = Local or locally
Le = Leaf or leaves
M = Mottling
Ma = Masking of symptoms
Mv = Midvein
N = Necrosis or necrotic
Ny = Necrosis surrounded with yellow tissue
Pc = Pale centre
Pe = Petiole or petioles
R = Red
Ri = Ring symptoms
S = Systemic or systemically
Se = Segments of rings
Sp = Spots
St = Stem or stems
Str = Streak
Stu = Stunting
V = Virus recoverable
Va = Virus absent
Vb = Vein banding
Vc = Vein clearing
W = White
Wi = Wilting
Y = Yellowing
Yo = Young

A special case of the symptomatological study of virus infected plants consists of the detection of primary infection sites prior to the appearance of visible symptoms.

*Starch-iodine lesions on leaves of Samsun tobacco plants infected with tobacco mosaic virus**

The principle of the method lies in the fact that the transport of starch from virus induced primary infection sites is delayed as compared to healthy leaf tissues. *Nicotiana tabacum* cv. Samsun plants are placed in the dark for 12–16 hr

* Holmes (1931).

at 10 °C, 3 days after infection with tobacco mosaic virus (by that time no visible symptoms have yet appeared). After the dark treatment the inoculated leaves (which are expected to show the starch-iodine lesions) are detached and immersed for a few sec into boiling water. After that chlorophyll is extracted from the leaves with 96 per cent ethanol at 40 °C for about 1–2 hr. The white leaves are then rinsed with distilled water and immersed in a solution of iodine in potassium iodide (2 g KI + 1 g iodine in 300 ml distilled water). The excess of the staining solution is removed from the surface of the leaves by rinsing them with 30 per cent ethanol. Primary infection sites, invisible to the naked eye, will appear after this treatment as small blue spots.

HISTOPATHOLOGY

Histological changes in plants upon virus infection can be divided into the following groups: 1. production of abnormal tissues, 2. changes in conductive tissues, 3. changes in parenchymatic tissues and 4. intracellular inclusions.

1. Production of abnormal tissues. In tobacco rosette, a complex virus disease, abnormal tissues and giant cells are formed in the cortex and pith of the plant. This will cause the stem to split. In clover plants infection with wound tumour virus will upset the entire process of histological differentiation (Lee and Black, 1955). As a result of the abnormal development of procambial cells and redifferentiation of the phloem parenchyma, meristematic tumour cells are formed. These, in turn, will develop into pseudophloem, hypertrophic cells, hyperplasiae and parenchymatic tumour tissues.

2. Changes in conductive tissues. As is known, the virus causing Pierces' disease of grape is restricted to the xylem of the plant. It causes gummoses in the tracheal elements of the xylem. Those viruses which are specialized to the phloem, such as sugar beet curly top virus, will cause histological alterations in the phloem.

3. Changes in parenchymatic tissues are caused by viruses inducing mosaic symptoms. Sugar beet leaves, for instance, infected with sugar beet mosaic virus are characterized by isodiametric mesophyll cells in the yellow leaf areas and a lack of differentiation between palisade and spongy parenchymatic cells. The cells located in the yellow leaf areas hardly divide if at all after infection, and contain few chloroplasts. The chloroplasts are irregular in shape and, in the case of heavy infection, combine to form amorphous masses. In several cells nuclei are absent.

For more details concerning the histopathology of virus infected plants we refer to Esau's (1948) review.

4. Inclusions. As early as 1902 Ivanovskiï observed that tobacco leaves infected with TMV contained intracellular bodies which were absent from healthy leaves. He described two types of intracellular formations: amorphous and crystalline. Later investigations (cf. Cook, 1947) have pointed out that in several other virus diseases similar inclusion bodies occur. There are three types of inclusion bodies:

a) Intracellular crystals, b) intracellular amorphous inclusions, and c) intranuclear inclusions.

a) Intracellular crystalline inclusions are true three-dimensional crystals of different shapes. Goldin (1938) showed that these crystals are actual aggregates of virus particles. This was later substantiated by several workers (e.g. Brandes, 1956).

b) Intracellular amorphous inclusions (X-bodies) are spherical or ellipsoid formations. Their size varies from 5 μ to 30 μ in diameter (Bawden, 1950). They are of granular consistency and contain 1 or 2 vacuoles.

c) Intranuclear inclusions. It was Kassanis (1939) who first observed that tobacco etch virus induced the formation of crystalline inclusion bodies in the nuclei of tobacco cells. Sheffield (1941) did not exclude the possibility of these inclusions being virus aggregates. Goldin and Vostrova (1959) described a special strain of tobacco mosaic virus which induced the formation of crystalline inclusion bodies in the nuclei of tobacco cells. These inclusions can be stained with special staining techniques (Bald, 1949; Rawlins, 1957). Virus inclusion bodies as pointed out by Zech (1954), are connected with certain stages of virus multiplication. Due to methodological difficulties, however, the role of viral inclusions in the process of virus multiplication is not quite clearly understood yet.

Up to the present about 30–40 plant virus diseases have been found to be associated with the formation of inclusion bodies. A detailed description of plant virus inclusions is to be found in Goldin's (1954) book.

In order to study the histopathology of virus infected plants the methods to be used are those which are commonly employed in plant histology. Such methods will not be described in this book. We confine ourselves to the description of two techniques used in plant virology for the fixation and detection of viral inclusions in plant tissues.

Demonstration of amorphous inclusion bodies from epidermal cells of Solanum nudiflorum infected with a strain of tobacco mosaic virus (tomato aucuba virus)

Plants infected for about 10–28 days are used for the experiments. From the midvein on the lower leaf surface, epidermal strips are prepared. If viewed in an ordinary or rather phase contrast microscope, amorphous inclusion bodies of granular structure can be observed near the cell nucleus. A differential staining of these inclusion bodies consists of treating the epidermal strips with a solution of iodine in potassium iodide (2 g KI + 1 g I in 300 ml of distilled water): the cell nuclei will strain bright yellow whereas amorphous inclusion bodies will be of a yellowish brown colour.

Fixation of crystalline inclusions in the epidermis of tobacco leaves infected with tobacco mosaic virus

For the experiments, leaf hairs or epidermal strips from tobacco plants infected for about 2 weeks are used. With the fixation technique described below hexagonal TMV crystals can be kept unaltered for more than a year.

Freshly prepared epidermal strips are treated for 3–4 min in a Carnoy-fixative solution (60 ml absolute ethanol + 30 ml chloroform + 10 ml glacial acetic acid) and rinsed with 96 per cent ethanol. The preparations are stored in 96 per cent ethanol and placed in distilled water when viewed under the microscope.

PATHOPHYSIOLOGY

For more than fifteen years several attempts have been made to correlate external symptom development with some specific metabolic alterations in plant virus diseases. Despite a large number of papers published on this topic (cf. Diener, 1963) no such correlation has been found. At the present time the general concept prevails (Farkas and Solymosy, 1965) that metabolic changes observed in virus infected plants are secondary phenomena and reflect a general metabolic disturbance which is far from being specific to virus infection, but takes place under the effect of a number of factors inducing stress. The story of controversial ideas concerning the possible mechanism of the hypersensitive reaction will serve as an illustration of the difficulties encountered in relating symptom development to what is believed to be specific metabolic changes.

Hypersensitive reaction which leads to the formation of necrotic local lesions (Holmes, 1929) is the most characteristic form of resistance to virus diseases in plants. Two problems connected with the hypersensitive reaction are to be distinguished:

1. What is the mechanism of necrotic local lesion formation? and
2. Why does lesion growth stop after a certain time?

Since the hypersensitive reaction appears in the form of a rapid necrobiosis it was thought that the dying of the cells was the cause of virus localization. By the end of the fifties, considerable interest was shown in the biochemical mechanism of necrotic local lesion formation mostly based on results of studies with hypersensitive fungus-host combinations. The idea which served as a working hypothesis was that the balance of oxidizing and reducing systems active on polyphenols may be upset in the diseased area, resulting in the accumulation of toxic polyphenol oxidation products which killed the host tissues. At that time it was already known that upon virus infection the respiration rate of the host increases (Yamaguchi and Hirai, 1959). It was tempting to suppose that part of the respiratory increase may be due to increased polyphenoloxidase activity.

Detailed studies along these lines revealed that, indeed, virus-induced necrogenic reactions are accompanied by an increase in polyphenoloxidase activity (Solymosy et al., 1959; Solymosy, 1960; Kikuchi and Yamaguchi, 1960; Farkas et al., 1960). The hypothesis of Solymosy et al. (1959) at that time was that due to the increased activity of polyphenoloxidase, polyphenol oxidation products (i.e. quinones) accumulate and, being more phytotoxic than polyphenols, lead to cell destruction. This hypothesis seemed to be supported by the finding that treatment with fairly high concentrations of reducing agents such as ascorbic acid, glutathione or cysteine actually reduced the number of necrotic lesions to a considerable extent without seriously interfering with virus multiplication.

The finding of Parish et al. (1965) that the administration of various phenols resulted in the darkening of existing local lesions and that lesions which have formed on ascorbic acid infiltrated leaves are larger and less pigmented than those in the control was in line with this hypothesis. But this was (in accordance with the above authors' claim), no proof for a causal relationship between increased polyphenoloxidase activity and tissue necrotization.

Solymosy et al. (1959) were also unable to demonstrate unequivocally a gradual increase in polyphenoloxidase activity prior to lesion formation (which would be an important prerequisite of the existence of a causal relationship). This suggested that the mechanism of lesion formation is not so simple as it might appear.

Later several reports appeared indicating that necrotic local lesion formation was based on more complicated processes than just increase in polyphenoloxidase activity and that increased polyphenoloxidase activity alone cannot be the cause of local lesion formation.

For example simple wounding seems to increase polyphenoloxidase activity (i.e. wounding induces a rise in enzyme activity and not the other way round). Jockusch (1966) cut uninfected *N. tabacum* cultivar Java plants above the 11th leaf and found that polyphenoloxidase activity rose by a factor of 50 to 70 within 3 days in the 11th leaf as compared to the control. It is known that tobacco plants which give necrotic lesions with a given strain of TMV at about 23 °C will show only chlorotic spots when kept at 35 °C. Now, enzyme activation by stem cutting could be induced in Java plants equally well at 23, 26, 29, 32 and 35 °C. Thus neither enzyme activation nor the enzyme itself is sensitive to high temperatures *in vivo* in contrast to the hypersensitive reaction. If the temperature sensitivity of the hypersensitive reaction were a property of a step following polyphenoloxidase activation in the necrotization process, an increased enzyme activity in virus-infected, hypersensitive tobacco varieties which were kept at high temperature would be expected. This, however, is not the case. These results by Jockusch (1966) support the view that the activation of this enzyme is a secondary phenomenon.

As to whether the reducing agents (ascorbic acid, glutathione and cysteine) used by Solymosy et al. (1959) for the inhibition of lesion formation are specific for the reduction of quinones to polyphenols, as originally postulated, the possibility of these reducing agents acting on some biochemical process other than this cannot be ruled out. Since the redox potentials of ascorbic acid, glutathione and cysteine are such that they will reduce ferricytochrome-c non-enzymically, it may be supposed that these reducing agents delay lesion development by inhibiting electron transport by the cytochrome chain just as cyanide and azide do (Merrett, 1962b). Thus if the energy for virus synthesis is provided by respiratory chain phosphorylation as suggested by Merrett (1962a) the above mentioned reducing agents may inhibit lesion formation by interfering with virus multiplication rather than with the oxidation of polyphenols to quinone.

To ascertain the possible role of ascorbic acid in the formation of necrotic lesions, Milo and Santilli (1967) found that in Pinto bean leaves inoculated with TMV, ascorbic acid concentration first increases and then decreases in the infected leaf halves. Since changes in ascorbic acid concentration do not run simultaneously with the process of necrotization it is hard to imagine that they may play an *in vivo* role in it. Another fact pointing to the aspecific nature of ascorbic acid oxidation is that it occurs after wounding, treatment with toxin, and after infection by pathogens other than viruses in a wide variety of plants.

The final conclusion concerning the role of polyphenoloxidase and of ascorbic acid in the necrotic process is that activation of polyphenoloxidase, increase in the quinone content, and decrease of ascorbic acid in the necrotized leaves are secondary phenomena rather than the primary cause of local lesion formation. One should rather think of a general disturbance of host metabolism manifesting itself in these conspicuous abnormalities.

Now we turn to the other question, i.e. why does lesion growth stop after a certain time?

The most logical approach to this problem is to look for some metabolic and/or structural alterations in the tissues surrounding virus induced necrotic local lesion which might point to the existence of defence reactions in these areas.

It was established several years ago that in the tissues around the necrotic lesions a number of enzymes were activated (Solymosy and Farkas, 1962; Farkas and Solymosy, 1962; Solymosy and Farkas, 1963). At that time a paper appeared by Yarwood (1960) and another one by Ross (1961) demonstrating that these areas are resistant to superinfection. Whether inhibition of lesion growth and resistance to superinfection are the result of the same metabolic events is not known. Based on some evidence, to be discussed later, one may suspect that they are.

At any rate, the experimental results reported so far point to the fact that in the tissues surrounding necrotic local lesions very peculiar metabolic and structural changes take place.

The activation of so many enzymes in the areas around necrotic lesions made Solymosy and Farkas (1963) suspect that it was due to a general enhancement of synthetic processes. There was a hint pointing to the correctness of this hypothesis because P^{32} was incorporated to a higher extent into the nucleic acid fractions of cells around the lesions than into the control (Farkas and Solymosy, 1962). The hypothesis was advanced that localized acquired resistance was associated with an increased protein synthesis which might interfere with the new information given to the cell by the viral RNA at reinfection (Solymosy and Farkas, 1963).

Recently two independent lines of evidence seem to support this view.

1. Israel and Ross (1967) have shown that mesophyll cells in a narrow band surrounding necrotic local lesions exhibited structural features suggestive of metabolically active tissues. Compared to normal mesophyll cells, cells within this zone had more cytoplasm, more ribosomes and membrane-bounded organelles, and a very marked system of extensive continuous elements of endoplasmic reticulum with which many of the ribosomes were associated. Compared to chloroplasts in normal mesophyll cells, those in this zone were more ameboid, had more ribosomes, large starch grains and lipid globules, and were more numerous. The authors suggest that the structural alterations are reflections of induced physiological changes related to the known capacity of these cells to resist virus infection.

2. The most convincing evidence for the existence of active synthetic processes in the zone around necrotic local lesions seems to come from Loebenstein et al. (1968). They found that the development of localized acquired resistance to TMV in *N. tabacum* cultivar. Samsun NN was inhibited in the presence of actinomycin D. These results suggest that the transcription mechanism of the cell from DNA to RNA has to operate so that localized acquired resistance may develop. This is the key reaction to synthetic processes. These authors refer in their paper to the finding that administration of actinomycin D to leaves bearing local lesions resulted in an increase of lesion size. This effect of actinomycin D together with its effect on reducing acquired resistance strengthen the idea that acquired resistance and the localization of the virus in a hypersensitive host are caused by similar mechanisms.

At present it thus seems that some factor spreads from the lesion to the surrounding tissues inducing an increased rate of synthetic processes, and that increased synthetic activity of the cells somehow inhibits establishment of viruses.

If this is so indeed one might except, that

1) some factors known to induce increased synthetic activity of the cells will at the same time increase resistance to virus infection and that
2) tissues characterized by intensive synthesis will be relatively resistant to virus infection.

As to the point of increasing virus resistance by means of inducing increased rates of synthetic processes the results of Király et al. (1968) are indicative. Cytokinins are known to exert a stimulatory effect on plant RNA and protein synthesis. Király and co-workers have found that treatment of tobacco or bean leaves with kinetin or benzyladenine resulted in a decrease of the number of local lesions. Some of their additional findings which are in line with the hypothesis that the rate of synthetic processes in the cells is inversely correlated with susceptibility to virus infection can be briefly summarized as follows:

1. Removal of the terminal bud of Pinto bean leaves, shown to arrest senescence, resulted in a decrease of the number of local lesions produced by infection with TMV.

2. Pinto bean leaves treated with chloramphenicol, an inhibitor of protein synthesis, showed more local lesions upon infection with TMV than did control

leaves. In addition the diameter of local lesions developing on chloramphenicol-treated Pinto bean leaves was greater than that on the control leaves, again pointing to a similarity in the mechanism of inducing resistance and reducing lesion growth.

3. Heat treatment of the leaves of Pinto beans increased the number of local lesions induced by TMV.

As to the second point i.e. that relative resistance to virus infection is a property of tissues characterized by intensive synthesis one has to refer to the well-known fact (cf. Crowley and Hanson, 1960) that meristematic tissues are resistant to virus infection.

There is another interesting phenomenon in plant virology which might have some bearing on the metabolic aspects of virus resistance in plants. This is the so-called systemic acquired resistance and the production of an antiviral factor (or interferon like substance) in virus-infected plant cells exhibiting necrotic local lesions or showing symptoms of systemic infection.

Ross (1961) has shown that inoculation of half-leaves of *N. tabacum* cultivar Samsun NN plants with TMV induced a high level of resistance to TMV in the opposite half leaves. The lesions formed on these halves were consistently only one-fifth to one-third as large in diameter as were lesions in susceptible half-leaves, again pointing to a similarity of the mechanism affecting susceptibility of infection sites and lesion growth. The resistance induced by TMV was not specific to TMV, but also for the other viruses tested (TMV, turnip mosaic virus, tobacco and tomato ringspot viruses). Another important point in Ross's experiment is that no resistance was induced by viruses that do not cause necrosis.

In later experiments it was shown (cf. Ross, 1965) that systemic acquired resistance is induced by some substance which is translocated. At the same time in tissues exhibiting systemic acquired resistance a protein-like substance was found by Loebenstein's group which resembled interferon in many of its properties (cf. Loebenstein et al., 1965).

The question arises whether there is any connection between systemic acquired resistance and production by the plant of an interfering substance or substances. For the time being it is impossible to answer this question. It would be interesting to see whether or not leaves treated with the hypothesized interfering substance will develop resistance before inoculation.

The fact that in tissues showing systemic acquired resistance no increased respiration and no increased polyphenoloxidase activity were found suggests that the mechanism of localized and systemic acquired resistance may be different.

In summary we may conclude that the hypersensitive reaction of plants to virus infection involves 1. the formation of necrotic local lesions; 2. localized acquired resistance; 3. systemic acquired resistance and; 4. the production of an interferon-like substance.

1. The mechanism of necrotic local lesion formation is not understood. Although lesion formation is associated with increased polyphenoloxidase activity this is most probably not the cause of the necrotic process.

2. Localized acquired resistance seems to be accompanied by an increased synthetic activity of the cells and most probably is actually the result of it. So far there is no indication that localized acquired resistance could be induced by effects other than virus infection.
3. Systemic acquired resistance on the other hand seems to be highly unspecific. As shown by several workers it can be induced by a number of quite unspecific factors, such as native protein of TMV (Loebenstein, 1960; 1962) saprophytic bacteria (Klement et al., 1966), heat killed cells of *Pseudomonas syringae* (Loebenstein and Lovrekovich, 1966) foreign nucleic acids or yeast-RNA (Gicherman and Loebenstein, 1968), as well as foreign proteins (Király and Klement, 1968; cf. Király, 1968). Thus systemic acquired resistance is not virus-specific with respect to induction.
4. The production of an antiviral factor has been shown so far to be induced by virus infection in general. This does not mean that it will not be formed whenever systemic acquired resistance is operating in the cells.

To correlate these four seemingly so different phenomena, it may be hypothesized that in plants any long lasting damage, which may be called stress—be it multiplication of plant viruses or injection of foreign macromolecular substances—synthetic processes are activated which lead to virus resistance. One aspect of this is localized acquired resistance and the other one systemic resistance based on the production of interferon-like substances.

For studying the pathophysiology of virus infected plants, plant physiological and biochemical methods are used. However, it is not aimed here to give a description of these methods.

3. ENVIRONMENTAL FACTORS INFLUENCING HOST-VIRUS RELATIONS

Environmental factors which influence host-virus relations can be divided into three groups: Physical factors, chemical factors and biological factors.

PHYSICAL FACTORS

Light intensity. Shading or darkening of the test plants prior to mechanical inoculation usually leads to increased susceptibility. This is partly connected with the fact that the cuticle of plants kept at reduced light intensities is thinner than that of plants grown in the light. Therefore it is easier to wound such leaves. This, however, is certainly not a full explanation of the effect of darkening on virus susceptibility because tomato bushy stunt virus, for instance, not only causes more local lesions on plants subjected to preinoculation darkening than in the controls but is also translocated faster (Bawden, 1959). In addition, a pre-inoculation dark treatment is also effective if applied only for 24 hr, and this certainly does not result in basic structural changes in leaf tissues. It may be postulated that preinoculation photoperiod treatment of plants leads to a depression of synthetic processes which in turn will lead to increased susceptibility.

Temperature. Temperature has a different effect on the susceptibility of acceptor hosts depending on whether heat treatment is applied before or after mechanical inoculation. Kassanis (1952) has shown that if test plants are kept for 1 to 2 days at 36 °C their susceptibility, as judged from the number of local lesions to the 5 mechanically transmissible viruses tested, increases. On the other hand, keeping the plants at 36 °C after inoculation leads to an inhibition of virus multiplication. The rate of inhibition was different with the various viruses tested.

Temperature also influences the resistance of plants to viruses. Kassanis (1952) has shown that *Nicotiana glutinosa* plants which normally react with local lesions to infection with TMV will exhibit systemic symptoms if kept at 36 °C. The resistance of a spinach variety to cucumber mosaic virus is also temperature dependent. Spinach plants which are resistant to the virus at 16–20 °C will show systemic necrosis when grown and infected at 28 °C (Holmes, 1954).

In some cases high temperature can be used to cure virus infected plants (heat therapy). Until now heat therapy has been used with success mainly against virus diseases causing yellows (peach yellows, aster yellows, potato witches' broom, potato leaf roll) as summarized by Bawden and Pirie (1952). The present status of heat therapy of plant virus diseases has recently been summarized by Roland (1957).

It may be hypothesized that high temperatures similarly to reduced light intensities decrease resistance to virus diseases by suppressing synthetic processes in plant cells.

CHEMICAL FACTORS

Chemicals have an effect on both the virus and the host plant. They influence the symptoms and susceptibility of the host on the one hand (Effect of chemical factors on the host plant), and on virus multiplication on the other (Effect of chemical factors on the virus). The separation of these two effects is artificial and is justified only from a didactical point of view. Since plant viruses are devoid of their own metabolism, different chemicals will always act by altering host metabolism even if this effect of chemicals is reflected in changes in virus concentration. Chemicals are often used as chemotherapeutants. It has to be stressed, however, that chemotherapy of plant viruses does not mean actual curing of infected plants but rather a drastic inhibition of virus multiplication.

The effect of chemicals on the host can be evaluated on the basis of the severity of symptoms, percentage of diseased plants and number of local lesions.

Böning and Diercks (1955) have shown that chloride-containing fertilizers increase the severity of symptoms produced by potato leaf roll virus and potato virus Y in potatoes. Sulphates have a similar effect. In contrast, overdoses of Ca lead to a reduction of symptom severity. Spraying with 10 per cent sucrose solutions enhances yellows symptoms in sugar beet.

Effect of chemical factors on the virus. Glucose, sucrose and a number of organic acids have been reported to stimulate virus multiplication and glycine, methionine and histidine to inhibit virus multiplication (Ryzhkov, 1957). In contrast to an earlier hypothesis Ryzhkov (1959) has demonstrated that there is no parallelism between the action of metabolites and antimetabolites on respiration and virus multiplication.

Chemotherapy. As mentioned before, chemotherapy of plant virus diseases does not mean actual curing in most cases. Chemotherapeutants can be found among both inorganic and organic compounds. Some antibiotics have also been found to exert a therapeutic effect on virus diseased plants.

Inorganic compounds. Stoddard (1947) was able to cure peaches from X-disease by using $ZnSO_4$. Thomas and Baker (1949) have successfully used $ZnCl_2$ and $CaCl_2$ against carnation mosaic disease. Nothing is known about the mode of action of these chemicals.

Organic compounds. Woods (1940) treated TMV infected tobacco leaves with sodium cyanide and found a well pronounced inhibitory effect on virus multiplication. Since sodium cyanide is known to be a respiratory poison, Woods (1940) suggested that this compound inhibited virus multiplication by blocking respiratory processes. Takahashi (1948) was successful in inhibiting virus multiplication by the administration of malachite green. He assumed that this compound decreased the rate of virus multiplication by inhibiting the activity of dehydrogenases.

Among organic compounds special attention has been given to purine and pyrimidine derivatives. Since viral nucleic acids contain purine and pyrimidine bases it is very likely that derivatives of these compounds will interfere with virus multiplication when introduced into virus infected plants.

Commoner and Mercer (1952) demonstrated that thiouracil (a pyrimidine derivative) strongly inhibits virus multiplication. This effect of thiouracil is reflected both in decreased virus concentration and in a delay of the appearance of TMV induced amorphous inclusions (Hirai, 1958). The inhibitory effect of thiouracil can be neutralized by the simultaneous administration of uracil. Matthews (1956) found that thiouracil is incorporated into TMV-RNA to replace uracil. Thiouracil-containing virus particles are presumably non-infectious. In addition to the effect on the virus particles themselves thiouracil alters the nucleic acid and protein metabolism of the host. Porter and Weinstein (1957) found that in tobacco plants treated with thiouracil the amount of amino acids, amide and ammonia nitrogen increased as compared to untreated controls. This points to the fact that protein synthesis is inhibited due to the disturbance of nucleic acid metabolism. In a trial to use thiouracil as a chemotherapeutant against tobacco mosaic virus in tomato, Milinkó (1961) found that in addition to its inhibitory effect on the multiplication of TMV, thiouracil also caused a serious disturbance in host metabolism. Several other pyrimidine-derivatives have been reported to inhibit virus multiplication (Ulrychová-Zelinková, 1960).

Among substituted purines 8-azaguanine was first shown to inhibit virus multiplication (Matthews, 1951). 8-azaguanine, like thiouracil seems to be incorporated into viral RNA (Matthews, 1953).

Antibiotics. Most of the antibiotics used with success against bacteria and fungi proved to be ineffective against viruses. Manil (1947) was unable to inhibit TMV infection by treating the plants with actinomycin, actinomycetin, penicillin, streptomycin, and tirotrycin. Aureomycin and terramycin proved equally inactive against plant viruses.

There are, however, some antibiotics which are effective against plant viruses. Such are citovirin, noformycin, naramycin (an actidione analogue) and mitomycin C. These antibiotics inhibit virus multiplication. Their mode of action is not known yet except that of mitomycin C which was shown by Hirai and Shimomura (1960) to inhibitit ribonucleic acid synthesis.

The above antibiotics should be clearly distinguished from antibiotics isolated from culture filtrates of *Trichothecium roseum* and *Neurospora sitophila.* These antibiotics inhibit the establishment of virus infection rather than virus multiplication. Milk is also a typical inhibitor of infection. It can be used successfully to prevent TMV infection of tomatoes in the greenhouse (Newell, 1954; Milinkó, 1961).

BIOLOGICAL FACTORS

From what has been said above it can be seen that physical and chemical factors which influence host metabolism will act on both host susceptibility and rate of virus multiplication. The reaction of the host to virus infection will be altered also when the plant has been infected with another pathogen (fungus, bacterium or virus) prior to or simultaneously with infection by a virus. For these agents the collective term biological factors can be used. In the present chapter we shall consider to what extent a previous or simultaneous virus or fungus infection influences host susceptibility, symptoms or the process of virus multiplication in the case of a challenge inoculation with a virus.

Viruses. Here one has to distinguish between the interaction of different viruses and the interaction of different strains of the same virus.

In the case of interaction between different viruses both antagonistic and synergistic (additive) effects are encountered. These effects operate both with simultaneous infection and with infections carried out at different times. The effect may be on the susceptibility of the host, on symptom development or on virus concentration.

An antagonistic effect between different viruses was reported by Price (1935). He found that on *Zinnia* plants which had been infected either with tobacco mosaic virus or with tobacco ringspot virus, cucumber mosaic virus caused fewer local lesions than on healthy *Zinnia* plants. Tobacco plants infected with tobacco ringspot virus showed milder symptoms upon TMV infection than healthy tobacco plants. An antagonistic effect on virus multiplication was reported by Bawden and Kassanis (1945). A previous infection of tobacco plants with tobacco etch virus inhibited subsequent multiplication of either potato virus Y or *Hyosciamus* mosaic virus. An antagonistic effect with simultaneous infection was described by Ross (1950). He found that when potato seedlings which are immune to potato virus X and react to potato virus Y with the formation of local lesions are infected with sap containing both potato virus X and potato virus Y, fewer necrotic lesions were formed than upon infection with potato virus Y only. The extent of inhibition was proportional to the amount of potato virus X in the sap.

If tobacco plants infected with potato virus Y are inoculated with a necrotic strain of potato virus X the number of local lesions will be higher than on comparable healthy tobacco plants. It has been demonstrated that in potato and tobacco plants doubly infected with potato virus X and potato virus Y the concentration of the former was higher than in plants containing only potato virus X. In the case of simultaneous infection of tobacco plants with TMV and potato virus X local lesions will form whereas either of these viruses if introduced alone into tobacco plants will induce systemic symptoms.

Different strains of the same virus do not exhibit additive effect, the effect is always antagonistic. The antagonistic effect between different strains or different viruses is called interference. In a strict sense the term interference is used to denote the inhibitory effect of one virus strain on the multiplication of another strain in the case of simultaneous infection. A special case of interference is called cross protection (premunity). Cross protection takes place when infection with a mild strain of a virus (e.g. a strain causing mild mosaic symptoms) protects a plant from becoming infected with a severe strain of the same virus (e.g. a strain causing bright yellow mosaic symptoms or necrotic local lesions). Premunity was discovered by Thung in 1931. He found that when tobacco plants infected with the common strain of TMV are superinoculated with a white strain of the same virus, the symptoms of the challenge virus will fail to appear.

Interference has been interpreted in several ways (Bennett, 1953). The most acceptable hypothesis on the cause of interference seems to be that of Bawden (1950). According to this, different strains of the same virus require the same precursors (or rather sites of attachment) for their multiplication. When a given strain of a virus is introduced first into plant cells it will use up the precursors needed (or exhaust the sites of attachment), consequently another strain of the same virus introduced later into the same cell will be unable to multiply due to a lack of precursors (or attachment sites). This hypothesis which clearly points to the cellular nature of premunity seems to be supported by Köhler's (1955) finding. In TMV infected leaves the virus particles are unevenly distributed between yellow and green leaf areas. In the yellow areas virus concentration is much higher. If mottled leaves of White Burley tobacco plants infected with the common strain of TMV are superinoculated with the para strain of TMV (which causes necrotic local lesions in White Burley tobacco), the local lesions will appear only in the green leaf areas. Thus, there is a correlation between virus concentration in the cells and the extent of cross protection.

Premunity tests are widely used to establish the degree of relationship between virus strains. It has to be stressed, however, that with certain virus strains cross protection is not complete. In some extreme cases, as for example with strains of sugar beet curly top virus, there is no premunity at all (Bennett, 1953).

According to Wu and Rappaport (1961) interference between virus strains in the case of simultaneous inoculation is based on the phenomenon of competition for potential infection sites.

Fungi. Yarwood (1951) found a synergistic effect between bean rust (*Uromyces phaseoli*) and TMV. In rust infected bean leaves the relative concentration of TMV was found to be 10,000 times higher than in healthy bean leaves. Furthermore, the bean variety Bountiful, which is normally resistant to TMV became susceptible when rust infected. On the other hand Wilson (1958) claimed that there was cross protection between the bean rust fungus and tobacco mosaic virus in beans. According to Blumer et al. (1955) there is an antagonism between cucumber mosaic virus and *Erysiphe polyphaga.* With virus infected cucumber leaves *Erysiphe polyphaga* was shown to attack first of all the green leaf areas. Fewer haustoria were found in the yellow leaf areas than in the green ones. Under natural environmental conditions cucumber plants showing signs of mildew disease were infected with cucumber mosaic virus to a lesser degree than healthy cucumber plants. According to the above authors conidial suspensions of *Erysiphe polyphaga* have an inhibitory effect on cucumber mosaic virus *in vitro.*

CHAPTER V

GENETICS OF PLANT VIRUSES

The amino acid sequence of proteins is ultimately determined by the nucleotide sequence of the nucleic acid which contains the genetic information for the particular protein under consideration. Nucleic acids code for proteins. Since the biosynthesis of TMV particles is initiated and coded for by TMV-RNA, there must be a close correlation between the nucleotide sequence of TMV-RNA and the amino acid sequence of TMV protein. The simplest method of demonstrating this specific correlation between the two structural components of TMV would involve the separate analysis of the nucleotide sequence and amino acid sequence and the establishment of regularity in the correlation of the two kinds of sequences. The determination of the complete amino acid sequence of TMV protein has already been achieved (Anderer et al., 1960; Tsugita et al., 1960). The determination of the nucleotide sequence of TMV-RNA, however, is not possible at present due to methodological difficulties. Therefore, the establishment of the correlation between the two sequences has to be approached in an indirect way: changes should be induced in the RNA chain of TMV, i.e. a mutant has to be produced, the mutants obtained in this way have to be available in large quantities, and the amino acid sequence of the protein part of these mutant strains analyzed. The final step in this type of experiment consists in the establishment of specific amino acid replacements in the protein part of the mutants with reference to well defined nucleotide changes induced experimentally in the parental TMV-RNA.

The most accurate method of inducing definite changes in nucleotides of an RNA chain consists in the oxidative de-amination of adenine, guanine and cytosine by treatment with nitrous acid. Gierer and Mundry (Mundry and Gierer, 1958; Gierer and Mundry, 1958) have treated isolated TMV-RNA or intact TMV with nitrous acid and used such treated preparations for the infection of Java tobacco plants. This cultivar is known to react to infection with the original (untreated) TMV inoculum by local chlorotic spots. On infection of Java tobacco plants with the nitrous acid treated TMV-RNA preparations, however, local necrotic lesions also appeared in addition to chlorotic spots. Mundry (1959) by subjecting the results to a detailed kinetic analysis, came to the conclusion that nitrous acid induced an actual mutation of the virus and that this mutation was a consequence of the oxidative de-amination of a single nucleotide residue in the viral RNA chain. When these so-called nitrite mutants were produced in large quantities and their amino acid composition determined it was found (Wittmann, 1960, 1961) that in most cases there was an amino acid replacement. A comparison of the expected transformations of nucleotide bases upon treatment with nitrous acid (A → G and C → U) with the amino acid replacements revealed that these replacements were compatible with the nucleotide composition of the triplet code words established from analyses of results obtained in cell-free protein synthesizing systems of *E. coli* with synthetic nucleotides.

In practice the following criteria for strain relationship can be used as summarized by Knight (1963):

1. Positive serological cross-reactions
2. Positive cross-protection tests
3. Similarity in host range
4. Similarity in method of transmission
5. Similarity in response to genetic changes in the host
6. Similar resistance to destruction of infectivity by chemical and physical agents such as heat, dessiccation, hydrogen ion concentration, irradiations, pressure, chemicals, etc.
7. Coincidence of specific chemical and physical properties:
   a) Same size and shape of particles
   b) Same proportions of protein and nucleic acid
   c) Same proportions of nucleotides in nucleic acid components
   d) Same nucleotide sequences except for mutationally introduced differences
   e) Same kind of N-terminal and C-terminal residues in protein subunits
   f) Same number of terminal residues and, hence, same number of subunits in protein
   g) Same number of amino acids in protein subunits and the same sequences of amino acids except for mutationally introduced differences

From a methodological point of view the study of the genetics of plant viruses coincides with virus identification. Virus identification does not involve any special techniques in addition to those described so far.

If the virus strain is to be identified, the so-called cross-protection test is used. From a methodological point of view a strain which induces necrotic local lesions should be used for superinfection. In such a case the cross-protection test will last for a shorter time than if the indicator strain induces systemic symptoms. If no necrotic strain is available for superinoculation in premunity tests a strain causing conspicuous symptoms differing very much from the strain to be tested should be employed.

The process of virus identification includes the following steps:

Establishment of the symptoms on the main host

↓

Transmissions to the host from which the virus has originally been isolated by means of

↙ Grafting | ↓ Dodder transmission | ↓ Mechanical transmission | ↓ Insect transmissions

Insect transmissions: ↙ Aphids ↘ Leafhoppers

Aphids: ↙ non-persistent ↘ persistent

Leafhoppers: ↙ persistent

↓

Transmission of the virus using the most efficient method to other host plants and description of the symptoms produced

↓

Screening of host plants to find a differential test plant

↓

Determination of the stability of the virus to physical factors in vitro

↙ Thermal inactivation point | ↓ Dilution end point | ↘ Ageing in vitro

↓

Serological properties of the virus

↓

Morphology

↓

Tests for possible soil transmission and seed transmission

Cross-protection test

For the test, nine plants are taken and divided into three groups each comprising three plants. One group is infected immediately with the virus strain to be tested. The other group is infected with the same strain three weeks after the inoculation of the first group. After an additional period of three weeks, all nine plants are superinoculated with the indicator strain causing necrotic local lesions. If the virus tested is a strain of the same virus as the indicator strain, the three control plants inoculated only with the indicator strain will exhibit many necrotic local lesions whereas the three plants which have been infected for three weeks with the virus strain to be tested will show fewer local lesions. On those plants which have been infected for six weeks with the virus strain to be tested prior to the inoculation with the indicator strain only very few or no lesions at all will be produced. The reason for this phenomenon is that in the plants which have been infected for six weeks the virus strain to be tested is much more evenly distributed than in those infected for only three weeks, and will thus induce a higher degree of cross-protection.

CHAPTER VI

NOMENCLATURE AND TAXONOMY OF PLANT VIRUSES

As early as the beginning of plant virus research a need for the nomenclature and taxonomy of plant viruses emerged. Due to a lack of appropriate criteria serving as a basis for plant virus classification, plant virus nomenclature has been worked out on several occasions independently from virus taxonomy. Recently, however, owing to the achievements in plant virus serology, electron-microscopy and last but not least virus biochemistry, an increasing trend of convergence occurred with respect to problems of virus nomenclature and systematics. From a didactical point of view virus nomenclature and taxonomy will be surveyed separately, the cases, however, in which the two problems are intervowen will be pointed out.

1. VIRUS NOMENCLATURE

The simplest form of virus nomenclature (vulgar nomenclature) consists of naming the virus on the basis of the most conspicuous symptom produced in its main host (e.g. tobacco mosaic virus). This type of nomenclature is widely used at present (Anonymous, 1950).

Another kind of nomenclature is the so-called "numbering" system of nomenclature. Here the word virus and a number is added to the vulgar (Johnson, 1927) or Latin (Smith, 1937) generic name of the main host (e.g. tobacco virus 1 or *Nicotiana virus 1*). This system of nomenclature is out of use now.

The principle of the third system of nomenclature is that in naming a virus in addition to the main host of the virus in question, some of its additional properties are also given. For instance tobacco mosaic virus can be named *Nicotiana virus altathermus* (Bennett, 1939) based on its high temperature resistance, or *Virus Nicotianae perdurans* based on its high degree of *in vitro* stability. This system of nomenclature, although potentionally applicable to classification, is not used at present.

The binomial nomenclature uses a term for the genus and a term for the species to name plant viruses similarly to Linné's system of nomenclature. The principles of this type of nomenclature have been laid down and the system of classification based on these principles worked out by Holmes (1948). Presently binomial nomenclature is used in addition to the vulgar nomenclature. Another binomial nomenclature was developed by Ryzhkov (1952).

2. VIRUS CLASSIFICATION

Plant viruses can be classified on the basis of two principles: biological properties of the virus (symptomatology, histotropism, transmission), and physicochemical properties (resistance to environmental conditions, shape, size and antigenic structure). In many cases both these criteria have been combined in the classification of plant viruses.

Quanjer (1931) based his system of classification on virus induced histological changes. The following groups were distinguished: nonnecrotic mosaic viruses, phloem necrosis viruses, acronecrosis viruses, and acropetal necrosis viruses.

Johnson and Hoggan (1935) divided plant viruses into two main groups: insect transmitted viruses, and viruses which are not transmissible by insects. Further subdivision of these two main groups was based on the resistance of viruses to environmental factors *in vitro*.

Holmes (1948) used both symptomatological and biological properties of plant viruses for their classification and introduced at the same time a binomial nomenclature mentioned above. McKinney (1944b) reported a similar system but went further in subdividing the individual taxonomical groups according to modes of transmission. Ryzhkov's (1952) system is based on the mode of transmission, shape and crystallization of plant viruses.

Schramm's (1954) system is based exclusively on the morphology and Chester's (1937) classification on the antigenic structure of plant viruses.

An up-to-date combination of the morphological and serological properties of plant viruses comes from Brandes and Wetter (1959). This system, however, has been used so far only for the classification of rod-shaped viruses. A recent trend of virus systematics is represented by the classification elaborated by Hansen (1956). This is based on the histotropism of plant viruses.

CHAPTER VII

CONTROL OF PLANT VIRUS DISEASES

To find appropriate control measures against a given virus disease the biological properties of both the virus and the host have to be taken into consideration. This means that the actual measures to be taken for the control of the virus diseases will vary to a great extent with different host-virus combinations. Therefore, it is impossible to go into details in this respect. Instead, a general outline of the theoretical background of control measures against plant virus diseases will be presented.

Control measures against plant virus diseases can be divided into the following groups (Gäumann, 1951):

1. Preventive control measures. This aims at the prevention of the disease and can be carried out in either of two ways: infection prophylaxis and disposition prophylaxis.

Infection prophylaxis is designed to prevent the virus from coming into contact with the host. There are two ways of achieving this aim: a) Preventing the virus from coming into contact with the healthy crop (measures against mechanical transmission, soil transmission and insect transmission), b) Preventing the virus from infecting the offspring of virus diseased plants (measures against seed transmission).

Disposition prophylaxis is intended to decrease plant susceptibility to virus infection. This can be achieved in either of two ways: influencing disposition factors (e.g. fertilization) and breeding for resistance.

2. Therapy. This procedure is designed to cure already diseased plants. Whereas heat therapy has been used with success against several plant virus diseases, chemotherapy has so far yielded little success. This is certainly due to the biochemical mechanism of host-virus combination.

APPENDIX

SOME REPRESENTATIVE PLANT VIRUS DISEASES

In the following some representative virus diseases of important crops will be described. This chapter does not aim to compile a comprehensive list of the most important virus diseases of crops. It rather enumerates a few virus diseases to show what the experimental object is to which the principles outlined in the foregoing chapters are to be applied.

TOBACCO

MOSAIC DISEASE

a) Symptoms (Fig. 7)

On the inoculated leaves usually no local lesions are produced. If temperature and light intensity are high, small circular slightly chlorotic spots may appear on the inoculated leaves. The first signs of systemic infection appear in the form of vein clearing of the youngest leaves. Vein clearing is accompanied by mosaic mottling and often by leaf deformations.

b) The pathogen

Tobacco mosaic virus (*Marmor tabaci* var. *vulgare*, Holmes; *Nicotiana virus 1*, Smith).

1. Morphology. Rod-shaped particles 300 mμ long and 15 mμ in diameter.
2. Thermal inactivation point. In undiluted sap 93 °C upon a heat treatment for 10 min.
3. Dilution end point. With crude sap 10^{-6}.
4. Ageing *in vitro*. Upon storage at room temperature the virus-containing sap retains only about two per cent of its original activity in 1 month. In desiccated plant tissues, however, the virus remains active for many years.
5. Serology. Using the usual immunization techniques a very active antiserum can be prepared against tobacco mosaic virus.
6. Transmission. The virus is easily mechanically transmissible. No unequivocal proof of successful transmission by aphids is available. The virus is soil transmitted. Seeds of tobacco do not seem to transmit the virus. It is transmitted, however, through seeds of tomato and red pepper.
7. Strains. There is a wide range of strains of tobacco mosaic virus. Those mostly used in the laboratory are the so-called ribgrass mosaic viruses which cause,

1

2

3

Fig. 7. Symptoms caused by tobacco mosaic virus on a systemic host (*Nicotiana tabacum* cultivar Samsun) [1], and on two local lesion hosts (*Nicotiana glutinosa* and *Nicotiana tabacum* cultivar Xanthi-*nc*.) [2 and 3] (Courtesy of Dr. R. Gáborjányi, Research Institute for Plant Protection, Budapest) in contrast to the other TMV strains, local lesions in *Nicotiana sylvestris* and the so-called para strain which induces necrotic local lesions in White Burley tobacco.

8. Differential test plants. All strains of tobacco mosaic virus cause necrotic local lesions on *Nicotiana glutinosa* and *Nicotiana tabacum* cultivar Xanthi-*nc*.

NECROSIS DISEASE

a) Symptoms

Symptoms induced by natural infection and artificial infection are different. Under natural conditions tobacco plants become infected through the lowest leaves which are in contact with infested soil. Those leaf parts which are in direct contact with the infested soil die. In some cases the virus is able to infect tobacco plants systemically. Upon artificial inoculation a large number of necrotic local lesions are produced. These lesions tend to coalesce.

b) The pathogen

Tobacco necrosis virus (*Marmor lethale*, Holmes; *Nicotiana virus* 11, Smith).

1. Morphology. The virus particles are spherical, about 20 mμ in diameter.

2. Thermal inactivation point. When heated for 10 min the virus loses its infectivity between 72 °C and 85 °C.

3. Dilution end point. With crude sap it is between 10^{-4} and 10^{-6}.

4. Ageing *in vitro*. Upon storage at room temperature the virus-containing sap loses its infectivity by the 20th day.

5. Serology. With the usual immunization techniques tobacco necrosis virus yields active antiserum.

6. Transmission. The virus is easily transmitted mechanically, but not at all with aphids. It has recently been shown that zoospores of the fungus *Olpidium brassicae* act as vectors. The virus is soil transmitted and this is the most common way of spreading in nature. No seed transmission has been reported.

7. Strains. Several strains of tobacco necrosis virus have been described. It is interesting to note that some strains do not show a serological relationship with each other. It seems therefore that tobacco necrosis virus is a collective species and its name reflects the type of symptoms rather than the taxonomic position of the pathogen.

8. Differential test plants. The virus causes comparatively large necrotic spots on tobacco and small necrotic local lesions on *Nicotiana glutinosa*. *Chenopodium amaranticolor* also reacts with the formation of small necrotic local lesions.

TOMATO

MOSAIC DISEASE

a) Symptoms

The most common pathogen of tomato mosaic disease is tobacco mosaic virus. The type of symptoms is determined by the strain of tobacco mosaic virus which infects the plant. With the so-called ringspot strains there is usually only mild mottling and a limited reduction of growth. The fruits remain symptomless. With the so-called yellow strains, however, bright mosaic symptoms are produced, there is a fairly pronounced inhibition of growth and the fruits show yellow rings or spots. When tomato plants are infected in an early stage of development, yield losses as high as 30 per cent may result.

b) The pathogen

The properties of the pathogen (*tobacco mosaic virus*) are described on p. 104.

ASPERMY DISEASE

a) Symptoms

The most characteristic symptom consists of the inhibition of the growing point of the main stem. On the other hand the axillary shoots grow fast and this gives the plant a characteristic bushy appearance. The top leaves are distorted and bear mosaic symptom. The formation of so-called enations is characteristic. Seed formation in fruits set after infection is suppressed.

b) The pathogen

Chrysanthemum aspermy virus (tomato aspermy virus).

1. Morphology. The virus particles are spherical, about 35 mμ in diameter.
2. Thermal inactivation point is between 55 °C and 60 °C upon a heat treatment for 10 min.
3. Dilution end point. In summer 10^{-3} and in winter 10^{-4}. This difference is due to the fact that at reduced light intensities the virus reaches higher concentrations in the host than at high light intensity.
4. Ageing in vitro. In expressed sap the virus retains its infectivity for 24 to 48 hr.
5. Serology. The virus is a weak antigen. To prepare antiserum against the virus, special techniques have to be employed.
6. Transmission. The virus is transmitted both mechanically and by means of aphids. It is a non-persistent virus. Seed and soil transmissions have not been described.
7. Strains. Three main groups of strains have been isolated so far with respect to the symptoms produced in *Nicotiana glutinosa*. The mild strain causes weak mosaic symptoms without leaf deformation on this particular plant species.

The common strain induces the usual leaf deformations and enations. The severe strain produces necrotic symptoms. Some authors claim tomato aspermy virus to be related to cucumber mosaic virus.

8. Differential test plants. *Nicotiana glutinosa* is a useful test for distinguishing the above mentioned three strains of the virus. *Datura stramonium* and *Tetragonia expansa* react to the infection with this virus with the formation of local lesions. Red pepper can be used to distinguish tomato aspermy virus from cucumber mosaic virus. Whereas infection with most strains of cucumber mosaic virus will result in a bushy appearance of red pepper, but aspermy virus causes only mosaic symptoms in this particular host.

CUCUMBER

MOSAIC DISEASE

a) Symptoms

The symptoms consist of leaf mottling and blistering. The leaves are reduced in size. The internodes of diseased plants are shorter than those of healthy ones. It is characteristic that cucumber plants infected with cucumber mosaic virus bear fewer flowers and fruits than healthy cucumber plants. The fruits which are set are distorted and show mosaic mottling.

b) The pathogen

Cucumber mosaic virus (*Marmor cucumeris* var. *vulgare*, Holmes, *Cucumis virus 1* Smith).

1. Morphology. Virus particles are spherical, 35 mμ in diameter.

2. Thermal inactivation point. In undiluted sap the virus loses its infectivity between 60 °C and 70 °C.

3. Dilution end point. With crude sap this is 10^{-4}.

4. Ageing *in vitro*. Upon storage at room temperature the virus loses ts infectivity in about 1 to 2 days.

5. Serology. Active antiserum has been prepared so far only with purified cucumber mosaic virus as the immunizing agent.

6. Transmission. The virus can be transmitted both mechanically and by aphids. It is a non-persistent virus. Some unconfirmed data refer to seed transmission in very low percentages.

7. Strains. The Pierce's indicator strain causes local lesions in *Zinnia elegans* in contrast to the other strains of this virus. Therefore, it can be used with success in cross protection tests. Strain Y causes systemic symptoms in *Vigna sinensis* in contrast to the other strains of this virus.

8. Differential test plants. The virus causes local lesions in *Vigna sinensis*, *Amaranthus caudatus*, *Chenopodium quinoa* and *Chenopodium hybridum*.

POTATO

MOSAIC DISEASE

a) Symptoms (Fig. 8)

The leaves show mosaic mottling. The degree of severity of mosaic symptoms varies with different potato varieties.

Fig. 8. Potato seedling infected with potato virus X (Courtesy of Dr. K. Schmelzer, Institute of Phytopathology, Aschersleben, Germany)

b) The pathogen

Potato virus X (*Marmor dubium* var. *annulus* Holmes, *Solanum virus 1* Smith).

1. Morphology. The virus particles are thread-like, 520 mμ long and 10 mμ in diameter.

2. Thermal inactivation point. There are two groups of strains which can be distinguished on the basis of their serological properties. One group has a thermal inactivation point of 68 °C and 69 °C whereas the thermal inactivation point of the other group lies between 72 °C and 74 °C.

3. Dilution end point. With crude sap it lies between 10^{-5} and 10^{-6}.

4. Ageing *in vitro*. Depending on the virus strain studied the *in vitro* stability of this virus varies from some weeks to one year.

5. Serology. Using conventional procedures of immunization active antiserum can be prepared against potato virus X.

6. Transmission. The virus is easily transmitted mechanically. No insect has been shown to transmit the virus. It is not transmitted by the seed or through the soil.

7. Strains. Depending on the severity of symptoms 6 different strains have been described. On the basis of their serological properties and thermal inactivation point they can be combined to form two major groups of strains.

8. Differential test plants. *Gomphrena globosa* reacts with brownish red necrotic local lesions.

LEAF ROLL DISEASE

a) Symptoms (Fig. 9)

The symptoms vary to a considerable extent with different potato varieties. The most characteristic symptom with all varieties is the rolling of the leaves.

b) The pathogen

Potato leaf roll virus (*Corium solani* Holmes, *Solanum virus 14* Smith).

1. Morphology. The virus particles are more or less hexagonal and are 23 mμ in diameter.

2. Thermal inactivation point.	Mechanically not transmissible consequently no way of testing these properties has been available
3. Dilution end point.	
4. Ageing *in vitro*.	

Fig. 9. Potato seedling infected with potato leaf roll virus (Courtesy of Dr. J. Horváth, Research Institute for Plant Protection, Budapest)

5. Serology.–

6. Transmission. The chief insect vector is *Myzus persicae* but other species of aphid also act as vectors.

7. Strains. On the basis of severity of reaction of Physalis floridana at different temperatures and of the degree of cross-protection the presence of four distinct leafroll virus strains has been established.

8. Differential test plants. The most widely used differential host is *Physalis floridana* because it is very susceptible to infection with potato leaf roll virus and exhibits characteristic symptoms consisting of stunting, chlorosis and leaf rolling.

STREAK DISEASE

a) Symptoms (Fig. 10)

The symptoms of different strains of potato streak disease are very divergent. As a rule slight mosaic, mottling of the leaves or acropetal necrosis is visible. In addition, black local lesions or vein necrosis appears on the leaves. Stem necrosis and necrosis in the vascular system may also occur.

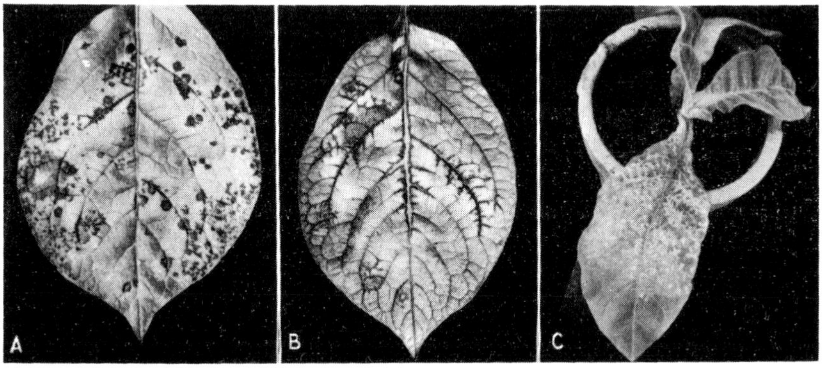

Fig. 10. Reaction of potato virus Y on test plants. A and B: *Solanum demissum* A6 hybrid. C: *Nicotiana tabacum* cultivar White Burley (Courtesy of Dr. J. Horváth, Research Institute for Plant Protection, Budapest)

b) The pathogen

Potato virus Y (*Marmor upsilon*, Holmes; *Solanum virus 2* Smith).

1. Morphology. The virus particles are long and flexuous thread shaped 730 mμ long and 15 mμ in diameter.

2. Thermal inactivation point. The undiluted sap loses its infectivity between 56 and 74 °C depending on the strains (cf. Horváth, 1967).

3. Dilution end point. In the expressed sap it is 10^{-4} to 10^{-5}.

4. Ageing *in vitro*. The crude sap retains its infectivity for 6 to 18 days.

5. Serology. Active antiserum exists against the pathogen.

6. Transmission. The virus can be transmitted both mechanically and by aphids. Seed transmission is also possible.

7. Strains. The usual grouping of the strains of potato virus Y is as follows: Normal strain Y^N, the strain causing the browning of midribs in tobacco Y^R, the strain Y^C (not aphid-borne), and the anomalous Y^{An} strain (cf. Horváth 1967).

8. Differential test plants. The virus causes local lesions on *Lycium halimifolium*, *Solanum demissum* Bastard A 6 and *Physalis floridana*. *Nicotiana glutinosa*, *Nicotiana tabacum* (differential host for the determination of strains) and *Datura innoxia* are systemic hosts. *Datura stramonium* is immune to the infection of potato virus Y.

BEAN

SOUTHERN BEAN MOSAIC DISEASE

a) Symptoms

The symptoms caused by southern bean mosaic virus are dependent on the bean cultivars. On some cultivars the virus induces large local lesions. Those cultivars which do not give localized symptoms produce systemic mottling or mosaic. In a few cultivars a systemic necrosis is produced.

b) The pathogen

Southern bean mosaic virus 1, *Bean mosaic virus 4*, Smith.

1. Morphology. The virus particles are spherical about 25 mμ in diameter.
2. Thermal inactivation point. The thermal inactivation point of the virus lies at about 95 °C.
3. Dilution end point. 1 : 500,000.
4. Ageing *in vitro*. The undiluted sap loses its infectivity in about 10 months.
5. Serology. Active antiserum can be prepared against the pathogen.
6. Strains. There are two different strains of southern bean mosaic virus. They differ in the size of local lesions and in the severity of symptoms.
7. Differential test plants: Certain cultivars of beans on which different symptoms are produced. Some solanaceous plants are susceptible such as *Nicotiana tabacum*, *Nicotiana glutinosa*, *Capsicum annuum*, *Datura stramonium*, as well as many species of *Leguminosae* as *Pisum*, *Vicia*, *Lupinus*.

TURNIP

YELLOW MOSAIC DISEASE

a) Symptoms

Systemic symptoms (vein clearing, yellow areas and mosaic mottling on the leaves) develop on *Brassica* species in two weeks. Flower breaking may also occur.

b) The pathogen

Turnip yellow mosaic virus (*Crystallococcus cruciferi* Ryshkow).

1. Morphology. The virus particles are almost spherical and measure 20 mμ.

2. Thermal inactivation point. The virus becomes inactivated between 70 and 75 °C.

3. Dilution end point. Using crude expressed sap positive infections are obtained at dilution 10^{-5} but not infrequently at dilutions of 10^{-6}.

4. Ageing *in vitro*. The inoculum may be uninfectious after a few weeks.

5. Serology. Using conventional procedures, active antiserum can be prepared against turnip yellow mosaic virus.

6. Transmission. The virus is transmitted by biting insects grasshoppers, earwigs and mustard beetles. Flea beetles are common vectors in the field, however aphids or other sucking insects are unable to transmit the virus.

7. Strains. The common strain is serologically related to Northumberland virus and Edinburgh strain. These strains react differently on cruciferous test plants. The thermal inactivation point of the two England strains is higher than that of the common one being between 75 to 80 °C.

8. Differential test plants. On all *Brassica* species mosaic mottling develops, except on *Brassica chinensis* which produces yellow variegation. Red local lesions appear on the inoculated leaf of *Cleome spinosa*, followed by a systemic mosaic.

PART II

BACTERIOLOGY

INTRODUCTION*

Bacteria are an important group of plant disease-causing organisms. In 1878–1880, during the great age of classical bacteriology when many important human pathogens were still unknown, the American plant pathologist Burrill (1880) was the first to prove that bacteria could cause plant disease. Today, the number of known bacterial plant diseases is over two hundred. Some of these cause considerable economical loss. For example, we have, in Hungary, bacterial diseases of the bean (*Pseudomonas phaseolicola*, *Xanthomonas phaseoli*, *Corynebacterium flaccumfaciens*), and the bacterial blight of pepper (paprika) and tomato, (*X. vesicatoria*). *P. mors-prunorum*, and *P. syringae* cause serious problems in orchards. In the U.S.A. and recently in Europe *Erwinia amylovora* destroys apple and pear trees. In tropical and sub-tropical areas, the damage caused by *P. solanacearum* to *Solanaceae* and banana plantations, or by *X. malvacearum* in cotton fields is significant. The pathogenic bacteria belonging to the group *E. carotovora*, which has spread almost all over the world, causes soft rot in the field and post-harvest in storage and crop losses regularly.

Approximately 15–20 per cent of the 1500 bacterial species known in bacteriology at the present time are phytopathogenic. Hence the aim of this section is to discuss some practical aspects of phytobacteriology and methodology.

Some methods in connection with bacterial plant pathogens have been published in the books of Dowson (1957), Izrailsky (1960), Stapp (1958, 1961) Ercolani (1968), Beltukova et al. (1968) and in the First Workshop on Phytobacteriology edited by Goodman (1967). In this work we tried to collect those methods used generally in our daily laboratory practice. The part on bacteriology is especially useful for beginners and those inexperienced in this field although it may prove valuable to experienced scientists as well.

* The following abbreviations are used: *A.* (*Agrobacterium*) ; *C.* (*Corynebacterium*); *E.* (*Erwinia*); *P.* (*Pseudomonas*); *X.* (*Xanthomonas*).

CHAPTER I

GENERAL CHARACTERISTICS OF PLANT PATHOGENIC BACTERIA

Schizomycetes, or bacteria, are unicellular organisms, visible under the light microscope. They may be motile if flagellated. Reproduction is mostly asexual by simple cell-division, fission forming colonies on solid media. All plant pathogenic bacteria are rod-shaped. Their length varies between 1–3 μ. The structure of a rod-shaped bacterial cell is shown in Fig. 1.

The cell substance consists of the protoplasm, which includes the cytoplasm, the nuclear substance (nucleus) and different granules, which are probably reserve nutrients. The cytoplasmic area is rich in RNA, the nucleus in DNA. With the electron microscope, one can differentiate macromolecular bodies about 200 Å in diameter which are RNA-protein particles, the ribosomes. The ribosomal fractions contain many enzymes which are necessary for protein synthesis. The bacterial cells do not contain a well defined nucleus, although cells of higher plants and animals do. However, within the cytoplasm it is often possible to differentiate such "chromatin bodies" which may be considered to be equivalent to the nucleus.

The cell substance is covered by the cytoplasmic membrane, which is surrounded by a more rigid cell wall, giving shape to the cell. Nutrition takes place through the semi-permeable membrane, through which the metabolites formed in the cytoplasm also pass.

Since the plant pathogenic bacteria do not form spores, the bacterial capsule or *slime substance* assumes greater importance. The capsule-enclosed bacteria are resistant to external influences (heat, light, radiation, chemicals) and in this way they maintain their viability for long periods of time (Hedges, 1926, Burkholder, 1945, Leach *et al.*, 1957 and Wilson *et al.*, 1965). Rosen (1936, 1938) pointed out, for example, that the exudate of the pathogen *E. amylovora* preserved the viability of the bacterial cells for 610 days

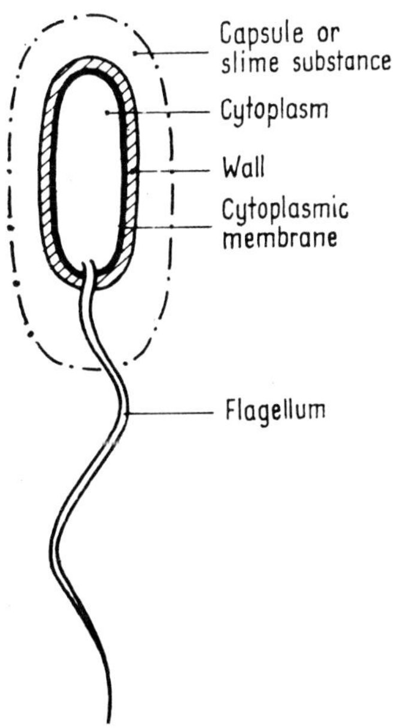

Fig. 1. Anatomic structure of a bacterium

in drying conditions; without exudate, the pathogen survived no longer than one month. The capsule material may interfere with the serological investigations, because the somatic (and surface) antigens are covered by the slime material (Elrod and Braun, 1947, Lovrekovich and Klement, 1965).

Most plant pathogenic bacteria are motile. The *flagellum* or *flagella* are responsible for the motility of the bacteria. These are extremely thin hair-like appendages (0·1 μ wide) which are not visible under the light microscope with simple investigative methods, but become apparent only with special staining procedures. The arrangement and number of the flagella are species variables, accordingly the bacterial cell may be *monotrichous* if a single flagellum is attached to one end of the cell; *lophotrichous* if more than one flagella comes from one pole; and *peritrichous* if the cell is flagellated all around (Figs 2, 3, 4). The presence, number and arrangement of the flagella characterize the different bacterial genera. For instance, the members of the phytopathogenic *Xanthomonas* are monotrichous, pseudomonads are lophotrichous, and the members of the genus *Erwinia* are peritrichous. Among the species of the *Agrobacterium* and *Corynebacterium* there are both flagellated and non-flagellated types (see Fig. 7).

In a suitable medium, the bacterial cell will start to reproduce. This may take place asexually by simple binary fission. However, as it has been shown during the past two decades, some species have a sexual mode of reproduction. The star formation occurring in agrobacteria is also considered to be one of the (simple) forms of sexual reproduction (Stapp, 1957, 1958).

The dynamics of bacterial reproduction under controlled constant conditions, may be characteristic of the species. Generally the soil-borne saprophytic bacteria reproduce quickly, while plant pathogens reproduce more slowly. This fact is very

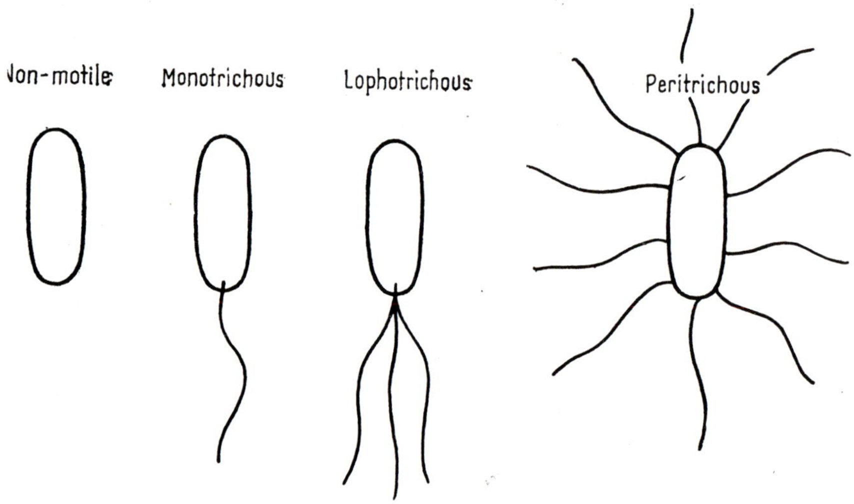

Fig. 2. Flagellation of bacteria

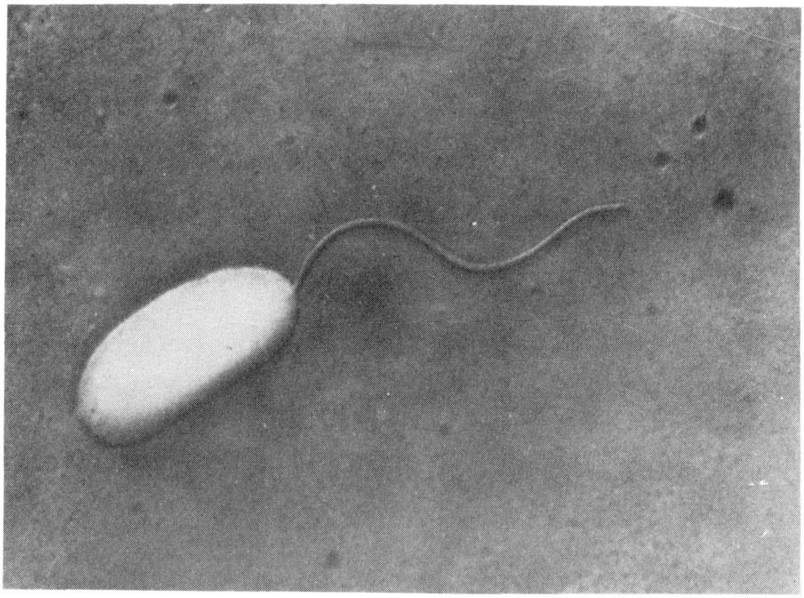

Fig. 3. Monotrichous flagellation of *Xanthomonas carotae* (Prepared by Z. Klement and B. Lovas)

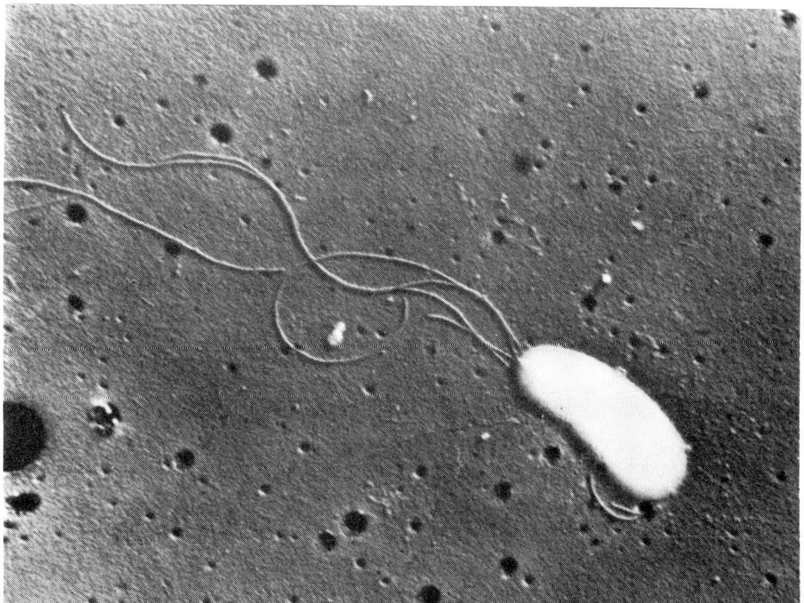

Fig. 4. Lophotrichous flagellation of *Pseudomonas lachrymans* (Prepared by Z. Klement and B. Lovas)

important, because saprophytic forms, frequently present in infected plant tissue can inhibit the pathogen when isolation in an artificial medium is attempted.

All plant pathogenic bacteria can be easily cultivated on media generally used in bacteriology. If, for example, one follows the reproduction of bacteria in liquid medium, the following stages can be differentiated (Fig. 5). The life of

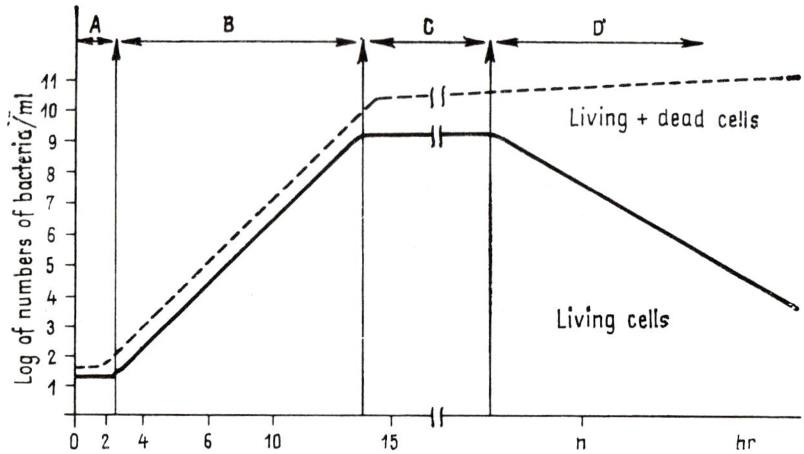

Fig. 5. Bacterial growth curve (A) lag phase; (B) log (logarithmic) or exponential phase; (C) stationary phase; (D) death or decline phase

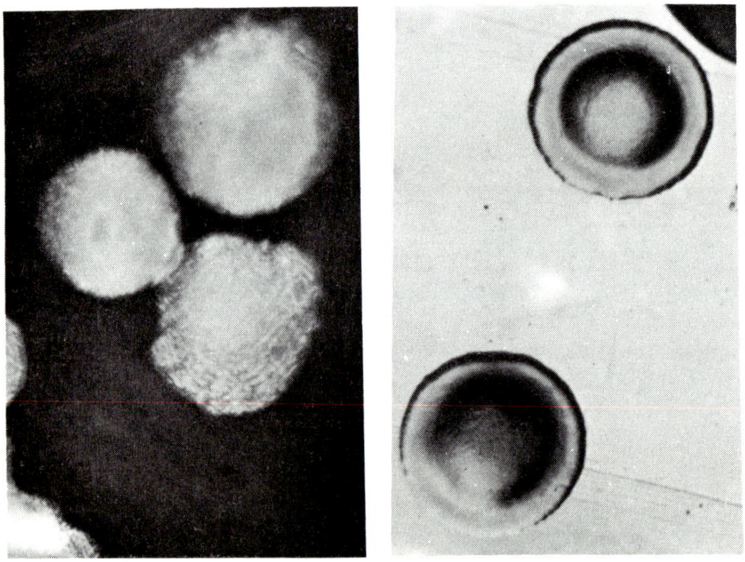

A B

Fig. 6. A) 48 hr-old colonies of *Pseudomonas syringae* on nutrient agar plate in transmitted light. B) Colonies of *Pseudomonas oryzicola* in throw light the culture begins with the *latent phase*. This is also called the "*lag*" *phase*. In this phase, reproduction has not yet begun following the inoculation of the media. The bacteria inoculated into the medium become acclimatized to the new environment and, following an increase in their size, initiate the *logarithmic* or "*log*" *phase*. During this phase, the number rises geometrically. Due to the exhaustion of some nutrients and the presence of inhibiting factors (bacterial metabolites), the rate of reproduction is reduced and a *stationary phase* follows, which in turn is followed by a phase of decline or *death phase*. Phytopathogenic bacteria begin to reproduce three or four hours after inoculation and, depending on the species, reach the stationary phase within twenty-four to seventy-eight hours.

The bacteria produce colonies on solid media (Fig. 6). The type of colony is to a certain extent characteristic of the bacterial species or strain. The colonies grow on the surface of the solid medium in lentiform or with slightly undulating edges. Many phytopathogenic species are chromogenic, a feature which may be used for their taxonomic characterization (Fig. 7). The phytopathogens can produce two kinds of pigments: water insoluble (non-diffusible into the medium),

Genus	Motility	Colony on nutrient broth agar	Gram stain	Type of disease (with few exceptions)
Agrobacterium	Motile or non-motile	White	−	Hypertrophy
Corynebacterium	Non-motile or motile (2 spp.)	Cream	+	Wilt vascular diseases
Erwinia	Peritrichous	White	−	Soft rot; Die-back
Pseudomonas	Lophotrichous (More than one flagella)	Greyish white transparent and diffusable green fluorescent pigment	−	Leaf spot; Die-back; Wilt
Xanthomonas	Monotrichous	Yellow	−	Leaf spot

Fig. 7. Some morphological and pathological properties of plant pathogenic bacteria

and soluble. Most of the xanthomonads, e. g. produce a non-diffusible pigment, giving the characteristic yellow colour to the colony. The pseudomonads on the other hand, produce mstly a green-fluorescent pigment diffusible into the medium (see page 146). Chromogenesis of the phytopathogens is summarized in Fig. 7.

The reproduction of bacteria in a given medium is influenced by the external environment. Therefore, temperature, pH, composition of the media and the oxygen requirement of the bacteria must be considered.

Regarding the temperature requirement of the bacteria, we usually differentiate between the minimum, optimum, and maximum growth temperatures and the so-called thermal death point. The bacterial cell is able to survive any low temperature. The reproduction of phytopathogens begins as low as 5–10 °C. The optimal growth temperature is between 25–30 °C in contrast to human pathogenic bacteria (37 °C). The plant pathogens cease to reproduce between 33–40 °C and the reproducing cells are destroyed at about 50 °C in 10 min.

The rate of growth is also influenced by the pH of the media. The phytopathogens develop well on neutral or slightly alkaline media (pH 77·2). This property may be used for separation from the microscopic fungi, i.e., the fungi are not or are only slowly developed on slightly alkaline media.

The bacteria are classified into *autotroph*, *heterotroph* and *paratroph* nutrient types according to the material from which they obtain their C and N requirements. The autotrophs are capable of building their bodies and metabolites from inorganic compounds. The heterotrophs receive their required energy for vital processes from the decomposition of organic materials. In this case, products having energy (which is normally liberated during decomposition) are necessary primarily for the movement and assimilative processes of the bacterial cell and lastly for its growth and reproduction.

The phytopathogens, with the exception of one or two, are heterotrophs. These grow only if organic materials are available. The C-requirement of many phytopathogenic bacteria may be supplied from only one source e.g., sugars or proteins, and the N-requirements from inorganic salts. Paratrophic, i.e., true obligate pathogens reproducing in the host tissue only, cannot be found among the plant pathogenic bacteria.

All phytopathogenic bacteria are *aerobic*, though some are *facultatively anaerobic*. They develop very rapidly on the surface and in aerated and shaken cultures. But in plant tissue, presumably, they cannot come in contact with air to a satisfactory extent, therefore, some species also grow in the absence of molecular oxygen (facultatively anaerobes). The *E. amylovora* develops aerobically only on media containing, e.g., mineral salts, glucose and nicotinic acid; but if malic or fumaric acid is added, it is able to grow anaerobically too.

CHAPTER II

THE CLASSIFICATION OF PHYTOPATHOGENIC BACTERIA

The classification of bacteria and other lower organisms is being perfected only very slowly, and even today it cannot be said to be complete. The reason for inadequate classification is the lack of morphological and pathological characteristics of the microorganisms. The faulty description of the same bacteria under different names stems from these difficulties.

Generally the most accepted system is contained in the 7th edition of Bergey's Manual of Determinative Bacteriology, published in 1957, which is followed in this book too. In this work we will limit ourselves only to the classes, orders, families and genera which contain phytopathogenic bacterial species.

Protophyta (unicellular plants):

Class II: *Schizomycetes* (fission fungi, bacteria)
- Order I: *Pseudomonadales*
  - Family 4: *Pseudomonadaceae*
    - Genus 1: *Pseudomonas*
    - Genus 2: *Xanthomonas*
- Order IV: *Eubacteriales*
  - Family 2: *Rhizobiaceae*
    - Genus 2: *Agrobacterium*
  - Family 4: *Enterobacteriaceae*
    - Genus 6: *Erwinia*
  - Family 12: *Corynebacteriaceae*
    - Genus 1: *Corynebacterium*
- Order V: *Actinomycetales*
  - Family 3: *Streptomycetaceae*
    - Genus 1: *Streptomyces*

There is no doubt that the nomenclature of phytopathogens used at the present time is out of date and needs revision from both bacteriological and pathological points of view. The struggle over contradictory opinions is still continuing. According to one point of view, the many species in one genus should be drawn into a few species (Graham, 1964; Stolp, Starr and Baigent, 1965; Lelliott, Billing and Hayward, 1966; Lelliott, 1967; Mandel, 1967; De Ley, 1968). The other, which follows the pathological point of view, would like to keep the classification used at the present time, with some modifications.

The most characteristic property of the phytopathogens is their ability to cause disease. Many research workers have been confused by the identification of the host range of bacteria by the atypical necrosis which appeared (hypersensitive reaction, see page 153). They considered such plant species as hosts, though this is not so in nature. The host range depended on which inoculation method was used and how many plant species and cultivars were investigated. The investigations carried out so far (Klement, 1968) have convinced us that the host spectrum of most plant pathogenic bacteria is much narrower than it was believed until now, and it is restricted to one or only a few plant species.

If one agrees with the fact that the pathogenicity of phytopathogenic bacteria is perhaps an even more important factor of classification than, for example, the action on carbon compounds or the capacity for breaking down proteins, then this important characteristic must not be forgotten when revising the classification of phytopathogenic bacteria. Furthermore, if we agree that the phytopathogenic species should be restricted to a few species within the same genus, we should not neglect the consideration that the naming of the bacterium should reflect its pathological characteristic also. This is an accepted thing in the classification of numerous human pathogenic bacteria (*Mycobacterium tuberculosis*, *M. bovis*, *M. avium*).

CHAPTER III

TYPES OF SYMPTOMS

The symptoms caused by bacteria are often similar to those incited by infections of fungi or sometimes of viruses. For example, the fungi belonging to the *Sclerotinia* and *Pythium* spp. cause a soft rot like that caused by some *Erwinia* species. The canker by *Nectria* species is similar to the symptom produced by the *P. mors-prunorum*. It is generally characteristic to symptoms of bacterial origin—in contrast to viral and fungus diseases—that the attacked tissues are water-soaked and when observed in transmitted light are greasy in appearance; quite often bacterial exudate occurs on the diseased parts.

The following types of symptoms can be differentiated: spot, die-back, soft rot, wilting, tissue proliferation (see Figs 8 and 9).

We speak of *spot diseases* when the necrosis of tissues is localized (Fig. 10). Many diseases of leaves belong to this group e.g., the wildfire of tobacco (*P. tabaci*), the blight of tomato, paprika (*X. vesicatoria*) etc. In the early stages of bacterial infection, the spots are of a darker green than the uninvaded tissue. The spots are mostly surrounded by small veins, which is why they have angular forms, as in angular leaf spot of cucumber (*P. lachrymans*). At first these spots appear water-soaked or greasy. The spots are often surrounded by a pale yellow zone (halo blight) originating from diffusion of the bacterial toxins into the tissues around the spot. It may happen that the spots fall out so the leaves become full of holes (leaf scar).

Local spots usually having irregular circular forms can appear on fruits (e.g., scab spot of tomato, Fig. 11). The spots formed on the stem and leaf stalk are usually elongated.

The *die-back of shoots* occurs if the necrosis completely encircles the shoots; in this case the tissues which are above the invaded part wilt. Wilting of shoots occurs in the case of *E. amylovora*, which causes fireblight disease on apple and pear shoots; and *P. syringae* and *P. mors-prunorum*, which infect stone fruits. Although under such circumstances the pathogen could also invade the vascular bundle, the disease still does not become systemic.

In the case of *soft rot*, the pathogenic bacteria producing pectin-decomposing enzymes break down the pectin which keeps the plant cell walls together. By this means, the tissues will be disorganized, and the attacked tissues become softened. The most typical case is the soft rot of onions and tubers, but the succulent plant tissues and stems are also infected by these bacteria. The rot is completed by the

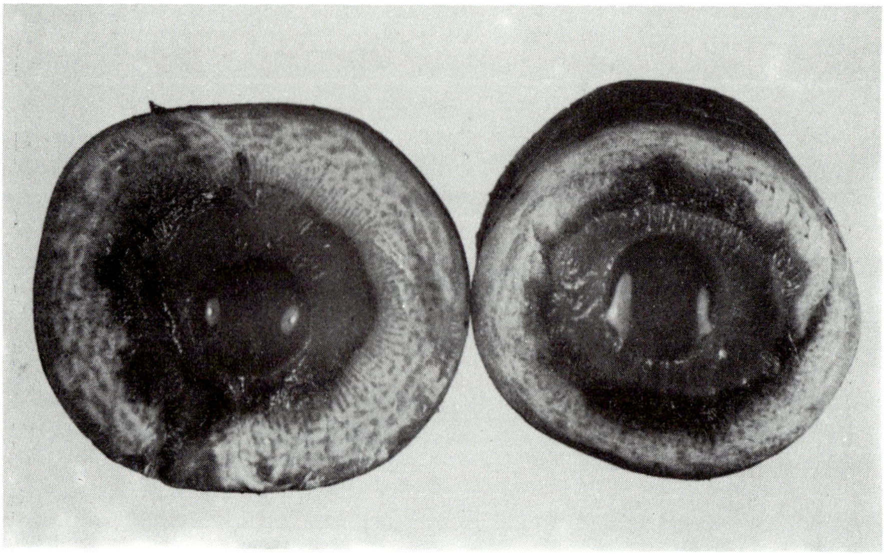

1

2

Fig. 8. Different types of symptoms induced by plant pathogenic bacteria: 1) Soft rot (*Erwinia carotovora*) on potato tuber. 2) Wilt (*Corynebacterium michiganense*) on tomatoes.

3

4

5

Fig. 8 (cont). 3) Spot disease (*Pseudomonas lachrymans*) on cucumber leaf. 4) Die-back (*Erwinia amylovora*) on apple shoot. 5) Crown gall (*Agrobacterium tumefaciens*) on apple tree

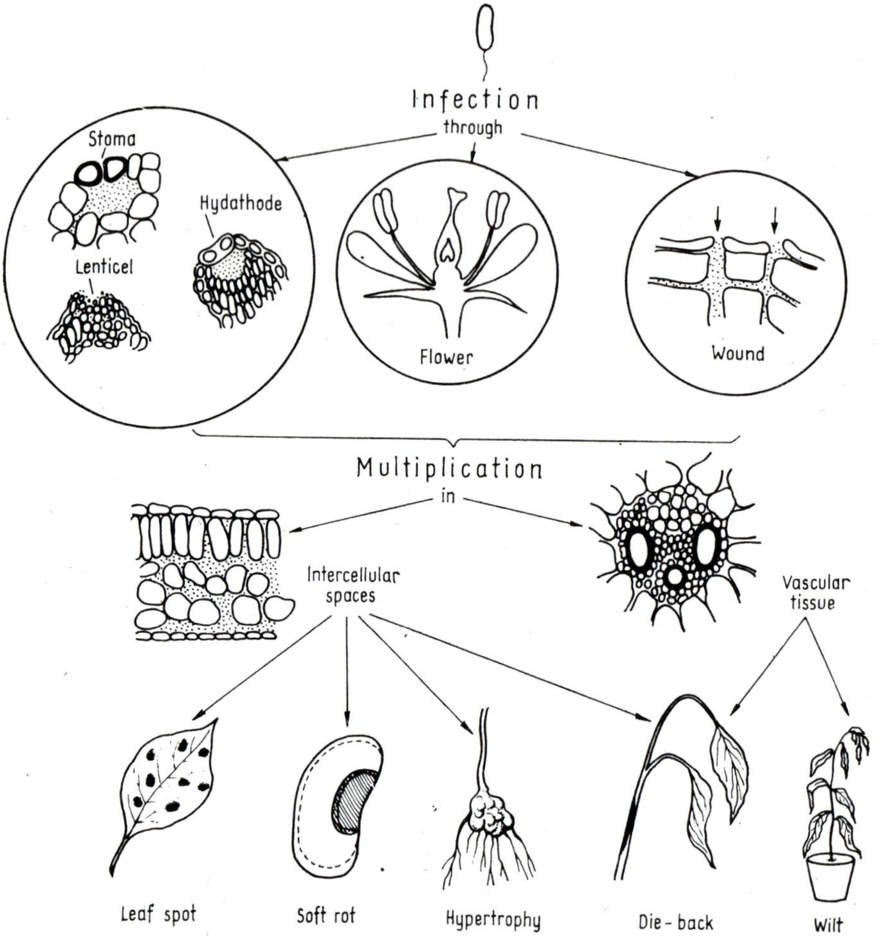

Fig. 9. Modes and places of infection, multiplication of plant pathogenic bacteria and types of symptoms

1

2

3

Fig. 10. Spot disease symptoms on leaves:
1) Shot holes on peach leaf (*Xanthomonas pruni*). 2) Angular leaf spot of hemp (*Pseudomonas cannabina*). 3) Striped spot on millet leaf (*Pseudomonas panici*),

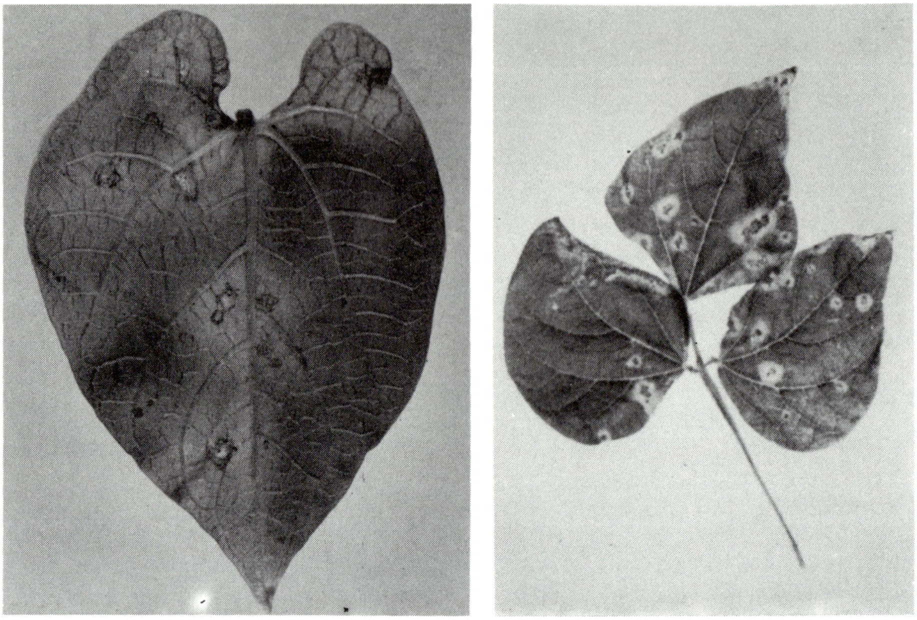

4 5

Fig. 10 (cont.). 4) and 5) Halo blight of bean, toxic effect round the spots

secondarily colonized bacteria and fungi. But at that stage the rotting has already become "putrescent rot".

Wilting occurs if the bacteria attack the vascular bundles. Since the water and nutrient transportation is blocked, the plant will wilt. If the bacterium penetrates from the vascular bundles into the parenchymatous tissues surrounding them and pushes its way to the surface, canker is formed (e.g., tomato wilt, *C. michiganense*).

Sometimes the symptoms of spotting and wilting appear together, as in the case of *P. phaseolicola* or *X. phaseoli*, which cause the grease spot of bean.

Some pathogens stimulate the meristematic tissue to grow abnormally, and in this way *proliferation of tissues* (crown-gall and hair root) are formed.

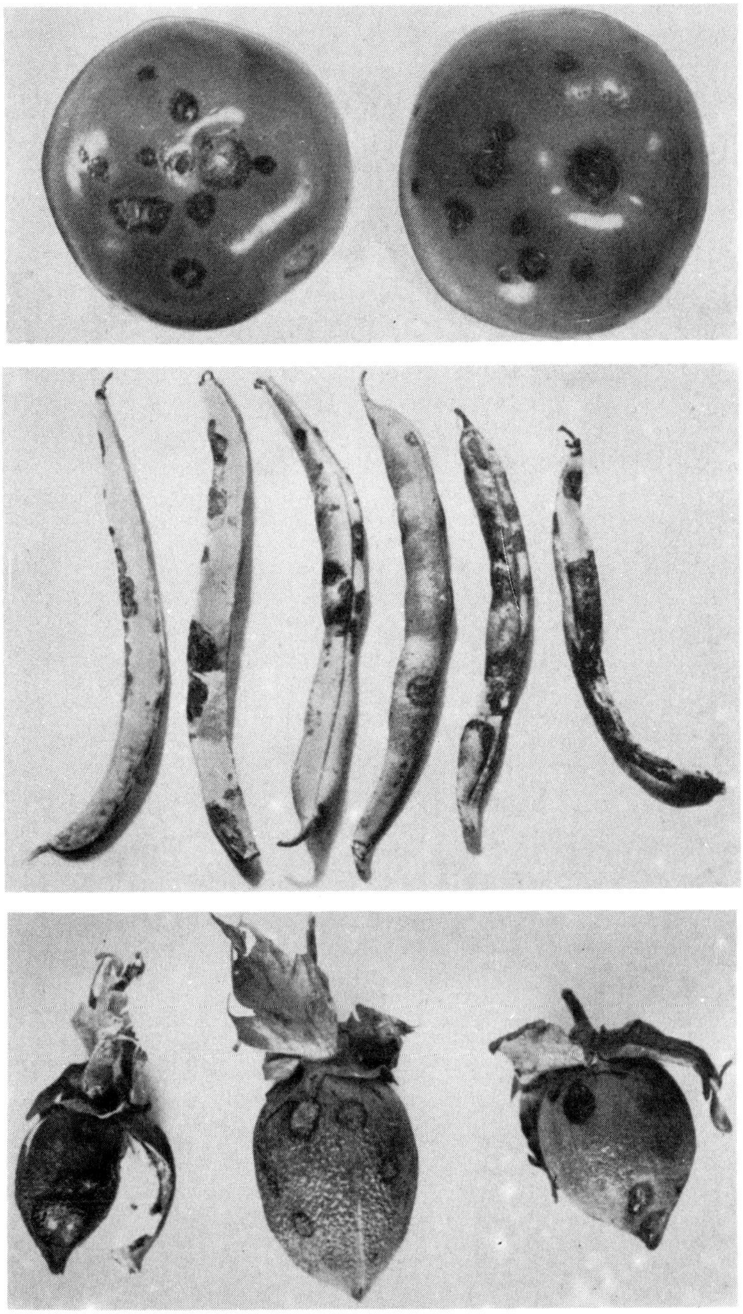

2

3

Fig. 11. Different types of spots on fruits, bulbs and pods: 1) *Xanthomonas vesicatoria* on tomatoes. 2) *Pseudomonas phaseolicola* on bean pods. 3) *Xanthomonas malvacearum* on cotton.

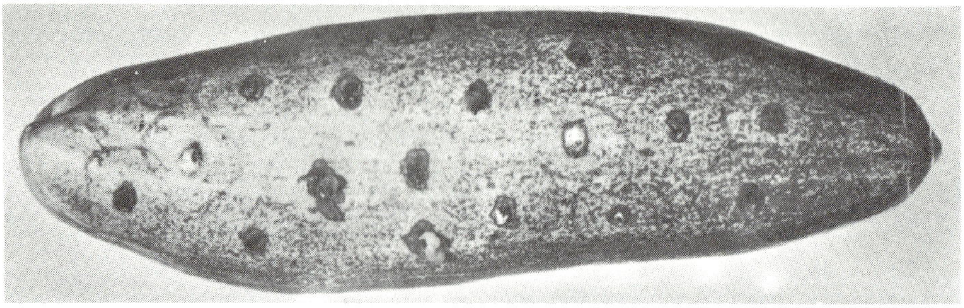

4

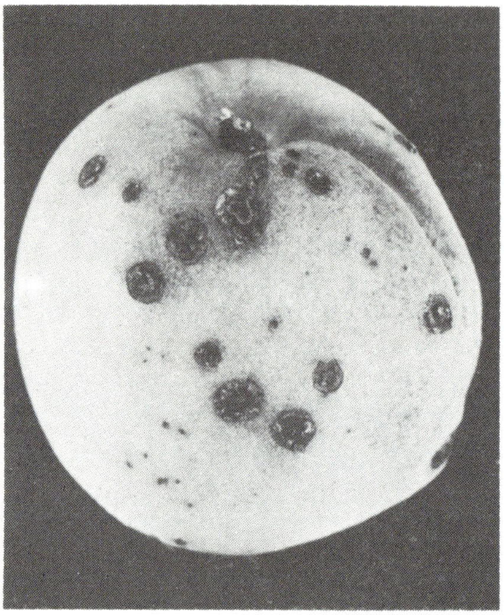

5

6

Fig. 11 (cont.). 4) *Pseudomonas lachrymans* on cucumber fruit. 5) *Xanthomonas pruni* on peach. 6) *Pseudomonas marginata* on *Gladiolus* bulb

CHAPTER IV

ISOLATION OF THE PATHOGEN

Conclusions as to the pathogen cannot always be drawn simply from the symptoms of disease. The same or similar symptoms may appear in species, which often are systematically far from each other. In the case of bean bacteriosis, three different bacterial species (*P. phaseolicola*, *X. phaseoli*, *X. phaseoli* var. *fuscans*) bring about the same symptoms. Recognition of the symptoms is not enough when diagnosing the disease. Isolation and identification of the pathogen should also be carried out. In fact, in many cases one should be convinced of the pathogenicity of the isolated strain.

It is best to follow Koch's postulates, which can be used on plant pathogenic organisms with the following modifications:

1. Recognition of the bacterium with microscopic examination in the infected tissue.
2. Isolation of a pure culture of bacteria.
3. Inoculation of the susceptible host plant to obtain the original symptoms.
4. Re-isolation of bacteria from the inoculated tissue.
5. Comparison of the original and reisolated bacterial strains, identification of the bacterium.

When *collecting the diseased material*, one should take care in order to get a successful isolation of bacteria. It is well known from practice that the plants sent in for investigation from farms are usually extremely damaged. The isolation of the pathogen from such plants is only possible either with great difficulty or not at all, because the tissues are overgrown by the secondary saprophytic microflora.

The bacteria can be successfully isolated from fresh plant material which demonstrates the initial symptoms of the infection. Plastic bags are not suitable for transportation and storage of such materials because the number of permanently present saprophytic organisms will increase under the closed and humid conditions. A dried material is always more suitable for this purpose than a too wet one. The sample transported and stored between paper layers is most suitable for laboratory cultivation.

It is not advisable to use chemicals for the *surface sterilization of the leaf* during examination of the leaves, because these materials can easily diffuse into the tissues and kill the bacterial cells. The most suitable substance for the disinfection of the surface is hypochlorite solution or another disinfectant. The most common method is to immerse the material to be investigated into 0·5 per cent Clorox (Na-hypochlorite) (1 : 9 solution) for two min. One should then swill the material thoroughly in distilled water and proceed with the preparation. In the case of leaves it is better to wash only the surface of the leaf blade with running tap water.

The isolation *from the leaves* is as follows: the fresh translucent leaf spots together with the healthy tissue parts are cut out with a sterile knife and put onto the surface of the slide which has been sterilized by flaming. One drop of sterile tap water or peptone water is pipetted onto the slide and the spot is cut with a sharp knife. The bacteria will diffuse from the spot into the peptone water after one or two min. The preparation covered by the cover slide is examined with more than 400 times magnification, but it is best to work with 800–1500 times magnification.

On the place of the cut, a bacterial cloud will diffuse from the infected tissue. For finer details, oil immersion or phase contrast set-up are used. The application and description of these is found in any microbiological manual.

It is especially important in the case of *vascular bundle diseases*, diseases of the stem or the transporting vessels, that a sample be taken from a part which shows the symptoms at an early stage. These stems are cut into 3–4 cm pieces. They are then split in length and the washed out prepared vascular bundles are macerated. From more fleshy stems, roots or tubers the bacteria can be obtained by squeezing the stem with the fingers so that the bacteria from the vascular bundles together with the sap is pressed out.

Bacteria from *tumours and woody tissues* can be obtained in the following way: a small piece of woody tissue which has been surface washed is dipped into alcohol and the alcohol in it is lit. While the alcohol burns off in 1–2 sec, the surface becomes sterile. However, 0·5 per cent Clorox solution can also be used for surface sterilization. The bacterium prepared from macerated tissue of discoloured or fresh tissues is then ready for examination.

The same procedure can be followed in the case of *seed investigations* if the pathogen settles under the seed coat (e.g., in the case of bean diseases). However, if the bacterium is adhering to the surface of the seed coat (with pea and tomato diseases) surface sterilization cannot be used. Isolation of bacteria from tumours is discussed in more detail on page 232.

MICROSCOPIC EXAMINATION OF THE DISEASED PLANT

Simple microscopic examination in the case of bacterial diseases does not provide sufficient information. Often such examination of plant material cannot even determine whether the disease is of bacterial origin. The reason for this is that frequently bacterial cells can be confused with plant cell pieces, starch or chlorophyll parts. The bacterial cells contained in the plant often do not have the forms of the bacteria cultivated by the media, but are rather smaller and of irregular rod shape. The bacteria found in the preparation from older spots lose their active motility. Probably only the Brownian movement is seen, making recognition of the structure of bacterial cells even more difficult. With young water soaked spots it is easy to recognize the bacterial cells. A correct microscopic picture in a later stage of the disease is greatly hindered by the presence of saprophytic microorganisms, because they cannot be distinguished from pathogens under the microscope. Therefore, microscopic examination only serves to give general information. It may be of more significance if tissue staining is done. With this procedure, the bacteria become differentiated in colour from the healthy and diseased tissue parts.

MICROSCOPIC EXAMINATION OF DISEASED TISSUES BY MEANS OF STAINED SECTIONS (DIFFERENTIAL-STAINING)

With suitable plant material, such as the stem or leathery leaf, sections can also be prepared free-hand if a sharp razor is available. Fine sections are prepared by hand from the diseased tissue placed between elder pith. The razor's edge is moistened with alcohol and the sections are placed in water. Stoughton's (1930) method for differential-staining.

1. Staining with carbol thionin for 5 min (Thionin blue 0·1 g in 100 ml of 5 per cent phenol containing distilled water).
2. Washing in water.
3. Transferring to 95 per cent alcohol.
4. Differential staining in Orange G for a few min (Orange G saturated in absolute alcohol).
5. Washing the tissues well with absolute alcohol.
6. After transferring to xylol, covering the section with a cover slip and mounting it in balsam.

In the stained preparation, the bacteria are deep blue, the cellulose cell wall yellow or green, and the lignified tissue light blue.

GENERAL LABORATORY PROCEDURES

Sterilization of equipment

The glassware, such as flasks, Petri dishes, tubes, pipettes, etc., which are used in the laboratory should be well washed in a detergent and rinsed out with tap water and finally with distilled water. The test tubes and the flasks should be fitted with cotton plugs and either covered with aluminium foil or with metal caps. Before the pipettes are sterilized, the end which is placed in the mouth should be plugged with cotton in such a way that the cotton in the pipette does not protrude or touch the mouth. This is necessary to prevent unsterile air from getting inside the pipette. Also, the cotton plug should not be moist. Pipettes and Petri dishes should be wrapped in wrapping paper or placed in pipette and Petri dish boxes.

Glassware is sterilized in an oven at 180 °C for 2 hr. To prevent damage, containers or equipment containing rubber or plastic parts should be sterilized in a steam autoclave or in a pressure cooker for 20 min under 1·5–2 atm.

Preparation of media

The phytopathogenic bacteria and fungi (with the exception of obligate pathogens) can be well cultivated on the generally used bacterial and fungal media. The nutrient requirements of a good number of pathogenic microorganisms are satisfied by carbon and nitrogen sources as well as inorganic salts. Some microorganisms, however, require certain vitamins or other growth substances besides the nutrients mentioned.

The media are prepared from different materials either from one's own preparation or a factory preparation (powder preparations). In the latter case, the amount of distilled water prescribed by the factory is added to the powder medium; then it is boiled and sterilized. Almost all phytopathogenic bacteria develop well on the generally used nutrient broth agar media, but potato dextrose agar used for fungus cultures (page 284) can also be used with success. When there is no factory preparation, the nutrient agar medium is prepared in the following way: 25 g non-fatty, minutely cut or minced horse meat or beef is placed in a 2 l flask to which is added 10 g peptone and 5 g NaCl in 1 l water, this is cooked for three hr. After it is cool, it is filtered firstly through a muslin cloth and then through cotton and supplemented to 1 l. After the pH is adjusted to 7·2, the medium is sterilized in the autoclave for 20 min at 1·5 atm. Solid media are mostly used for isolation and cultivation of the pathogens. The jellifying material is generally prepared from a sea alga, so-called agar-agar, which is added in 1·5–2 per cent amounts to the medium. The characteristic property of this curdling material is, that it dissolves at 100 °C and solidifies at 40 °C. In contrast to this, gelatin liquefies at 28–30 °C. If unwashed agar is used, then the impurities in the medium can be eliminated by centrifuging. If the medium is sterilized in the autoclave in a bigger glass container and then is allowed to cool in the autoclave for a night, the impurities will fall to the bottom of the solified medium in a thin layer which is easily seen. This layer can be eliminated by tipping out the solid lower part of the medium and cutting the impurities out with a knife.

Adjusting the pH of the medium

The bacteria multiply optimally in a slightly alkaline medium, while the fungi usually grow better under acidic conditions. The reaction of the medium can be adjusted to the required pH level before sterilization with indicator paper. The optimal pH requirement of bacteria is about 7·2 and that of the fungi about 5·5–6·0.

To acidify or alkalize the medium to the desired level, one adds HCl or NaOH solution drop by drop keeping in mind that 1–2 min are necessary for the exact estimation of the final pH. It is advantageous to adjust the pH in a warm medium. It should also be considered that with every sterilization the pH will be less by 0·1–0·2 (i.e., it is acidifying).

The changing of the pH precludes certain microorganisms from the medium. For example, if one wishes to isolate fungi from soil or heavily infected plants, bacterial development may be inhibited by adjusting the pH of the medium at pH 5·5. The fungi still grow well at this pH level, but the bacteria do not. In the same way fungi may be eliminated by alkaline media (pH 7·5–7·8).

Many bacteria may also be excluded from the medium with the addition of antibiotics. Generally the presence of 100 ppm sterile (filtered) streptomycin will inhibit growth of most bacterial spp.

Sterilization of media

Care has to be taken with the sterilization of media containing sugars, vitamins or growth substances. It is good to sterilize these on three consecutive days, fractionally steaming at 100 °C. The commonly used nutrient broth and nutrient agar or potato-dextrose agar media are sterilized in the autoclave at 1·5 atm for 20 min.

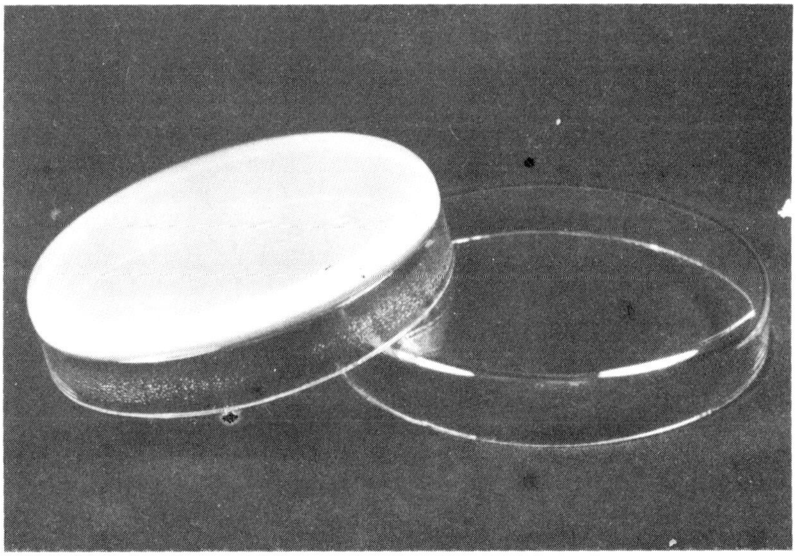

Fig. 12. Position of the Petri dish for drying surface of the agar plate

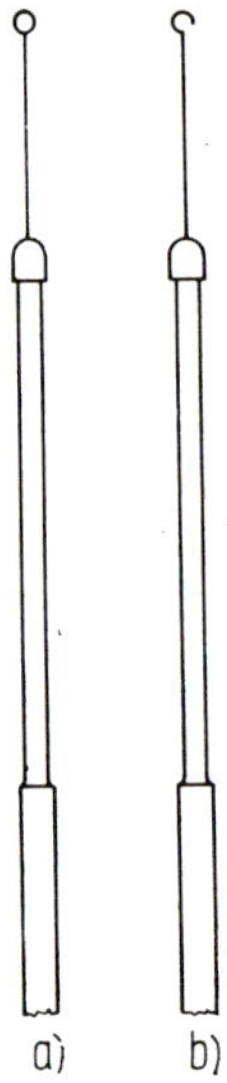

Fig. 13. Shaping of the inoculating loop *a)* the correctly formed loop for keeping of the bacterial suspension; *b)* the incorrectly formed loop

Preparation of pure cultures

The pathogen is isolated on solid media. After the medium is poured and solidifed in Petri dishes, it should be dried a little before streaking out the bacteria. This eliminates the condensed water which precipitates on the surface of the media. This condensed water makes it impossible to obtain isolated colonies, because the bacteria are swimming in it. The surface of agar plates can be dried at 50 °C for half an hour in a position as seen in the figure (Fig. 12). The bacteria are streaked out with a wire loop (Fig. 13) onto the agar plate. A suspension is prepared on a sterile slide from the infected plant tissue with the methods described earlier (see page 136).

In order to obtain separated colonies it is best to streak out the bacterial suspension according to the figure (Fig. 14) (Dowson, 1949). With the wire loop

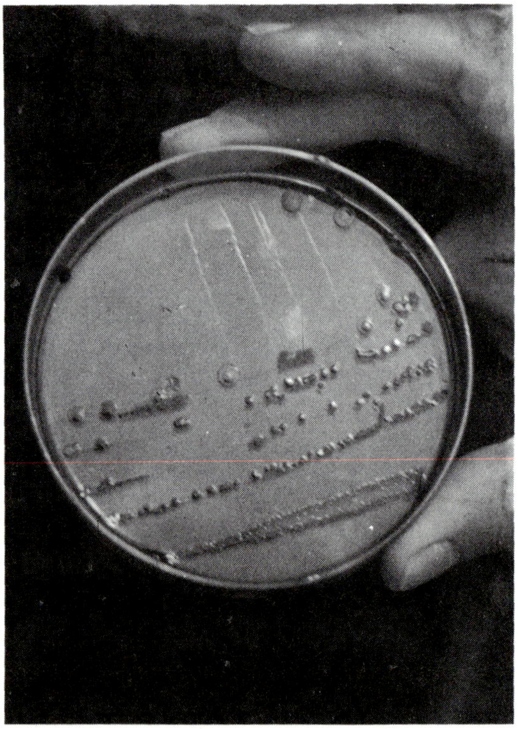

Fig. 14. Isolation of bacteria by streaking out a suspension on a nutrient agar plate

which is dipped in the bacterial suspension, five lines are drawn at first from left to right. The wire loop is placed in the flame, the Petri dish is turned right by 90°, and five more streaks are made beginning from the second streak right to left according to the figure. The Petri dish is turned 90° right again and with the enflamed loop the bacteria are further streaked from right to left according to the figure. In this way such a dilution is reached that the bacteria grow in separate colonies. After 48–72 hr of incubation at 28 °C, different types of colonies are transferred to tubes each containing an agar slope. The bacterial strains originating from diseased tissues are investigated by generally used cell and colony morphology, biochemical and pathological methods.

CHAPTER V

IDENTIFICATION OF THE PATHOGEN

Because of the morphological characteristics of bacteria, the same methods cannot be used for their study as those used in mycology. Since the determination of fungi and higher plants is based on morphological characteristics, slight morphological differences in bacteria make it impossible to use the same methods. Certain phytopathogenic bacteria have such a close relationship to each other that one cannot distinguish them with classical identification methods.

The following investigations are made for the determination of the pathogenic bacterium (Fig. 15):

1. Pathological test for picking of the pathogen isolate.
2. Morphological investigation of the bacterial cell and colony.
3. Determination of physiological and biochemical properties of the bacterium.
4. Pathological investigations for determination of the host range.

PATHOLOGICAL TEST FOR PICKING OF THE PATHOGEN ISOLATE

It is best to carry out a rapid simple pathological examination, because then only the strains which have been proved to be pathogenic need to be biochemically tested. The pathogenicity of the strains can be proved easily within 24 hr by the Klement's (1963) injection infiltration method (see page 159). This is based on the hypersensitive reaction of plants (see pages 153 and 195).

The bacterial strains originating from the isolation plates can be suspended in sterile tap water. Bacterial cell number of these suspensions can be adjusted to 10^{7-8} cells/ml densitometrically. These suspensions are injected into the intercellular space of an intact tobacco leaf with a syringe. This method enables us to inject different suspensions into intervenal sections of the same tobacco leaf (Fig. 16).

Since only plant pathogenic (non-saprophytic) bacteria are able to induce necrosis or disease in tobacco this method can be used for the rapid selection of the pathogenic isolate. Only the isolate which induced necrosis on the indicator plant should be investigated.

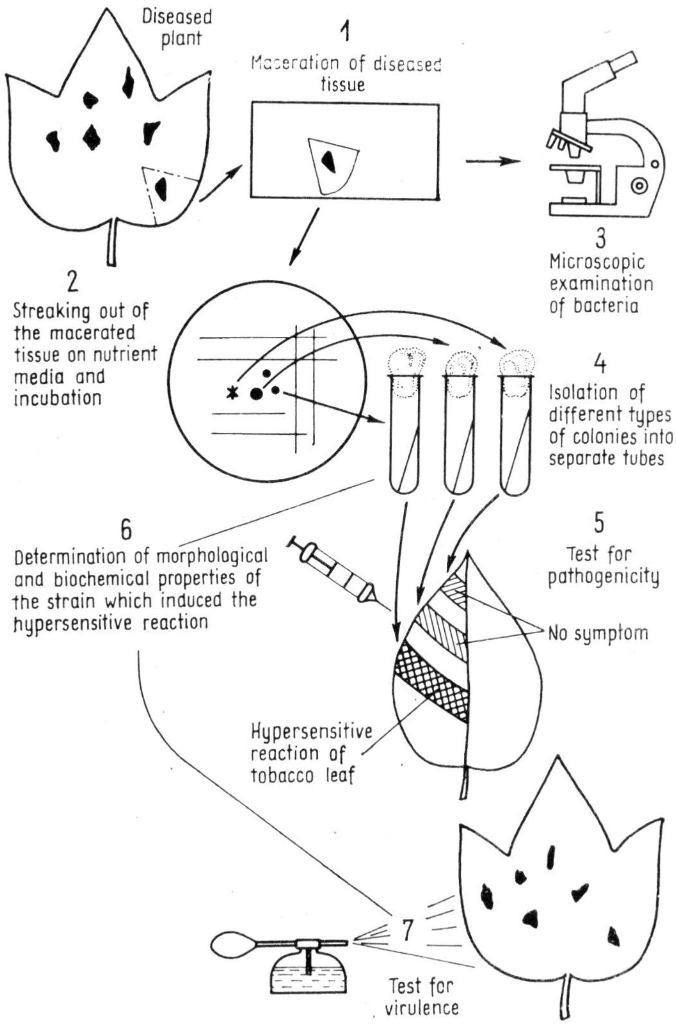

Fig. 15. Schematic course of isolation and determination of plant pathogenic bacteria

CELL AND COLONY MORPHOLOGY

The microscopic investigation of bacteria in the living state has numerous advantages, but usually does not give enough information. For example, movement of the bacteria is observed in a native state, because the flagella staining procedures used at the present time are too complicated for routine series tests.

Fig. 16. Injection infiltration of tobacco leaf by hypodermic needle

Hanging-drop examination

Motility can be easily observed by microscopic examination of a hanging-drop preparation. A drop of 10–24 hr old bouillon culture is placed on a cover slip, and the cover slip is turned with a quick movement onto the hollow-ground part of the slide in such a way that the drop does not touch the hollow part of the slide. It must be emphasized that the motility of the bacteria can only be observed in young cultures because the energy of the flagella movement is exhausted in older cultures.

Flagella stain

Most phytopathogenic bacteria are motile with polar or peritrichous flagella. The position and number of flagella may be characteristic for the bacterial genus (sse Fig. 7). The flagella are of such a small width that they cannot be resolved under the light microscope. For this reason they are not visible without staining. The essence of the staining procedure is to make salts or other compounds adhere to the flagella, enlarging their diameter so that they are observable under the light microscope. Flagella staining is a procedure which needs the greatest patience and one has to make the preparation with special care. Many methods are used.

We supply here the latest procedure, which Shaffer (1967) cites from Blendon and Goldberg's (1965) work.

The culture to be stained should be on slants and 12–18 hr old. The slants are flooded with water and slightly agitated (not shaken) and the cell suspensions are put in clean test tubes. These tubes are allowed to stand for a half to one hour, while stains A and B are prepared.

Solution A

tannic acid	5 g
ferric chloride · $6H_2O$	1·5 g
distilled water	100 ml

Mix and add:

1 per cent sodium hydroxide, 1 ml
40 per cent formalin (make this using 4 ml formaldehyde and 6 ml water), 2 ml

The final pH of the solution should be 1·5 or 1·8.

Solution B

silver nitrate	2 g
distilled water	100 ml

Make up this solution, then reserve 10 ml for back titration.

To the major portion of silver nitrate solution, add concentrated ammonium hydroxide by drops until a heavy precipitate appears; keep adding ammonium hydroxide until the precipitate disappears, then stop. Now back titrate by drops using the reverse nitrate until a slight precipitate reappears. Check the pH, which could be below 10. If the pH is not low enough, add more reserve silver nitrate until a pH of 9·8 is reached. (This pH is critical.) The staining must be done within 4 hr, otherwise the pH changes and cannot be re-altered.

From the bacterial suspensions one loop of suspension is transferred to an alcohol-cleaned slide and diluted with one loop of sterile water. To the air dried slide solution, A is added and allowed to stain 6–8 min. Be sure to get the cells to be stained from the top of the liquid, because the more motile cells are at the top. The stain is poured off and the slide is thoroughly washed with distilled water. Reagent B is added for 30 sec and then washed with water. The slides are allowed to air dry and are examined under the microscope.

Gram stain

The Gram stain is a well utilizable method in the determination of bacteria. Among the phytopathogens the *Corynebacterium* species are Gram-positive. The stain is carried out in the following way: The bacterial suspension is smeared on the degreased slide. After drying in air, the bacteria are fixed by drawing the Bunsen flame three times across it. The smeared and fixed preparations are stained with freshly filtered carbol water gentian violet (4 g gentian violet is dissolved in 100 ml 96 per cent alcohol: this stock solution is diluted by 5 per cent carbol solution prepared with 90 parts by volume distilled water). The stain solution is then poured off after 2–3 min without water rinsing and the Lugol solution is poured over the preparation (1 g iodine, 2 g potassium iodine, 300 ml distilled water). Two or three minutes after the Lugol solution has been poured over the preparation, absolute alcohol is dropped onto the slide which is held in a sloping position. Initially, the alcohol is stained blue-lilac, but it will become colourless. This is the sign that the elements of the preparation, which do not stain according to Gram, have already given off their colour. The treatment with alcohol must not be continued further because the Gram-positive bacteria sooner or later lose their colour. After treatments with alcohol, the preparation is rinsed with water and dried between filter paper. The Gram-negative organisms are colourless or pale blue. As a result of counter-staining with dilute water fuchsin solution, the Gram-negative bacteria turn red.

Colony-morphology

The colour, form and consistency of the colony, as well as the colour of the stains which have diffused from the colony into the medium, are observed for the determination of bacteria. The form and colour of the colony and the importance of these characteristics in the identification were discussed earlier (see page 123) and summarized in Fig. 7. But it should be noted that an effective description of the colour of the colony or pigments which diffused into the medium depends on knowledge of the exact composition of the medium, because the medium strongly influences the stain producing materials. For example the green fluorescent pseudomonads' stain production is well observable on King's medium. This medium is composed of the following: Peptone 2 per cent, agar-agar 1·5 per cent, glycerine 1 per cent, K_2HPO_4 (anhydrous) 0·15 per cent, $MgSO_4 \cdot 7\,H_2O$ 0·15 per cent and distilled water in the required quantity. The pH of the medium should be adjusted to 7·2. The fluorescent stain produced in the medium after growth of the colony is observable especially with UV light.

BIOCHEMICAL CHARACTERISTICS OF BACTERIA

For the determination of the biochemical activity of phytopathogenic bacteria the following main biochemical tests are described: 1. Action on carbon compounds, 2. nitrate reduction, 3. gelatin liquefaction, 4. action on peptone (ammonia, hydrogen sulphide and indole production), 5. hydrolysis of starch, 6. action on litmus milk, 7. acetylmethyl carbinol production (Voges–Proskauer reaction), 8. oxidase test.

It is rather important to separate the bacteria diagnostically which are morphologically the same on the basis of the carbon compounds they decompose and it is necessary to state whether or not they will cause the production of acid and gas from carbon compounds. Organic acids may be produced during the bacterial decomposition of the different sugar and sugar-like alcohols added to the peptone water or synthetic nutrient solution. These acids change the colour of the indicator which is added to the medium. If, however, the reaction takes place during the production of the gas and acid, organic acids are further decomposed and can be retained in the gas bubble form by using a little tube (Durham's tube), and turning it down with its d fitopen ented into the liquid medium. Most of the phytopathogenic bacteria do not produce gas during sugar decomposition, with the exception of a few species of the genus *Erwinia.* Identifying the bacteria, the following compounds are tested: xylose, arabinose, glucose, mannose and galactose as monosaccharides; among the disaccharides, raffinose, sucrose, maltose, lactose; and among the polysaccharides, insulin and dextrin ; among the alcohols, glycerine (trihydric), mannite, dulcite and sorbite (hexahydric); and finally salicin and aesculin glucosides.

Regarding the carbohydrate decomposing capacity of the same species, different authors have reached varying results. The reason for this is, that some of them used peptone water as a basic nutrient solution, while others used synthetic or other compositions of basic nutrient solutions. The phytopathogens generally produce little acid. The alkaline materials originating during peptone decomposition neutralize this slight acid production and this may cause different results. Therefore, it is best to use the following basic nutrient solutions:

NH_4Cl	1 g
KH_2PO_4	4 g
K_2HPO_4	4 g
$MgSO_4 \cdot 7H_2O$	0·2 g
$FeSO_4$	0·028 g
$ZnSO_4$	0·003 g
nicotinic acid	0·010 g
phenol red	0·018 g
distilled water	1000 ml

First nicotinic acid is dissolved, then the other components are added after being mixed, the solution is autoclaved at 121 °C for 5 min.

The members of pseudomonads produce only a little acid from carbohydrates. In the case of some *Pseudomonas* species a simpler basal medium is used containing less buffer salts. 1·0 g $(NH_4)H_2PO_4$, 0·2 g KCl, 0·2 g $MgSO_4 \cdot 7H_2O$, 1000 ml distilled water. Bromthymol blue should be made up as 1·6 per cent alcoholic solution, 1 ml of which should be added to 1000 ml of basal medium.

Adjust the pH to 7·0 by the addition of 0·1 N-sodium carbonate of which abou 6 ml will be required. This basal medium is autoclaved at 121 °C for 5 min.

The different carbon compounds are dissolved separately in distilled water. Then the solutions are either filtered through a sterile bacterial filter, or fractionally sterilized for three consecutive days at 100 °C for 20 min. At a higher temperature the sugars became caramelized more quickly. After this, the filtered or sterilized sugar solution is added aseptically to a basic nutrient medium. If one also wishes to examine the gas production, then a small tube (Durham's tube) is placed into the nutrient medium with the opening turned down (Fig. 17). It is advisable to heat the prepared tubes again at 100 °C for a few min. The tubes should be allowed to stand for a few days so that they should be really sterile.

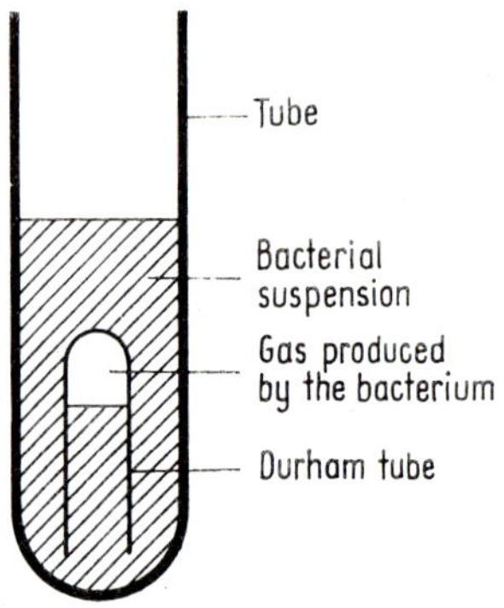

Fig. 17. Placing of Durham tube in a broth culture

The tubes are inoculated by dropping 0·1 ml of the homogeneous suspension of the bacteria to be investigated into the tubes. In the case of acid formation, the colour of the medium becomes pink or yellow in the presence of bromthymol blue. The tubes are kept for three weeks in an incubator at 28 °C.

Nitrate reduction

The nitrate test is important as a determinative method for testing bacteria. With this, depending on the pH, the nitrates take up the hydrogen which has been broken down during the life processes. As a result, nitrite or ammonia is produced, or the whole process continues until free nitrogen formation. Nitrate broth, having the following composition, is used for the investigation:

peptone	10 g
KNO_3 (nitrite-free)	1 g
distilled water	1000 ml

This is steamed till dissolved, then distributed and autoclaved for a moment at 121 °C. The test tubes are examined on the first and fifth day after the inoculation.

The investigation is carried out as follows: A few drops of reagents a) and b) should be dropped into the broth. Nitrite is present if the liquid is a distinct pink or red.

Reagent a) 8 g of sulphanilic acid is dissolved in 1000 ml of 5 N acetic acid (40 per cent by volume acetic acid). Reagent b) 5 g of naphtylamine in 1000 ml of 5 N acetic acid.

If the red colour does not appear, then zinc powder should be added to the broth so that the original nitrate is reduced to nitrite. Red colour appears if the nitrate has not been reduced to nitrite by the bacterium.

Gelatin breakdown

By means of the enzyme organization of bacteria, peptides of smaller molecular weight are formed having a partial breakdown of protein molecules; or they are broken down to amino acids which build up proteins. Such a proteolytic enzyme liquifies the gelatin. The gelatin liquefaction can be used for bacterial determination.

10–12 per cent good quality gelatin is added to nutrient broth and dissolved by steaming. The pH is adjusted by KOH to 7·2 because of the low pH value of the solution. For three consecutive days the gelatin is fractionally sterilized for 20 min at 100 °C. The test tubes, kept at 25 °C, are inoculated by dipping the pointed tip of the wire loop in the bacterial suspension and then vertically stabbing 1–2 cm into the tube containing the set and slanted gelatin. When the proteolytic enzymes of the bacteria breakdown the gelatin, it becomes liquefied in 2–3 days after the inoculation.

Ammonia production

Ammonia formation is of general significance in all those microorganisms using amino acid or similar compounds for bacterial cell building. Ammonia in the culture containing peptone usually accumulates because of deamination of amino acids, and/or amides (splitting off of the NH_2 group). The bacteria, in the absence of protein, utilize amino acids for building up their cells. Since more carbon than nitrogen is required to build the cells and produce energy, ammonia accumulates in the medium. In the presence of utilizable carbohydrates, a protein saving effect takes place whereby the organism utilizes the amino acids for building up the cell rather than for energy. In this case a smaller amount of ammonia accumulates.

The investigations are carried out on peptone water broth. One per cent peptone and 0·5 per cent NaCl are dissolved in warm distilled water and the pH adjusted to 7·2. Then the peptone water is poured into test tubes and sterilized in an autoclave for a moment at 121 °C. Ammonia production is tested with Nessler reagent. The culture is examined 2–3 days after the inoculation. When equal amounts of the Nessler reagent are added to the culture, a yellow-orange coloured precipitation indicates the presence of ammonia. If there is no reaction, the test should be repeated in 5 days.

Preparation of the Nessler reagent: In 250 ml distilled water, 62·5 g KI are dissolved. About 10 ml of this solution is reserved, and a cold saturated solution of $HgCl_2$ is added to the main portion until a distinct slightly red precipitation remains. Add the reserved KI solution and additional $HgCl_2$ solution stirring continuously until a permanent slight precipitate is formed. Then 150 g KOH are dissolved in 150 ml distilled water. This is cooled and added to the above solution. It is diluted to 1 l with distilled water and allowed to stand for 1 week and then decanted for use.

Production of H_2S

The bacteria inoculated in the peptone water (see p. 149) produce hydrogen sulphide from sulphur-containing amino acids. The hydrogen sulphide turns into lead sulphide if the indicator paper is placed above the medium. Dry strips of filter paper are previously moistened in a saturated solution of lead acetate. These strips of paper are inserted in the tubes and held by the plugs above the culture without touching the medium. If H_2S is formed the paper strips will turn black in 3 to 7 days.

Indole production

With the aid of certain bacterial enzymes under aerobic conditions and in the absence of sugar, indole is produced as a product of tryptophane breakdown. Indole formation occurs during the nitrogen metabolism of the organism. The nitrogen compounds containing a tryptophane group of higher carbon atomic number (peptone) are broken down into simpler indole compounds.

The 1 per cent tryptophane and 0·5 per cent NaCl containing broth are inoculated with the organism to be tested. Strips of dry filter paper which had been previously dipped in a saturated solution of oxalic acid are inserted in the test tubes. The paper strips are secured by the plugs, taking care that they do not touch the tryptophane broth. In the presence of indole the paper strips will turn pink.

Hydrolysis of starch

Certain phytopathogens are able to hydrolyze starch. Enzymes hydrolyze starch to sugars of lower carbon number. Among these enzymes, the most important are α- and β-amylases. This process is easily demonstrable on a medium containing soluble starch with iodine reaction (Lugol solution). To the nutrient agar, 0·2 per cent soluble starch is added. The bacteria are streaked with a wire loop onto agar plates and in 3–5 days the culture is flooded with Lugol solution. With a negative reaction the medium becomes blue. However, around the colonies, if the bacterium in question has hydrolyzed the starch, the blue colouring does not occur (Fig. 18).

Lugol solution

iodine	1 g
potassium iodine	2 g
distilled water	300 ml

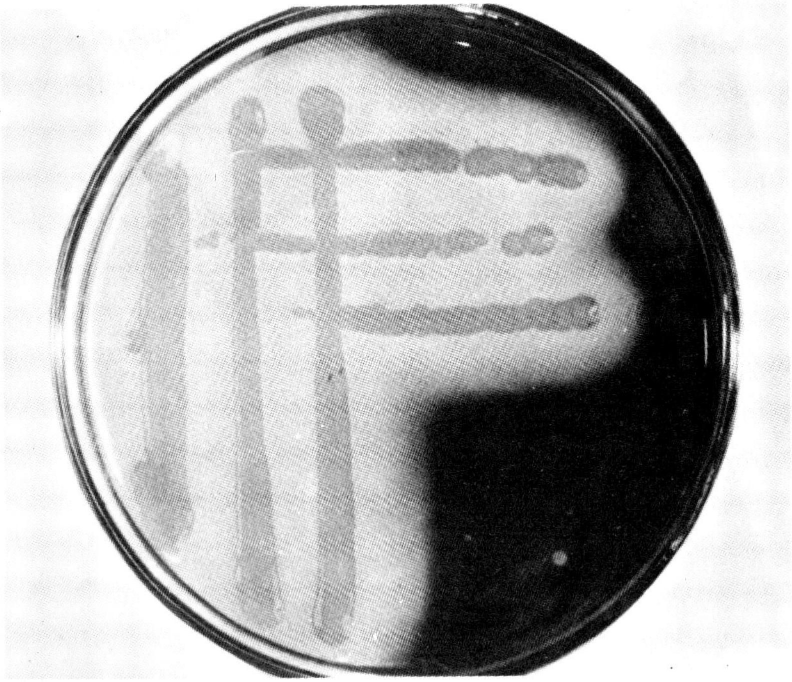

Fig. 18. Starch hydrolysis of *Xanthomonas vesicatoria* (tomato strain)

Action on litmus milk

Litmus milk contains five important materials suitable for the determination of bacteria. These are, lactose in a larger amount, and glucose (after sterilization) in a smaller amount, two milk proteins and litmus. During milk decomposition we are interested in whether the microorganisms to be examined make the milk acidic, alkaline, coagulated, peptonized or reduce the litmus.

If the bacterium utilizes only glucose, which is present in small amounts, then little acid is formed and litmus added to the milk colours it pink.

If it also attacks lactose present in a bigger quantity, then more acid is formed, and a deeper pink appears.

If the bacterium only breaks down the lactalbumen, then ammonia or basic amines are formed which turn litmus blue.

With caseinogen decomposition two possibilities may take place. A larger amount of lactic acid formed from lactose in the course of the bacterium's life processes, break the physico-chemical balance of the caseinogen. The otherwise homogeneous caseinogen is in combination with calcium which is removed by the acid to form calcium lactate and leaves behind the casein which flocculates to form a semisolid mass, the curd. This effect (called coagulation) may be characteristic for the bacterial species. Above the coagulated proteins there is a clear fluid. Curdling can take place without acid formation also if the cultures produce a curdling enzyme.

Peptonization is the opposite of this, and it occurs when the enzymes of the bacterium utilize casein as a carbon source and hydrolysis reduces it to polypeptide units. During peptonization the milk becomes clear and transparent.

Skim milk is used for the experiment. Thus either powdered milk is used or the butter fat is removed from whole milk by centrifugation. 1–2 per cent litmus solution is added to skim milk. Tubes containing litmus milk are sterilized by steaming for 20 min on three successive days.

Litmus solution is prepared as follows: 80 g litmus granules are ground in a mortar and boiled in 150 ml 40 per cent alcohol solution for 1 min. After the decantation of this liquid the remaining precipitated litmus particles are boiled again with 40 per cent 150 ml alcohol for 1 min. After decanting, one drop of *N* HCl is added with agitation to the litmus solutions until the liquid becomes purple.

Acetylmethyl carbinol formation

The essence of this test (Voges–Proskauer reaction) is that the acetylmethyl carbinol produced by the bacterium, in the presence of alkali, oxidizes to diacetyl, which is red.

The test is done in the following way: To one, or two-day-old nutrient broth culture containing glucose and peptone, one drop of yellow ferric chloride solution and 1·5–2 ml 20 per cent sodium hydroxide solution are added. In the case of positive reaction the culture turns red in 2–3 hr.

Oxidase reaction

Oxidase reaction is also used for taxonomical purposes. One platinum loop bacterial cell from slant agar is placed onto a filter paper impregnated with 1 per cent aqueous solution of tetramethyl-p-phenylene-diamine-dihydrochloride. The reaction is positive if a red colour appears during 10 sec. The reaction is delayed positive if the red colour appears in 10–60 sec (Steel, 1961).

The oxidation of phenylene-diamine is catalyzed by traces of irons, therefore, it is important to use a platinum loop. The reagent may lose its sensitivity and change its colour during storage. The solution can be preserved in a refrigerator in a glass plugged bottle in the dark.

CHAPTER VI

INOCULATION TESTS TO DETERMINE THE HOST RANGE

After acquainting oneself with the morphological and biochemical characteristics of bacteria, one should also become aware of the pathological characteristics of the strain to be investigated. In the case of different disease types, various inoculation methods have to be applied. Before we go into the details of inoculation methods, it is important to discuss the principles generally followed.

Phytopathogenic bacteria, as opposed to numerous pathogenic fungi, do not secrete substances which dissolve the cuticle. It follows from this that they can only penetrate into the plant tissue by a passive way. Some bacteria having a higher pathogenicity infect the plants through the openings, the stomata of leaves, lenticels, hydathodes and other green parts. Others enter the tissues through damages or wounds. In certain unusual cases, the pathogens penetrate through the flower or stigma.

The bacteria which have entered the tissues multiply either in the intercellular spaces or in the vascular system. In contrast to numerous fungi, the bacteria do not penetrate the cell wall or grow in the host cells.

Experiments have proved that most phytopathogenic bacteria are able to produce symptoms on most plant species. The symptoms, however, may be atypical and typical. *Typical symptoms* are referred to when exactly the same alterations occur on an artificially inoculated plant as appear in nature. *Atypical symptoms* are those which occur when a rapid necrosis forms at the place of the inoculation or immediately around it. This necrosis generally appears earlier than the typical symptom of the disease. The atypical necrosis is never water-soaked but is rather of a dry nature. Today we already know that this atypical symptom is the defence reaction of the host (hypersensitive reaction in more detail later) to the attack of the pathogen (Klement, 1968).

The appearance of atypical symptoms has misled many research workers. When a bacterium in question caused some symptoms in a test plant it was described as a host while under natural conditions it would never have become diseased. This would not be such a serious problem if it had occurred in only a few papers and then could have been corrected. However, these mistakes have been introduced by various authors of manuals. Thus there is chaos at present in the determination of the host range of plant pathogenic bacteria. It can be seen, therefore, that in different works and manuals, the host range of some bacteria is rather wide. This often depended on how many plant species were included in the experiment. However, we never find in nature that, for example, *P. tabaci* will infect wheats and other grasses, as is mentioned in manuals.

The question arises as to where these mistaken conclusions came from, and how this can be prevented in the future. The mistake stems from the fact that the atypical symptom was confused with the typical symptoms of the disease. This arose because, on the one hand, the inoculation methods used were not suitable for inducing the typical symptoms: on the other hand, the defence reactions of plant were not known well enough (page 195). Since the atypical symptoms were considered to be typical symptoms of disease, such plants were wrongly included among the host range. In fact, these plants actively resist the pathogens (hypersensitive reaction).

Atypical necrosis, the hypersensitive defence reaction of the plant, however, does not always become visible. This depends on the number of bacterial cells which have entered the plant tissues. Since few bacteria penetrate the intercellular spaces of the plant, the defence reaction takes place only around the point of the infection when examined microscopically. Take the following case: Only a few bacterial cells penetrate the stoma through the substomatal cavity (Fig. 19). Here the bacteria which begin to multiply, immediately induce the hypersensitive reaction in cells of the incompatible host, which is in contact with the bacterium. These few plant cells will die in 8–10 hr and in this way the infection becomes localized. This occurs in nature, or in artificial inoculations with incompatible host–parasite relationships when the inoculum contains few bacteria. In this relationship, the plant remains disease-free since the reaction appears only in micro dimensions. However, if the same bacteria are put into the intercellular spaces, in a higher number with infiltration, the many little spots appear in the form of continuous necroses.

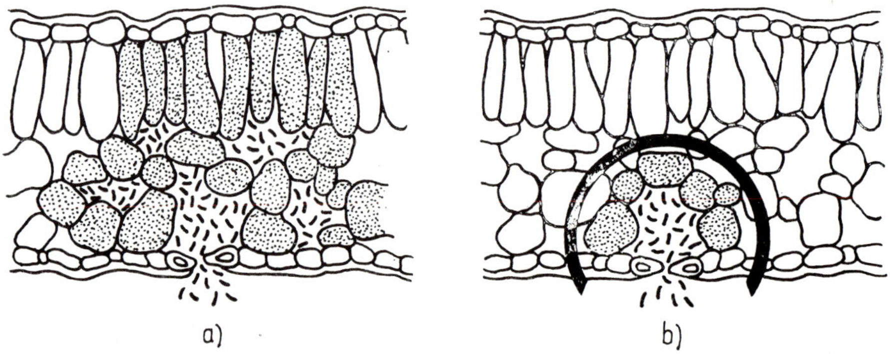

Fig. 19. Fate of infection in compatible and incompatible hosts. *a)* distribution of bacteria in a water soaked leaf tissue; *b)* localization of infection by the hypersensitive necrotic reaction

Hov car we prevent these mistakes during our experiments? A suitable method and a determined bacterial number should be used for artificial inoculation. As it has been proved (Klement, 1968) in the case of pseudomonads, 5×10^6 cell/ml

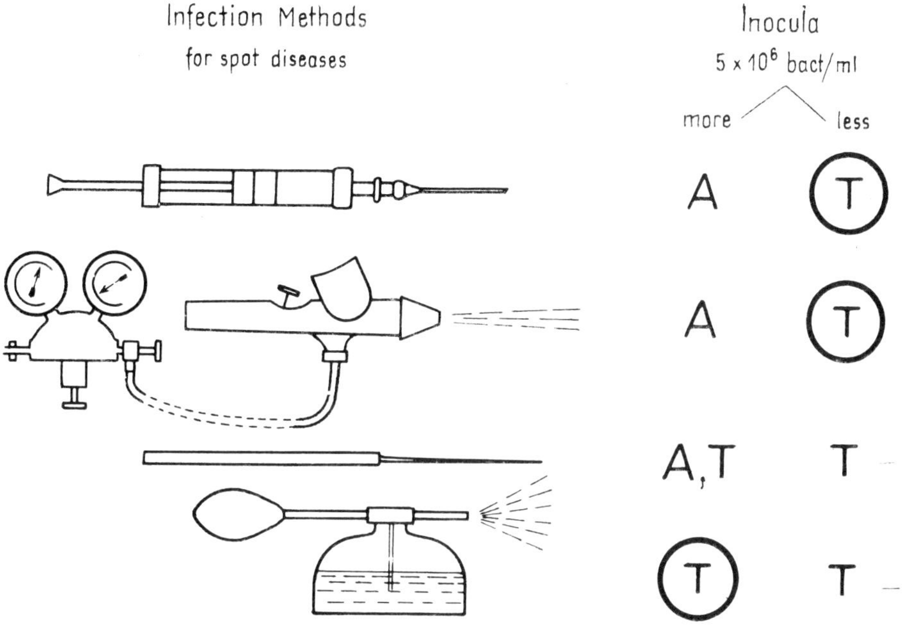

Fig. 20. Different symptom types obtained by using various inoculation methods and different inoculum concentrations. Circles indicate proper techniques. A: atypical symptom, T: typical symptom, –: no symptom

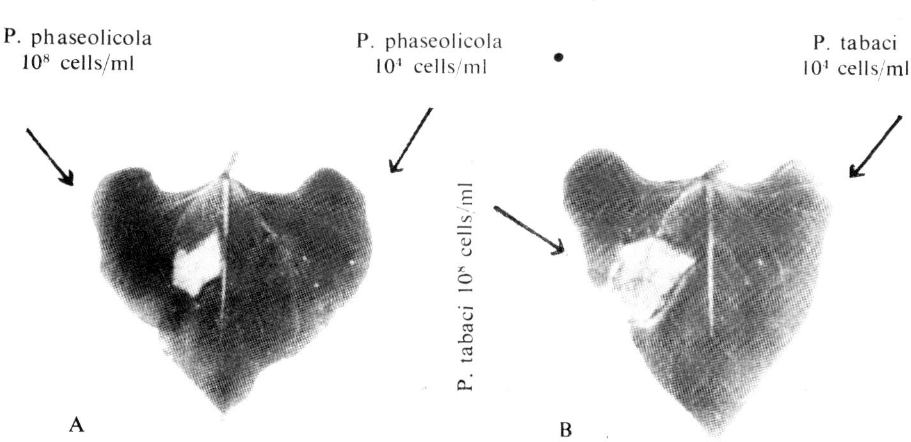

Fig. 21. Role of the concentration of inocula in A) a compatible (bean leaf injected with *Pseudomonas phaseolicola*) and B) incompatible (bean leaf injected with *P. tabaci*) host-pathogen relationship taken on the sixth day after inoculation

or more than this number of bacterial cells are required so that all the cells of the plant tissue can be in contact with the pathogen.

Thus, using an inoculum of high cell number, e.g., applying one of the infiltration methods, we can only reach atypical symptoms (Fig. 20). In order to avoid the formation of atypical symptoms, the appropriate cell number of the different inoculation methods should be used. These can be observed in Fig. 21. When using different inoculation methods, one must note the inoculum concentration marked by a circle on the table. Otherwise one may count on getting atypical symptom formation.

PRETREATMENT OF PLANTS FOR INOCULATION

It is advisable when inoculating to use the same plant species and cultivars as were used in isolating the pathogen. This is especially important if one wishes to prove the virulence of a still unknown pathogen. In the case of well known bacterioses, the so-called indicator plants, may also be used for artificial inoculation. With bacteria causing soft rot, onions, carrots and cucumber pieces may be used. However for the investigation of spot-causing bacteria, young, well growing plants are used for artificial inoculation. With spot diseases the bacteria penetrate the intercellular spaces through the stomata; therefore, it is important that the stomata be open during inoculation. For this reason the plant can be put in a humid half-shaded place under a bell jar or moist chamber for 24 hr before inoculation. The humidity may be increased by lining the wall of the bell jar or inoculation chamber with moist filter paper.

PREPARATION OF THE BACTERIAL SUSPENSION AND CHOOSING OF THE INOCULATION METHOD

A 24–48 hr slant agar culture is used. The culture is washed with sterile water and the suspension is adjusted to a determined bacterial cell number (page 161). The cell number of the inoculum depends on the following inoculation methods.

Spraying of bacterial suspension onto the surface of the plant

In the case of spot disease, the inoculation method which best resembles those of natural conditions is the spraying of leaves with bacterial suspension. Twenty-four hours before inoculation the plants have to be placed in a moist chamber to allow the stomata to open (Fig. 22). The under surface of leaves is sprayed with a low pressure sprayer. In this case more than 10^6 cell/ml inoculum may be used. With this method we cannot get atypical symptoms because only a few bacterial cells are able to penetrate the intercellular spaces through the stomata. In the case of incompatibility, the bacteria only express their activity around the stomata, and as seen before, the atypical symptom develops in micro sizes invisible to the naked eye. In the case of compatibility the pathogen may increase to a great extent, which may lead to the characteristic symptoms of disease (Klement, 1968).

A B C

Fig. 22. Angular leaf spot symptoms 3 days after inoculation A) preinoculation treatment in moist chamber B) non-treated C) preinoculation wilting (After Williams and Keen, 1967)

After inoculation the plants can again be placed in a moist chamber. Williams and Keen (1967) when investigating the angular leaf spot of cucumber (*P. lachrymans*) found that only a few minute spots formed if the plants were placed in relative humidity of 80–90 per cent for 48 hr after inoculation. Typical angular spots developed in large numbers if the relative humidity was 95–100 per cent (Fig. 23).

Use of high pressure inoculation equipment

At high pressure spraying with the help of the apparatus (atomizer) seen in the figure (Fig. 24), the bacterial suspension is applied with 1·5 kg/cm^2 pressure onto the lower surface of the leaf. As a consequence of pressure, the bacterial suspension is pressed through stomata into the intercellular spaces. It should be noted, however, that with the application of a pressure higher than 1·5 kg/cm^2 the epidermis will be damaged and atypical necroses will appear. Using this method the cell number of the inoculum cannot be more than 5×10^6. If we use a higher cell number inoculum, the atypical necroses may develop on plants which are not hosts.

A B

C D

Fig. 23. Angular leaf spot (*Pseudomonas lachrymans*) symptoms 3 days after inoculation on leaves incubated at A 85, B 90, C 95 and D 100 per cent relative humidity after inoculation (After Williams and Keen, 1967)

Fig. 24. Infiltration of bean leaf by high pressure atomizer

Injection of bacterial suspension into the intercellular spaces with a syringe

This method consists of injecting bacterial suspensions into the intercellular spaces of leaves with a syringe (Fig. 16). In this way it is possible to introduce a definite number of bacteria uniformly throughout the intercellular spaces without, ostensibly damaging the leaf tissue (Klement, 1963). The injection is carried out on the under surface of the leaf; the most appropriate syringe is a 25 gauge hypodermic needle, 5/8 size pin: with this method tobacco, paprika and tomato leaves or bean pods and paprika fruits can be well injected. Generally, the leaves which are lower and older can be more easily injected than younger ones. With practice the injection infiltration method can be carried out easily. It is important to hold the needle in the correct position (Fig. 25). The suspension, if carefully injected, may reach the whole leaf blade. With this method it is possible to examine the effect of different bacterial suspensions on one leaf. In this case it is advisable to use intervenal parts of the leaf. In this way, the effect of 10–12 different bacterial strains on the same tobacco leaf can be investigated at the same time. The water will evaporate and the leaf regains its original state in one half or one hour. Undetached leaves may be used with this method.

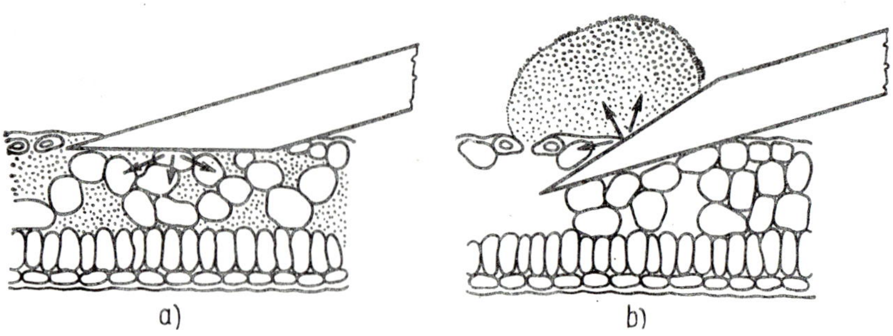

Fig. 25. Correct a), and incorrect b) holding of the hypodermic needle during the injection infiltration of the leaf

Rubbing of leaves

Small wounds are made on the epidermis by rubbing the leaves. A little carborundum powder sprinkled on the surface of the leaf, is a method particularly effective with monocotyledonous plants which are covered with waxy substances. After this, the bacterial suspension is carefully spread over the surface of the leaf, either with the finger or with a brush.

Pricking

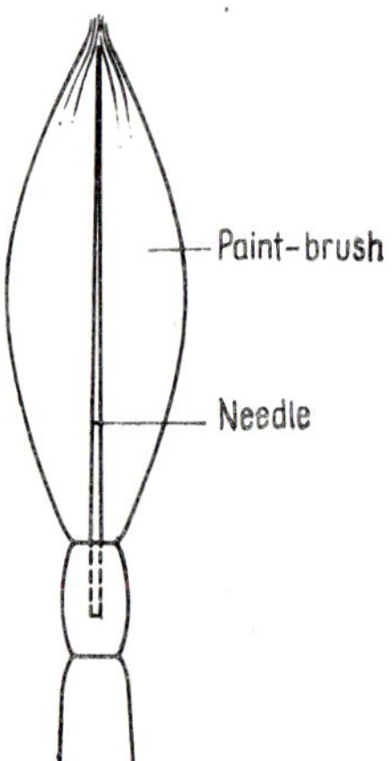

Fig. 26. Location of the inoculation needle in brush puff

The stem or more fleshy parts of the plant are pricked or damaged with a needle or scalpel dipped in bacterial suspension. With bacteria causing wilt or soft rot the pricking of the stem, or in the latter case the damaging of the fleshy parts of the plant, seems to be the most appropriate method. In this case brown necrosis around the pricking cannot be mistaken for wilt or the typical symptoms of soft rot. With this method, the typical symptoms of disease can be obtained regardless of the bacterial cell number of the inoculum. If we wish to inoculate more plants, it is advisable to fasten the needle onto a brush puff (Fig. 26). The infection of the needle is insured by keeping the brush puff in the bacterial suspension for a long time. In the case of wilt disease, the young plants are pricked in the area between cotyledon and the first leaf. In the case of die-back diseases, young shoots or seedlings are inoculated about 2–3 cm from the tip. The bacterial suspension can also be introduced into the shoots by an injection needle.

Inoculation of cut tissues by bacterial suspension

With wilt diseases the roots of the plants are cut back to 1–2 cm, and immediately placed into the already prepared bacterial suspension. The plants are kept in the suspension for 2 hr and then placed in the soil.

In wilt diseases, the petioles of young leaves or shoots are broken off from the stem. The bacterial suspension is placed onto the fresh wound surface with a pipette or brush.

In the case of bacteria causing soft rot, the bacterial suspension is dropped onto the surface of the washed, disinfected and freshly cut pieces of carrot, potato, cucumber or onion. The plants treated in this way are placed in a moist chamber for 48 hr after infection (see pages 223 and 224). The inoculation methods (in order of importance) which can be used for different disease types are as follows:

Spot diseases:	spraying
	tissue infiltration by high pressure atomizer
	pricking
	tissue injection infiltration method by syringe
	rubbing on the surface of the leaves
Die-back of shoots:	pricking
	injection
Wilt diseases:	pricking
	immersing of the cut surface of the plant wounds into bacterial suspension (root inoculation)
	dropping of bacterial suspension onto the cut surface of plant wounds
Soft rot:	dropping of bacterial suspension onto the cut surface of the plant wound
	pricking
Hypertrophy:	pricking, cutting (wounding)

DETERMINATION OF INOCULUM CELL NUMBER

The cell number of a bacterial suspension can be determined in two ways: turbidimetrically and by plate-count method.

The advantage of the first method is that with the help of the quick and directly measurable density of the suspension, the bacterial cell number of the suspension can be adjusted. Its disadvantage is that the number of the living cells is counted with the non-living cells. Since the cell size of the different species is not identical, it is necessary to calibrate the densitometer to the bacterial species to be examined. The advantage of the plate-count technique is that it only determines the living bacterial cells and that the cell number can be determined more exactly. Its disadvantage is that it is slower and that the result can be obtained only in one or two days after the suspension has been used. In this way we can only calculate the bacterial cells that were used. Its further disadvantage is that it requires considerably more work and laboratory equipment.

Turbidimetric measurements

a) *McFarland's scale* (as shown below) is used for the tubes prepared according to the table and plugged with rubber plugs so that they can be used for years to compare the bacteria in question. Using the bacterial suspension tube, and the McFarland's scale tube placed side by side in transmitted light, one can directly conclude the density of the suspension.

Tube	ml of 1 per cent $BaCl_2$	ml of 1 per cent H_2SO_4	Bacterial number/ml
1	0·01	9·99	30,000,000 = 3 $\times 10^7$
2	0·05	9·95	150,000,000 = 1·5 $\times 10^8$
3	0·1	9·9	300,000,000 = 3 $\times 10^8$
4	0·15	9·85	450,000,000 = 4·5 $\times 10^8$
5	0·2	9·8	600,000,000 = 6 $\times 10^8$
6	0·3	9·7	900,000,000 = 9 $\times 10^8$
7	0·4	9·6	1200,000,000 = 1·2 $\times 10^9$

b) *For photoelectric measurements*, the densitometer, colorimeter or spectrophotometer are used. The measurement is more exact in this way than when the turbidimetric scale is used. The essence of this is, that if a tube containing bacterial suspension is put in the path of a beam of light, only one part of the beam can reach the photocell and turbidity can be read on a suitable scale (Fig. 27). From this one can calculate the density of the suspension, that is, the cell number of the suspension. At measurement, the scale pointer of the apparatus has to be set to zero with the uninoculated medium. For example, in the case of bouillon suspension one has to determine firstly the density of bouillon and to set the equipment to zero with this. Then the measurement of the bacterial suspension follows.

Since the shape and size of the bacteria is variable depending on species, the photometer should be calibrated for every bacterial species. Calibration is done in the following way: the density of the bacterial culture multiplied in bouillon will be determined with the plate-counting technique. Simultaneously with the measurements, the bacterial cell number estimated with the aid of the colony and density values gained from the photometer are summed up in a graph (Fig. 28).

In certain tests, such as the determination of the bacterial cell number of shaken or continuous cultures, a 100 ml Erlenmeyer flask constructed with a tube in the side is very useful. When using continuous cultures, it is not then necessary to obtain samples in separate tubes, for the flask can be tilted. Thus the cell number can be controlled without loss of the sterility (Fig. 29).

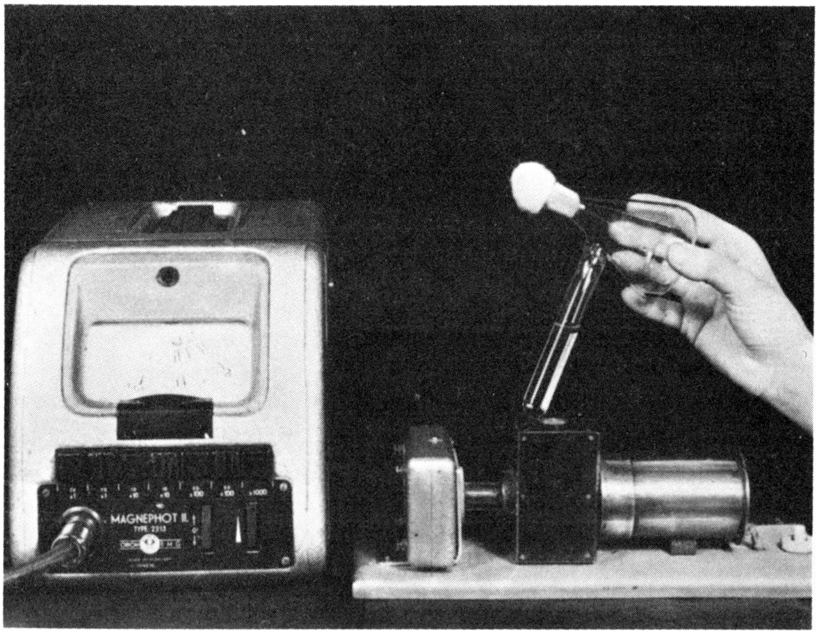

Fig. 27. Determination of turbidity (or growth) of bacterial suspension by densitometer

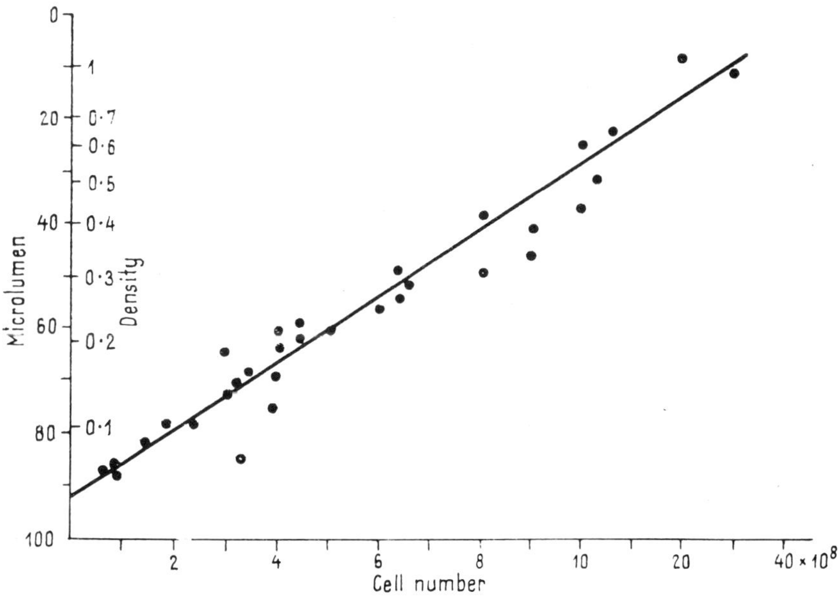

Fig. 28. Calibration curve of the density of a *Corynebacterium flaccumfaciens* suspension

Fig. 29. Erlenmeyer flask adapted for measuring bacterial growth

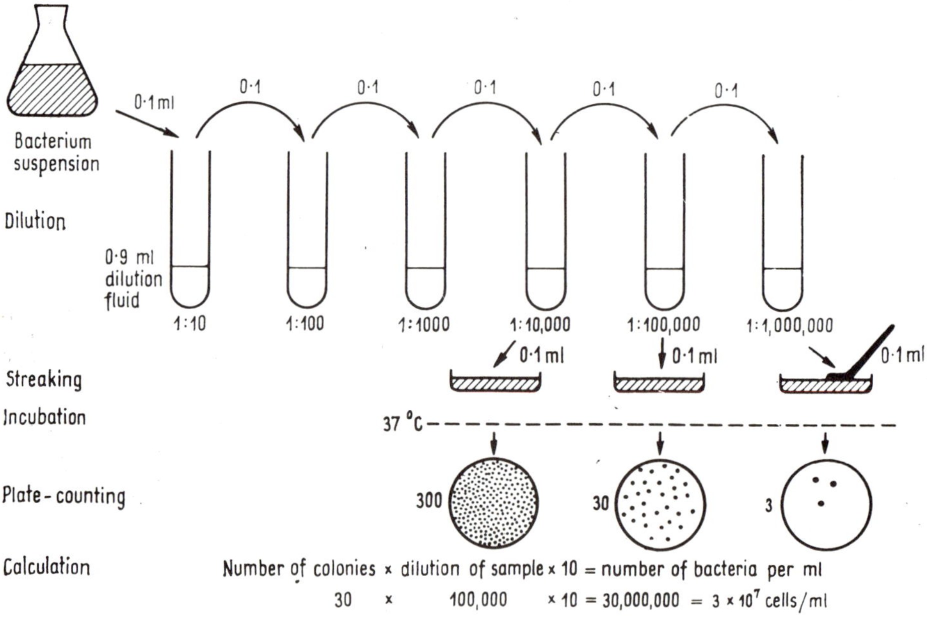

Fig. 30. Scheme of the dilution row and calculation of living bacterial cells

Dilutions are made from the bacterial suspension in question. The samples taken from certain tubes of the dilution row are spread onto the agar plate. From the number of developing colonies one can conclude the bacterial number of the original suspension. There are four steps. The first one is the preparation of the dilution series; the second one is the spreading of the bacteria onto the agar plate; the third is the incubation and the fourth the colony count and calculation.

The scheme of the preparation of the dilution series is shown in Fig. 30. The dilution can be conducted in sterile tubes or in hollows of plexiglass plate. With the use of the Takátsy's dilution spiral loop, the dilution can be done more quickly and exactly (Fig. 31). The sterilization of the spiral loops is completed simply by flaming the loops. Their use is more advantageous than pipettes, because suspension will adhere onto the wall of the pipette. Dilution can be done the same way as with the tubes, with the difference that here the tubes are positioned in hollow plexiglass of 1·5–2 ml depth. Plexiglass is sterilized with UV light (for more than 30 min, at a distance of 30 cm).

The tenfold dilution row is prepared in such a way that transfers of 0·1 ml are made first from the bacterial suspension in question to the sterile tubes or holes

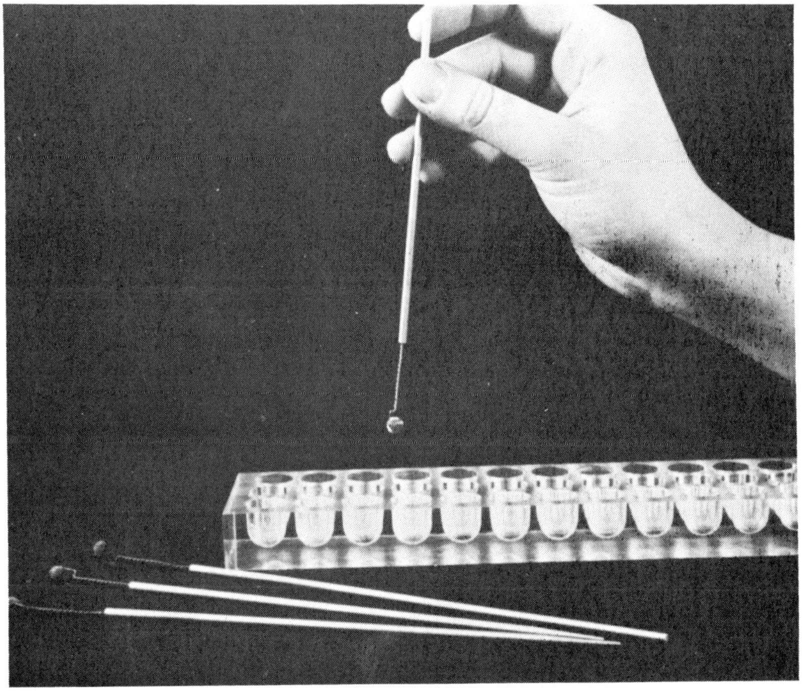

Fig. 31. Method for dilution of bacterial or phage suspension by the Takátsy's dilution loop in plexiglass

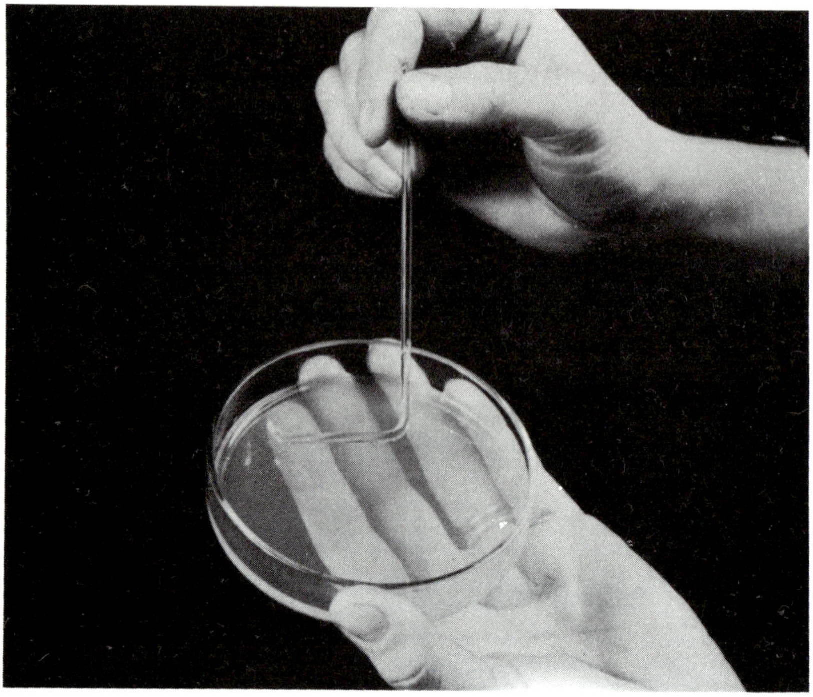

Fig. 32. Streaking of the bacterial suspension by a glass rod on to the surface of the agar plate

of plexiglass plate, containing 0·9 ml dilution liquid (sterile water or broth), and then from each tube into the next tube. If dilution is carried out with a pipette we have to take care that every step of dilution should be done by a sterile pipette.

Then 0·1 ml suspension is placed onto the surface of the prepared agar plate surface (see later) with a pipette from the last members of the dilution row. This is spread over on the surface of the agar plate by a sterile glass rod (Fig. 32). The glass rods are sterilized by momentarily holding them across a flame after dipping them in alcohol. If spreading over the Petri dish is begun from the highest dilution end (with the smallest cell number), the same pipette and glass rod may be used for the work.

On plates kept at 28 °C the colonies develop during 48 hr. The bacterial number of the suspension at the beginning may be concluded from the number of the colonies on the basis of the equation indicated in Fig. 30. If greater exactness is required, dilution and spreading over the plate is repeated three times.

CHAPTER VII

MAINTENANCE AND PRESERVATION OF BACTERIAL CULTURES

Phytopathogenic bacteria generally retain their virulence for a long time under laboratory conditions, but some species lose their viability relatively quickly. The pseudomonads generally can keep their viability for many months without transfer, whereas the xanthomonads die in one or two weeks, on commonly used bacteriological media. Certain xanthomonads are especially sensitive, e.g., *X. phaesoli*, var. *fuscans* or *X. vesicatoria*, which die in three days on nutrient agar medium. The xanthomonads produce acids from sugars which are in the medium, and in this way the pH becomes very low. For this reason, on Starr and his collaborators' advice, it is best to maintain xanthomonads on agar containing calcium carbonate, which will neutralize the acid produced in the culture.

This medium contains 1 per cent yeast extract, 2 per cent agar-agar, 2 per cent glucose, 2 per cent precipitated $CaCO_3$ and distilled water to volume.

Generally, the phytopathogenic pseudomonads, erwinia and agrobacteria, should be transferred onto fresh media every 3–4 weeks, while it is advisable to transfer the xanthomonads and corynebacteria more frequently, every 1–2 weeks.

After cultivation, the cultures are stored in a refrigerator at 3–6 °C.

PRESERVATION OF CULTURES BY OVERLAYING UNDER MINERAL OIL

Cultures can be preserved for longer periods if sterile paraffin oil is poured on the surface of the agar slant cultures. The oil must cover the slant completely; to ensure this the oil should be about 1–2 cm above the tip of the slanted agar. The oil closes the bacteria from the air and in this way their metabolism will be slowed down. With this method, the cultures can be maintained for 1 year without subculturing.

LYOPHILIZATION

The most effective method for maintaining bacterial strains is the rapid drying in a frozen state. Different apparatuses are available for this purpose. The essence of the method is that the bacterial suspension will suddenly freeze due to a high vacuum ($10^{-3, -4}$ Hg mm) and after this will continuously dry in 5–6 hr. The absorbed gases will evolve from the liquid on the effect of the vacuum. This should be accompanied by boiling. Bubble-formation causes radical change in the material, therefore, the gases should be released in such a way so as not to be followed by frothing. This can be done in two ways. The suspension of the bacteria is placed in small vials which are then immersed in a mixture of dry ice and alcohol (—78 °C) before lyophilization; or the material is centrifuged under high vacuum. The exhausted steam is precipitated with dry ice, or it is absorbed by phosphor pentoxide. In this way the steam cannot go into the pump and its efficiency will not be decreased.

Generally, bacterial suspension washed from 24-hr agar culture is used for lyophilization. If the agar culture is washed with physiological salt solution or bouillon, the preservation will not be satisfactory. If, however, a protective colloid is applied, 50–80 per cent of the bacterial cells will survive the trials of the lyophilization process. Moreover, the further suspendability and maintenance will be incomparably better. According to one group of research workers skim milk, to others, sterile filtered inactivated serum, is considered suitable as a protective protein. In our laboratory lyophilization with milk has been used successfully for years. Before use fat is removed and the skim milk is placed in the autoclave at 100 °C and sterilized on three consecutive days. The bacterial suspension prepared with the milk is measured with a Pasteur pipette into a sterile lyophilization ampulla. It is advisable to contract the middle of the neck of the ampulla by holding it over a flame before the sterilization and filling of the ampulla. With this method, the scalding of the ampullas can take place without their removal under the origina vacuum. Not more than 0·5 ml of bacterial suspension should be measured into one ampulla. The lyophilized bacteria could be stored in room temperature, or more precisely at a low temperature (0–5 °C). Lyophilized cultures are able to preserve their viability for 15–20 yr.

The opening of the lyophilized cultures is as follows: the outside of the ampullas is disinfected with alcoholic cotton wool and is rubbed with an ampulla file. Sterile physiological water or bouillon is dropped by means of a sterile Pasteur pipette onto the preparation. From this suspension the bacteria will be streaked onto the agar plate with a wire loop (page 140). Colonies generally develop more slowly from lyophilized cultures.

COLLECTIONS OF PHYTOPATHOGENIC BACTERIA

For breeding for disease resistance as well as for comparison of bacterial strains authentic bacterial strains are required. These can be obtained at the following places.

National Collection of Plant Pathogenic Bacteria, Plant Pathology Laboratory, Ministry of Agriculture Fisheries and Food, Harpenden, Hertfordshire, England.

International Collection of Phytopathogenic Bacteria, Department of Bacteriology, University of California, Davis, California. U.S.A.

The American Type Culture Collection, 12301 Parklawn Drive, Rockville, Maryland 20852, U.S.A.
Bacteriology Department of the Federal Biological Institute, Berlin-Dahlem, Königin-Luise-Str. 16–18 G.F.R.
Culture Collection of Microorganisms, Faculty of Sciences, J. E. Purkyne University, Brno, Czechoslovakia.

CHAPTER VIII

SEROLOGY

Such a close relationship exists between certain phytopathogenic bacteria that often they cannot be differentiated from each other even with traditional identification methods (Stapp, 1961; Starr, 1959). Some species differ from each other in biochemical properties only as slightly as the different strains of the same species. It is obvious, therefore, that merely a biochemical and morphological knowledge of the bacteria, besides the pathological investigation, are insufficient. These have to be completed by serological and sometimes phage sensitivity tests.

As mentioned previously, the problem of diagnosing the pathogens of bacterial diseases is not entirely solved. Generally, diagnosis of the disease is based on the investigation of disease symptoms and on the cultural properties and biochemical activities of the bacteria. These tests are rather lengthy, especially in the case of phytopathogenic bacteria; often they require several months' work. In practice, these methods cannot be applied in routine tests. Therefore, it was necessary to apply such methods as serology from medical microbiology so that each unknown pathogen can be identified in a relatively short time. However, to do this, the antiserum of the bacterium to be investigated is needed.

Furthermore, serological methods can be used for information on the relationship between certain bacteria, and thus provide information on the classification of these organisms. Unfortunately, there are only a few publications in the area of phytobacteriology dealing with serology; therefore, the methods which may be applied in phytobacteriology are not yet entirely worked out.

When antigens are introduced parenterally into an animal organism, it reacts, after some time, with produced antibodies. The antibodies are globulin, originating in the active mesenchyma cells, which are most easily detectable in blood-serum. The serum containing the antibodies is known as the immune serum.

A property of antibodies is their ability to react with the antigens, which induce their production. They are able to react with similar antigens, too. The demonstration of the reaction between the antigen and antibody is an extraordinarily sensitive and specific test, which is the basis of serological methods.

The antigens may be proteins or protein-polysaccharide complexes. The bacteria consist mainly of substances with antigenic characteristics; therefore, serological methods can be applied for rapid detection and identification of bacteria.

The bacteria homogenized in a physiological (0·85 per cent) sodium chloride solution, remain in a homogeneous suspension for a relatively long time. As a result of mixing with the serum containing antibodies specific to the bacteria under discussion, the bacteria will be clumped in the suspension. This reaction is known as *agglutination*. If the immune serum is mixed with a soluble antigen (e.g., cell extract), then the antigens react with the antibodies, giving a precipitate. This reaction is the *precipitation*.

Agglutination and precipitation take place in two phases. In the first phase, the antigen is combined with the antibody. In the course of combination, high molecular aggregates are formed. In the second phase, physicochemical processes follow, which help to precipitate the aggregation. Lipids and electrolytes are necessary for the precipitation of aggregates consisting of antigens and antibodies. The serum contains the lipids needed for the reaction. The hydrophobia of the antigen-antibody complex is increased by the lipids. Precipitation is promoted in this way. The presence of monovalent cations is also required for the formation of the precipitate. The electric charge of the bacteria, or of the colloidal extract obtained from them is negative. This charge renders bacterial cells and colloidal particles to remain homogeneously dispersed in the liquid. When the antigen unites with the antibody, the negative charge of the antigen decreases. This decrease of charge is intensified by the monovalent cations, and in this way they help the precipitation of the antigen-antibody complex.

The bacterial cell is not a homogeneous antigen, but is composed of many antigens forming a so-called antigenic mosaic. The antigens of the bacterial cells are distributed in the cell and cell wall, as well as in the flagella. It can be shown that only certain antigens take part in the serological reaction. If the bacteria, having flagella, are boiled before agglutination, the antigens, which are proteins in the flagella, will be denatured.

Bacterial (somatic) antigen with a homologous antiserum form a fine agglutination, because the antigens contained in the flagella are eliminated by boiling. This type of reaction is called *O-agglutination*. As opposed to this, the living bacteria are precipitated by the immune serum in such a way that the cells having flagella form a larger clump, the so-called *H-agglutination*.

Certain atomic groups of the bacterial antigen have an antigenic effect independently. Every atomic group having an antigenic effect can produce its antibody and react upon it. The antibody should be thought of as an imprint of the antigen. The very high sensitivity of the serological reactions is also obvious for the fact that not only the structure of the whole antigenic molecule, but the number and the spatial arrangement of the atomic groups are also reflected in the antibody. Two different bacteria have different antigenic mosaics. Thus, serum produced against one bacteria has no antibody against another one. But two related bacteria may contain similar antigens. The serum produced against one bacterium reacts with the antigen of another one, so that common antigens are contained by both. The degree of relationship can be expressed quantitatively with the agglutination and precipitation titres.

Agglutination titre is a dilution of the immunological serum, which agglutinates with a determined quantity of bacterial cells. However the *precipitation titre* is a dilution of the antigen which still gives a visible reaction with a well-defined quantity of serum.

The identity and systematization of the bacteria with serological methods is based on the identification of the common antigens of the related bacteria. The antigen structure of bacteria is determined by their chemical composition; for this reason these methods give a relatively stable basis for the identification and systematization of these organisms.

PREPARATION OF THE ANTIGEN FOR IMMUNIZATION

The bacteria is suspended in saline from the 24 hr slant agar culture and the density adjusted (see page 161). For this purpose the densitometer or McFarland scale is used. 10^9 cells/ml bacterium suspension is the most appropriate for immunization. It is not necessary to kill the phytopathogenic bacteria with heat or in any other way. However, the antigens must be prepared for the agglutination test or agar-gel diffusion method, which will be discussed later.

IMMUNIZATION OF THE RABBIT

Healthy, well developed, about 2–3 kg rabbits are the most suitable for immunization. It is not advisable to use too young, very old or pregnant rabbits.

Before immunization one has to take a sample from the blood of the rabbit to determine whether its blood serum contains non specific antibodies which could react with the bacteria. The collection of the blood sample takes place as follows: the hair is cut with fine scissors from the side of the marginal ear veins and then the skin is disinfected with alcohol. Then the marginal ear vein is cut across with a sterile sharp scalpel or a razor blade. The dripping blood is caught in a centrifuge tube; 1–2 ml blood is enough for the above test. The blood is allowed to stand at room temperature for one hour; then it is centrifuged by a low speed of centrifuge revolutions. The titre of this normal serum will be determined (see page 175) to see whether the serum of the untreated rabbit does or does not agglutinate the bacteria. If the serum agglutinates the bacteria higher than 1 : 40 – 1 : 80 dilution, then the animal is unfit for immunization.

Before any injection is made one has to prepare an immunization plan. In the case of phytopathogenic bacteria the following immunization method was used successfully. Before the immunization, the ear on the site of the marginal vein was disinfected with alcohol. The rabbit was injected (into the marginal ear vein) with 1, 2, 2, 4 ml bacterial suspension every 3 days. A two or five ml injection syringe and a 25–27 gauge size needle were used. It is important that syringe and needle be clean and sterile and that the needle be sharp. The ear is held firmly and the needle inserted through the skin into the vein and parallel to the course of flow. After 6 days of the last injection one has to take a sample for serum titration in the way described later. If the serum titre is higher than 1280 or 2560, the rabbit will bleed to death.

It could happen that certain phytopathogenic bacteria cause shock effect in the rabbit. As regards some strains of *X. vesicatoria*, this shock effect can be avoided only if the first injection is carried out with 0·1 ml, then 0·3, 0·5, 1 and 2 ml bacterial suspensions (Lovrekovich and Klement, 1965).

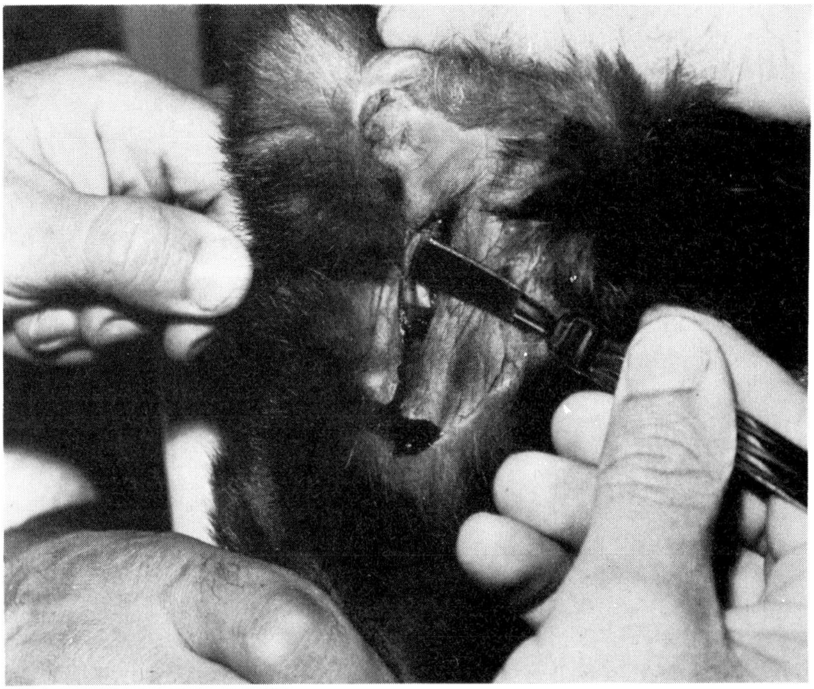

Fig. 33. Preparation of carotid artery of rabbit to obtain immunological serum

It is possible for the rabbit to be bled to death in several ways. Larger amounts of blood can be obtained from rabbits by performing a cardiac puncture. In our laboratory the animal is bled to death through the carotid artery. The site of the cut is disinfected after the rabbit is placed on its back on the autopsy board, and a hole in the skin of the neck is made with scissors in the vicinity of the oesophagus (Fig. 33).

After preparing the carotid artery which is located near the oesophagus the vein is cut and the blood is collected through a sterile funnel in a sterile calibrated cylinder. One can get 60–80 ml of blood from a rabbit, and about half of the blood will be antiserum.

The blood is allowed to cake in the calibrated cylinder at 37 °C for two hr and then one should put it in the refrigerator overnight. Before placing it into the refrigerator, it is necessary to separate the blood plasma from the wall of the cylinder with a glass rod, because the plasma does not separate the serum from itself. The next morning, the opalescent serum, separated from the plasma, is drained off. If it still contains a few blood-corpuscules, then it is necessary to centrifuge it. Phenol in 0·5 per cent final concentration will be added to the serum to inhibit the multiplication of microorganisms. The contamination can also be avoided with the addition of 0·1 per cent merthiolate. Serum dosage in sterile tubes can be maintained without any significant titre decrease in a refrigerator for years.

AGGLUTINATION TEST

Slide agglutination test. Agglutination can be produced on a clean glass slide or in test tubes. For quick information the glass slide agglutination test is used. Bacteria are suspended uniformly using a wire loop in a physiological saline solution which is dropped on the slide. Then a loopful of immunological serum is mixed with the suspension. The slide, held between the fingers, is moved carefully in order to set up contact between antigen and antibody and form the agglutination. The reaction is visible within 3–5 min in transmitted light. The bacterial cells which were originally in a uniform suspension become clumped, in other words, agglutinated (Fig. 34).

The tube agglutination is a more sensitive method, than a slide agglutination. The agglutination test is carried out in the following way: 10–12 agglutination or Wasserman tubes are put in the test-tube rack; 1·8 ml physiological saline solution is measured into the first tube, and 1 ml into each of the other ones (Fig. 35) Then 0·2 ml antiserum is pipetted into the first tube. The liquid is sucked up with the pipette and released several times to produce perfect mixing. After mixing,

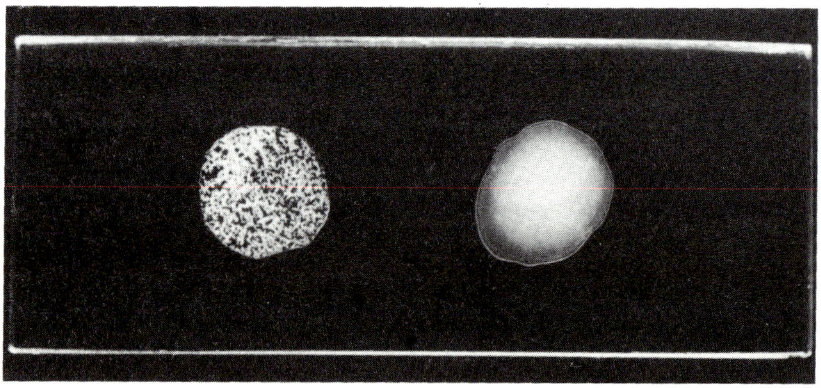

Fig. 34. Slide-agglutination test. Left positive, right negative reaction

1 ml is transferred from the first tube into the second tube, then after mixing this tube, from it into the third and similarly into the remaining tubes. The same pipette is used for the dilutions. The control tube contains no serum. When the half-dilution series is completed, 1 ml homogeneous suspension of the bacterium (10^9 cells/ml) taken from a young culture in physiological sodium chloride

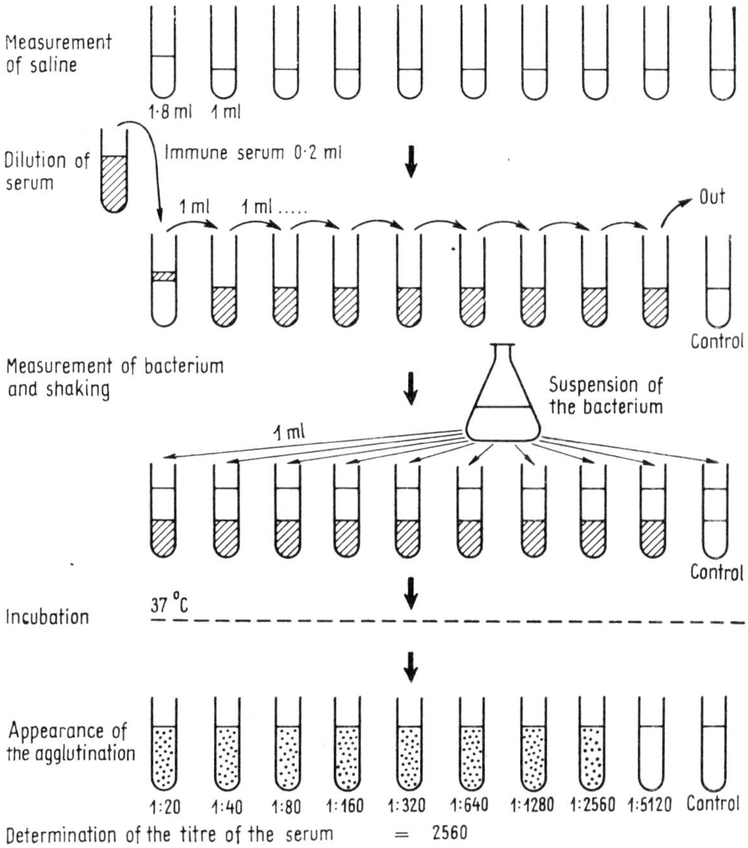

Fig. 35. Tube-agglutination and establishing the titre of the immunological-serum

solution is measured into the tubes. After being mixed, the tubes are placed in a water bath at 37 °C for two hr and then the agglutination titre can be read. The last dilution when there is still agglutination present, indicates the titre of the serum (Fig. 36).

It has been shown by Klement and associates (Klement and Lovrekovich, 1960; Lovrekovich et al., 1963) that the thermostabile antigens are more specific than the thermolabile ones in the case of the phytopathogenic pseudomonads and xanthomonads. Therefore, if one wants to determine the relationship of different isolates or species, it is advisable to coagulate the thermolabile antigens. This can be done by keeping the antigens (bacterial suspension) in a boiling water bath for two hr.

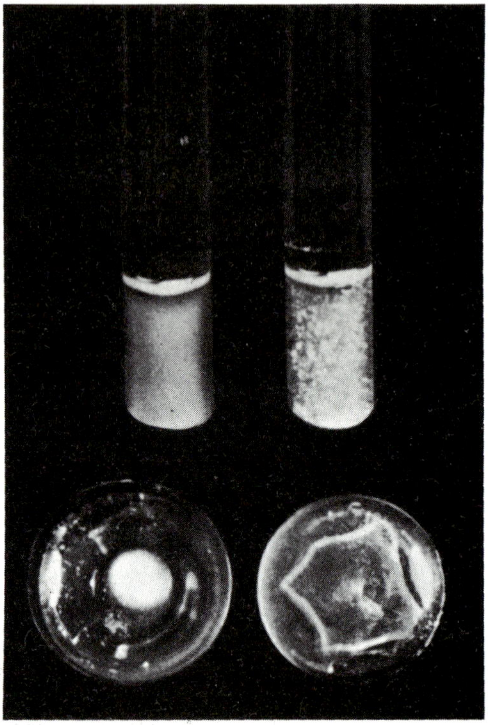

Fig. 36. Tube-agglutination. Negative (left) and positive (right) reaction. Upper: test tubes 2 hr after the test; lower: reaction on the bottom of test tube after 24 hr

PRECIPITATION TEST IN AGAR-GEL

There are very close relationships among phytopathogenic bacteria. For example, pseudomonads and xanthomonads cannot be differentiated from each other with the conventional identification methods, if the host plant is unknown. Simple serological tests do not bring the expected results, because the antigenic structures of species belonging to the genera mentioned above may be very similar.

For this reason it is necessary to apply more sensitive serological tests. The best method for the identification of the phytopathogenic bacteria and the determination of their relationship, at present, is the agar-diffusion test (Fig. 37). This method was devised by Ouchterlony in 1958. It has an advantage over the methods mentioned above, in that the various antigens and antibodies can be separated and compared directly. The agar-diffusion test is based on the precipition reaction, which is conducted in agar in a Petri dish. Bands of precip-

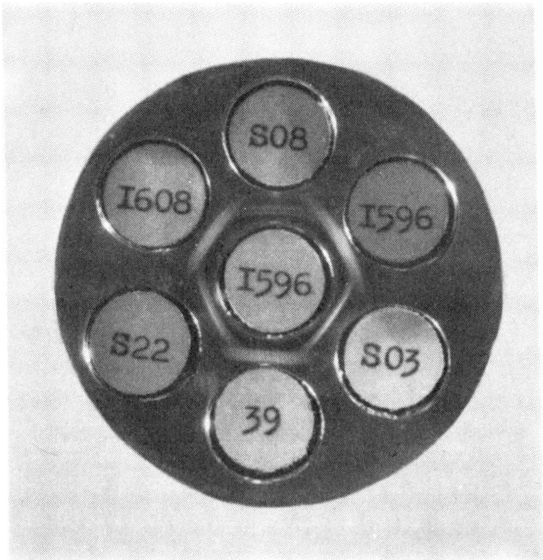

Fig. 37. Agar-diffusion test. Cross-reaction of pepper strains (I 596, I 608, 39) and tomato strains (S 03, S 22, S 08) of *Xanthomonas vesicatoria*. The isolates from pepper and tomato contain related antigens but they are not identical (Lovrekovich and Klement, 1965)

itate form in the agar between the wells in which there is an appropriate concentration of homologous antiserum and antigen. The diffusion speed of the different antigens and antibodies in the agar is not the same. For this reason, if the investigated organisms contain different antigens, several bands are formed accordingly. The antigenic structure of the bacteria under discussion can be determined from the number of precipitation bands formed.

The antiserum is prepared in the same way as in the agglutination test, but it is advantageous to try to produce a titre of serum as high as possible. The agglutination titre of antiserum for the agar diffusion test is ideal if it is higher than 1 : 2560. A large quantity of antigen is needed for the precipitin reaction. Therefore, it is advisable to grow the bacteria in Roux bottles containing 8 ml nutrient agar. After 24 hr incubation at 28 °C, the bacteria are washed from the surface of nutrient agar media with physiological saline solution and their density is adjusted so that the suspension contains $10^{10,-11}$ cells/ml. If one is comparing many strains of bacteria one should take care that the suspension of strains under comparison is uniformly concentrated. To exhaust and destroy the nonspecific antigen components, the suspensions must be boiled in the autoclave at 121 °C for two hr (Lovrekovich and Klement, 1961, 1965).

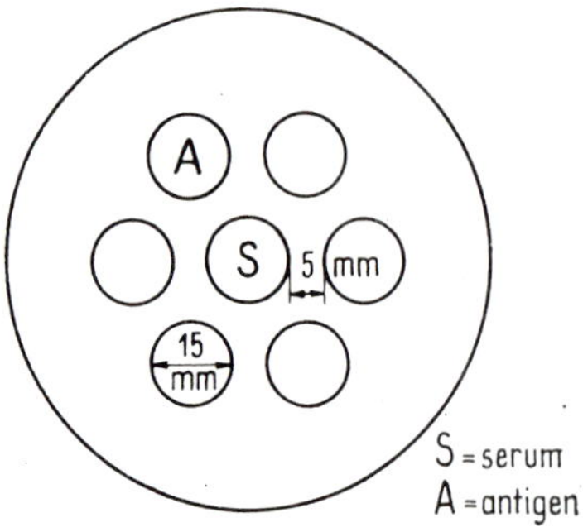

Fig. 38. Diagram for cutting of wells in the agar gel for Ouchterlony test, A: antigen, S: serum

The diffusion plates are prepared by Ouchterlony (1958) methods, 15 g washed agar is melted in 1000 ml physiological saline solution, 10 ml 1 per cent merthiolate and 10 ml 0.1 per cent Orange-G or trypan blue solution will be added to the agar before it solidifies. To inhibit bacterial growth or contamination, merthiolate should be added. Staining makes the precipitation bands more obvious either for visual or photo evaluation.

The pH of agar must be adjusted between 7 and 8, 15 ml agar is poured into a 10 cm diameter Petri dish. According to the figure (Fig. 38) wells are made in the agar plate with a cork borer 15 mm in diameter. Around the central well 6 wells are made 5 mm from the edge of the central well, and 0·2 ml antiserum is measured into the central well. Similarly, 0·2 ml of the antigens will be dropped into the wells around the well containing serum. The plates are placed under humid condition.

At a temperature between 5–10 °C the appearance of the precipitation bands needs more time, but the bands will differentiate from each other better. When the incubation is conducted at a higher temperature, between 24–28 °C, the bands of precipitate appear earlier, but often a few bands diffuse into each other, which can disturb the evaluation.

By comparing the number of places of bands, one can draw conclusions about the antigenic structure and the relationship of the investigated bacterial strains (Feinberg, 1957). Antigenic structure of two bacteria can be said to be identical only if the bands of precipitate, formed by the two antigens to be compared and the antibody, fuse at their junction. If the number of bands is the same, but the fusion produces crossing, or spur formation (Fig. 37), the investigated bacteria cannot be considered to be identical.

ADSORPTION TEST

The adsorption test serves as a comparison of the antigenic mosaics of bacteria. The essence of the method is to add surplus bacteria to the antiserum prepared from a well-known bacterium. If the bacteria is identical to the homologous one with which the serum was prepared, all the antibodies will be adsorbed by the antigens of the bacteria.

The bacteria, which are combined with the antibodies centrifuged from the serum, remove the antibodies with them. On completion of the adsorption test it is rather essential to add the bacteria in excess to the serum to be adsorbed, because otherwise antibodies remain in the serum after centrifugation. The serum treated in this way loses its ability to combine with the homologous bacterium. This negative reaction allows us to draw the conclusion that the two bacteria are identical.

If bacteria are added in excess to the immunological serum whose antigenic mosaic is not identical with the homologous bacteria, then the reactivity of the centrifuged serum will be lost in proportion to the common antigens contained by the two bacteria. With the appropriate selection of the antigen used for the adsorption of the serum, highly specific, so-called *factor-serums* may be produced affecting only one antigen of some bacteria.

RAPID SERODIAGNOSTIC TESTS TO DEMONSTRATE BACTERIAL INFECTIONS FROM PLANT MATERIALS

A practical method for the demonstration of bacterial infection in bean seeds

The bacterial diseases of bean seeds cannot in all instances be identified on the basis of symptoms. In the case of slight infection, the spot can hardly or not at all be detected on coloured seeds.

The essence of the procedure (Lovrekovich and Klement, 1963) may be outlined as in Fig. 39. From the seed lot an average sample containing 100 or more

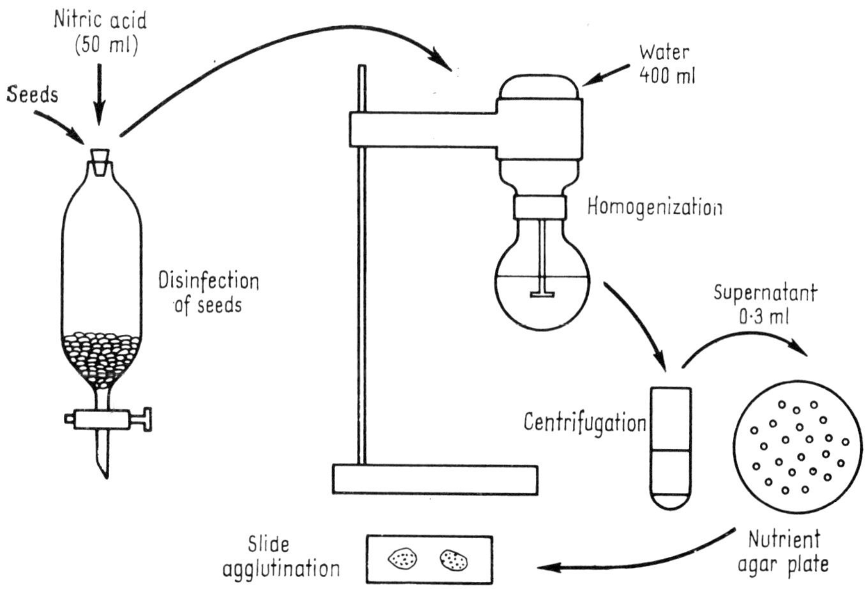

Fig. 39. Rapid serodiagnostic test for demonstration of bacterial infections (*Pseudomonas phaseolicola*, *Xanthomonas phaseoli*, *X. phaseoli* v. *fuscans* and *Corynebacterium flaccumfaciens*) from bean seeds (Lovrekovich and Klement, 1963),

seeds is taken. Because of saprophytic bacteria living occasionally in great numbers on the seeds it is necessary to disinfect the surface of investigated seeds. The seeds are placed into a separatory funnel of 150 ml capacity where the surface of seeds becomes sterile by shaking in about 50 ml of concentrated nitric acid for 2 min. After this, the seeds are washed with running water for 2 min. A suspension of bean seeds is prepared by the addition of 400 ml sterile water in a "Waring Blendor". After centrifuging about 50 ml of the homogenized liquid at 1000 rpm for 3 min the bacteria remain in the supernatant. 0·3 ml of the supernatant is spread on a nutrient agar plate. If the sample contains the pathogen, the developing bacteria will be agglutinated with immunological serum on glass slides, using the slide agglutination method.

A rapid serodiagnostic test for potato ring rot

Ring rot control of potato caused by *C. sepedonicum* is based on a rigid field inspection in conjunction with a seed certification program. Strobel and Rai (1968) reported a rapid serodiagnostic test to detect a systemic phytotoxic glycopeptide produced by *C. sepedonicum* in potato tissue (see page 198). The presence of wilt toxin is specifically related to the presence of the pathogen in the host

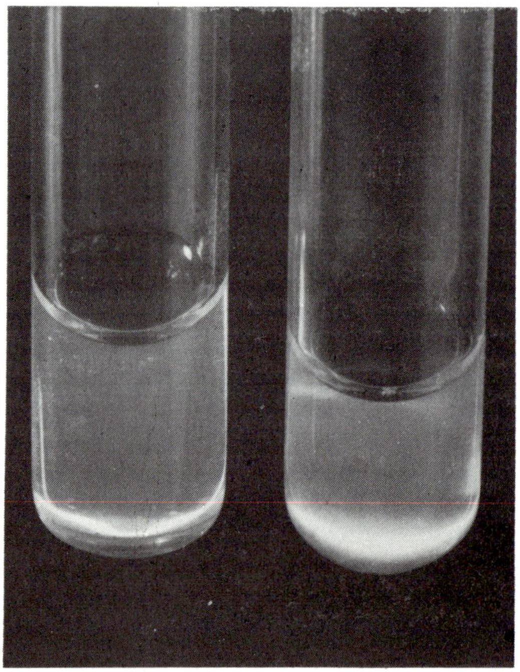

Fig. 40. Serodiagnostic test for potato ring rot (*Corynebacterium sepedonicum*). Left: Processed plant extract of a non-infected plant, right: processed plant extract of a ring-rot infected tuber (Courtesy of G. A. Strobel, and P. V. Rai)

tissue. The preparation of antigen, antiserum and the tissue extract is described by Strobel and Rai (1968) as follows: the toxins of *C. michiganense*, *C. insidiosum*, and *C. sepedonicum* were serologically related. The antiserum prepared against *C. michiganense* had titres of 1 : 16, the highest of any of the three antisera. It is for this reason that the antiserum preferred in this test was produced from the toxin of *C. michiganense*. Cultures of *C. michiganense* were grown in still cultures at 23 °C in 1 l batches of 1 per cent yeast extract dialysate, 1·5 per cent glucose, and 0·5 per cent calcium carbonate for 3 weeks prior to harvesting the crude toxin. Bacterial cells were removed from the cultures by centrifugation at 20,000 g for 30 min. The supernatant was passed through 20 cm × 2·5 cm columns of Dowex 50 (200–400 mesh-X 8, H^+ form) and Dowex 1 (200–400 mesh-X 8, formate form), respectively, per litre of solution. After reducing the volume of

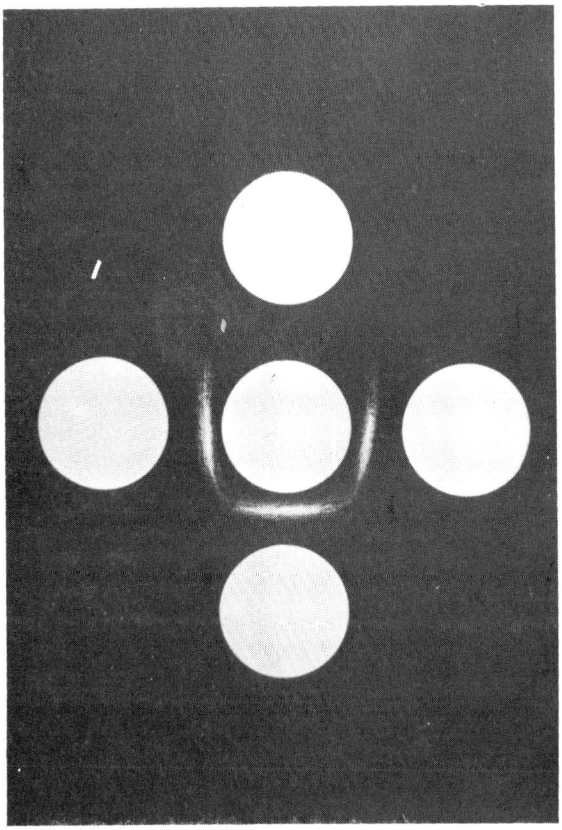

Fig. 41. Gel diffusion test for determination of the toxin of *Corynebacterium sepedonicum*. Middle: antiserum to *C. sepedonicum* pure toxin; top: buffer; right and left: pure toxin; bottom: purified toxin isolated from plants infected with *C. sepedonicum* (Courtesy of G. A. Strobel)

the solution by flash evaporation to less than 100 ml, 6–8 equal volumes of either cold acetone or 95 per cent ethanol were added and a precipitate resulted. This precipitate was collected by a low-speed centrifugation and dried in a vacuum desiccator over P_2O_5. This precipitate was then referred to, as the crude toxin.

An antiserum to the toxin was prepared by injecting 1 ml of an emulsion containing 10 mg of the toxin intramuscularly into the hip muscle of rabbits. The injected antigen was made by dissolving the toxin in phosphate buffered saline (pH 7·0, 0·01 *M* phosphate, and 0·15 *M* NaCl) and emulsifying in Freund's complete adjuvant. Two primary injections of toxin, each of 10 mg, were made at 7-day intervals. Serum was collected 28 days after the final injection and stored at —20 °C until used.

Toxin was extracted from 10 g of fresh tissue taken either from the vascular area of the tuber or from whole stem tissue and grinding it in a mortar with a pestle with 5·0 ml of distilled H_2O. The crude juice was then decanted into a small column (1·0 cm × 10 cm) containing 1·0 cm of glass wool, 1·0 cm of washed sand and 1 cm of Dowex 50 (200–400 mesh-X8, H^+ form). To the effluent from the Dowex 50 column, four volumes of cold acetone were added, and the precipitate pelleted by a slow-speed centrifugation. The supernatant liquid was then discarded and the pellet taken up in 0·5 ml of distilled H_2O. Once again the suspension was centrifuged and the supernatant liquid carefully removed with a pipette and 0·1 ml was placed into a small test tube (6·0 mm × 50 mm). To this solution 0·01 ml of undiluted antiserum was added. If a distinct heavy cloudiness appeared in a solution within 30 sec this was considered as a positive reaction; no cloudiness occurred in a negative reaction (Fig. 40).

Figure 41 shows the agar-gel diffusion test of the *C. sepedonicum* toxin. It seems that both the purified toxin isolated from plant infected with *C. sepedonicum* and the pure toxin of *C. sepedonicum* gave the same precipitation lines.

CHAPTER IX

BACTERIOPHAGES

In 1915, the English bacteriologist F. W. Twort observed that some bacterial cultures became lysed. Two years later, Felix D'Herelle, a Canadian bacteriologist who was working at the time at the Pasteur Institute of Paris observed the same phenomenon. Initially, different opinions were formed about the nature of the lysis. The controversy was finally resolved by electron microscopic examinations. It was demonstrated that small particles called bacteriophages, which are similar to virus particles, are liberated from the lysed bacteria. Delbrück and Luria have confirmed the viral nature of the phages. According to them the following are characteristic of phages: 1. when separate from bacterial cells they exhibit no metabolism, 2. they only multiply in cells which are sensitive, 3. their size is the same as that of viruses, 4. they have a large degree of host specificity, 5. they are nucleoproteins, 6. their lyzogenic property is identical to that of latent viruses.

Most bacterial viruses are shaped like sperm, with round or polyhedral heads and a hollow tail. The chemical investigation of phages demonstrates that they are made up of 40–50 per cent nucleic acids and 50 per cent proteins. They can also be characterized on the basis of protein and nucleic acid function. It has been proved experimentally that only the nucleic acid moiety of the phage can penetrate the bacterial cell. The protein coat has no special significance in phage reproduction. The whole phage can be imagined to be a small syringe which empties its contents, the nucleic acid (DNA) into the bacterial cell; but the syringe itself, the protein coat, remains outside the cell.

The process of phage reproduction consists of three phases (Fig. 42): 1. the adsorption and the injection of the DNA of the virus head into the bacterial cell, 2. the vegetating phase or synthesis of new virus particles within the cell, 3. lysis of the bacterium and the liberation of the newly formed phage particles.

1. The adsorption and injection of DNA of the phage head into the bacterial cell (Fig. 43).

Brownian molecular movement promotes the contact of bacteria with phage particles. When the two collide, the phage particles adsorb the cells. More phage particles may be adsorbed at the same time but only one phage particle can inject the host, as in the case of sperm cells. This phenomenon is one example of interference. It has been demonstrated with the electron microscope that the phage adsorbs to the bacterium by its tail. The electron microscopic picture also confirms

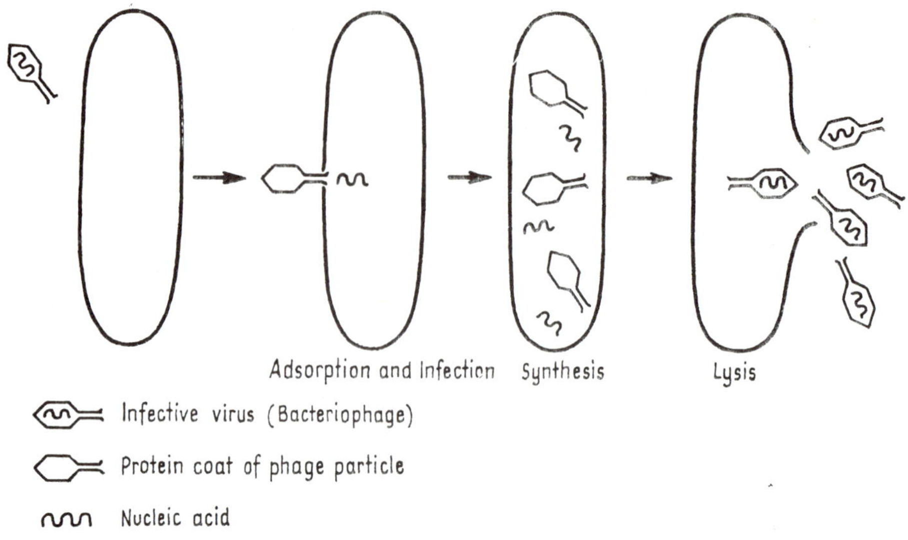

Fig. 42. Process of bacteriophage reproduction

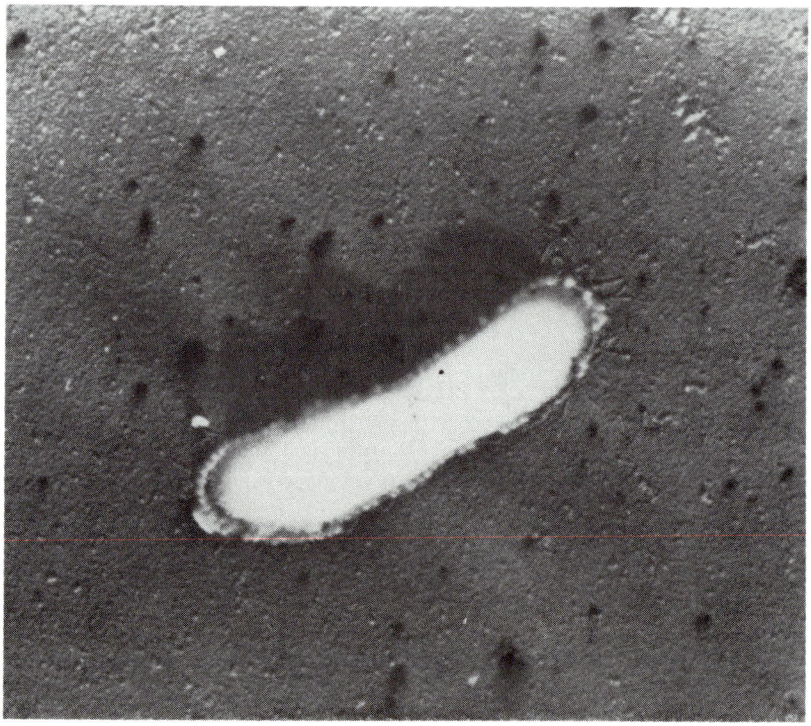

Fig. 43. Adsorption of phage particles to a cell of *Corynebacterium flaccumfaciens* (Z. Klement and B. Lovas, 1959, *Phytopathology* **49**, 107–112)

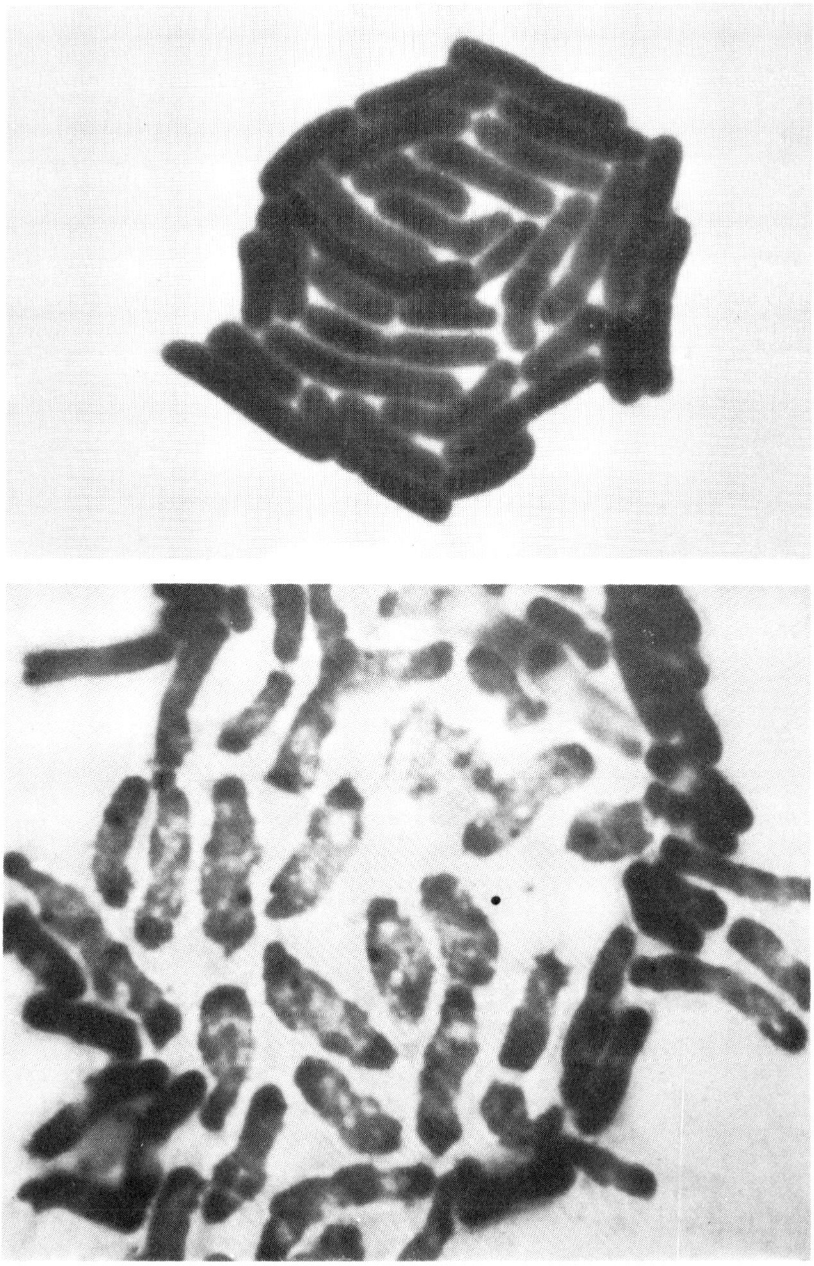

Fig. 44. Synthesis of phage in the cells of *Corynebacterium flaccumfaciens*. Above: microcolony produced by uninfected bacteria. Below: microcolony infected with phages. Prepared 60 min after inoculation. (Z. Klement and B. Lovas, 1959, *Phytopathology* **49**, 107–112)

that only the nucleic acid from the head penetrates the cell, and the external protein coat remains outside the bacteria.

2. The vegetative phase of reproduction within the cell (Fig. 44).

In this phase the multiplication of the bacterial cells is inhibited and the cells gradually become swollen. The cell structure is reorganized and gradual protein and nucleic acid synthesis begins immediately, serving to build the phage particles. The bacterial cells infected by the phage no longer divide. The original respiratory enzymes continue to operate regardless of the infection; however, the energy is used not for the building up of the new bacterial cells, but rather for the synthesis of phage particles.

3. Dissolution of bacterial cells and the release of the new phage particles (lysis) (Fig. 45).

When the synthesis of new phage particles is completed, the bacterial membrane ruptures and releases the virus particles. Ten to 300 new phage particles are liberated from one bacterial cell. This multiplication, however, greatly differs from the mode of reproduction of the living cells, the bacteriophage reproduction does not take place through division but through synthesis.

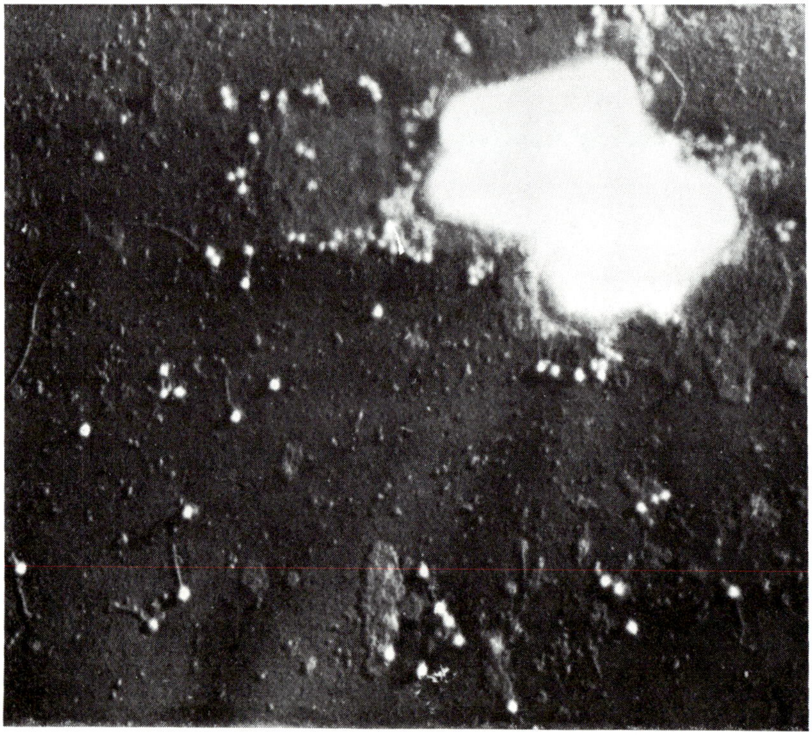

Fig. 45. New phage particles from the lysed cell of *Corynebacterium flaccumfaciens*. (Z. Klement and B. Lovas, 1959, *Phytopathology* **49**, 107–112)

The multiplication described above may be observed on both liquid and solid media. In liquid media inoculated with bacteria, the bouillon changes when multiplying bacteria are affected by phages. After some time (latent period), the bouillon suddenly becomes clear which is a sign that the bacterial cells have become lysed (lysis).

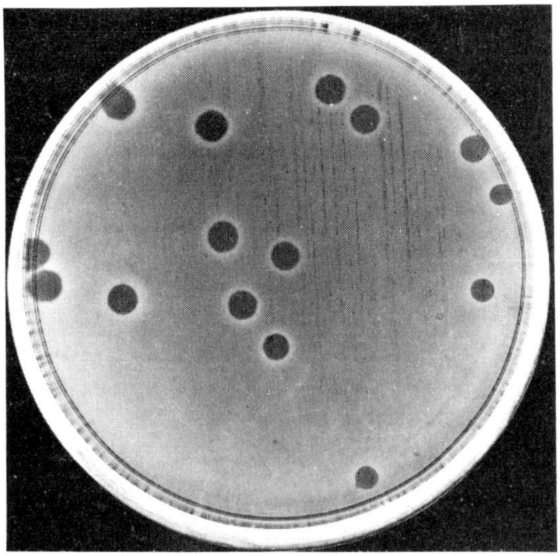

Fig. 46. Plaques of a phage for *Xanthomonas phaseoli var. fuscans*

The phage particles are of such a small size that they are invisible under the light microscope, but their presence can be well recognized by laboratory techniques. Phage and bacteria may be introduced into a surface layer of nutrient agar. After 12–24 hr incubation, the continuous bacterial layer is interrupted by clear, round areas of lysis produced by the phage (plaque) (Fig. 46). The plaque originates in the following way: the bacterium infected by a phage particle soon lyses; the phage particles released from it infect the surrounding healthy and multiplying bacteria, which also become lysed. The phage generations display their lysic effect in a homogeneously growing bacterial layer, then a regular round area free of bacteria appears. Since it is supposed that every plaque actually originated from the affect of the progeny of a single phage particle, the number of phage particles in a suspension can be easily and precisely determined by the plaque-count.

PLAQUE-COUNT TECHNIQUE

From the phage suspension a tenfold dilution row is prepared in sterile tubes. The dilution is carried out by measuring 0·9 ml sterile dilution fluid (bouillon, physiological salt solution or peptone water) into every tube in the dilution row.

The dilution is made by pipette or Takátsy's dilution loop, as was described earlier in relation to the bacterial cell number determination (see page 164–166), by transferring 0·1 ml of the suspension into each tube. After the dilution, 0·1 ml

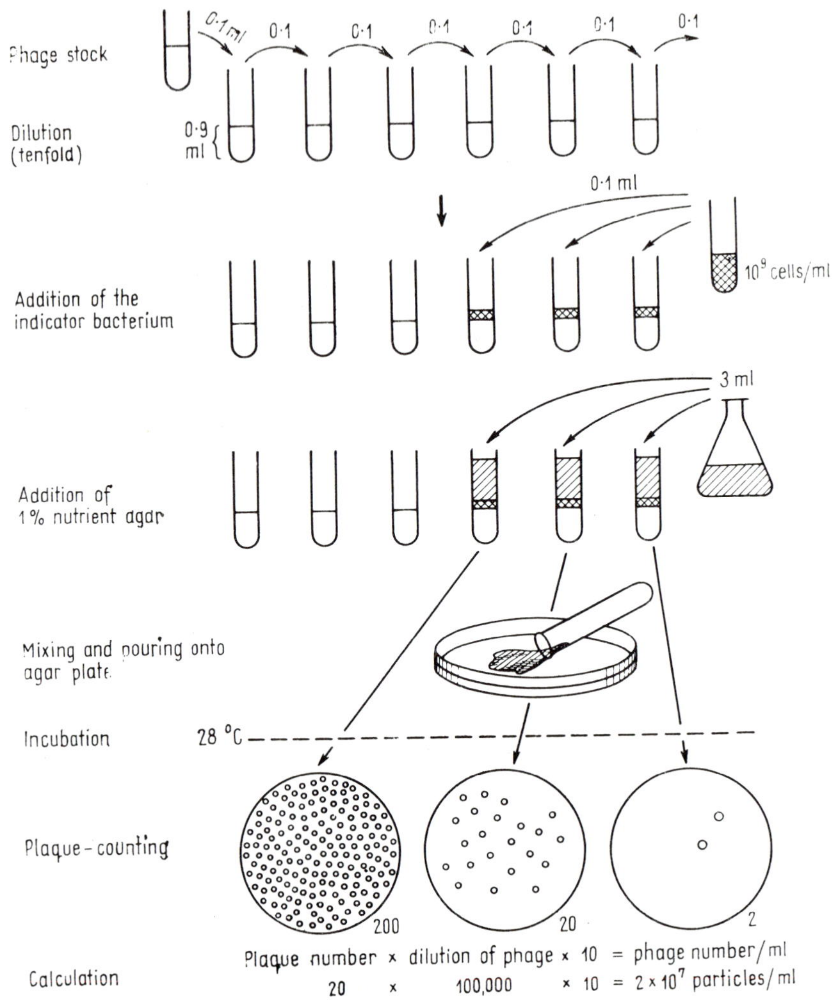

Fig. 47. Diagram of the plaque-count method for determination of phage number

10^9 cell/ml indicator bacterial suspension is placed in every tube. Three ml of melted 1 per cent nutrient agar medium, which was prepared beforehand and kept at 55–60 °C, is measured into every tube of the dilution series. The contents of the tubes are shaken separately and poured onto previously prepared 2 per cent agar plates and evenly dispersed. The plates are incubated in an incubator at 28 °C. After 24–48 hr plaques form on the plates in relation to the dilution (Fig. 47).

USE OF BACTERIOPHAGES IN PLANT PATHOLOGY

After the discovery of bacteriophages great hopes were set all over the world that with their help — in the fields of both medical and plant pathology — a successful control of bacterial diseases would be achieved. Experiments carried out brought temporary disillusionment due to inadequate biological knowledge. They did not take into consideration the fact that the phage always becomes a victim of the antibodies which are present in the body. The second stage of research began only in 1940, and today the results achieved are very significant. This progress is due first to the fact that bacteriophages are excellent models for virus research. Important results have been achieved with phages in the field of genetics because they are easy to use in culture. This work has greatly advanced our views on viruses.

Numerous bacteriophages of plant pathogenic bacteria have previously been described. Information on bacteriophages already isolated, can be found in the works of Stolp (1961) and Okabe and Goto (1963). The phages for phytopathogenic bacteria can be specific or polyvirulent. Observations up to the present (Klement and Lovrekovich, 1961), reveal that xanthomonad bacteriophages are very specific, while phages lysing pseudomonad are mostly polyvirulent. In the phages of xanthomonads, the degree of specificity extends to varieties and sometimes to strains. The specific phages are more often used for identification of bacteria, while the polyvirulent phages are used for the determination of the bacterial relationships.

ISOLATION OF PHAGES FROM INFECTED PLANTS

The bacteriophages of the phytopathogenic bacteria can be isolated either from the soil or from infected plant tissue (Fig. 48). The phages isolated from soil are mostly polyvirulent, while the phages received from the strongly infected plant tissues are generally specific.

The heavily infected plant material is homogenized in a mortar or a homogenizer together with bouillon. One cannot use sterile methods with phage isolation because the disinfection of the surface of the plant causes too much damage. The homogenized plant material is put into an Erlenmeyer flask with liquid to five times its volume. Although plant homogenate contains many phytopathogenic bacteria, it is necessary to inoculate with fresh indicator bacteria. If a strep-

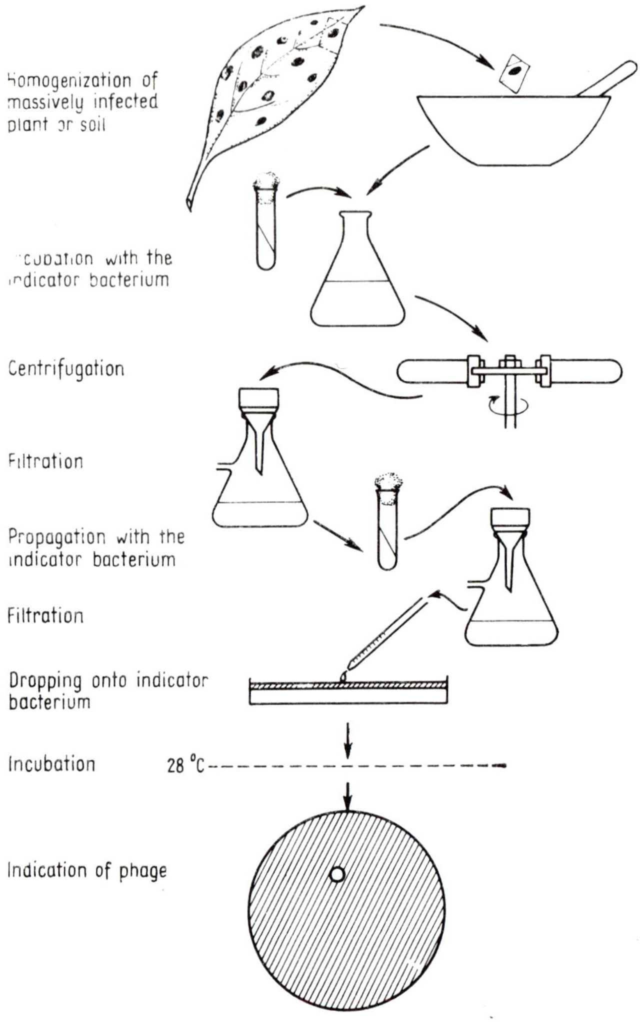

Fig. 48. Diagrammatic representation of phage isolation of plant pathogenic bacteria from diseased plant or soil

tomycin resistant indicator bacterium is used then one has to add 100 ppm streptomycin to the plant homogenate. In this way, one can filter out the majority of bacteria not required. The homogenate is incubated in an incubator at 28 °C for 48 hr. After incubation, this material is centrifuged at 3000 rev for 30 min and the supernatant is either filtered through bacterial filter, or the bacteria are killed with chloroform. The chloroform destroys only the bacteria. The supernatant is put into a test tube with chloroform (1–2 ml to 5 ml supernatant) fitted with a rubber plug and then shaken. After one half or one hour's shaking, the chloroform will separate out to the bottom of the tube. The liquid above the chloroform is then pipetted into a separate sterile tube.

After this, one can investigate the liquid for the presence of active bacteriophages. The procedure is to transfer one drop of this fluid by pipette or wire loop onto the plate containing indicator bacterium. This method will be described later (page 192). If the fluid contains phages, the bacteria will not grow where the drop was placed (Fig. 48).

ISOLATION OF PLAQUES

The liquid originating from plants in which there are bacteriophages should be tested to see if the isolate contains one or more phage types. This can be done with the help of the plaque-count method (page 187). The aim of the method is to reach a phage dilution producing separated plaques on the indicator plates. From this size, transparency and sharpness of the edge of the plaques one can conclude that the liquid contains one or more phage types.

The different plaque types are cut from the agar plate with a sterile loop and put into the fresh broth culture. In this culture the young cells are lysed by the phage particles which have diffused from the agar and their number is increased. After 24 hr of incubation, this culture is filtered by bacterial filter. The filtrate keeps its activity at 1–5 °C in a refrigerator for a long time. With this method one cannot reach a high titre phage suspension. If we wish to receive higher titre than 10^8 phage/ml, it is advisable to multiply the phages on agar layer plates. In this case the agar layers containing the phage particles are collected and homogenized in 4 ml broth/plates and this fluid is filtered through a bacterial filter. The titre of phages is determined by the plaque-count method.

DETECTION OF AN UNKNOWN PATHOGEN BY A KNOWN PHAGE

One can determine the pathogen in the infected plant tissue by applying specific phages directly or indirectly. Katznelson and Sutton (1951), then Katznelson, Sutton and Bailey (1954) worked out a method by which one can quickly and positively determine the infection of the plant material, the species of pathogen and the degree of infection. Bacteriophages of known titre should be added to the homogenate of the infected tissue. During incubation, an increase in phage titre shows that the plant tissue contains bacterial cells. This method is diagrammed in Fig. 49.

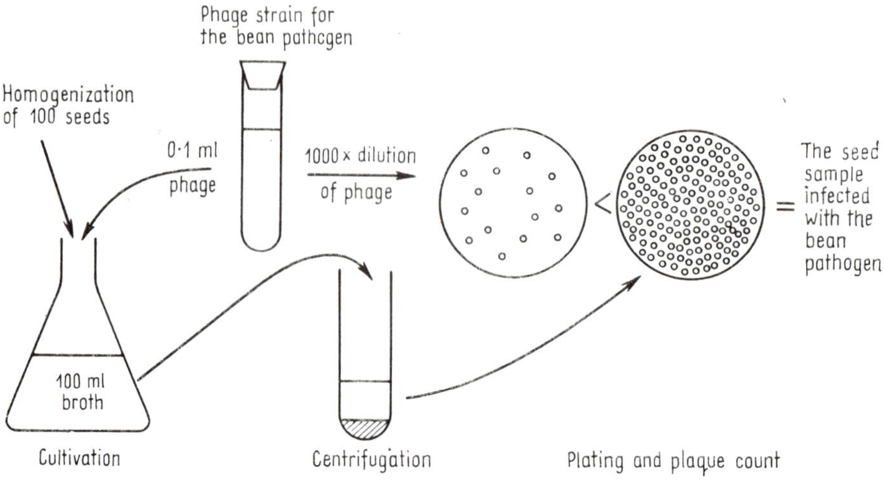

Fig. 49. Scheme of the Katznelson's method for detection of bean pathogens by phages

An alternate method may also be used: the bacteria are isolated from the infected plant tissues and these isolates are separately plated onto nutrient agar in Petri dishes. Then a known bacteriophage is dropped onto the agar culture (see the next paragraph); where bacteriolysis appears after a short incubation, the bacterium in question is identical with the bacterium specific to the phage.

DETERMINATION OF THE RELATIONSHIPS OF BACTERIA ON THE BASIS OF PHAGE SENSITIVITY

Only the polyvirulent phages are suitable for this purpose. The investigated bacteria are suspended in the nutrient agar previously melted to 50 °C. This is layered onto the surface of a solid medium containing 2 per cent agar. After the upper layer solidifies, the known indicator phages are dropped onto it. The bacterial strains have to be examined on separate plates. In the case of identical or related bacteria, lysis will occur. This is only a preliminary investigation serving to gain information which has to be supplemented by pathological and serological tests.

CHAPTER X

THE FATE OF THE PATHOGEN IN THE PLANT, THE PLANT'S DEFENCE MECHANISMS AGAINST BACTERIAL INFECTION

The nutrient requirement of plant pathogenic bacteria is very simple. It can be satisfied by common nitrogen and carbon sources. These bacteria can be grown well on cultures containing carbohydrates, or proteins and simple salts. They do not need any special materials, although certain amino acids greatly aid their growth. Those organic and inorganic compounds (Starr, 1959) are available in plants which satisfy the nutrient requirements of phytopathogenic bacteria.

Phytopathogenic bacteria enter the plant through the stomata or through wounds, and multiply in its intercellular spaces or conducting vessels. It has been shown experimentally (Klement, 1965) that the fluid in the intercellular spaces and in the conducting vessels contains an abundance of all the nutrients necessary for bacterial growth. This fluid can be used as a culture medium to support the proliferation of congenial, uncongenial and saprophytic bacteria in vitro. Despite the presence of all the necessary nutrients, phytopathogenic bacteria are able to multiply for only short periods of time in uncongenial hosts and resistant plants. The question of host specificity, e.g., why plant pathogenic bacteria are limited to their own hosts comes sharply into focus (Schnathorst, 1966).

Research workers who are studying the physiological and biochemical reasons for the disease resistance most often differentiate two main resistance types. We also find it advisable from a didactic point of view to group the known types of protectives which appear with bacterial diseases. These are:

Preformed resistance
Induced resistance

premunity   hypersensitivity

PREFORMED FACTORS OF RESISTANCE

"Preformed factors of resistance" are those that exist in the plants prior to infection and are independent of infection.

Garber (1956, 1961) has developed a "nutritional theory for resistance". This theory assumes either that (a) the nutrients necessary for the pathogen are absent from the resistant host tissues; or (b) as a result of the inhibitory effect of some other compounds in the plant, the required nutrients are unavailable to the pathogen.

Although this research is of interest, it fails to explain why certain pathogenic bacteria are unable to proliferate sufficiently and therefore, do not cause disease in either uncongenial hosts or resistant plants. Although nutritional mutants are likely to occur spontaneously in nature, it is very probable that such mutants will be eliminated from the original population by pressure from natural selection.

Garber and Heggestad (1958) and Garber and Schaeffer (1957) carried out these experiments with mutants requiring different amino acids.

More recently the works of Hildebrand and Schroth (1963, 1964) as well as Schroth and Hildebrand (1965) have created greater interest. They demonstrated that the aqueous extracts of pear tissue are highly toxic to *E. amylovora.* The inhibitor was purified and identified as hydroquinone.

The physico-chemical condition of the plant may also influence infection. In connection with this, Garber (1961a) showed that the buffer capacity of the plant influences the development of *P. tabaci.*

Although experimental data has recently accumulated linking substances in plants to resistance, this data is as yet not too convincing. The fact that a resistant plant contains some substances which are toxic to the pathogen, or that the rate of production of the toxic substances is correlated with the degree of resistance, may be little more than coincidental. In order to causally relate a substance to resistance, its activity in vivo must be conclusively demonstrated. It would seem essential to demonstrate that virulent and avirulent strains are differentially suppressed by the toxic substance in question. As yet such information has not been revealed. It seems that somewhat more definite information on the subject of disease resistance to bacterial pathogens can be obtained by studying induced defense reactions.

FACTORS OF INDUCED RESISTANCE

Those new or liberated compounds forming in the plant which were not present before infection and which are effective against pathogens produce induced resistance. Two types of induced resistance will be discussed: premunity and hypersensitivity.

Premunity

Premunity is regarded as a nonspecific acquired immunity which manifests itself in plants pretreated (inoculated) with one bacterium that "immunizes" or protects the plants from infection by another bacterial pathogen.

The existence of premunity because of previous infection by plant bacterial pathogens was established in 1965 (Lovrekovich and Farkas, 1965). Washed and heat killed cells of *P. tabaci*, the agent of the wildfire disease of tobacco, were introduced into the intercellular spaces of one half of a tobacco leaf; the opposite half of the leaf served as a control and was infiltrated with water. One day after this pretreatment the entire leaf was inoculated with living cells of *P. tabaci*. In the half-leaves pretreated with water, symptoms appeared two days after inoculation which led to a collapse of the inoculated tissue in six days. In the half-leaves which had been pretreated with a suspension of killed bacteria, symptoms failed to appear by the sixth day and bacterial multiplication was also strongly inhibited. The action of killed cells of *P. tabaci* did not seem to be specific since similar inhibitory effects could be achieved with killed-cells of some other plant pathogenic bacteria.

Later it was proved that not only plant pathogens but the saprophytic *P. fluorescens* induced a protective effect. In plants pretreated with bacteria, the protective effect was not only effective against bacterial infection but, as demonstrated by Klement, Király and Pozsár (1966) inhibited viral infection too. In summary, it can be stated that not only the pathogen but saprophytes are capable of inducing such a protective reaction. Furthermore, not only the living but also the killed microorganisms induce this kind of reaction. This proves the unspecificity of the inducing factor and the assumed formation of the antimicrobial "compounds". Similar phenomena have been reported by Crosse (1965) and Goodman (1965, 1967).

Hypersensitive reaction

The occurrence of hypersensitivity in the case of fungal and viral diseases has long been known; however, only recently has its important role in bacterial diseases been clarified. Klement and co-workers (1964) proved that all uncongenial pseudomonads in foreign hosts induce a hypersensitive reaction. Exactly the same type of reaction appears if the virulent bacteria meets with is resistant host plant, (Stall and Cook, 1966; Kennedy and Cross, 1966, and Klement and Goodman, 1967) or if the sensitive host comes in contact with the pathogen which has lost its virulence (Klement and Goodman, 1966). Therefore, it can be seen that the hypersensitive reaction comes about in every incompatible hostparasite relationship.

Host-parasite relationships

Combinations	Hypersensitive reaction	Typical disease symptom
Virulent bacteria – sensitive host plant	–	+
Avirulent bacteria – normally sensitive host plant	+	–
Normally virulent bacteria – resistant host plant	+	–
Pathogenic bacteria – nonhost plant	+	–
Saprophytic bacteria – plant (all)	–	–

The process of the reaction occurs in the following way. A foreign pathogen such as *P. syringae* penetrates the intercellular spaces of the tobacco leaf the bacteria are able to begin to multiply, but this ceases in a relatively short time 8–10 hr. At the same time, the plant cells around the site of infection suddenly die and in this way the infection becomes localized (Fig. 50). When more than 5×10^6 bacterial cells/ml of inoculum are injected then the whole inoculated tissue dies 8–10 hr after the infection. If the inoculum is lower than this the hypersensitive reaction likewise forms; however, due to the minuteness of the necrosis, it is impossible to see with the naked eye.

The hypersensitive reaction works similarly in natural conditions. One or two pathogenic bacterial cells can penetrate the substomatal cavity in a passive way through the uncongenial hosts's stomata where they begin to reproduce. Around the bacterial cells which are beginning to reproduce, only a few host cells die (Fig. 19). This, at the same time, localizes the infection and so the plant remains practically symptom-free. This local infection probably does not develop to such a degree as the viral necroses; the necrosis appears so quickly (8–10 hr) that in such a short time the bacteria are unable to increase to a degree so as to be able to induce necrosis on a bigger tissue area (Fig. 49). If, however, we work with a higher bacterial number, the many little local lesions which are invisible to the naked eye get so close to each other that the whole tissue dies. Therefore, it appears that if more than 5×10^6 bacteria/ml inoculum are introduced into the tissues, the hypersensitive reaction (necrosis), besides the natural environmental conditions, appears in every case in a visible form.

Fig. 50. The hypersensitive reaction of tobacco leaf induced by plant pathogenic bacteria. More than 10^7 cells/ml bacteria were injected into the intercellular spaces of the leaf

The hypersensitive reaction so far has not been reproduced with bacterial filtrate, heat killed, or bacterial cells sonically disintegrated or killed with antibiotics. Similarly, efforts to extract substances which may instigate the hypersensitive reaction from the plant infected with incongenial bacteria have been unsuccessful (Klement and Goodman, 1967).

It is therefore obvious that only the living bacterial cell is capable of induction. The question arises as to how much time the bacteria require to induce the hypersensitive reaction. Klement and Goodman (1967) proved that 20–25 min is a sufficient reaction time. In the 5–7 hr (latency period) following this, the already begun reaction may continue further without the presence of living bacterial cells. The death of the plant cell follows in the 8th or 9th hour. This process is relatively quick, going into completion in 1–2 hr. This quick death of the cells is a consequence of a sudden change in the permeability of the cell wall. As a result of the permeability change the ions liberated from the cells can be measured electronically.

PRIMARY AND SECONDARY FACTORS OF PATHOGENESIS

Although a pathological process cannot be imagined without the combined presence of host and pathogen, we will try to examine the problem from the bacterial aspect.

The old question, what determines pathogenicity still cannot be answered correctly. We know of different enzymes and toxins which take part in the formation of the disease; however, these in themselves cannot be held responsible for inciting the disease. It is the inducing ability of the hypersensitive reaction of bacteria which accentuates the secondary role of the metabolites of bacteria in pathogenesis. Until recently, studies of the relationship between the host plant and the bacterium have been primarily concerned with the bacterial products formed during the multiplication of the bacteria in vitro and in vivo. It has been assumed that these metabolites play the decisive role in pathogenesis. Starr (1959) has grouped these products as follows: (a) wilt toxins, (b) necrotizing toxins, (c) factors inducing hypertrophy, (d) rotting enzymes. Let us take some examples.

Wilt toxins and enzymes

Earlier it was thought that wilt was a consequence of the xylem vessels being blocked by the bacterial cells. Now it is an accepted fact that the microbiological products which Gäumann calls "wilt toxins" also take part in wilt. Certain polysaccharides which are produced by phytopathogenic bacteria are toxins which inhibit the transpiration process. A connection was found between the cellulase enzyme produced by the bacteria and wilt. Some enzymes attack the pectin and cellulose substances of the plant's vascular bundles, thereby liberating certain gum substances which block the plant's water conveying vessels.

P. solanacearum which causes wilt in numerous plants is a typical wilt causing microorganism. Buddenhagen and Kelman (1964) summarized the effectiveness of the bacterium as follows: "Movement of water in vessels declines as viscosity of vessel fluid increases because of formation of extracellular slime surrounding masses of bacterial cells. Secondary causes of reducing water movement include tylose formation and vessel collapse, but their role has not been clearly defined. Other factors, including the action of cellulolytic enzymes and possible pectic enzymes may be involved but are not considered to play an essential role in wilting per se."

Phytotoxic polysaccharides were isolated from *C. insidiosum*, *C. michiganense* and *C. sepedonicum* by Rai and Strobel (1967). The host specificity of the aqueous solution of these three toxins was tested with various plants. The toxins caused wilting in various test plants indicating their nonspecificity. Strobel (1967) investigated properties of the purified phytotoxic polysaccharide produced by *C. sepedonicum*. The purified phytotoxic glycopeptide has a molecular weight of 21,000, and it is water soluble. Although the general concept of bacterial polysaccharide action is that plugging of the vascular elements occurs resulting in retardation of water and nutrient flow (Husain and Kelman, 1958), Strobel and Hess (1968) suggest that the primary mode of action of the wilt toxin (glycopeptide) is to destroy the integrity of cellular membranes which results in a loss of water from the cells. In serological studies, the toxin was proved to be antigenic. On this basis Strobel and Rai (1968) worked out a rapid serodiagnostic test for potato ring rot (see page 180).

Averre and Kelman's (1964) investigation clearly proves that *P. solanacearum* can only express the characteristics causing wilting if the plant protective reaction does not prevent bacterial growth. The wilt toxins have no primary role in pathogenesis; they can only function when the pathogen has settled and multiplied in the plant.

Toxins causing necrosis

With many bacterial diseases the necrotic spots are surrounded by a chlorotic halo. The best example for this is the *P. tabaci* which causes "wildfire" disease of tobacco. The small spots of the leaves are surrounded by chlorotic halos of one and one-half cm in diameter (Fig. 51). These halos are free of bacteria. With the bacteria-free filtrate of the *P. tabaci* culture, exactly the same toxic symptoms may be produced not only on the tobacco but on numerous other plants too. The bacterium itself is host specific but its toxic product is not. Braun (1955) demonstrated that the wildfire toxin originates from a still unknown amino acid. Investigations carried out with the *Chlorella* unicellular alga have proved that L-methionine inhibits the toxic effect of *P. tabaci* toxin. From this it was supposed that the toxin inhibits the methionine metabolism of plants. This is also supported by the observation that a synthetic antimetabolite of methionine, sulphoximine creates a chlorotic halo similar to the *P. tabaci* toxin. Since the toxic effect is not specific, the question as to which factors are responsible for the specificity of the pathogen remains unanswered.

A very interesting problem of phytobacteriology is that connected with tumours. The most characteristic plant tumour is caused by *A. tumefaciens*. Braun (1962) studied the problem more intensely when he found an analogy in many respects between plant and animal cancers. Moreover, it was proved that the cancerous growth free of bacteria can be planted in healthy plant parts. The bacteria causing tumours, the host-parasite relationship, and the numerous consequences of the host's reactions were investigated (cf. Wood, 1967). As opposed to this, the role of *A. tumefaciens* in the induction of the normal host cells into the tumorous cells is still not entirely known. According to the Braun school, tumour formation has two phases: (1) The host cells changing into tumorous cells (2) stimulation of the tumorous cells with indole acetic acid and other growth factors which lead to the process of tumour formation. Researchers are of the opinion that the change of such pathologic plant cells originates by upsetting the balance of numerous physiological factors. This is induced by the "tumour inducing principle", which is elaborated by *A. tumefaciens*. This so-called "tumour inducing principle" (TIP) which is uncharacterized as yet, thus transforms the cells. It has been found, that only those host cells which have been wounded ("conditioning") can be transformed into tumour cells. The tumorous cells differ from the healthy ones in that their growth is not regulated, in other words, they are able to grow unlimitedly. Tissues formed from such cells are not under the control of the organism therefore, when tumorous cells are inoculated into healthy hosts or on tissue culture medium these tissues develop abnormally. This second phase is independent of the pathogenic bacteria. Here the presence of abundant auxins and other growth regulating substances is necessary for the develop-

Fig. 51. Effect of "wild-fire" toxin on tobacco leaf caused by *Pseudomonas tabaci*

ment of the tumour (the continuous cell multiplication and growth). There is a suitable balance in the normal plant metabolism between auxin production and decomposition. In tumourous tissues the balance is upset.

Enzymes causing soft rot

The plant cells are glued together by a pectin substance, the so-called protopectin, which is insoluble in water. One of the groups of phytopathogens decomposes the pectin substances between the cells with enzymes. This causes the disintegration of the cells and tissues. It has been determined that the *E. caratovora* strains causing soft rot produce pectin methyl esterase and polygalacturonase. These, in a way still not clarified completely, are able to decompose the pectin substance which normally keeps the cell walls together. The polygalacturonase enzyme is present in all bacteria causing soft rot, including many non-phytopathogenic bacteria. The other enzyme, the pectinmethylesterase, is only produced by some phytopathogens. Therefore, at the present time the unique role of the pectin decomposing enzymes in bacterial pathogenicity is not certain.

Concerning our knowledge of bacterial products, we may say in summarizing that these materials can be formed and be effective only if the pathogen has avoided the plant's protective reaction and in this way insure its multiplication in the host plant.

As mentioned earlier, the induction of the hypersensitive reaction is a general ability of phytopathogenic bacteria in incompatible plants. Disease in the host-parasite relationship can only develop, therefore, in the extraordinary case where the pathogen is unable to promote a hypersensitive reaction in the host. In this case the pathogen in the plant may increase and during multiplication produce the secondary factors, the different enzymes and toxins, until finally the typical symptoms of disease are developed. In the opposite case the hypersensitive reaction comes about and inhibits further multiplication of the pathogen, and in this way the secondary factors – enzymes and toxins – cannot be formed and typical symptoms of the disease do not appear.

Our present knowledge on the factors of pathogenicity of phytopathogenic bacteria may be summarized as follows:

	Infection
	↓
Primary factor (specific)	Absence of the inducing ability of the protective reaction
	↓
Secondary factors (unspecific)	Formation of toxins and enzymes
	↓
	Development of the disease

It should also be mentioned that the primary factor (the absence of the inducing ability of the hypersensitive reaction) is specific, that is, the hypersensitive reaction does not take place in the case of a compatible host-parasite relationship; while the secondary factors are unspecific (the toxins and enzymes) because their effect is not host specific.

BACTERIAL CELL NUMBER DETERMINATION FROM THE LIVING PLANT

It often happens that the process of bacterial multiplication in the plant has to be followed (Fig. 52). The multiplication of the bacterium in plants may be used as an indicator in studying the host-parasite relationship. The plate-count technique described earlier may be used for this purpose (see page 165). First any contaminations on the surface of the leaf blade are washed with running water, and then leaf discs (about 1 cm^2) are cut with a cork borer. These are placed in a mortar and with the help of a little sterile sand are macerated. Towards the end

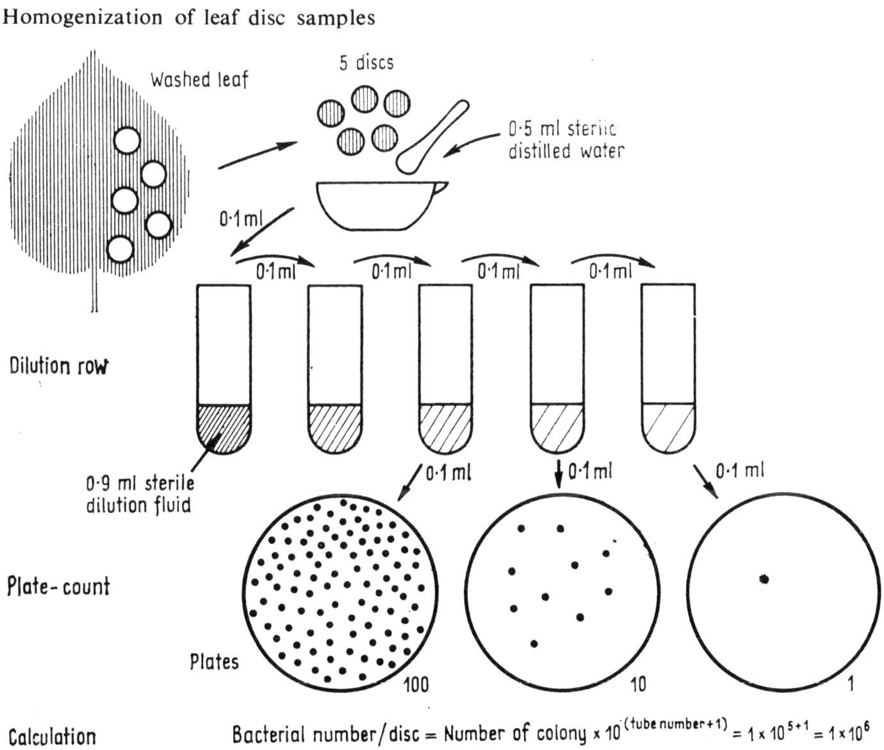

Fig. 52. Steps for determination of bacterial number from the infected leaf discs

of maceration 0·1 ml distilled water per leaf discs are added to the mortar. From the suspension prepared in this way a tenfold dilution series is set up. With the plate-count techniques (see page 165) the bacterial cell number of 1 cm^2 surface of the leaf can be determined on the basis of samples taken after inoculation at different times, and the curve of bacterial multiplication in the leaf tissue can be obtained.

THE POSSIBILITIES OF PLANT INFILTRATION

Most pathogenic bacteria multiply in the intercellular spaces of plants. In the investigation of the host-parasite relationship it is necessary to introduce a determined number of bacterial cells into the intercellular spaces. There are three possibilities for this: (a) infiltration by vacuum, (b) high pressure spraying and (c) injection. In the case of (a) intact plants cannot be used, but only cut plant parts.

On infiltration by vacuum the leaf or leaf part to be infiltrated should be placed in a vacuum glass containing the bacterial suspension. The air from this vacuum glass should be drawn out until bubbles emerge from the stomata of the plant part. With the slow replacement of the normal air pressure the bacterial suspension streams in through the stomata in place of the air which has left the intercellular spaces. If the plant part does not remain submerged in the suspension, the infiltration will not be perfect.

In the case of (b) the bacterial suspension is applied on the under surface of the leaf. As a result of pressure the bacterial suspension is pressed through the stomata into the intercellular spaces. For a detailed description see page 157.

Injection-infiltration method (c) has been described on page 159. In this case intact plants (tomato, tobacco, pepper or cotton leaves) can be used.

The further multiplication of the bacteria introduced into the tissues by means of high pressure or injection may be stopped at any time. This may be done by re-infiltration of different antibiotics. In the case of phytopathogenic bacteria the introduction of a solution containing 300–500 ppm streptomycin has proved to be the most successful. By changing the concentration of streptomycin there is a possibility of reaching a static effect on the bacterium.

CULTURE FILTRATE PREPARATION, FILTRATION, CENTRIFUGATION, AND WASHING OF BACTERIAL CELLS

If we wish to investigate the bacterial metabolites free of cells which have diffused into the medium, or if we wish to sterilize the liquid without heat, filtration is applied.

In our laboratory a *glass filter* (G-5 filter) is used (Fig. 53), but different asbestos, collodium or other filtering surfaces may also be used. The fluid to be filtered or the bacterial suspension is pressed through the filter either with pressure or sucked through with a vacuum. The disadvantage of the latter, although it is more often used, is that one must take care to prevent the filtered liquid from forming bubbles. Special care must be taken to prevent the microorganisms from being sucked back (as a consequence of the vacuum) into the filtered substance after filtering is finished. By placing a filter tube containing cotton wool before the glass (Fig. 54) this danger can be reduced to a minimum. In this case the glass filter and the tube with cotton must be sterilized in an autoclave at 2 atm for a few min.

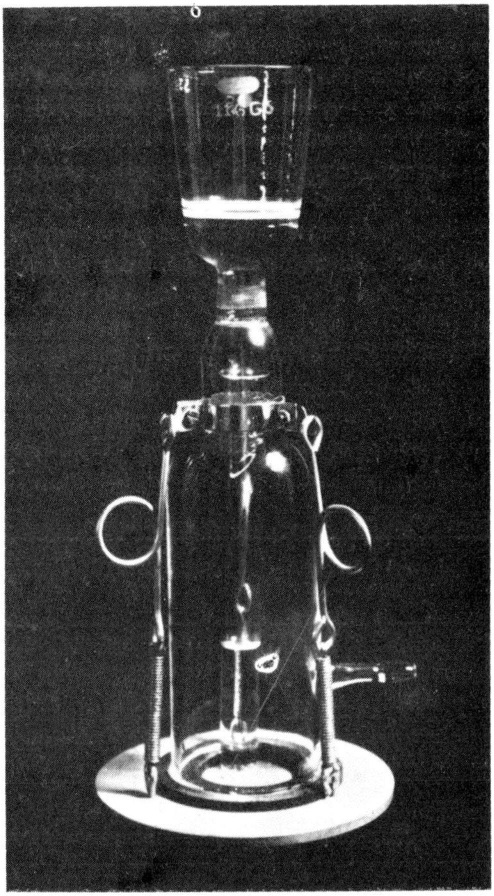

Fig. 53. Bacteriological filter of the sintered-glass type

It is advantageous in the preparation of culture filtrates to centrifuge the bacterial suspension before and then filter only the supernatant. In this way the filter surface will not become blocked. The glass filter should be washed after use.

Cleaning of the bacterial glass filter: Pour some hot H_2SO_4 with some KNO_3 into the glass filter 2–3 mm high up. Make some vacuum until 2–3 drops of H_2SO_4 come through the filter. After this, leave it to penetrate for a night. On the next day the rest of the acid is drawn through the filter by vacuum. After this treatment the glass filter is washed through with distilled water three times. It is necessary to check the acidity of the glass filter after cleaning. Mix up some $BaCl_2$ with a little distilled water and drop in a few drops of the washing water coming through the glass filter. If a strong precipitation appears you have to wash through the filter with distilled water once more until the precipitation disappears.

It is often necessary to separate the bacterial cells from the bacterial products which have diffused into the medium. The bacterial cells are separated by centrifugation. Depending on the bacterial size, at 6000 rev/min speed, the phytopathogenic bacteria settle in 20–30 min. However, it should be remarked that the supernatant never becomes sterile with centrifuging.

So as to be entirely free of media adhering to the bacterial cells, the bacterial cells have to be washed many times. The centrifuged bacterial cells are resus-

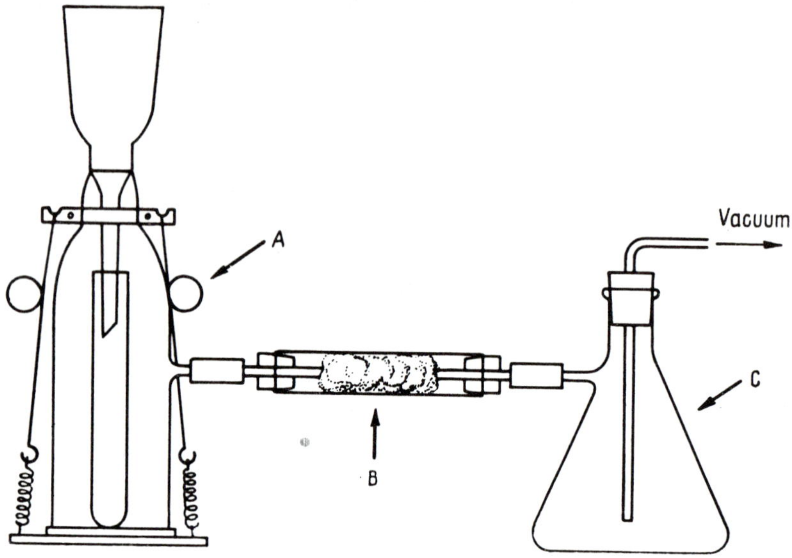

Fig. 54. Junction of the bacteriological filter and vacuum. A) glass filter, B) sterile cotton in a glass tube, C) vacuum flask

pended and again centrifuged either in distilled water, physiological saline solution, or in buffer solution. This procedure is repeated 2–3 times as is found necessary.

OBTAINING FLUID FROM THE INTERCELLULAR SPACES OF THE PLANT

Bacteria that cause leaf spotting diseases multiply within the intercellular spaces of the parenchymatous tissue of leaves and not within the host cells per se. Presumably, the intercellular fluid provides a medium for growth of the pathogen. Also, it may be assumed that toxins, enzymes, and other bacterial products that play a role in pathogenesis are formed by bacteria in these places, and that some of the defence reactions of the host are activated against the bacteria here as well.

The juices of the leaf as obtained by freezing, pressure, or homogenization do not possess the same biological and biochemical value as the intercellular fluid. Therefore, to study precisely those events that occur or are engendered by

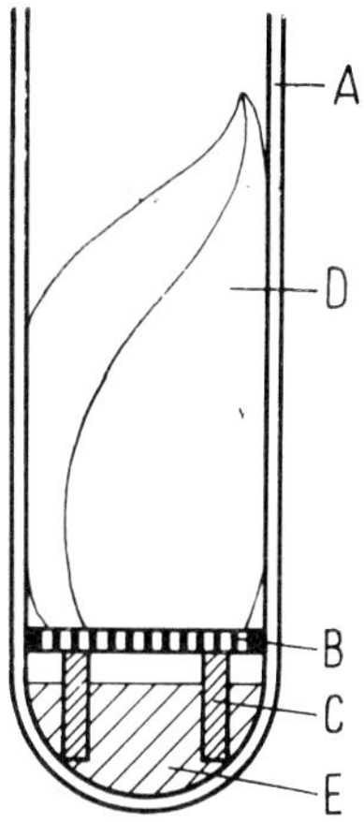

Fig. 55. Apparatus for obtaining fluid from the intercellular spaces of foliage: A) centrifuge tube, B) silver screen, C) glass cylinder, D) infiltrated leaf roll, E) fluid from the intercellular spaces after centrifugation

phytobacterial pathogens in the intercellular spaces, the following method was developed.

The basis for this simple method is that the intercellular fluid may be pulled out of the leaves by centrifugation after the leaves have been infiltrated with distilled water or a buffer solution. Considerable dilution of the endogenous intercellular fluid may result from this procedure.

The method reported by Klement (1965) was developed with tobacco plants (*Nicotiana tabacum*), although other plants may be similarly treated to obtain intercellular fluids.

The intercellular spaces of leaves still attached to the tobacco plant may be flooded with the test fluid with a fine hypodermic needle (see page 159). Detached leaves and leaf parts may be infiltrated under vacuum, then water on the surface of the leaves is removed by blotting with filter paper.

To facilitate withdrawal of the fluid from the intercellular spaces, leaves are cut lengthwise, parallel to the main veins. The main vein is generally removed, which exposes the ends of numerous large lateral veins.

The filtrated leaves are rolled (Fig. 55) and placed in a 50 ml centrifuge tube over a nonreactive disc of silver screening somewhat smaller in diameter than the centrifuge tube. The leaves and screening are supported by a plexiglass cylinder 10 × 18 mm, which prevents the leaf and screening from moving into the centrifugal fluid. The leaves are centrifuged at 1120 RCF (relative centrifugal force) for 10 min. The fluid is driven from the intercellular spaces and collects in the bottom of the centrifuge tube. With this method 1–1·5 ml fluid from one tobacco leaf 200 cm^2 in dimension are obtained.

The fluid from the intercellular spaces of a tobacco leaf is clear and slightly yellowish in colour. A slight precipitation occurs on standing and precipitation is strong when the fluid is heated at 60 °C for 10 min.

CHAPTER XI

SPREAD OF THE PATHOGEN IN THE FIELD

Generally speaking, the appearance of the disease in the field has two sources, the infected plant material and the infested soil (Fig. 56).

The pathogen makes its way to the field with the infected seed, infected seedling, or shoot. In the case of the seed being infected, the pathogen shelters itself either by adhering to the seed coat (e.g., *P. pisi*) or lodging among the seed hairs (*X. malvacearum*) or under the seed coat (*C. flaccumfaciens*). Information on the location of the bacterium is important in choosing the appropriate method of seed dressing as well as the compound for the seed treatment. For example, there is no use in treating bean seeds with non-diffusive chemicals because these cannot destroy the bacteria sheltered under the seed coat.

With diseases of woody plants, the bacteria overwinter in dead infected shoots and branches, from where they emerge in spring in the form of bacterial slime. The yearly process can be easily followed in the case of *P. mors-prunorum* (see later on page 297 and Fig. 59) *P. syringae* or *E. amylovora*.

The appearance of the disease in spring in many cases is begun by bacteria originating from contaminated frames of seedling bed, from contaminated manure or from seeds infected with bacteria. It would be cheaper and more successful to control the disease if the prevention is begun in the seedling bed and if care is taken that only infection free seedlings get into the field.

The disease in the field can also originate from the soil which has been contaminated with pathogenic bacteria in previous years. Numerous plant pathogenic bacteria are able to remain in the soil for longer periods, sometimes many years, without multiplying. They reach the soil by means of withered plant parts which have fallen on or into the soil during the previous year. The observance of crop rotation has an important role in the case of infested soils.

The disease developed from the seedling or the young plant infected by the soil may spread to the neighbouring plants. Generally the epidemic begins from such small centres of infection and the disease spreads over the whole field under favourable conditions (Fig. 56).

If the disease centre is present, the further rapid spread of the disease is aided by the environmental conditions. Thus rainy, humid, warm weather has an important role in the epidemic spread of the bacterial diseases. However, for the development of other bacterioses, the grease spot of the pea (*P. pisi*) and bacteriosis of fruit trees (*P. syringae*) cool, rainy weather is favourable. The pathogenic spread is aided by strong winds combined with rainfall. It has been observed that tobacco "wildfire" disease spreads more rapidly in the direction of the wind. The rain drops rebounding from the soil may carry the pathogens in the soil onto the leaves.

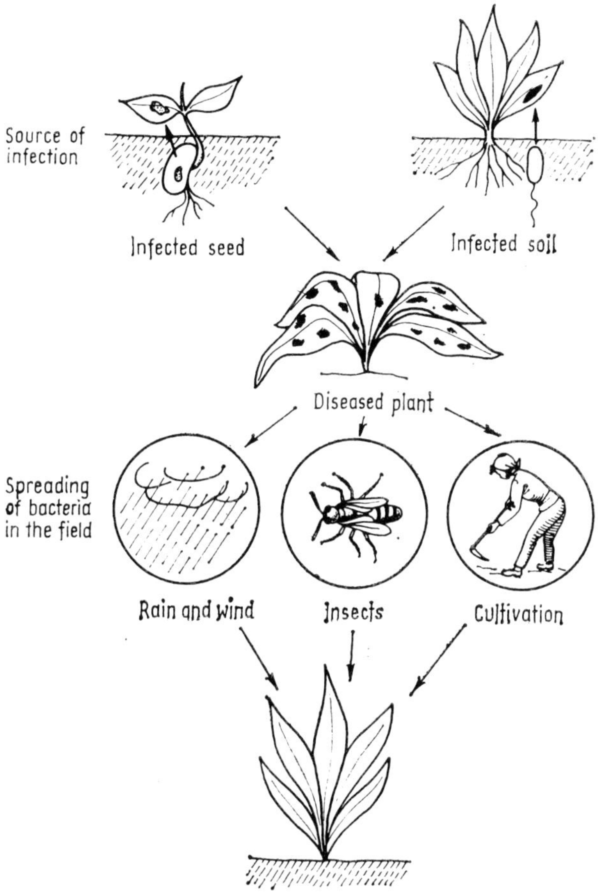

Fig. 56. Source of infection and spreading of bacteria in the field

ARTIFICIALLY INDUCED EPIDEMICS IN THE FIELD

In the course of experiments it is often necessary to carry out inoculation in the plant beds, or in the field. Since the plant pathogenic bacteria are only able to infect the plants in humid conditions, great attention should be paid to ensure the appropriate external conditions and the choice of time suitable for inoculation.

If we wish to observe the susceptibility of plants in the field, then we cannot trust natural infection but must create an epidemic artificially. This may be done in two ways. Either infected seedlings are put among healthy plants or the pathogen is sprayed onto the plants. The former is the better method since the infected seedling, during the course of cultivation, is constantly forming a centre of infection. Artificially spraying the bacteria depends for it success upon the prevailing weather conditions during and after the spraying.

For inoculation in the seedling-bed or in the field, large amounts of pathogens are required. The bacteria are cultivated in Roux bottles or Kolle flasks. Media containing 2 per cent agar is poured into the flasks to make a $^1/_2$ cm layer on the bottom of the flask. After sterilization and after the solidification of the nutrient agar, the dishes are inoculated with a bacterial suspension washed from a 24 hr agar slant culture, using 2 ml in each vessel. The suspension should be spread evenly on the surface of the medium. After a 24–48 hr incubation the bacteria are washed from the surface of the agar plate with water of 25–30 °C and adjusted approximately to 10^7 cell/ml. The suspension is sprayed onto the under surface of the leaves with a clean sprayer free of chemicals.

Four to 5 hr before the seedling bed inoculation, the plants are sprinkled with water and the cover clip is slightly shaded. As far as possible the under surface of the leaves should be sprayed and afterwards the seedlings should be slightly smoothed over. With this method the plants become clustered and therefore the leaves which have not been sprayed also come in contact with the bacteria. After inoculation for at least 24 hr, the spray-like watering of the seedlings is repeated every 2–3 hr to insure appropriate humidity. After inoculation it is again advisable to use a slight shading.

When inoculating the plants in the experimental field, it is important to choose a time suitable for the inoculation. The inoculation should be carried out in the early hours of the morning when the plants are still covered with mildew and the stomata are open. If this is impossible then it can also be done in gray or cloudy weather. The plants should be sprayed from below with the bacterial suspension. If dry weather follows the inoculation, the plants should be frequently sprayed with water.

In the case of vascular bundle disease, the plants are pricked between the cotyledon and first leaf with a needle dipped in a bacterial suspension. To avoid having to dip the needle in the bacterial suspension after every prick, it is advisable to place a needle in the middle of the brush (Hedges, 1926) dipped in the bacterial suspension, which insures the constant contact of the needle with the bacteria (Fig. 26).

CHAPTER XII

CONTROL OF PLANT PATHOGENIC BACTERIA

In controlling bacterial plant diseases, we are faced with the same difficulties as in viral diseases. The bacteria multiplying inside the tissues can hardly be reached with chemicals or antibacterial substances which have been used until now. When the disease occurs in clusters, overcoming it is in most cases unsuccessful. Therefore, we must not depend entirely on defensive substances but preventive measures should be taken so that the pathogen can be kept far from the plant.

In some closed areas the pathogen may be kept away by strict quarantine regulations. In this way *P. tabaci*, the pathogen which had spread all over the world causing tobacco "wildfire" disease, was successfully kept out of Australia for a long time.

In preventing diseases from spreading by seed, it is important to be sure that the seed to be planted is healthy. The bacterial spot disease of the bean was eradicated in the U.S.A. by growing the seed to be planted in dry areas of California. With this kind of control the damages caused by the pathogen have been reduced to practically zero.

With seedlings the disease is often taken into the field in the infected plants. The disease spreads in the seedling bed because of humid, warm conditions. It is for this reason that we must constantly watch the seedlings and if the slightest sign of the disease is noticed on the under surface of the leaf, then the seedlings should not be planted. It is best to constantly treat the plants in the seedling beds with bactericides. The application of streptomycin in the field is not economical but through its use a large number of seedlings can be relatively protected. The spraying should be repeated every 5 days. Seeds coming from infected areas should not be used because though the seed appears healthy externally, inside the seed or on the surface of the seed the bacterium may begin infection.

Among the generally used plant protective agents—especially in the case of seed treatment—are substances containing 0·001 per cent mercuric chloride and, for preventive spraying, different copper preparations. It should be noted however, that contradictory results are often published in the literature in connection with these agents (Dimond, Stoddard and Chapman, 1952).

The most generally used antibiotic against bacterial diseases is streptomycin. Moreover, a mixture of streptomycin and oxytetracycline (terramycin) has also been used successfully A considerable advantage of the antibiotics as opposed to other plant protective agents is that they have a good absorbing ability and they translocate in the tissue. These advantageous characteristics are of significance particularly with bacterial diseases, since the bacteria which have increased in the tissues cannot be reached with the generally used plant protective agents which lack an absorbing capacity. The streptomycin of 200–400 ppm concentration does not kill the bacteria but only inhibits their multiplication or is bacteriostatic. Some days after spraying, the antibiotic content in the plants gradually decreases. Therefore, repeated spraying is necessary. Generally the most suitable schedule is weekly spraying. When spraying is neglected for longer than 10 days, a successful result cannot be insured.

There are also some disadvantages in applying streptomycin. Even more publications appear which describe the appearance of streptomycin resistant strains. In the course of our experiments we experienced that for one bacterium, every 10–1000 millionth cell is resistant to streptomycin or operates with spontaneous resistances. The progeny of these cells will all be resistant.

It can be imagined therefore, that the streptomycin resistant strains also develop and multiply in nature. Antibiotic mixtures such as agrimycin containing 10 per cent terramycin in addition to the streptomycin may be applied.

APPENDIX

SOME EXAMPLES OF BACTERIAL DISEASES

In this section we must restrict ourselves to discussing one or two characteristic diseases from among the five types of symptoms mentioned in the general part (spot diseases, die-back, wilt, soft rot and hypertrophy). This mode of discussion was chosen because the isolation and technique of artificial inoculation varies with the disease type. We shall introduce important diseases in which the applied methods may be used in other bacterial diseases.

SPOT DISEASES

BACTERIAL BLIGHT OF TOMATO AND PEPPER (PAPRIKA) (*X. VESICATORIA*)

The bacterial spot of tomato and paprika is caused by *Xanthomonas vesicatoria* (Doidge) Dowson.

The symptoms on the paprika differ from those on the tomato. In the case of *paprika*, mainly the leaves and the peduncle become infected; the fruit remains practically symptom-free. First small, irregular, often slightly swollen, dark green, moist spots appear on the underside of the leaf, which later spread to become angular spots. The middle of the spot becomes dafed and dries out and is surrounded by a narrow dark-brown border (Fig. 57). The spotted leaves turn increasingly yellow and then fall off. Thus strongly infected plants become defoliated. The pathogen also infects the peduncle of the flower. The flowers and the young fruits together with the attacked peduncles fall off. A significant part of the damage occurs because the paprika plants which have lost their leaves shed most of their flowers and in this way their yield is greatly reduced.

The disease occurs in the leaves, shoots and fruits of the *tomato*. Damage is most significant on the fruit, where brown spots appear. The disease is quite obvious whether it is on green or on red fruits, in spite of the fact that the pathogen can only infect young fruit. On green fruits, first tiny dark green, brown or black round bulging spots appear (Fig. 58). Later these spread and coalesce. Due to the attacked and lacerated epidermis and cuticle, the developing fruit may crack, providing the opportunity for attack by secondary organisms. Such fruits may rot while still on the plant.

Fig. 57. Symptom of *Xanthomonas vesicatoria* on pepper leaf

At first small, water soaked, green spots appear on the leaves and then turn brown: they are most obvious on the underside of the leaf. The spots which occur on the stem, the peduncle and the shoots are similar to the leaf spots but a little more elongated. On older stem parts cancerous wounds may arise which, with time, become suberized. Attack of the flower peduncle frequently results in loss of the flowers.

The pathogen reaches the intercellular spaces through the stomata of the leaves. The multiplying bacteria cause blistering which in time cracks and the bacterium again reaches the surface. From here the splashing rain and insects convey the bacterium to a healthy plant.

The *pathogen* is a rod shaped 1·0–1·5 × 0·6–0·7 micron in size with a polar flagellum. It is Gram-negative. Colonies appear on the third day on nutrient broth agar medium and are pale yellow, lens-shaped. The gelatin is moderately liquefied. Peptonization begins in litmus milk producing a weakly alkaline reaction without coagulation. All strains produce considerable amounts of hydrogen sulphide. Ammonia forms in the cultures in quantities varying according to the strains. But indole formation cannot be demonstrated. The nitrates are not reduced to nitrite. From among the different carbon compounds acid is formed without gas from glucose, galactose, levulose, saccharose, raffinose, glycerine, maltose and mannose. Acid formation does not occur in the media containing dulcite, mannite, xylose and arabinose. Strains originating from tomato and paprika behave differently on nutrient agar containing soluble starch. All strains isolated from paprika will not hydrolyze starch. One group of isolates of tomato does not hydrolize starch, while the other does strongly. Therefore, this may be used for differentiating the strains of *X. vesicatoria* isolated from paprika and tomato.

Fig. 58. Development of the symptom of bacterial blight of tomato fruits by *Xanthomonas vesicatoria*

Pathological, biochemical, serological and phage sensitivity tests (Lovrekovich and Klement, 1965) have proved that *X. vesicatoria* is not a uniform species. Three biotypes can be distinguished. One type only infects the paprika, the other only the tomato, and the third attacks both (Klement and Goodman, unpublished work). Lovrekovich and Klement (1965) as well as O'Brien and co-workers (1967) proved that the strains isolated from the paprika also differ in antigenic structure from tomato isolations. On the basis of phage sensitivity, the three different biotypes were distinguished (Lovrekovich and Klement, 1965). In contrast to these Dye, Starr and Stolp (1964) were unable to differentiate the isolates of paprika and tomato.

Isolation of the pathogen

The pathogen may be successfully isolated from leaf spots or fruit spots. As far as possible young, still grey-green spots should be used in isolations. It is not advisable to sterilize the surface of the leaf before, because there is a danger that the pathogen may also be killed. However, it is necessary to rinse the surface of the leaf with running water. The cut out leaf spot together with a healthy tissue part is dropped on a sterile slide and macearated in sterile water. A wire loop is used to streak liquid from the macerated tissue part onto the broth agar plate (see page

140). The yellow colonies develop on the third day in an incubator at 25–28 °C. Strains isolated from the paprika form colonies similar to those formed by isolates of tomato. Thus differentiating them on the basis of colony morphology is impossible.

Artificial inoculation

Paprika and tomato may be successfully inoculated in the greenhouse and the field. For greenhouse inoculation we may follow two procedures. The leaves are infiltrated either with a syringe or by spraying. In both cases we have to be careful because the paprika easily sheds its leaves. If we choose infiltration, then the leaves of the paprika, as of the tomato, may be well infiltrated with an injection infiltration technique (see page 159). In this case we may use a bacterial suspension of below 10^6 cell/ml because if greater amounts are used we may count on atypical symptoms.

The other method of artificial inoculation is low pressure spraying of the underside of the leaves with a bacterial suspension. For spraying, the inoculation suspension must be higher than 10^7 cell/ml. For artificial inoculation in the greenhouse, the plants should be placed in a moist chamber for 24 hr before and after inoculation. The spots on the leaves appear earlier and in greater numbers if we can maintain the temperature in the glasshouse at 28–30 °C. The spots become obvious in about 6–10 days depending on the temperature.

For the field inoculation, two procedures may be followed. An epidemic may be achieved either with the planting of infected seedlings or the spraying of the pathogen. Both methods are described in the general part (see page 207).

Sowell (1960) found a *Capsicum annuum* line resistant to *X. vesicatoria*. According to Cook and Stall's (1963) investigations, this resistance can be attributed to a single dominant allele (P.I. 163/92). Stall and Cook (1966) have further shown that resistance in these plants can be explained by the development of the hypersensitive reaction.

HALO BLIGHT OF BEANS

(*P. PHASEOLICOLA*)

The symptoms of halo blight disease caused by *Pseudomonas phaseolicola* (Burkholder) Dowson are exactly the same as those of *Xanthomonas phaseoli* (E. F. Smith) Dowson and *Xanthomonas phaseoli* var. *fuscans* Burkholder (Starr et Burkholder) and somewhat like those of *Corynebacterium flaccumfaciens* (Hedges) Dowson. In the field, their discrimination is impossible. It can be assumed that in all probability the disease extends throughout all countries where beans are cultivated (Zaumeyer and Thomas, 1957).

This is a leaf and pod disease. The lesions on the leaflets are of different sizes and darker in colour than the basic colour of the leaves; they are water soaked, mostly irregular in shape with a chlorotic halo (Fig. 8). In time the spots spread and become dry and fragile. If one or more such lesions coalesce into a larger spot, the leaf turns yellow and shrivels. Leaves which have been strongly infected become torn by pelting rain or wind. If the infection reaches young developing leaves, they will grow one sidedly.

The most serious damage of bacterial spotting is to be found on pods where first water soaked grease spots can be seen around the stomata. These later develop into bigger grease spots which are generally circular. The spots initially are grey-green; later their edges become slightly reddish. A yellowish-white bacterial exudate oozes to the surface of the spots. This exudate dries out to a shiny mass in dry weather. When the lesions get older they lose their moist characteristic, dry out and become slightly hollow. But this is never so deep as in the case of anthrachnosis.

The pathogen may also attack the parenchyma and vascular bundles, but spotting is more characteristic of the disease. In the case of systemic infection, the higher, younger leaflets form mosaic-like symptoms and develop asymmetrically. The pathogen is not present in the mosaic-like leaflets (Waitz and Schwartz, 1956); thus the symptom arises rather from the toxic effect.

Goss (1940) investigated the effect of temperature changes and determined that the pathogenic symptoms develop in 6–10 days at 24–28 °C, 2–3 days later at lower temperatures. The typical halo blight develops mainly at 20 °C and below. At 24 °C a still weak chlorotic halo appears. Between 28–32 °C the halo disappears, but many more spots arise than at lower temperatures.

The disease is spread by seeds. The symptoms are also characteristic on the seed. The cotyledon and the first leaves are infected from the bacterial exudate which is sheltered under the seed coat. The bacterium can spread from plant to plant from an infected centre in different ways. Rain combined with wind dissolves the bacterial slime on the surface of the plant and washes it onto other parts of the plant or onto neighbouring plants. The bacteria are also spread further by insects. The pathogen adhering to the body of insects is conveyed from one plant to another.

The plant is most susceptible when the transport of nutrients is the most intensive. Gloyer (1924) determined two susceptible phases of plant growth. The first one is from germination to the appearance of 1 or 2 trifoliates; the second from pod formation to ripening.

Under natural conditions *P. phaseolicola* penetrates the tissues through the stomata of the leaf. It floods the intercellular spaces and causes partial dissolution of the middle lamella.[1] The toxic production of *P. phaeseolicola* was observed by Skoog (1952). Toxin formed in a broth culture only at temperatures below 20 °C.

The pathogen *P. phaseolicola* is a 1·3–3·2 × 0·4–0·6μ-sized polar flagellated rod-shaped bacterium. It is Gram-negative without spores. On nutrient broth agar the colonies develop on the third day. The colonies are white, opalescent, transparent, regular lens form, shiny and smooth. In contrast to other authors, we could not observe green fluorescense staining in the medium. According to Burkholder it does not liquefy gelatin, but according to others and our investigations, it does so slowly. It does not hydrolyze starch. Litmus milk turns an intensive blue without curdling or peptonization. In peptone water, hydrogen sulphide does not form, but a very small amount of ammonia is produced. It does not reduce nitrates. It forms acid in a synthetic basic medium from glucose, galactose, arabinose, xylose, raffinose, saccharose, mannose and fructose. Acid does not form from lactose, maltose, dextrin, dulcite, mannitol, asparagine, glycerine and salicin.

Differentiation of the four bean pathogens

As mentioned earlier, the four bacterial pathogens *P. phaseolicola*, *X. phaseoli*, *X. phaseoli* var. *fuscans* and *C. flaccumfaciens* cause pathologically similar symptoms. This is especially the case for the first three as whole differentiation in the field is impossible. Since we cannot conclude from the pathological symptoms as to the pathogen it is necessary to do selective investigations as followss

Outline for differentiating the bacterial pathogens of beans

Isolated bacterium

- Colour of colony: yellow
  - Browns the nutrient medium
Xanthomonas phaseoli var. *fuscans*
  - Colour of medium unchanged
    - Gram-positive
Corynebacterium flaccumfaciens
    - Gram-negative
Xanthomonas phaseoli
- Colour of colony: white
Pseudomonas phaselicola

The pathogen is isolated in the same way as is *X. vesicatoria*. We may also use Groth's (1965) method as follows: The diseased leaf is held above the nutrient agar. A fresh leaf spot is pierced with a pointed needle so that the tip of the needle touches the surface of the agar plate. With this method, the needle touches only a few saprophytic microorganisms which are on the surface of the leaf. After they are pricked, the bacteria are streaked over the surface of the agar plate with a sterile glass rod. The colonies develop into well distinguishable forms on the third day. We can prove the pathogenicity of the isolated strains by infiltration of the bacterial suspension (10^{7-8} cell/ml) into the tobacco leaf. (See general part, page 153) A hypersensitive reaction (necrosis) of the tobacco leaf appears 24 hr after injection.

An improved rapid technique for the pathogenicity test has been offered by Fenwick and Guthrie (1969). Beans are inoculated in the staple (crookneck) stage of growth. At this stage beans are placed on paper towels on a greenhouse bench and inoculated by (a) stabbing twice through the cotyledons and hypocotyl and flooding with the broth culture of the bacterium using a sterile pipette; or (b) by stabbing the seedlings with a dissecting needle immersed into the bacteria growing on agar plates. The seedlings then are placed on damp soil in clay pots and covered with soil. Upon re-emergence (3–5 days) the plants exhibit seed-borne halo blight symptoms in hundred per cent.

The virulence of the pathogen may vary according to strains. Small and Worley, (1956) suggest a differentiating medium with which the virulence of the pathogen can be distinguished. The virulent bacterial colonies are red colour on potato-glucose agar containing 0·005 per cent 2,3,5-triphenyl-tetrazolium chloride while the weakly virulent ones are lightly coloured or remain white.

Two biotypes of *P. phaseolicola* were found and called race 1 and race 2. The Red Mexican U.I.3. cultivar is susceptible to race 2, while race 1 infects all other cultivars. Resistance of the Red Mexican to race 1 is due to a monogenic dominant gene.

Schuster (1955) offered an inoculation method suitable for investigations of resistance to bacterial spot of the bean. Plants growing at 21 °C were inoculated with high pressure spraying equipment in 2–3 leaf stages. The concentration of the inoculum was adjusted to about 8×10^6 cell/ml. At this bacterial concentration, the typical water soaked lesions and the chlorotic halo occurred around the lesions only on the susceptible cultivars. It should be noted, however, that it is more advantageous to work with 10^6 or 5×10^6 cell/ml inoculum. With the use of such an inoculum, the hypersensitive reaction (brown necrosis) does not appear on resistant cultivars. The symptoms on the susceptible (Red Kidney) cultivar appear after the second day, especially on younger plants. The more appropriate time for evaluation of the results is 10–14 days after inoculation.

DIE-BACK OF SHOOTS

BACTERIAL BLOSSOM BLIGHT AND BACTERIAL CANKER (*P. SYRINGAE* AND *P. MORS-PRUNORUM*)

Two pathogens are responsible for the bacterial blossom blight and canker of fruit trees. As summarized by Crosse (1966), bacterial blossom blight (*Pseudomonas syringae* van Hall) has to be distinguished from the bacterial canker whose pathogen is *P. mors-prunorum* Wormald. Although this separation is not accepted by every research worker, Crosse and co-workers, concluded that what we are concerned with is two well distinguishable bacterial species or pathotypes.

The *Pseudomonas syringae*, besides the stone fruit trees attacks pear, lemon, lilac also attacks non-woody plants. Reliable identification of bacterial strains of *P. syringae* originating from herbaceous plants has not yet been achieved. The *P. mors-prunorum* has, in the first place, spread mainly in Great Britain where significant damage is caused on stone-fruit trees, especially cherry trees.

Symptoms

The bacterial blossom blight and canker of fruit trees is a disease which proceeds in a typical cycle. During winter the bacterium multiplies in the woody tissues of shoots and branches while in the summer phase the leaves and other green tissues become infected. During the summer months the shoots and branches become resistant. This cyclical process is seen in Fig. 59. In autumn the pathogen penetrates the tissues through the leaf scar at the spot of the fallen petiole (Crosse, 1956). In the winter months, until the temperature is above 1–2 °C, the bacterium is capable of weakly multiplying. In spring, before and during budding, the pathogen commences sporadic multiplication and the bacterium breaks to the surface from

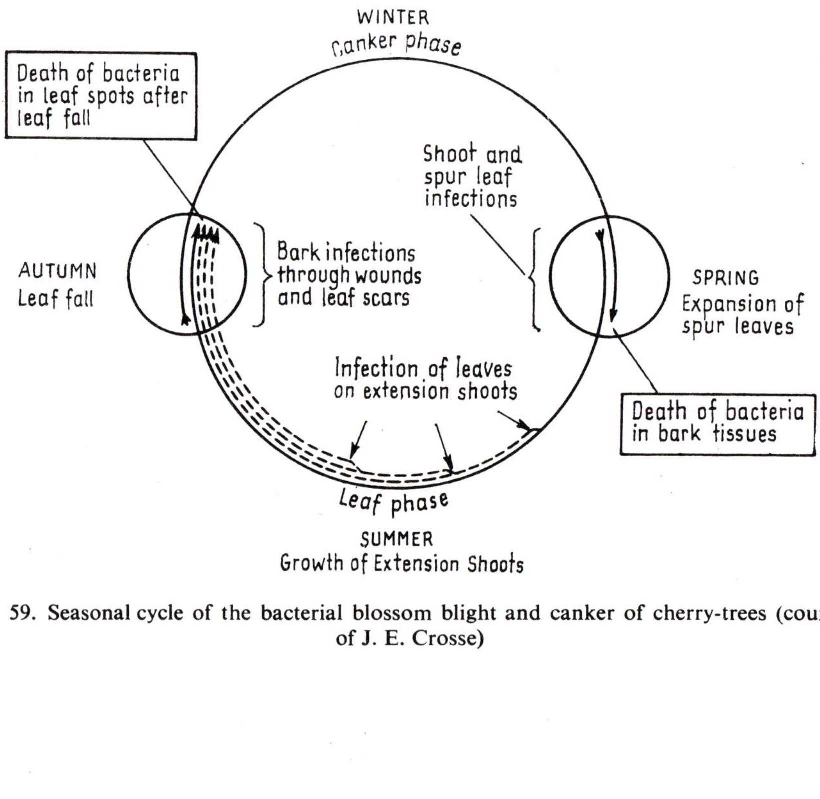

Fig. 59. Seasonal cycle of the bacterial blossom blight and canker of cherry-trees (courtesy of J. E. Crosse)

the infected buds or from blight formed in winter and early spring. Tiny red-brown spots which gradually fall out, are formed on newly developed leaves.

The *P. syringae* causes browning and blossom blight in the pear and cherry tree and also causes spotting on young fruits. Dark-brown or black blight is formed on the leaves as well as on young fruits. The shoots formed from infected buds gradually die and dry. In summer and autumn the pathogen, as demonstrated by Crosse (1963), English and Davis (1960) as well as Panagopoulos and Crosse, (1964), lives epiphytically on leaves of the pear tree. The pathogen cannot be isolated from the infected branches or the cankerous wounds developed in the summer months. The members of *P. syringae* group cause spot and die-back of many herbaceous plants as well.

Isolation of the pathogen

It is obvious from the previous discussion that the pathogen can be isolated in different ways during different seasons. Generally, the pathogen can be isolated successfully during the spring and autumn months. In spring it can be isolated from the spots formed on young leaves, while in the autumn months after the falling of the foliage, it can be isolated from the freshly inoculated shoots and branches (Wilson, 1936).

Isolation from the leaf spot should be made according to the procedure described in the chapter on spot diseases. However, another procedure has to be followed in isolating the pathogen from shoots and branch parts. It is necessary to sterilize the shoots and surface of branches, which is done by dipping a small piece of inoculated shoot in alcohol (96 per cent) and holding it for a moment above a flame. The isolate is plated by macerating the tissue which has been surface disinfected in the usual way and then streaking it on 5 per cent sucrose nutrient agar (Wormald, 1932). On this medium *P. mors-prunorum* grows in white colonies, while the colonies of *P. syringae* are pale yellow.

The culture characteristics and biochemical properties of *P. mors-prunorum* and *P. syringae* are similar. They do not reduce nitrate to nitrite, cause ammonia formation, and do not form indol or hydrogen sulphide. Starch is weakly or not at all hydrolyzed. Acid is formed without gas from saccharose, dextrose, galactose, levulose, glycerin and mannitol. Acid is not formed from maltose and lactose. *P. mors-prunorum* either does not cause a change in litmus milk, or causes slight curdling. *P. syringae* peptonizes milk and makes it alkaline. Garret, et al. (1966) carried out other differentiating tests for distinguishing the two pathogens.

Differences between P. syringae, and P. mors-prunorum

	P. syringae (pear, lemon, lilac)	*P. mors-prunorum* (plum, cherry)
Longevity on nutrient sucrose agar	8–14 days	2–6 days
Growth in nutrient sucrose broth	yellow	white
Proteolytic activity	+	−
β-Glycosidase activity	+	−
Tyrosinase activity	−	+
Use of *L*-leucine as solo C and N source	+	−
Use of tartaric acid	−	+
Use of lactic acid	+	−
Lesion produced on: pear fruitlets	black, sunken	nil
Cherry fruitlets	black, sunken	brown, small
Lemons	black, sunken	brown superficial very small, superficial

Reaction: + positive, − negative.

Artificial inoculation

The pathogenicity of *P. syringae* and *P. mors-prunorum* can be clearly demonstrated with the tobacco leaf injection infiltration technique. If a higher inoculum than 5×10^6 cell/ml is used, the hypersensitive reaction appears within 10–12 hr (see page 153).

The virulence of *P. syringae* is demonstrated on young, 10–20 cm long, developed peach shoots under greenhouse conditions. About 3 cm below the shoot tip, the stem is pricked with a needle which has been dipped in a bacterial suspension. Symptoms of the die-back of shoots appear 3–5 days after inoculation.

By using the method employed by Crosse and Garrett (1966), we may achieve a successful inoculation with *P. mors-prunorum* in the orchard. In autumn, from the beginning of October to the middle of November, before the fall of the foliage, the leaves are broken off from the one-year-old shoots. The leaf scar petiole is inoculated with 10^8 cell/ml bacterium suspension with a fine brush. The first symptoms in spring appear around the bud. The meristematic tissue around the bud dies. The results of artificial inoculation the following summer are particularly outstanding the following summer when the shoots turn yellow and die-back.

Crosse and Shanmuganathan (1961) carried out inoculation in the winter. They made a V incision on two three-year-old branches of six-year-old trees and smeared them with a bacterial suspension. They immediately stuck down the surface of the wound with a polyethylene strip. The degree of virulence was concluded from the length of die-back arising until April.

SOFT ROTS

BLACKLEG AND SOFT ROT OF POTATO TUBER AND OTHER PLANTS (*E. CAROTOVORA*)

Bacteria belonging to the group *Erwinia* (*Pectobacterium*) *carotovora* (Jones) Holland cause soft rot in numerous plants.

Symptoms

The blackleg of potato is probably the most important disease caused by soft rot bacteria but it may occur on many other herbaceous plants too. A dark brown, black colour appears on the base of the stem. This is from where the name of the disease originates "blackleg". The attacked plants are stunted, the foliage is pale green or yellow and often the upper leaflets curl downwards which resembles the viral leaf curl disease. If the black stem is broken in two, the inside of the stem is also dark brown and soft. Not all stems of the potato stock are infected.

The first appearance of the disease in spring originates from seed or infested soil. During the cultivation period the pathogen spreads from one plant to another through the wounds. The infected plants are vigorous for a long time, however, they produce infected tubers.

The soft rot of tubers may spread to one part of the tuber or to the whole to such an extent that only the corky epidermis keeps the decayed mass together. The infected parts on the cut tuber are at first colourless, however, soon

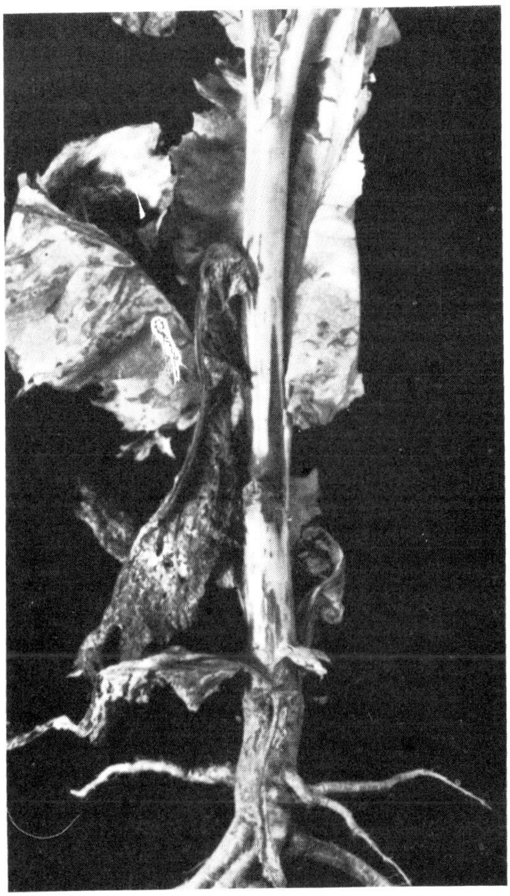

Fig. 60. Soft-rot disease of poppy caused by *Erwinia carotovora*

they turn pink, brown, red or brownish-black in contact with air. Such tubers only have weak odours, however, it often occurs that the tubers decay with odour due to secondary organisms.

The organisms of soft rot not only attack potato but many other herbaceous plants and crops too. We can often see the decay of tobacco, poppy (Fig. 60) and corn stems. During storage the pathogen also spreads on potato tubers, carrot roots, onions and other tuber plants.

Bacteria causing soft rot produce enzymes which decompose the pectic substances. Soft rot is the symptom that results from the destruction of the cellulose of cell walls or of the pectic substances in the wall matrix or in the middle lamella of plant cells. In this way the cells separate and the protoplast dies. The role in pathogenicity played by the enzyme decomposing pectin has not been determined, since many other saprophytic organisms are able to produce these enzymes (see page 200). Perhaps pectin methylesterase is the enzyme which can be regarded to be involved in the pathogenicity of bacteria (Smith, 1958a, b).

Lovrekovich et al. (1967) examined the biochemical factors of resistance in the potato tuber infected with *E. carotovora.* Pectinase produced by bacteria in the potato tuber tissues, induces polyphenoloxidase which moreover oxidises phenols. The oxidised phenols cause a black colour which at the same time prevents the further spread of the disease. Resistance to soft rot depends on how quickly the black barrier arising from the oxidation of phenols is formed. The bacterium while infecting, produces dehydrogenase enzymes too, which, however, inhibit phenol oxidation, that is the development of the barrier. From this, it follows that the dehydrogenase activity of *E. carotovora* may be one of the factors of virulence of the bacterium. Summarizing, the oxidation of phenols is inhibited by dehydrogenase production of the virulent bacterium, therefore, the barrier does not form in the potato tuber.

The pathogen

The nomenclature of bacteria causing soft rot is still incomplete. In the latest review Graham (1964) determines the following: "There is sufficient evidence to justify removing these organisms from the genus *Erwinia* Winslow et al. and placing them in the genus *Pectobacterium* Waldee retaining *Erwinia* for the non-pectolytic organisms." He recognizes only one species: *Pectobacterium carotovorum* which is divided into: *P. carotovorum*, *P. carotovorum* var. *atroseptica*, *P. carotovorum* var. *aroideae* and *P. carotovorum* var. *chrysanthemi*.

The same disease symptom can be produced by different bacterial species. Graham and Dowson (1960) found that temperature was a basic factor in the formation of the disease. Isolates from different host plants caused blackleg symptoms if the infected plants were kept at 25 °C or higher temperatures while other strains were able to produce the symptoms of blackleg at 19 °C or lower temperatures. Those belonging to the so-called high temperature group originate from subtropical countries and these could be listed under *E. carotovora* and *E. aroideae* species while the low temperature group came from temperate regions. In natural black leg infection the bacterium *E. atroseptica* was found to be responsible.

Symptoms similar to the blackleg disease caused by *E. chrysanthemi* which produces slow wilt of the carnation (Lelliott, 1956), the bacterial blight of chrysanthemum (Burkholder et al., 1953), the leaf rot disease of *Phylodendron* (Miller and McFadden, 1961) and the *Parthenium* disease (Starr, 1947), at temperatures above 25 °C.

The *E. carotovora* is a facultative anaerobic and rod-shaped bacterium with peritrichous flagella and is Gram-negative. The colonies on nutrient broth agar are round, smooth and slimy, they liquefy gelatin and coagulate milk but do not peptonize it. They reduce nitrate to nitrite, do not form indole. They form acid and gas from arabinose, xylose, glycose, levulose, saccharose, rhamnose, raffinose salicin and mannitol. They form acid from glycerine and arbutin. They do not hydrolyze starch.

Isolation of the pathogen

Generally it is very difficult to isolate the pathogen directly from the tissues suffering from soft rot, since the secondary organisms from the disintegrated tissues have increased considerably. Often the tubers infected by *Phytophthora infestans* also have the symptoms of soft rot because the secondary micro-organisms overgrow the attacked tuber tissues. Thus it is best to isolate the pathogen not directly from the tissues of soft rot but to inoculate a healthy potato tuber with the diseased tissue. In such cases the following procedure is advised: at first the healthy potato tubers are washed well with running water and placed for 15 min in an acidic mercuric chloride solution (1·7 g $HgCl_2$ + 10 ml HCl + 1 l H_2O). Then the tubers are rinsed with a 70 per cent ethanol. The dried tubers are cut longitudinally with a sterile knife. The cut tubers are placed in dishes 5 cm deep and 20 cm in diameter, with a top lined with double filter paper. In order to maintain the relative moisture content of 95–98 per cent the filter paper is moistened and care is taken that it be continually wet. A very dilute suspension is prepared from the infected tissues as thin as possible. The suspension is placed by using a pipette onto the horizontally placed potato half so that it should cover the surface of the potato thickly without flowing off the surface. The other half of the cut potato is used as a control, therefore, only sterile water is used. The covered dishes are placed in an incubator at 26 °C. After 48 hr the tubers are examined for soft rot. The rot of the tissues can be easily tested by piercing the potato tissues with a sterile loop. The rotten tissue is soft, the healthy one remains hard. The soft mass is separated by a sterile knife so that only a narrow zone remains between the healthy and diseased tissues. The bacteria may be streaked from this part onto the surface of the nutrient agar. If necessary this procedure may be repeated again.

The bacteria may be isolated directly from the stem of blackleg potato too. At the border zone of the soft and healthy tissue the bacteria can be easily isolated.

Further details in connection with the isolation of the pathogen may be found in the works of Stapp (1961) and Henniger (1965). Many have tried to apply selective media (Starr, 1947, Dowson, 1957, Paton, 1958, Noble and Graham, 1959). From among these we advise using Starr's (1947) method. 0·9 ml 1 *N* NaOH, 0·6 ml 10 per cent $CaCl_2 \cdot 2H_2O$ (freshly prepared), 0·1 ml 1·5 per cent alcoholic brom thymol blue and 1·05 g Difco yeast extract were added to 100 ml of double distilled water. 3·05 g sodium polypectate (No. 6024 Sunkist) was added slowly to the solution above with constant stirring, pressing out the lumps and bubbles with a spatula, then placing over boiling water and stirring until dissolved. The medium should have a grass-green colour at this point, about pH 7·3. It should be placed in a relatively large flask (mixture tends to foam!) and autoclaved for 15 min at 151°C. After autoclaving, the pH will be about 6·4. It must be poured immediately after the medium comes out of the autoclave because it sets quickly and cannot be melted again in a water bath after it gels. Alternatively, it may be distributed in culture tubes for "deeps" and autoclaved.

If we want to prepare a semi-solid agar pectate medium, then 10 ml of a sterilized (still molten) 3 per cent solution of agar in distilled water is added to 100 ml of the sterilized but still fluid pectate medium. The final agar concentration in the medium is thus 0·3 per cent, enough to keep the gel fairly stiff even after the pectate is liquefied. The pectinolytic bacteria will make sunken colonies on this medium.

Artificial inoculation

We have already described the method of artificial inoculation of the potato tuber. If necessary carrot or cucumber discs may also be used. However these methods are not appropriate for determining the degree of resistance of potato cultivars. The most suitable for this is the potato disc method (Henniger, 1965).

Before use, the potato tubers are disinfected in the usual way. Potato pieces are made by a sterile cork borer 8 mm in diameter and 5 cm long. The potato cylinders are immediately placed into test tubes containing 0·5 ml (5×10^5 cell/ml) bacterial suspension. Care has to be taken that the bacterial suspension should not get onto the wall of the tube. The potato cylinders will immerse 4–5 mm in the bacterial suspension. If resistance tests are carried out, it is important that the potato pieces should immerse equally deep in the bacterial suspension. The tubes after being plugged, are placed in an incubator at 15 °C. The extent of rotting can be evaluated after 5 days. Henniger (1965) found in his investigations that the degree of resistance changed proportionally with the height of rotting formed on the cylinder.

WILTS

BACTERIAL WILT OF BEANS
(*C. FLACCUMFACIENS*)

Bacterial wilt of beans is caused by *Corynebacterium flaccumfaciens* (Hedges) Dowson.

Symptoms

The pathogen infects the vascular system and the disease remains in the plant from the germination phase to the ripening of seeds. The inoculated plants do not recover from the disease; when there are unfavourable conditions for the pathogen, the symptoms temporarily disappear and the bacteria assume a latent form.

In spring, the source of infection is the infected seed. The strongly infected seeds either do not emerge or die in the cotyledon stage. Generally the plants which have been attacked are dwarfed and less developed. The leaves become spotted in a mosaic-like form at first and on larger areas they are discoloured to yellow. With certain cultivars, the leaves girdle to a spoon-like shape. With the progress of the disease the intervenal spaces become yellow, dry and papery, then turn light-brown—one of the most characteristic symptoms of the disease. Sometimes a yellow halo surrounds the necrotized tissues. A typical symptom is that the trifoliate leaves droop and wilt. This is in contrast to other bacterial diseases of the bean where spotting does not occur simultaneously on all the leaflets of the trifoliates. Water-soaked spots are not characteristic of wilt disease though this symptom generally occurs in bacterial diseases.

The bacterium causes a systemic disease of the plant. It attacks the vascular system. On the longitudinally cut stems of the plants the vessels are brown. This symptom is typical in the lower part of the stem. When the vascular system is attacked, the plant suffers from lack of water and nutrients and in this way is only able to resist dryness with difficulty. Therefore, in wet humid weather the disease is hardly noticed, but when the weather turns dry the infected plants suddenly wilt.

At the end of the season the diseased stems generally become dry two or three weeks earlier than the healthy ones. Such plants generally give 60–70 per cent less yield than the healthy ones.

The restricted transportation of water and nutrients naturally has a marked effect on pod formation. There are very frequently no external signs of the disease on the pods though all the seeds are infected. Spot formation on pods rarely occurs and the infection does not progress outside the pod but rather inside along the vascular bundles through the peduncle. Thus, if spot formation appears it is more likely to pass along the suture of pods.

The diseased plants carry infected seed although the signs of infection can hardly be seen. In the case of heavy attack the bacterium under the seed coat forms a mass. With white seed cultivars this is especially obvious.

Isolation of the pathogen

Isolation may be made from the stem or infected seed. When isolating from the stem, the infected stem or the points of branching are cut longitudinally. The browned vascular bundles are prepared with a sterile needle and macerated in one drop of sterile water on a slide. After this liquid is streaked onto an agar plate, colonies of the pathogen appear on the second and third day. The colonies are pale cream, of lens form and with intact borders. The pathogen can be maintained for a relatively long time in the medium without losing its virulence.

Identification of the pathogen

The bacterium is a short non-sporing rod with a rounded end of $0{\cdot}6–3 \times 0{\cdot}3–0{\cdot}5$ μ in size. Electron microscopic investigations indicate one polar flagellum. It is Gram-positive in contrast to the already mentioned three other bean pathogens. It does liquefy gelatin. Starch is weakly hydrolyzed. In litmus milk, peptonization commences on the fifth day and the milk turns pale pink. Nitrite is not formed from nitrate. Ammonia is formed but it could not be demonstrated in every case. Indole is not produced in peptone water culture. Hydrogen sulphide does not develop. Acid is produced from glucose, galactose, arabinose, xylose, raffinose, saccharose, maltose, mannitol, mannose, fructose, salicin and dextrin. A little acid is formed from lactose and dextrin after 15 days. Acid is not produced from dulcite and asparagine.

Artificial inoculation

This can be easily done with *C. flaccumfaciens* in the greenhouse as well as in the field. If the bacteria have been introduced into the vascular bundles by pricking, the wilt symptom often appears typical because environmental conditions are not as important as in the case of spot diseases.

On the basis of Hedges' (1926) idea, we successfully used a needle placed among the hairs of a brush puff (Fig. 26). This implement can easily be prepared. The metal clamp of an aquarel brush puff is loosened and a needle is secured among the hairs of the puff so that the tip of the needle is about 1–2 mm deeper than the top of the puff. A suspension of a 24 hr slant agar culture is used for inoculation. With one dipping, 50–60 plants may be inoculated, while without the use of the puff, the pin has to be dipped into the suspension after every second or third inoculation.

Inoculation is as follows: young bean plants with one or two developed trifoliates are pricked with a needle on the stem between the cotyledon and the first leaves.

The first symptoms of successful inoculation are the rotting and wilting of the youngest leaves. This gradually spreads from the top to the lowest leaves. Eight–ten days are necessary for the appearance of the first symptoms.

Resistance

In practice, all bean cultivars and types are susceptible to *C. flaccumfaciens*, although Rands and Brotherton (1925), who examined 663 cultivars from England, France and other places, found some to be relatively resistant. In the course of investigations in Hungary, the author (unpublished work) observed partial resistance in the Kidney Wachs and Beste van Allen cultivars in the field, with an artificial inoculation test (Fig. 61).

Fig. 61. Result of artificial field inoculation on beans by *Corynebacterium flaccumfaciens*

SOUTHERN BACTERIAL WILT OF SOLANACEAE, OTHER DICOTYLEDONEAE AND BANANA
(*P. SOLANACEARUM*)

In subtropical and tropical areas the most widespread wilt disease of economically important plants occurs in tobacco, tomato, potato, pepper, peanut, eggplant and banana. The pathogen is *Pseudomonas solanacearum* E. F. Smith, which, as we will see later is not a uniform species.

Symptoms

This bacterium typically causes tracheobacteriosis primarily attacking the vascular bundles. Generally, the browning of the vascular bundles is typical, although it has been observed that in some districts the vascular bundles of the attacked potato and tomato plants are not discoloured. If the infected stem is cut crosswise, a bacterial slime appears on the periphery of the cut vascular bundles. In the later stage of the disease the bacteria floods the lining tissues and the cortex, forming cavities. The infected stems suffer from lack of nutrients and water, turn yellow and gradually wilt and die. The presence of bacteria in the potato tuber can be demonstrated by pressing the potato tuber between the fingers. If bacterial slime exudes from the vascular bundles we may conclude that bacterial infection has taken place. In the case of the banana the infected fruit ripens and turns yellow earlier than the healthy one.

Isolation and identification of the pathogen

The pathogen may be isolated from the browned vascular bundles. The small pieces of brown tissue of the xylem are placed on a slide in sterile water. When bacteria are present a larger number of the bacterial cells diffuse from the vascular bundles and in a few minutes cloud the water so that it appears milky. This is typical of bacterial infection of vascular bundles in contrast to fusarial wilt where the water does not cloud. During incubation, the bacteria streaked on potato dextrose agar are kept at 32–35 °C in contrast to the usual 25–28 °C. A brown pigment forms on most media and on old cultures the medium becomes brown. Most strains colour litmus milk a deep blue and a slow peptonization begins.

The bacterium does not form acid from salicin but from lactose and maltose, a characteristic differentiating it from the other pseudomonads. Strains isolated from different geographical locations and host plants are not uniform in their biochemical, serological and phage sensitive characteristics; they differ from each other in many characteristics (Kelman, 1953).

Biotypes

Buddenhagen et al. (1962), as well as Okabe, Goto (1961) and Buddenhagen and Kelman (1964), separated three biotypes: race 1 attacks *Solanaceae* and other plants, race 2 the banana and *Heliconias*, race 3 only the potato. These races not only differ in their pathological characteristics, but in their biochemical properties, antigenic structure and phage sensitivity as well.

Artificial inoculation

The most suitable portion of the plant for inoculation is the upper 5–6 leaves of a 5-week-old tomato. Young tobacco, eggplant and potato plants can also be well inoculated.

From the gently clouded bacterial suspension spread from the 48–72 hr culture, one drop is placed with a pipette or an eye dropper on the axil of the third leaf counted from the top of the experimental plant. Then the stem is pricked but not pierced with a downward-sloping needle through the water drop.

After inoculation the plants are placed in a greenhouse where the daylight temperature is between 32–33 °C and the night around 25 °C. Within two days the first signs of wilt as well as yellowing and later stunting growth can be noticed on the inoculated plants.

Change of virulence and its demonstration

The *P. solanacearum*, after numerous transfers on standard media, seems to lose its virulence. The virulence of cultures can be maintained by preparing a suspension of the virulent culture in sterile water and storing it at 21 °C. In this way the culture can be maintained without transfer for 2–5 yr (Kelman and Person, 1961).

In cultures of *P. solanacearum*, the mutants which have lost their virulence develop relatively early. Kelman (1954) found a correlation between the colour of the colonies, occurring on the tetrazolium medium, and the degree of virulence. The medium consists of 1000 ml distilled water, 10 g dextrose, 10 g peptone, 1 g casaminic acid, and 18 g agar. This basic medium is poured in 200 ml quantities into flasks and sterilized the usual way in the autoclave. Before pouring the warm liquid medium out into Petri dishes, 1 ml of sterile 2,3,3,-triphenyl tetrazolium chloride (TTC) in 1 per cent concentration is added to the basic 200 ml medium. In this way the TTC concentration of the agar medium will be 0·005 per cent. The TTC solution is prepared with 1 g of TTC dissolved in 100 ml distilled water and sterilized by filtering (see page 202). This solution must be stored in a dark place until use.

A dilution of the bacterial suspension is prepared with sterile water and this diluted suspension is streaked onto the surface of TTC agar with a wire loop.

After a 36 hr incubation at 32 °C the colour of the colonies is examined in the light. The virulent or wild type colonies are of irregular shape, slimy, white with pale pink centres. The avirulent ones are small and round, cream and deep red with a thin lighter border. In 14–21 day-old 5 per cent tryptophane cultures more and more avirulent mutants will appear.

PROLIFERATION OF TISSUES

CROWN GALL
(*A. TUMEFACIENS*)

The *crown gall* caused by *Agrobacterium tumefaciens* Smith et Townsend Conn. is economically one of the most important and most studied diseases. This disease does not resemble other bacterioses, for here the bacterium induces an uncontrolled growth of plant tissue.

Tumour formations occur on many different plants. The best common hosts of the pathogen are the apple, pear and stone fruit trees, hop, grape, poplar, berry fruits as well as many herbaceous plants. The infected plants are mostly of weaker growth than those free of crown gall because one part of the nutrients is used up fðr tumour formation. The appearance of tumour is due to the uncontrolled multiplication of the cells (see page 199). In the tumour of herbaceous plants mainly parenchymatic cell proliferation dominates while in the woody plants the unsystematic development of the trachea or tracheids is a characteristic feature. On herbaceous plant stems the tumours are generally soft, on woody plants the tissues which at first were soft, gall-like become woody and hard. The size of the galls varies depending on the plant as well as the location. In most cases, especially in woody tissues, it arises around the root crown root. On herbaceous plants the tumourous tissue may also develop on the stem and leaves. With woody plants the young tumours at first, are grey-white in colour, have cauliflower-like forms and are soft. The gall finally disintegrates or turns brown and becomes woody.

A. tumefaciens is a typical wound parasite. It can only penetrate plants through the wounds.

An abundant nutrient supply is required for the development of tumours, therefore their development is only possible during the cultivation period.

There is disagreement in connection with the economic loss caused by crown gall. In practice the fruit tree grafts infected with *A. tumefaciens* have been destroyed. In contrast to this many observations have shown that the later development of trees is not influenced disadvantageously by crown gall (Riker et al., 1959). As opposed to this, crown gall may kill hop, grape and common berry fruits (Fig. 62).

The *A. tumefaciens* is a rod-shaped bacterium which often develops, in short chains. It is also flagellated, polar, although Dowson (1957) observed a few peritrichous flagella too. On this basis, he placed the bacterium in the *Erwinia* genus, however, this is not generally accepted. This pathogen is Gram-negative and aerobic. Small, circular, slightly elevated colonies are formed on nutrient agar. These are moist, shiny and transparent. It does not liquefy gelatin, coagulates milk but does not peptonize it. The nitrates are weakly or not at all reduced. Indole is formed but sometimes only weakly. Ammonia and hydrogen sulphide is produced in large amounts. Only weak acid without gas formation was observed from glucose, levulose, arabinose, galactose, mannitol and salicin. Starch was not hydrolyzed.

Fig. 62. Crown gall on vine-stock

The bacterium cells form star formations on media rich in carbohydrates but containing little nitrogen. Stapp pointed out this phenomenon in 1942. Stapp (1957) considered this star formation to be the preliminary sexual stage of the bacterium. This star cell formation is characteristic for *A. tumefaciens*.

Fe–Mn-carrot-juice for the detection of star-formation (Stapp and Bortels, 1931): Washed carrot is cut to small pieces and cooked with twice the amount of distilled water for 10 min. The cooked carrot is filtered through gauze. The filtrate is diluted with three times of the original amount of water. The pH is adjusted to 7·0–7·2 with Na_2CO_3 and add 0·01 per cent $MnSO_4$ and 0·01 per cent $FeSO_4$ of the volume. At first the $MnSO_4$ is dissolved in some distilled water and added to the carrot juice then the $FeSO_4$ solution separately. After shaking, this medium becomes dark brown. 5 ml portions are measured into test tubes and sterilized fractionally on three successive days. The tubes are inoculated with 20–24 hr old bacterial cultures and incubated at 26–28 °C. The star-formation is detected in the 24–36 hr old carrot-juice cultures under the microscope.

In this medium the following species are able to produce star forms:
Family of *Rhizobiaceae:*

Gen. *Rhizobium*	*leguminosarum*	a little star-formation
	phaseoli	
	trifolii	
	lupini	
	japonicum	
	meliloti	
Gen. *Agrobacterium*	*tumefaciens*	intensive star-formation
	rhizogenes	
	rubi (little)	
	radiobacter	

Family of *Pseudomonadaceae*

Gen. *Pseudomonas savastanoi* (after 6–10 days)
tonelliana (after 4–8 days)

Isolation of the pathogen

Often it is not easy to isolate *A. tumefaciens* from tumorous tissues. It is futile to even try to isolate the pathogen from old, large or woody crown galls. Frequently it is even difficult to isolate the pathogen from the soft woody tumours. Therefore, it is best to graft a small tissue piece from such a tumour onto another herbaceous test plant. A small V-shaped tissue is cut out from the soft tissue and this is immediately placed in the stem part of a fresh herbaceous plant in such a way that the tissue part is completely covered by the new host phloem tissue. Then the plants are placed in a moist chamber for 24–48 hr. Stapp suggests the following test plants: *Pelargonium zonale*, *Datura stramonium*, *Datura tatula*, *Lycopersicum esculentum*, *Chrisanthemum frutescens*, *Nicotiana tabacum*, *Kalanchoe digremontiana*, *Helianthus annuus*.

It is not likely that the tumour will develop on all cultivars, therefore, it is advisable to use different cultivars simultaneously.

In 2–4 weeks time, where larger tumours formed at normal callus formation, a small piece is cut with a sterile knife and a suspension is prepared with the addition of sterile water in a mortar. Before the utilization of the tumorous tissues, the surface is cleaned from impurities with plenty of running water. The macerated tissue suspension is streaked onto nutrient agar plates.

A. tumefaciens develops well on relatively simple media. If the medium contains amino acids, after longer cultivation the bacterium will lose its virulence on such media. This kind of medium helps the formation of R forms, which is at the same time accompanied by a reduction of virulence.

Several herbaceous plants which have been mentioned may be used. However it is best to use tomato and sunflower. The young stem of the growing plant is pricked with a needle dipped in suspension prepared from the investigated bacterium culture. After incubation the plants are placed in a moist chamber for 48 hr. The temperature is likewise an important factor in tumour formation. If for example tomato plants are kept at 28 °C after inoculation, vigorous tumours develop. At 31–32 °C the tumours no longer arise; but at temperatures below 28 °C the tumours develop too but much slower. The tumours appeared in 2–3 weeks on inoculated plants.

Lehoczky (unpublished data) uses sunflower seedlings for the detection of *A. tumefaciens.* Sunflowers 5–6 days old are grown on wet filter paper and pricked near the roots with a needle previonsly dipped in the bacterial suspension in question. Thickening and swelling occur at the place of inoculation.

PART III

MYCOLOGY

INTRODUCTION

Fungi cause the majority of "infectious" plant diseases. Throughout the world, they reduce the yield and quality of plant production and cause material economic loss. Knowledge of the morphology and taxonomy of plant pathogenic fungi is essential for the recogniton and successful control of plant diseases. In breeding for resistance, one also needs to know something about their physiology and life cycles. In phytopathological practice, as in breeding for fungal resistance in plants, the most commonly utilized processes are microscopic examination and the cultivation of fungi for identification and inoculation.

CHAPTER I

GENERAL CHARACTERISTICS OF FUNGI

Fungi are unicellular or filamentous multicellular organisms with true nuclei and, in most cases, with chitinous hemicellulose or cellulose cell walls. They have no chlorophyll, and are spore-forming thalloid organisms. They differ primarily in their mode of nutrition from other members of the living world, especially from autotrophic plants and from holozoic animals. The fungi resemble plants the most, because they have a cell wall and are generally unable to move actively. For this reason, they are often regarded as members of the plant kingdom. However, fungi are neither plants nor animals, but have developed during the course of time in an independent, phylogenetic line (Martin, 1960; Ubrizsy and Vörös, 1966). Most fungi can multiply both sexually and asexually. The systemization of fungi is based primarily on sexual reproduction, since this process is a rather constant and stable characteristic.

VEGETATIVE MYCELIUM

The somatic body of a unicellular fungus containing one or more nuclei is restricted by a cell wall, or it may be naked. The vegetative body of multicellular fungi is usually filamentous, and always has a cell wall. Each cell may have one or more nuclei. The filaments of fungi, the hyphae, are apical in growth and in mass are called the mycelia. In some groups of fungi, the hyphae are divided by septa; in others septa are formed only exceptionally, or not at all. Certain cells are connected by plasmodesmata through pores in the septa of multicellular hyphae. Mycelium not divided by transverse septa is called coenocytic mycelium.

Nutrient uptake is often carried out by morphologically undifferentiated hyphae growing into the substrate, or penetrating the host plant. In certain fungi, this is performed by modified hyphae called rhizoids (Fig. 1). The germ tube of some germinating fungus spores often forms characteristic, specially developed, adhesive discs called appressoria on the surface of the substrate. In the case of phytopathogenic fungi these are formed on the surface of the host plant. Obligate parasitic fungi, living on the surface of the host plant or in its intercellular spaces, take up nutrients through specific absorbing hyphae called haustoria, that penetrate the cells of the host plant. The haustorium is a branching or simple knob-like formation, insuring a large surface for the absorption of nutrients. Those phytopathogenic fungi having mycelium solely on the surface of the host plant, and meeting their nutrient requirements only through haustoria penetrating epidermal cells, are grouped together as ectoparasitic fungi. In contrast to this, the hyphae of endoparasitic fungi grow inside the host plant, usually in the intercellular spaces, and haustoria penetrate into the adjacent cells.

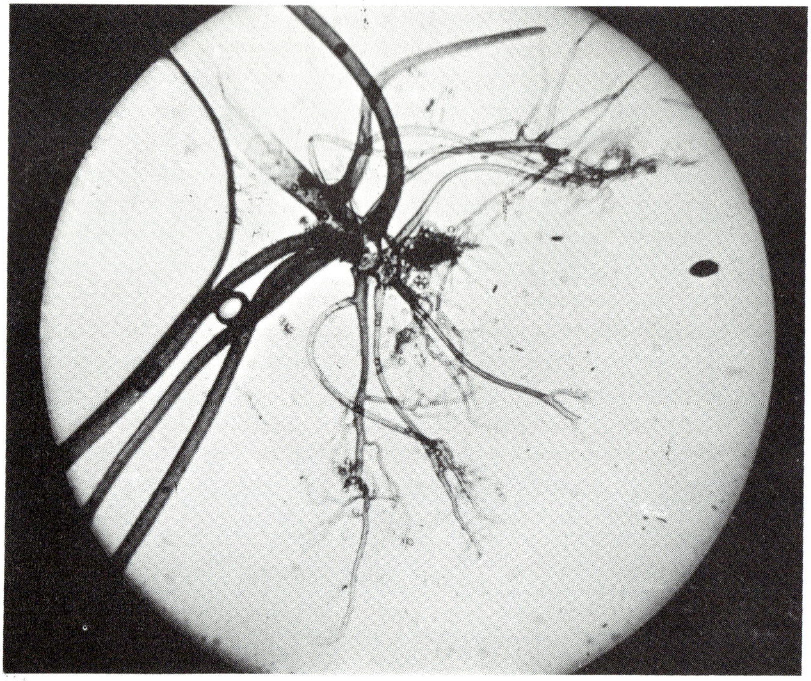

Fig. 1. Rhizoids (*Rhizopus stolonifer*)

In certain stages of their life cycle, the somatic body of most fungi is organized into fungus "tissues" composed of loosely or compactly interwined hyphae. Although never differentiated into true tissues typical of higher plants, two kinds of plectenchyma or organized fungus tissue, can be distinguished. The constitutive hyphal elements are still identifiable in the prosenchyma but not in the closely packed, isodiametric cells of the pseudoparenchyma that under the microscope remind us of the true parenchymatous cells of higher plants. Rhizomorphs, sclerotia, and stromata are somatic structures composed of these tissues. In some higher fungi, the mycelium forms compact, thick strands called rhizomorphs. These structures extend over long distances in the soil. The more compact, outer layer has a hard cortex and growing tip. The rhizomorph is a very resistant structure against unfavourable external conditions. Another resistant structure, the sclerotium, also formed from vegetative hyphae, is a hard, spherical to ellipsoid-shaped body that is rarely elongated (Fig. 2). It is divided into an exterior cortex that is harder and usually darker coloured than the internal, softer medulla part. The cells of the medulla are larger in diameter than the compact cells in the cortex, and reserve nutrients are stored in them. The hyphal mass that sexual or asexual reproductive cells are formed on or in, is called the stroma.

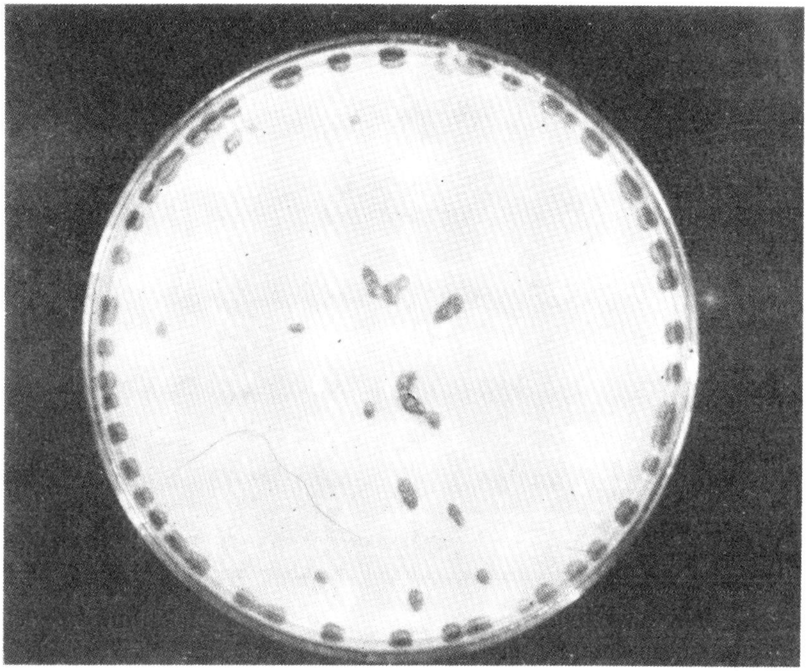

Fig. 2. Sclerotia (*Sclerotinia sclerotiorum*)

The mycelium and other vegetative structures may be extraordinarily variable in their formation, size, colour, etc. These will be discussed later in connection with some of the more important phytopathogenic fungi that produce them.

REPRODUCTIVE ORGANS

Fungi reproduce asexually and sexually. The latter is characterized by the union of nuclei and the appearance of sexual structures, but in the former there is no fusion of cell nuclei or sexual organs. Reproduction may be holocarpic or eucarpic. Mainly unicellular fungi are holocarpic because the whole somatic body is transformed into the reproductive structures. Most fungi are eucarpic in that the vegetative body remains functional while special reproductive organs or cells are formed on its surface during reproduction. Holocarpic reproduction is the more primitive, simpler type of the two. The uniform vegetative hyphae of eucarpic fungi are modified during reproduction. Hyphae are formed which are morphologically diverse in producing sexual organs or reproductive cells. This may be a consequence of the massing of hyphae, different types of stromata, and fruiting bodies.

ASEXUAL REPRODUCTION

In the broadest sense, one can term asexual reproduction as all processes where the formation of a new individual or colony is not preceded by a sexual process, that is, the fusion of nuclei. Thus asexual reproduction is the formation of new individuals by way of propagation, broken off pieces of hyphae or mycelia, sclerotia, rhizomorphs or special cells. In a narrower sense, reproduction is the spreading of differentiated reproductive cells specially formed for this purpose. The reproductive cells of fungi are called spores. The cells formed without sexual processes are called asexual spores. The spores are generally nonmotile, but many fungi produce motile swarm spores with one or two flagella. The asexual reproductive cells are generally formed in large masses. This guarantees the further spread of the species in the quickest way, and the formation of epidemics of phytopathogenic fungi. The asexual spores are unicellular or multicellular, colourless, hyaline, or they are variously coloured. The period of viability is generally short and the asexual spores very rarely survive overwintering or unfavourable conditions. Under suitable conditions, they germinate with one or more germ tubes, developing the mycelium of the new individual. Many reproductive cells are dispersed widely by the wind for great distances, but humans and animals help to spread them, too. Spores of some fungi are discharged by force. On the basis of their formation, one can differentiate other types of asexual reproductive cells: fragmentation, budding, constriction, and in multinucleated cytoplasm, the massing of a plasma membrane around the nuclei (free-cell formation).

The asexual reproductive cells originating by means of the fragmentation of hyphae are called arthrospores or oidia. These occur most commonly among yeasts (*Geotrichum*). Previously, the conidia of powdery mildew fungi wrongly were considered to be oidia. In budding, one part of the dividing nucleus passes into a protuberance of the mother cell. When the bud reaches the size of the parent cell, it separates or remains together by forming short budcell chains, or pseudomycelium. Most yeasts and the reproductive cells of some fungi reproduce by budding, e.g. the basidiospores of smut fungi by sporidia. The thalli of many higher, filamentous fungi that are produced under extreme conditions (e.g. very low pH, under anaerobic conditions, etc.) also form pseudomycelia consisting of bud cells (Figs 3, 4, 5).

The asexual reproductive cells originating usually on terminal hyphae are conidia. Most plant pathogenic fungi reproduce asexually by conidia. On the basis

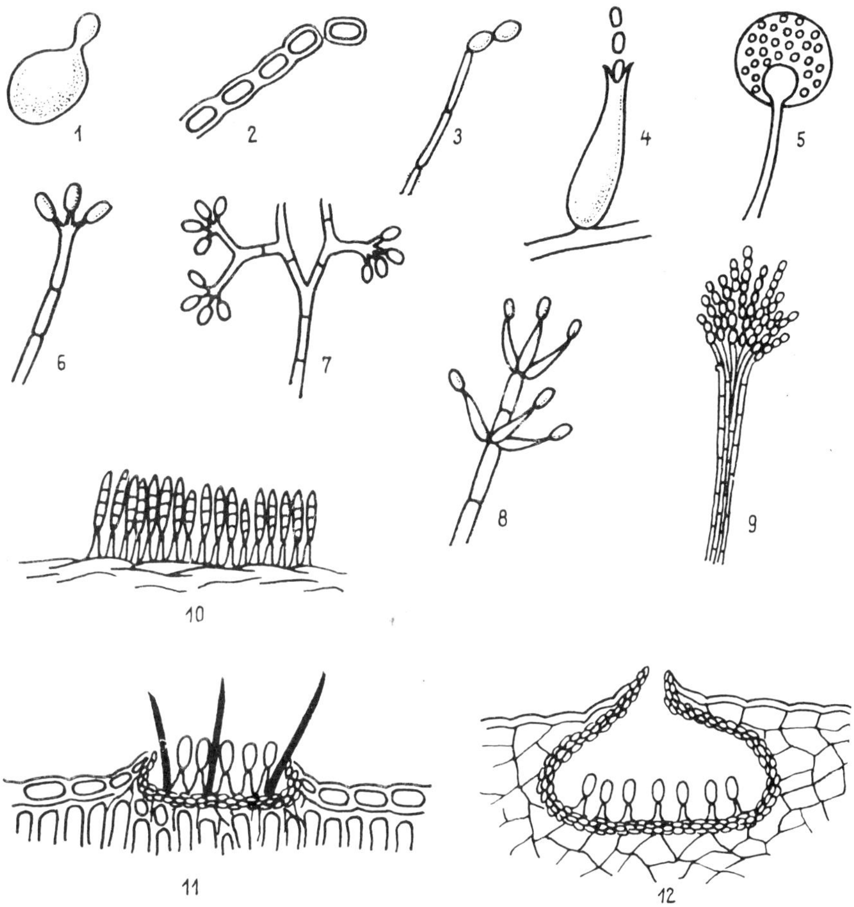

Fig. 3. Asexual spores and types of conidiophores

of their size, we can differentiate macro- and microconidia, and on the basis of their formation, exo- and endoconidia. Exoconidia are produced on conidiophores which are formed in various ways; however, endoconidia arise inside a hyphal tube or cylindrical shaped cell (in a narrow sense phialide or sterigmata) e.g. tube conidia in the case of *Thielaviopsis basicola* (Fig. 6). The conidiophore can be a simple hyphal side branch which does not differ from the vegetative hyphae, or it can be a morphologically differentiated, simple or branched formation such as found in dichotonously branching and verticillate conidiophores. Spore-forming cells or the apex of the conidiophores are sometimes swollen into a vesicle. The terminal, pointed cell of branched conidiophores on which the conidia are produced is the sterigmata, or in a broader sense, a phialide. The conidiophores are formed isolated or formed in groups. In the latter, the really long, compact

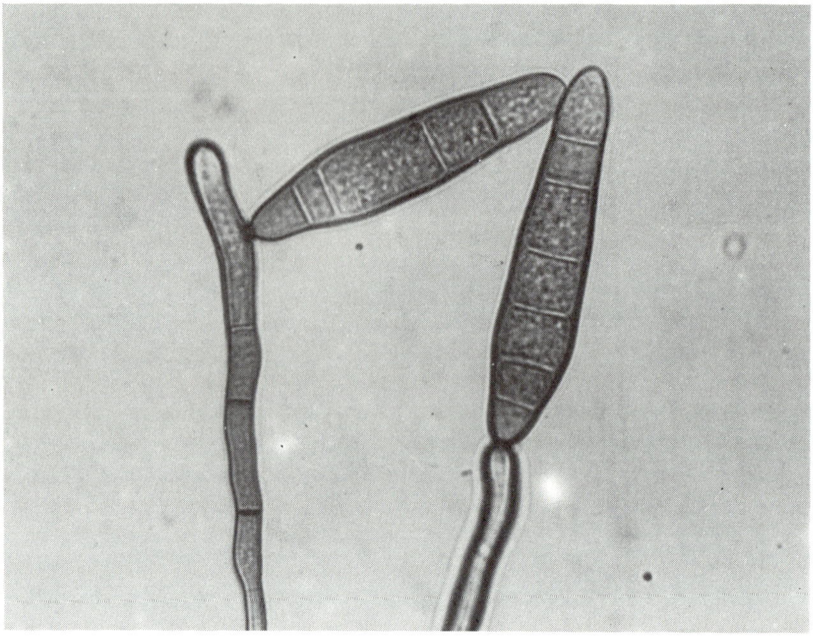

Fig. 4. Conidia cut off from conidiophores (*Helminthosporium turcicum*)

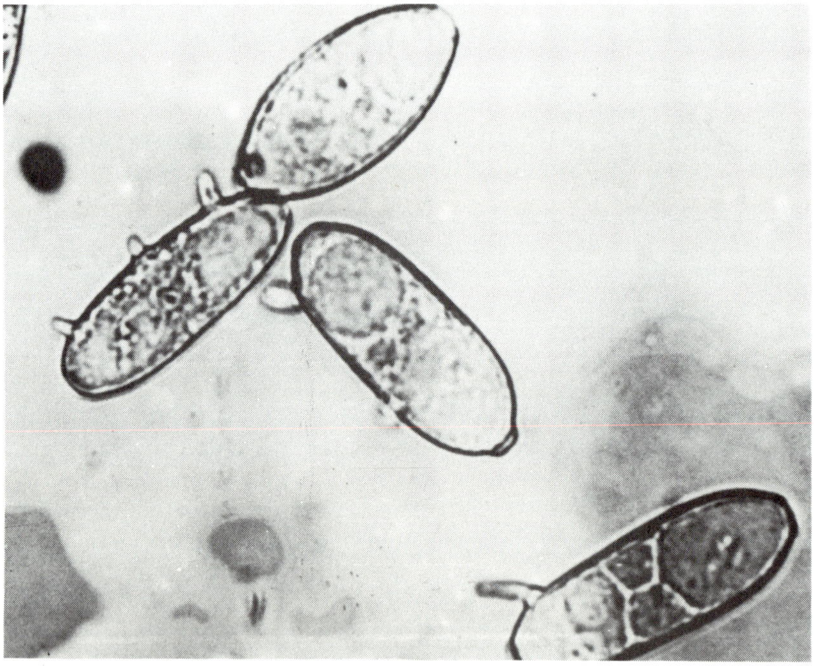

Fig. 5. Germinating conidia (*Erysiphe graminis*)

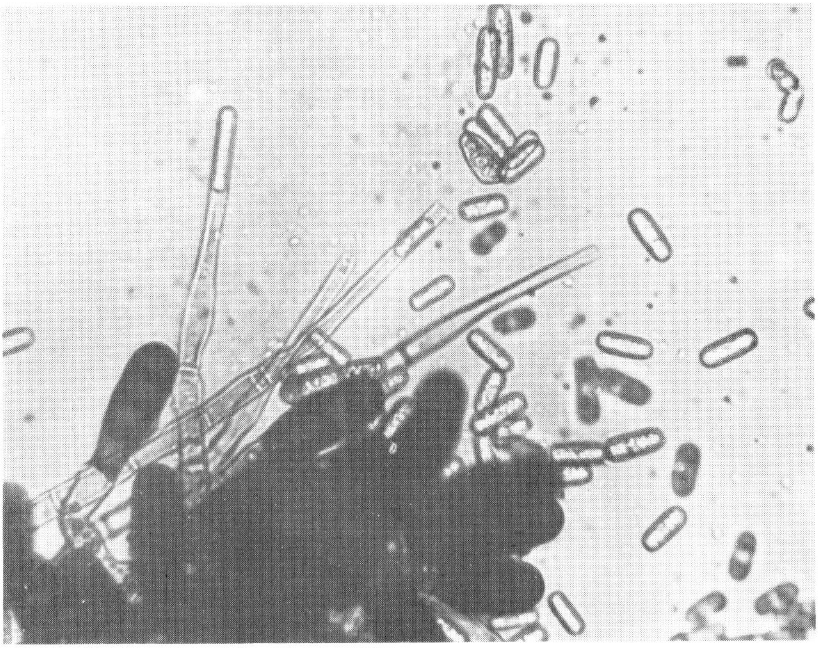

Fig. 6. Endoconidia (*Thielaviopsis basicola*)

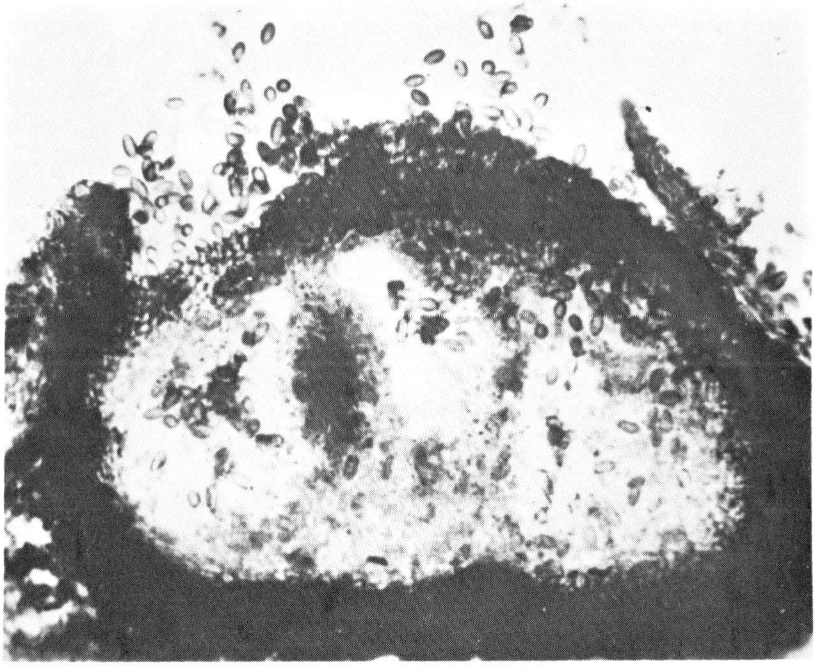

Fig. 7. Cross-section of a pycnidium (*Diplodia sp.*)

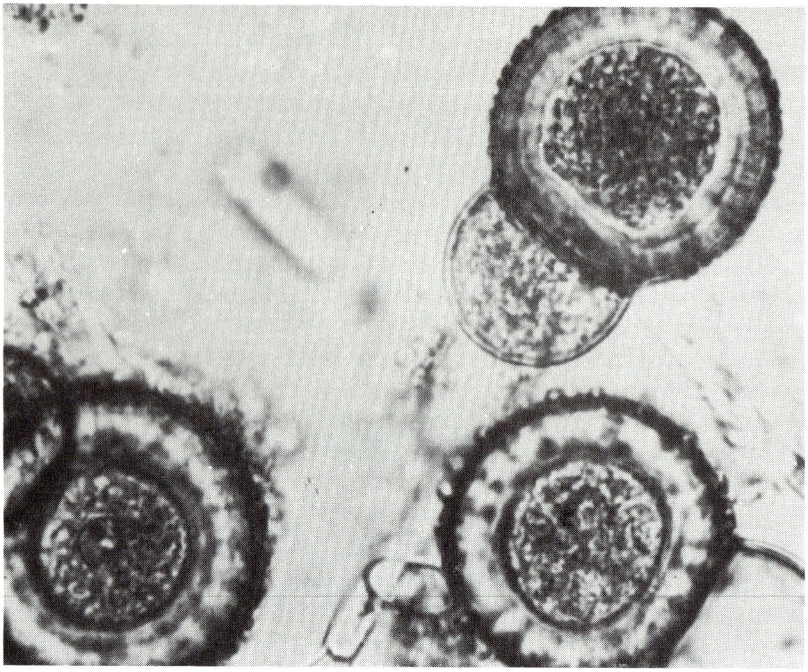

Fig. 8. Chlamydospores (*Mycogone rosea*)

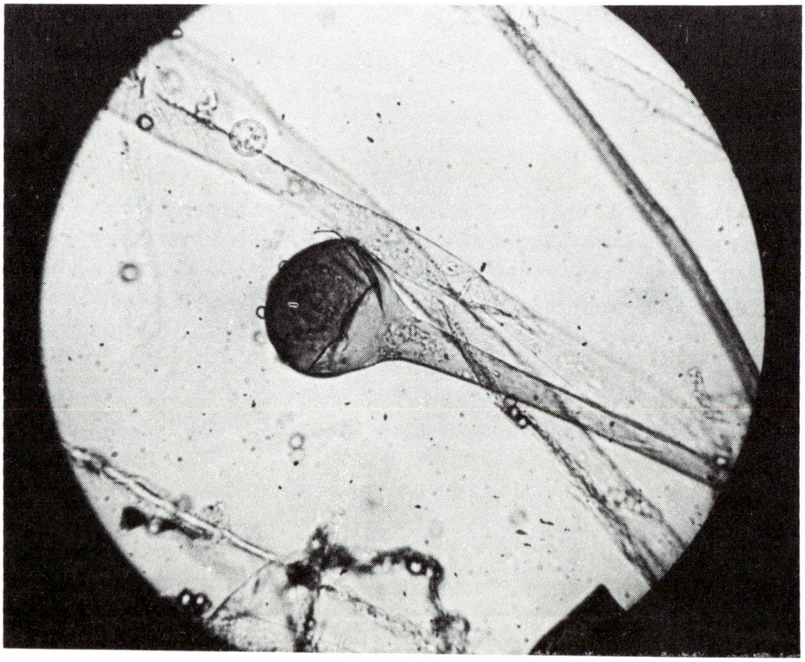

Fig. 9. Young sporangium (*Rhizopus stolonifer*)

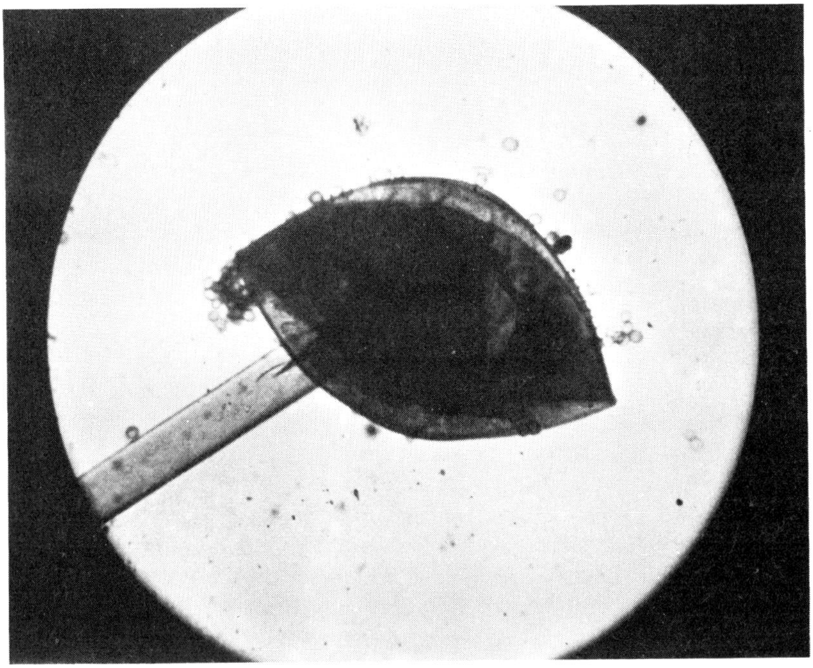

Fig. 10. Mature sporangium (*Rhizopus stolonifer*)

conidiophore fascicle is called a coremium or synnema, and the short palisade-like conidiophore mass is called a sporodochium. The acervulus is a sporodochium-like conidiophore group arising underneath the epidermis or cuticle of the host plant. Conidia are often formed inside globular or bottle-shaped fruiting bodies known as pycnidia (Fig. 7). Such asexual reproductive cells are known as pycnidiospores. The conidia are emptied when the pycnidium cracks irregulary at maturity, or upon ripening, opening by way of a split in the wall or through a pore formed earlier. If the apex of the pycnidium bearing the pore becomes elongated like a beak, it is called an ostiole. The pycnidium can be formed on the surface of the host or can be sunken or embedded in a stroma. If only a loose hyphal mass is formed under the pycnidium, it is referred to as subiculum (subicle).

In a broad sense, all terminal or intercalary, thick-walled resting, asexual reproductive cells are known as chlamydospores (Fig. 8). The chlamydospores which do not separate from the hypha but are channeled along a thick stalk are termed aleurispores. In the hyphal cells of a few fungi, resting spores formed by means of cytoplasmal aggregation are known as gemmae.

Asexual reproductive cells can be formed within sporangia too (Figs 9, 10). The sporangium is a multinuclear swollen growth in which at maturity the cytoplasm aggregates around the nuclei and the reproductive cells called sporangiospores arise by free-cell formation. In some fungi, the hypha, bearing sporangium (sporangiophore) extends as a columella inside the sporangium which contains many spores. The small sporangium with or without a columella is called a sporangiolum which contains but a few spores. The merosporangium is a cylindrical sporangiolum in which the spores settle into a single line.

Most spores are non-motile and are called aplanospores, but flagellated, motile reproductive cells are produced too. These are the swarm cells or zoospores occurring among the lower aquatic fungi which have not fully adapted to a terrestrial way of life. Knowledge of them is important both from systematic and phylogenetic points of view. The zoospores are formed inside a zoosporangium, and may be round, drop, or kidney shaped. The swarm spores of fungi can be mono or biflagellated. Even in the presence of a pair of flagella, one of them may be very rudimentary and easily missed, being barely visible in some fungi. Among the motile forms of fungi there are two flagellum types. The surface of the so-called "whiplash" type is smooth while the "tinsel" type is hirsute with fine fibrils resembling a feather. The submicroscopic structure of both flagella are almost identical in all living organisms higher than the bacteria. The flagellum consists of two central fibrils which are surrounded by nine others acting as a mechanical tube. The two central fibrils protrude from this cover in the "whiplash" type and the motility of the flagella is insured by this part of the flagellum. The fibrils constituting the flagella are made up of subfibrils. The central fibrils continue the base of the flagellum inside the zoospore in the form of a blepharoplast and the rhizoplast connects the cell nucleus to the flagellum. In the zoospores of some fungi, the cell nucleus is surrounded by a "nuclear-cap".

SEXUAL REPRODUCTION

Sexual reproduction is characterized by the fusion of sex cells (organs) and a subsequent reduction division. Sexual reproduction always takes place in a given phase of ontogeny, or life cycle. Sexual fusion in the lower fungi occurs in the same way as in other living organisms, that is with the fusion of two haploid nuclei having $1n$ number of chromosomes in sex cells or sex organs. Upon completion of the development of the somatic diploid part with $2n$ number of chromosomes, new homo- or heterogeneous sex cells of $1n$ number of chromosomes are formed again by reduction division or meiosis. This process becomes somewhat more complicated in the higher fungi. Here copulation of the sex organs brings together the cytoplasm and nuclei from separate parents into one cell but without fusion of the nuclei occurring. This is called plasmogamy. The binucleate cell containing one nucleus from each parent continues to grow and multiply by conjugate division. This stage is called the dikaryophase and takes place between plasmogamy and karyogamy. Fusion of the cell nuclei occurs in the organ which later produces sexual reproductive cells (asci or basidia). Immediately following fusion, reduction division takes place with the formation of new sex cells containing $1n$ number of chromosomes.

Generally the sex organs of fungi are called gametangia and the sex cells are called gametes. The fusion of morphologically identical male and female sex organs is isogamy. Copulation of gametes of morphologically different forms is heterogamy. In the latter case, the male sexual organ is the antheridium, and the female organ is the oogonium (*Oomycetes*) or ascogonium (*Ascomycetes*). Among the heterogametangia, the male one is always smaller and usually has an elongated shape. Sexual fusion in the lower fungi (*Phycomycetes*) occurs in many cases between motile sex cells or planogametes with flagella, but here too, both iso- and heterogamy can be distinguished (Fig. 11). In the course of changing from

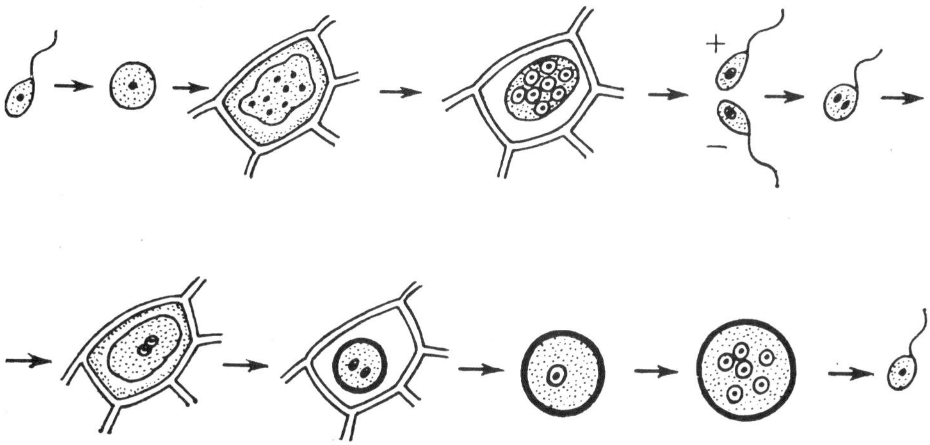

Fig. 11. Sexual process between planogametes. Diagram of the life cycle of a *Chytridiales* (*Olpidium*)

an aquatic life into a terrestrial one, the role of the planogametes is assumed more and more by non-motile aplanogametes. In the fusion of gametangia, either the male nucleus (nuclei) penetrates into the female sexual organ or the protoplasm of both sex organs is united through copulation. With some higher fungi that role of the male sex organ is filled by non-motile male sex cells, the spermatia. These are ordinarily of very small size and contain 1*n* number of chromosomes and are unable to germinate and grow independently. When these reach the female sex organ or its receptive hypha (trichogyne), they fertilize them. The following four types of sexual reproduction can be differentiated in the higher fungi that have changed to a terrestrial way of life. These divisions serve also as the basis for their classification.

1. *Oomycetes:* The smaller sex organ (antheridium) having one or more nuclei, grows towards the large female sex organ (oogonium) which is formed on coenocytic mycelium and contains one or more haploid nuclei. At the point of contact of the two sex organs the cell wall is dissolved and the protoplasm of the antheridium flows with the nuclei into the oogonium. The sexual process

takes place with the fusion of the different sex cell nuclei. The cytoplasm of the oogonium clustering around the nucleus containing 2*n* number of chromosomes constitutes the sexual reproductive cell or oospore. This process is characteristic of the *Oomycetes* class of fungi. It is obvious that here the cytoplasm

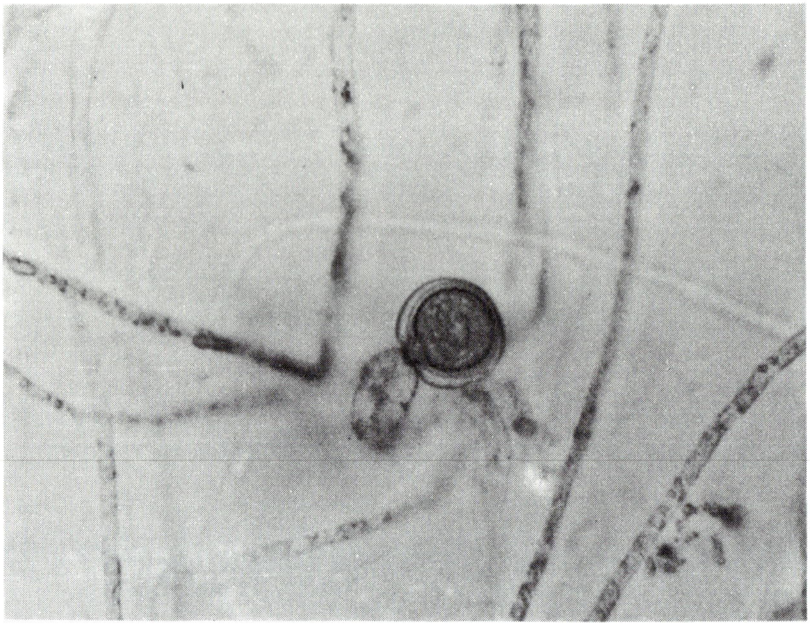

Fig. 12. Copulation between oogonium and antheridium (*Pythium sp.*)

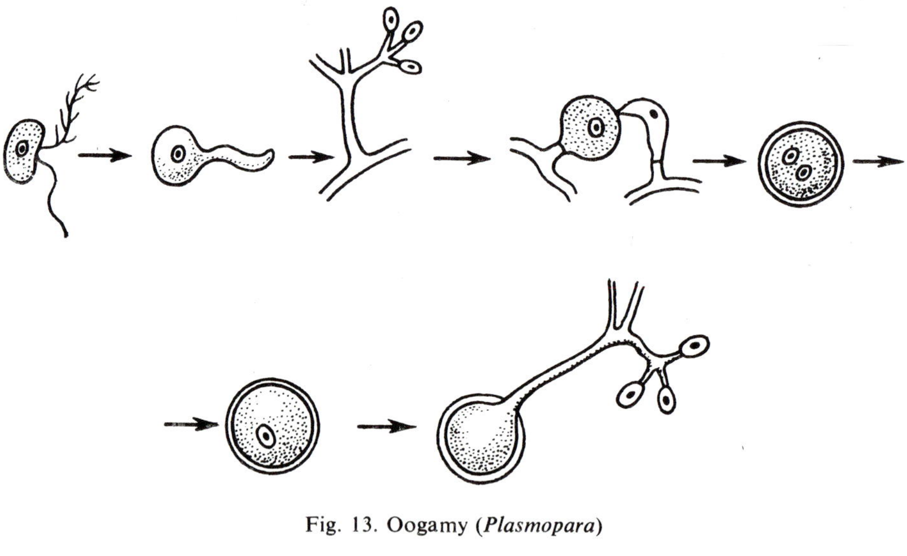

Fig. 13. Oogamy (*Plasmopara*)

and the nuclei of the sexual organs fuse at the same time and place and do not include an intervening dikaryon phase (dikaryophase) between the plasmogomy and karyogamy stages (Figs 12, 13).

2. *Zygomycetes:* Two morphologically similar but sexually different hyphae containing $1n$ chromosome number nuclei fuse in copulation (isogamy) by the dissolving of the contacting walls. The copulation bridge is separated from the copulation branches by a thick wall that forms around the protoplast of the zygote. The resting zygospore with $2n$ number of chromosomes is the product of the sexual process. In the type of sexual reproduction one finds in the *Zygomycetes*, no dikaryophase occurs (Figs 14, 15).

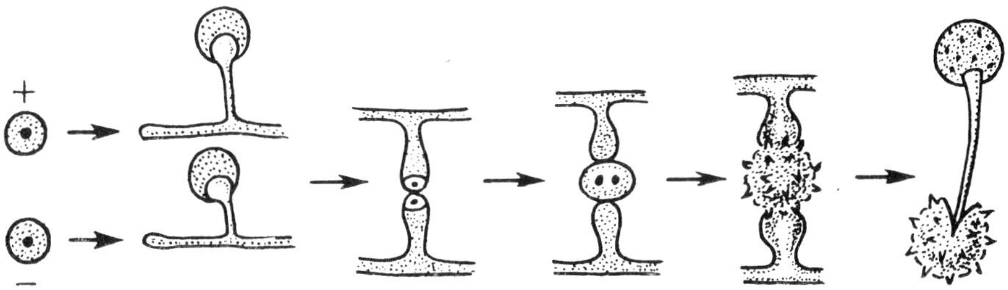

Fig. 14. Zygogamy (*Mucor*)

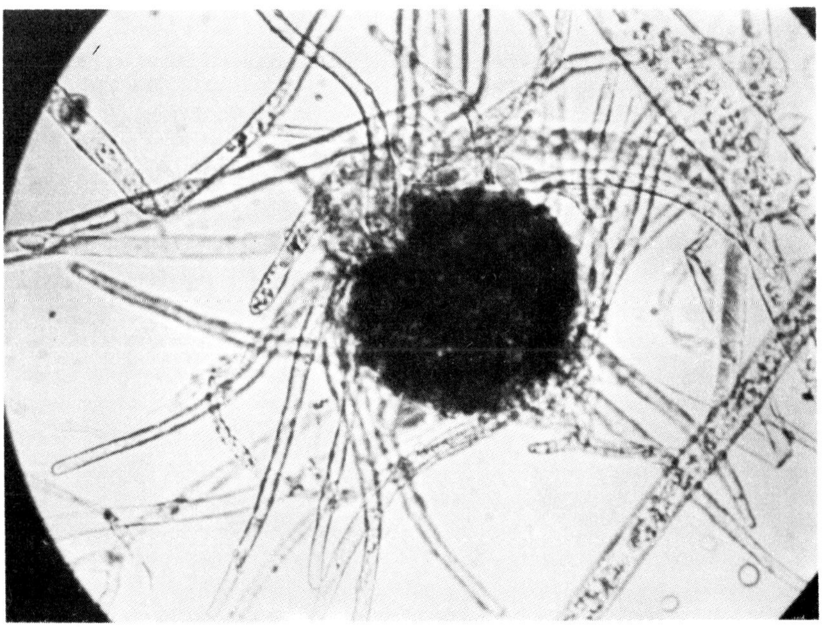

Fig. 15. Zygospore (*Absidia sp.*)

3. *Ascomycetes:* After the development of the haploid mycelium, the union of the protoplasm (plasmogamy) of two hyphal ends takes place. These hyphal ends are morphologically different (ascogonium and antheridium) or identical but of different sexual characteristics. The different sex cell nuclei do not fuse at this time. From the fertilized ascogonium, that was developed from the female hypha, the binucleate ascogenous hyphae develop. After a short development, clavate, clubshaped asci are formed on the end of the dikaryotic ($n + n$) hyphae and in which finally the fusion of nuclei occurs (karyogamy) (Fig. 16). The ascospores, the sexual reproductive cells, are formed after meiosis followed by mitosis. Usually 8 haploid ascospores are formed in one ascus. Thus in the course of the sexual processes characteristic of the *Ascomycetes*, one finds that a short dikaryophase occurs between plasmogamy and karyogamy.

4. *Basidiomycetes:* The dikaryophase starts very early by means of plasmogamy. Two protoplasts unite from different haploid hyphae (sometimes spores) so that the mycelium is binucleate at the beginning of the life cycle. The dikaryotic mycelium continues until knob-like basidia develop on the ends of the hyphae. The two nuclei in the terminal cell fuse (karyogamy), meiosis occurs, resulting in 4 haploid cell nuclei that migrate terminally inside the wall, while outside of the basidium, short projection-like sterigmata form. A small portion of the cytoplasm and the nuclei are extruded from the basidium to form the basidiospores

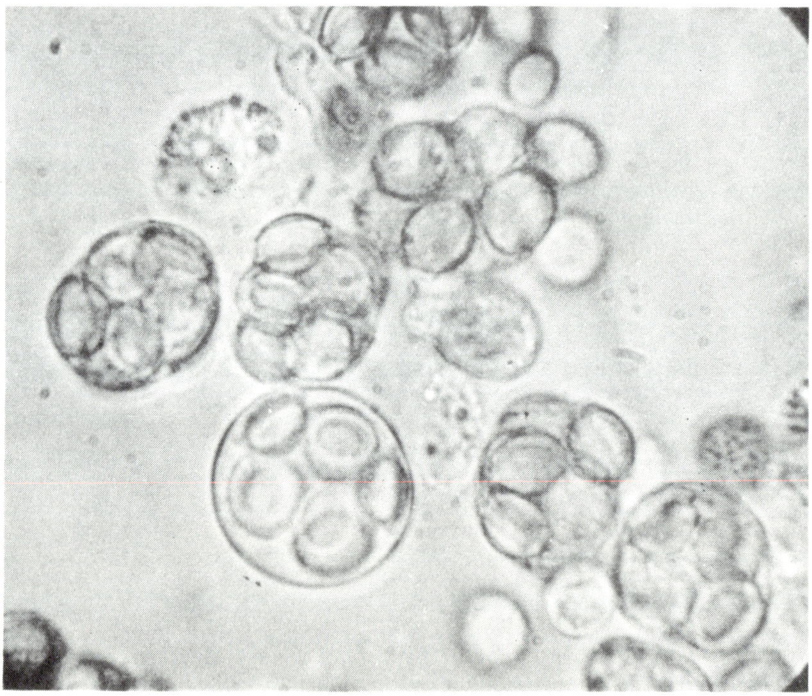

Fig. 16. Asci (*Aspergillus umbrosus*)

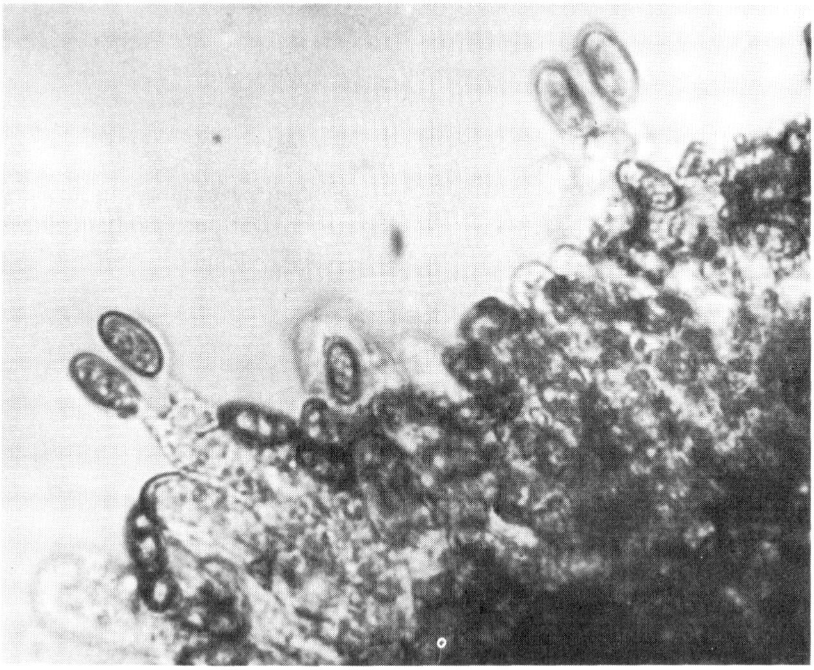

Fig. 17. Basidia (*Merulius lacrymans*)

or sexual reproductive cells characteristic of the *Basidiomycetes* Class. The dikaryophase characterizes the whole mycelium for most of the life cycle (Fig. 17).

The fusion of planogametes of aquatic fungi as well as the oogamy and the zygogamy of the *Oomycetes* and *Zygomycetes*, generally are morphologically simple processes. But this is not the case with the higher *Ascomycetes* and *Basidiomycetes* (Fig. 18). Their sexual processes often result in characteristic, differentiated fruiting bodies. In many cases the formation of the fruiting bodies occurs before the sexual process is completed. Formation of these bodies does not take place in the course of the life cycles of the *Endomycetales* and the parasitic *Taphrinales* in the *Ascomycetes* or the *Uredinales* and *Ustilaginales* in the *Basidiomycetes*. Many of the *Basidiomycetes* species forming fruiting bodies (e.g. *Agaricales*, *Gasteromycetes*) are but of secondary importance from a plant pathological point of view. Therefore, only the types of fruiting structures of the *Ascomycetes* will be referred to in detail. The classification and identification of the genera and species in the *Ascomycetes* happens to be based mainly on distinguishing characteristics of the fruiting bodies (Fig. 19). There are five general groupings based on these fruiting structures that are useful to phytopathologists: 1) *Hemiascomycetes*, naked asci produced without an ascocarp; 2) *Plectomycetes*, deliquescent asci produced in a cleistothecium; 3) *Pyrenomycetes*, persistant asci in a hymenial layer borne in a perithecium; 4) *Discomycetes*, asci arranged in a hymenium but borne in an apothecium; and 5) *Loculoascomycetes*, bitunicate asci formed within an ascostroma without any separating ascocarpic walls.

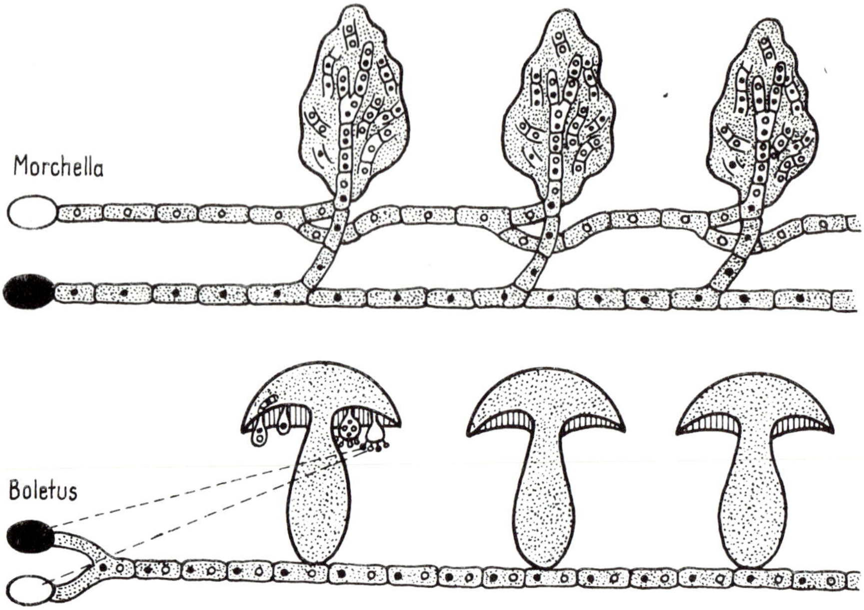

Fig. 18. Comparison of the development of an ascomycete (*Morchella*) and a basidiomycete (*Boletus*)

Phylogenetically, the simplest type of fruiting body in the *Ascomycetes* is the cleistothecium. The asci are arranged randomly in a globular or pear shaped fruiting body having a thin, membranous wall or a loose peridium. At maturity, the cleistothecium ruptures irregularly without a preformed opening, freeing masses of ascospores already liberated from the asci.

The second most simple fruiting structure is the type characteristic of the *Loculoascomycetes.* Here the typically double-walled (bitunicate) asci rest in cavities of the stroma (locules). The cavities bearing the asci have no independent walls so they are not really fruiting bodies like a perithecium. But the asci produced in a single cavity in a stroma is extraordinarily similar to true perithecia. The name applied to this structure is pseudoperithecium (e.g. *Mycosphaerella*).

The formation of true perithecia surrounded by well-developed, often hard peridia, represents a still higher stage of development (*Pyrenomycetes*). There is often an opening (pore or ostiole) at its apex. The asci are not arranged irregularly but actually form a continuous fertile layer, called the hymenium, at the base of the fruiting body. Asci may not be the only structures found in the hymenium, for sometimes paraphyses or sterile filaments originate in the bases lining the fruiting structure. Those sterile filaments not originating in the hymenium but

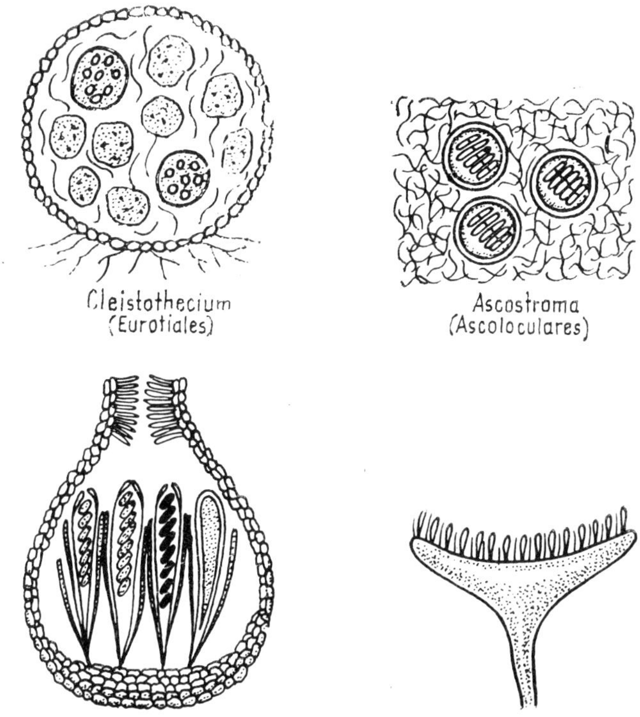

Fig. 19. Types of fruiting bodies of ascomycetes

rather from the outer covering wall of the perithecium extending into the hymenium are called pseudoparaphyses. The short sterile filaments found lining the inner part of the ostioles are called periphyses (Figs 20, 21).

The highest developed forms of the *Ascomycetes* are characterized by producing their ascospores in an open ascocarp or apothecium (*Discomycetes*). This type can be deduced evolutionally from the perithecium. The apothecium, at least at maturity, is similar to a perithecium made into an open saucer or cup-shaped fruiting body with its upper surface covered with a fertile hymenial layer. The apothecium may be sessile or have a long stalk (stipe).

In the classification and identification of the *Basidiomycetes*, basidium formation is the most decisive characteristic. As mentioned earlier, a fruiting body is not formed in the course of the ontogenesis of the phytopathologically important

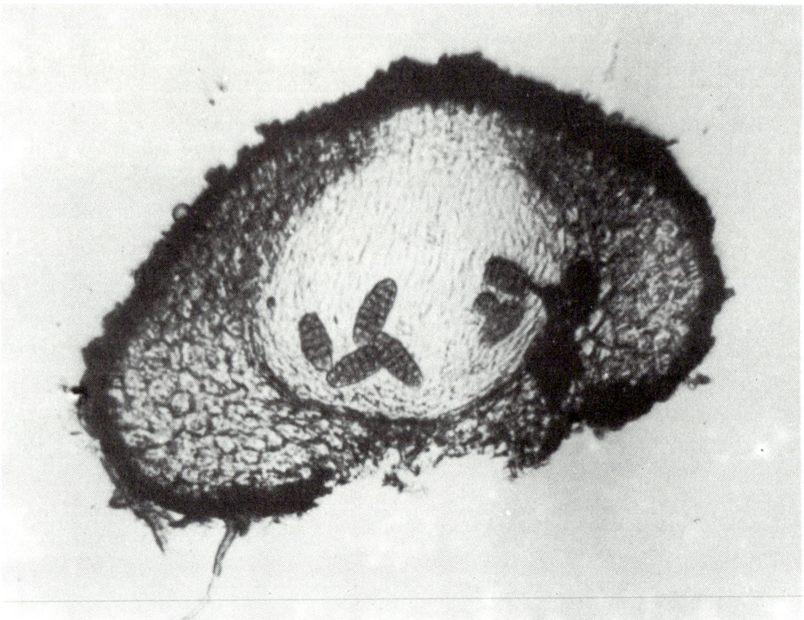

Fig. 20. Cross-section of an ascostroma (*Pleospora herbarum*)

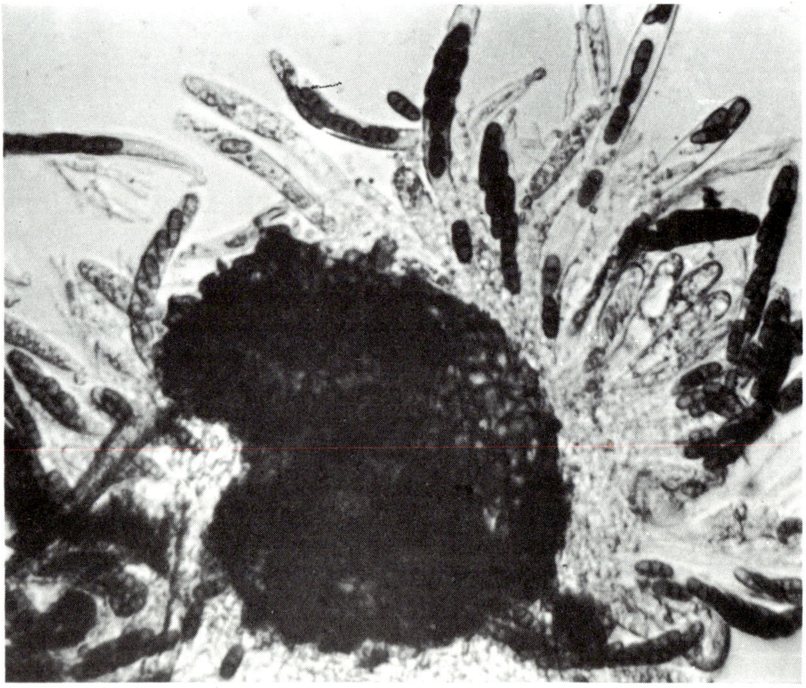

Fig. 21. Fruiting body and asci (*Leptosphaeria sp.*)

smut and rust fungi. Within the two groups, at the end of the dikaryophase, that cell from which the basidium will develop forms a resting spore. Such a thickwalled, encysted "basidium mother cell" is called a teliospore. For this reason the collective name of *Uredinales* and *Ustilaginales* is termed the *Telosporae* by some authors.

Two principal types of basidia are formed. The homo- or holobasidium is unicellular, while the hetero- or phragmobasidium is four-celled or multicellular, divided by longitudinal or transverse walls. The homobasidium can be cylindrical (stichobasidium) or clavate (chiastobasidium). There are two large groups of the homobasidial or nonseptate type: *Hymenomycetes* that form an exposed hymenial layer, and contain a number of plant parasitic forms, and the *Gasteromycetes* that never expose their hymenial layer, and contain no plant parasitic forms. Two types of heterobasidia can be distinguished. A septate basidium divided by longitudinal walls (*Tremellales*) or by transverse septa into a four-celled promycelium with each cell producing one basidiospore on a short sterigma (*Uredinales*) or one sessile, without sterigmata (*Ustilaginales*) (Fig. 22).

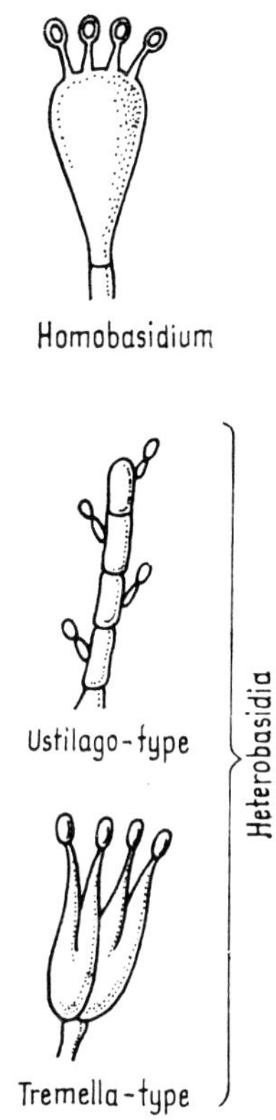

Fig. 22. Types of basidia

The following terms are used for the various regions of the basidium (Donk, 1954; Talbot, 1954). In the holobasidium type, the basidium body is the metabasidium, the basal, thick part of the sterigma is the protosterigma, and the tapering apical point is the spiculum. The same terminology also can be applied to the *Tremella*-type basidium septated by longitudinal walls. Concerning the septobasidium or septate basidium of the teliospore (smut and rust teleutospores) from which the basidium (or promycelium) is the metabasidium, the basal part of the sterigmata is the protosterigma, its tapering apex is the spiculum. According to other authors, the teliospore is the hypobasidium and the promycelium is called the epibasidium.

In connection with the dikaryophase, an additional phenomenon of simultaneous division in the binucleate mycelium that is important from both the ontogenetic and phylogenetic points of view should be considered. Further divisions of the dikaryotic cell following plasmogamy of two nuclei from different haploid hyphae take place in a characteristic way in the *Basidiomycetes*. From the two sexually different nuclei (*A* and *B* nucleus) one of them (*A*) moves into the basal end, the other (*B*) into the terminal end of the elongated apical cell of the growing hypha. At the beginning of conjugate nuclear division and at the same time as the separation of the cell nuclei is affected, a clamp connection arises to form a hooking-back extension on itself. The plane of nuclear division in the *A* nucleus is longitudinal, resulting in one of the daughter nuclei passing into one end of the elongated basal cell, the other into the terminal end of the basal cell. Meanwhile the *B* nucleus also will be elongated in nuclear division, but its spindle is oblique, so that one of the daughter nuclei passes to the apex of the basal cell, but the other into the clamp connection. After the complete separation of the daughter nuclei, the clamp connection bends back to the basal part of the dividing parent cell whereupon the point of contact, the cell wall is dissolved so that one of the

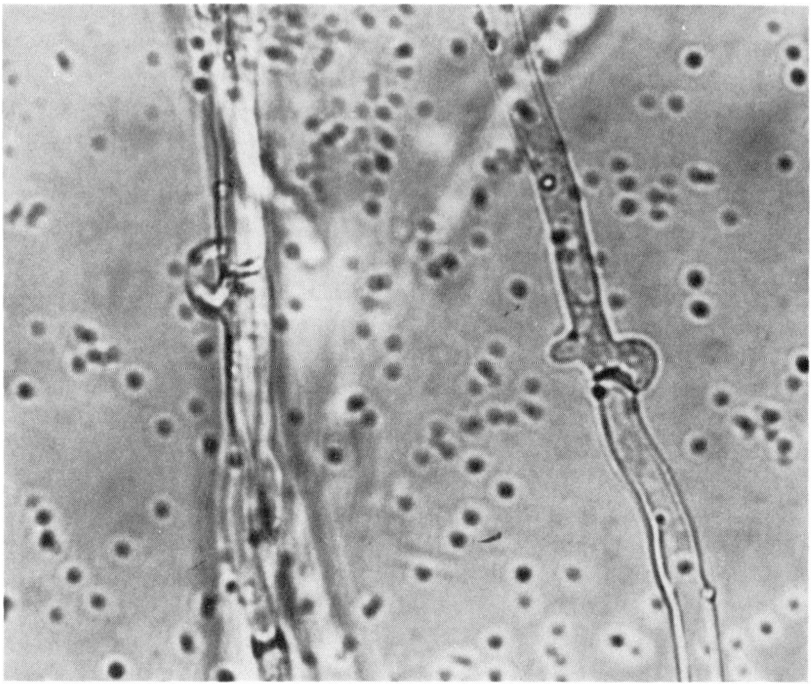

Fig. 23. Clamp connections (*Merulius lacrymans*)

daughter nuclei (*B*) will be paired with an *A* daughter nucleus located in the basal part of the parent cell. Immediately septa are formed at the middle of the elongated parent cell and the place of origin of the bent clamp connection. The conjugated division is finished with this, and in this way one *AB* nucleus pair is located in the new apical cell. This process is repeated in every division until finally at the end of the dikaryophase, karyogamy takes place in the developing basidium (Fig. 23).

A process homologous to clamp connection formation occurs in the case of the more highly developed *Ascomycetes*. Here, in contrast to the *Basidiomycetes*, the dikaryophase is restricted only to a few cells which are unable to carry on a separate existence either parasitically or saprophytically. At the time of cell division, the apex of a binucleate ascogenous hypha bends back on itself like a clamp connection, while the conjugate cell division of the nuclei take place as outlined above in the clamp connection formation of the *Basidiomycetes*. After the two septa are formed, the bent, transversely septated cell at the apex becomes the binucleate ascus mother cell, and the tip and basal cells that are uninucleate soon fuse without uniting to form another crosier hook from this newest binucleate basal part. The process is repeated over and over again until karyogamy occurs and the development of the asci is completed.

With certain fungi, male and female sex organs (sex cells) originate on the same thallus, but on others, male or female organs occur in separate individuals. In the first case, the hermaphroditic fungi are referred to as being monoecious, and in the latter case, dioecious. In those fungi in which the sexual organs cannot be separated morphologically, that is, the male and female sex organs (sex cells) are similar and not different from each other, one cannot speak about monoecism or dioecism in a morphological sense.

Those monoecic fungi which are self fertile are called homothallic species. In this case, a single individual is able to reproduce sexually. However there are self-sterile monoecic fungi in which the male sex organ of one individual can fertilize only the female sex organs of another individual but not its own female organs. These self-sterile monoecic fungi, as well as all dioecic species are referred to as heterothallic fungi. More types of heterothallism are known, but it is not our aim to give a detailed description of sexual compatibility here.

Among the fungi, genetic recombinations are insured, by the described sexual processes, but also by two more phenomena. These are heterokaryosis and parasexuality (see chapter on the variability of microorganisms page 291).

Many fungi have not only one, but more than one nuclei in their asexual reproductive cells and vegetative hyphal cells. If there are genetically different nuclei side by side in the same cell, this phenomenon is called heterokaryosis. A fungus in which this occurs is a heterokaryotic fungus.

The parasexual processes differ from sexual processes (plasmogamy, karyogamy, meiosis), but are equivalent to them. They take place at no certain definite period in the life cycle or portion of the thallus and can occur anytime and even in fungi reproducing sexually.

The general characterization of fungi would be incomplete without data referring to their ultrastructure. Those results described above are based on many decades of light-microscopic investigations, but the finer details of cell structure of fungi are beyond the resolving power of the light microscope. Electronmicroscopic studies of the cell structure of fungi hardly looks back on a past of ten years. It is natural therefore, that from the relatively little data accumulated during this time, one cannot generalize, even in retrospect. Nevertheless, there is no doubt that submicroscopic research already has discovered some very important facts from a general biological point of view, e.g., the almost identical pattern of flagella structure of various organisms in the living world, or the presence of lomasomes exclusively in fungous cells. An excellent summary of this research is given by Moore (1965). Here, only certain submicroscopic structures of fungi will be dealt with and these are as follows: cell division, septum types, lomasomes and penetration of haustoria. Research on the universality of similar flagellata structure in widely separated biological groups will be discussed in the chapter dealing with phylogeny and taxonomy.

Cell division and septum formation in bacteria, and in the *Actinomycetes*, and in reproductive cells of fungi originating from hyphal fragmentation, occur in such a way that the plasma membrane turns inside out at a definite place in the side wall in the form of a peripheral ring by extending in the direction of the middle of the cell. Thus at the place of a future septum, an almost parallel, double plasma membrane layer develops, meeting in the centre of the cell. This is dissolved at the point of contact and the two daughter cells become separate cells. Meanwhile, either the old cell wall separates along the perimeter penetrated or a new, inner, second cell wall layer is formed on the whole surface of both daughter cells. In the latter case, the side wall of the daughter cells or spores is double and the surface of the disjunction is limited by a simple cell wall. After the formation of the new cell-wall layer, the separation of the daughter cells is completed (Chapman, 1959; Moore, 1963).

The process outlined here thus means a complete and perfect cell division when the formed daughter cells actually separate from each other, but they may remain attached for a shorter or longer time together in chain formations (bacteria, *Streptomyces* spore chains, aeciospores of rust fungi, etc.). Other processes take place, however, at septum formation in the hyphae of higher fungi but not in the coenocytic ones (Moore, 1963; Moore and McAlear, 1961). In contrast to the general process of fision in bacteria, at the beginning of cell division, the centripetal growth of the plasma membrane starts with a wider ring of origin and the constriction becomes progressively thinner towards the centre. Viewed from the side, two wedge-shaped parts approach each other, limited by the plasma membrane, and which space later is filled with cell wall substances. The central portion of the diaphragm-like septum, meeting in the centre of the cell becomes completely closed, but thin. Afterwards it bulges so much that it breaks and a pore is developed in the middle of the septum. The septum pore formed in this way is large enough in the hyphae of *Ascomycetes* and in the non-dikaryotic *Basidiomycetes* that the cytoplasm of two neighbouring cells is not only connected, but their corpuscular structures, such as the cell nuclei, can penetrate the pores without difficulty. This means, therefore, that the septated hyphae of the *Ascomycetes* and mono-karyotic *Basidiomycetes* have functionally coenocytic mycelia that forms a continuous protoplasmic connection between the cells. It has been observed a few times (Moore, 1965) that the septum pore of *Ascomycetes*, e.g., *Mollisia* is closed by a plug of high density as shown by intense electron absorption which hinders the penetration large structures through the pores, e.g., cell nuclei.

A more advanced differentation of septum formation may be observed in the dikaryotic hyphae of *Basidiomycetes* examined with the electron microscope (Moore and McAlear, 1962; Moore, 1965). The formation of the septum begins in much the same way as that found in the *Ascomycetes*, but the middle of the septum apparently becomes inflated. The septum bulging outwards in all directions becomes barrel-like in form, but with open ends. A pore is formed by this doliform structure. This insures a cytoplasmic connection between the two cells, but holds the cell nuclei separated from each other. Both ends of the dolipore structure are covered by a two-layered membranous formation called parenthesomes, which keeps the corpuscular structures from one cell from wandering into another cell. The roughly hemispherical or semielliptic parenthesoma is closely connected with the endoplasmatic reticulum of the cell.

The relatively complicated dolipore septum found in *Basidiomycetes* so far has been observed only in dikaryotic mycelia. Thus, it may be supposed that this micromorphologic differentation is to insure the preservation of the dikaryotic state and parallels the similar functions of the clamp connection which could be followed even with the light microscope. The septum types outlined here are indirect proofs that the existence of the phylogenetic line of the *Phycomycetes-Ascomycetes-Basidiomycetes* is correct. The septa of *Ascomycetes* supplied with pores means a more advanced developmental stage in contrast to the hyphae of the *Phycomycetes* which are coenocytic and without septa and can be considered primitive. Further morphological differentiation is found in the doliporus type of septum in the dikaryotic mycelium of *Basidiomycetes*. The pore is covered by parenthesomes and does not functionally insure the coenocytic character of hyphae as do the continuous pores in the *Ascomycetes* and the non-dikaryotic *Basidiomycetes*. However, it must be mentioned that according to the latest available data, the septum pore structure of the rust and smut fungi, which are regarded at the present time to be the most highly developed *Basidiomycetes* are in fact extremely simple and have septa similar to those of the *Ascomycetes* (Ehrlich and Ehrlich, 1963; Moore, 1963).

Although there are many observations and much data has accumulated in connection with the fungous cell organelles (Moore, 1965), here we can only touch upon the formation of lomasomes. Lomasome formation occurs exclusively in fungus cells and is a characteristic of the fungi. The presence of lomasomes was first detected by Girbardt (1961) in the hyphae of a wood-destroying fungus and later was noticed in many other fungi as well (Moore and McAlear, 1961). Lomasomes settling in between the cell wall and the plasma membrane are composed of a loose, light electron absorptive, amorphous substance which is found on the side and cross-walls of hyphal cells. The mode of formation is unknown and also it is not known for sure that all lomasome-like formations are homologous. Their role is debated, but it seems likely that they come into existence by means of the turning inside out of invaginations formed in the plasma membrane and are a means of excluding certain substances from the cytoplasm. It is moreover possible that they are connected with some regular excretion processes (Moore and McAlear, 1961).

The penetration of host cells by parasitic and symbiotic fungi occurs by means of a modified hyphal side branch, that is called a haustorium. The haustorium may be a simple side branch which is not ramified, but is knob-like or tassel-like.

What ever shape the haustorium takes, it never causes injury to the plasma membrane of a compatible host. One can differentiate morphologically two types of haustoria. In the first haustorium type, the cell wall of the fungus is continuous, while in the other type, the haustorium has no cell wall for some distance from the site of penetration. An example of the continuous type is that of *Peronospora manshurica*, which causes the downy mildew disease of soybeans (Peyton and Bowen, 1963). Here intercellular mycelium entering the leaf sends haustoria into the cells adjacent to the mycelium and at the point of penetration, the host cell wall grows sheath-like around the haustorium. Between the wall of the haustorium and the plasma membrane of the host cell there is a relatively thick capsule, called the zone of apposition. The fungous cell wall is continuous without breaks even in the haustorium. The other type on the whole, has the same structural build up, e.g., *Puccinia graminis*, but the difference being that the fungus wall is absent from around the neck of the penetrating knob-like haustorium (Ehrlich and Ehrlich, 1963). In this region, the haustorium is limited only by the zone of apposition and of the fungus plasma membrane. The nutrient uptake of pathogenic fungi occurs in both haustorium types through the zones of apposition without any direct contact of the protoplasms of the fungus and host. Both the fungus and host plasma membranes are uninjured and, in fact, a zone of apposition occurs between them as well as a fungus cell wall, except in the neck region of the *Puccinia* type of haustorium.

The fungi show micromorphological characteristics in agreement with these two types of haustoria, not only in fungi obtaining nourishment in a parasitic way of life but also in the so-called symbiotic ones too. For example the development of fungus haustoria in lichen colonies (Moore and McAlear, 1960). The difference therefore between a parasitic and symbiotic way of life, based on our present knowledge, cannot be explained on the basis of microhistology alone.

CHAPTER II

SYSTEMATIC MYCOLOGY

NAMING OF FUNGI

The naming of fungi with bi- or trinominals has been continued since the time of Linnaeus in the same way as it is done with higher plants. So for example, the fungus causing cucumber anthracnose is designated as *Colletotrichum lagenarium*. One can find in descriptive or taxonomic works the author's name placed after the fungus name; i. e., the name of that person or persons who first described the species scientifically, or who put it in its right taxonomical place, or genus, which is still valid. So, *Colletotrichum lagenarium* (Pass.) Ellis et Halst. means that the fungus was first described by Passerini, but incorrectly named *Fusarium lagenarium*. Later, Ellis and Halsted recognized that Passerini's fungus belonged really to the genus *Colletotrichum* and for this reason it is the valid name and the right taxonomic place for the pathogen. *Fusarium lagenarium* Pass. is the obligate synonym of *Colletotrichum lagenarium* (Pass.) Ellis et Halst. Meanwhile, the pathogen of cucumber anthracnose was observed and described by others, too, who designated it by different names, e.g., *Colletotrichum orbiculare* Berk., *Colletotrichum oligochaetum* Cav., etc. These are also synonyms but they are not based on the original type specimen and type description, therefore they are called facultative synonyms. The mark of the obligate synonym is: $\equiv$, the facultative one is: $=$. On the basis of the author's or authors' name recorded after the species name, one can establish for certain the origin of the species, and how the taxonomic position and the name of the fungus under question was modified in the course of time. Over and above this, the use of the author's name can clear up mistakes originating from homonyms. This may occur when two authors describe two different fungi, having the same name, which causes much confusion.

The method of naming fungi that is prescribed by the *International Code of Botanical Nomenclature* uses rules and recommendations to insure that the names and descriptions of fungi will be uniform throughout the world. According to the rule of priority, the earliest name is valid, if it is done in accordance with these rules, and if it does not give rise to any misunderstanding. The starting points of fungus nomenclature are the works of Fries: Systema Mycologum (1821–1832) and Persoon: Synopsis Methodica Fungorum (1801). The fungus names published since 1935 are valid only if the publication contains descriptions in Latin too. The names of genera are written with the initial letter capitalized, and the names of species or lower taxa are written with small letters. The new, Latin name usually indicates the most important and typical characteristics of the fungus.

The names of the taxonomic categories are formed according to the Rules, too. The names of species, genera, subspecies and forms are formed according to the rules of Latin grammar. If the name of a species or lower taxa originates from the name of a person or host plant, it receives a genitive case ending, e.g. *Ustilago vavilovii*, *Ustilago tritici*. The ending for the names of taxa higher than genus and species are given in the following summary:

Phylum:	*Mycota*
Subphylum:	*Eumycotina*
Class:	*Phycomycetes*
Subclass:	*Syphomycetidae*
Collective order:	*Zygomycetes*
Order:	*Mucorales*
Suborder:	*Mucorineae*
Family:	*Mucoraceae*
Subfamily:	*Mucorideae*
Genus:	*Mucor*
Species:	*racemosus*

One has to keep in mind that the taxonomic categories listed here probably do not occur as such in nature. They consist of individuals in the living world which more or less resemble each other. These taxonomic groupings are made by human beings to facilitate their survey and to enable a systematic study to be made of them. Considering the unusual variability of living organisms, and the fungi in particular, and that there are many in-between forms existing among types, one is not surprised that not all individuals can be classified into presently accepted taxonomic categories. The taxonomic groups were made on the basis of the ontogenic and phylogenetic relationships of the fungi, and these reflect our present limited knowledge. It is obvious that with the widening of our knowledge, our opinions on the systematic relationships of fungi have necessarily changed in the past, and will continue to be changed by discoveries in the future.

THE PHYLOGENY OF FUNGI AND THE FORMATION OF THE PHYLOGENETIC SYSTEM

In the broad classification of living organisms only two kingdoms (regenum) were differentiated, the plant kingdom and the animal one. The great difference in the way the higher green plants and animals obtained their nutrition was so obvious that the classification of these organisms into one or the other kingdoms was no problem. But among the lower, little-differentiated organisms, three or more types of nutrition are distinguished: autotrophic, holozoic, and such heterotrophic organisms which can absorb their nutrients only from solutions. The latter type is called a chylotrophic organism. Such organisms cannot be classified arbitrarily into either the plant or animal kingdom or their classification is only very uncertain. This is the reason that recently three, and even four kingdoms were proposed (Copeland, 1956; Whittaker, 1959; Pénzes, 1962).

There is relatively little data concerning the descent and phylogeny of the fungi and so their phylogeny and phylogenetic system is based largely on theory. As in higher plants, the most important characteristic in the classification of fungi is their mode of sexual reproduction. So it generally has been accepted to divide the fungi, into the *Myxomycetes* and *Eumycetes* and the latter into classes: *Phycomycetes*, *Ascomycetes* and *Basidiomycetes*.

The fungi have traditionally been regarded since the time of Linnaeus and almost up to the present day to be organisms belonging to the plant kingdom. Most theories derive fungi from the algae by mono- or polyphyletic ways. Accordingly, the *Phycomycetes* are derived from the *Chlorophyta*, the *Ascomycetes* from the *Rhodophyta*. Some authors presume that there is such a close correlation between these two groups that they create alga-fungus categories on the basis that the fungi are only nutritional variants of algae (Clements and Shear, 1954).

Clearly, the fungi do not only differ from plants in their means of nutrition, but their cell organization is absolutely unique (Martin, 1955). An overwhelming majority of fungi have cell walls of chitin and hemicellulose in contrast to the cellulose walls, found in higher plants. Fungus cells have an endoplasmatic reticulum similar to those of plants. Ribosomes in fungus cells, if there are any, are never joined to the endoplasmatic reticulum, but are randomly arranged in the cell. Fungus mitochondria are irregular and less rigid than plant mitochondria. Lomasomes between the cell wall and the cell membrane occur exclusively in fungus cells. Cell division in the fungi is an extraordinarily simple process compared to plants and spindle formation is rare (Hawker, 1965).

The fine structure of flagella was not only the most important discovery from a phylogenetical aspect of fungi, but also was important from a general phylogenetic point of view. That flagella of motile forms of plants, animals and fungi are composed from eleven fibrilla, two in a central position and nine of the fibrilla forming a supporting tube can be recalled from our earlier discussion. Among fungi two types of flagella occur. The surface of the whiplash flagellum is smooth while the surface of the tinsel flagellum is feathery. An exception to this 2 plus 9 fibrilla structure is the cilia of bacteria, which consist of only a single fibrillum (Koch, 1956; 1958).

We can conclude therefore that the plants, animals and fungi which have originated from a proto-type more highly developed than bacteria were equipped with two flagella. According to this hypothesis, three groups of organisms developed parallel to each other but as independent phylogenetic lines (Ubrizsy and Vörös, 1966). The autothrophy of the ancient type may be explained by a chemosynthetic process which differed from present day photosynthesis.

The phylogeny of fungi may be derived on two side branches from the ancient prototype having two flagellata. The *Myxomycetes–Plasmodiophorales* line is formed by a reduction in the number of flagella. These forms have a naked thallus without cell walls. This evolutional direction resulted in a limited morphological differentiation. On the other branch, flagellum differentation progressed and these organisms evolved to a terrestrial way of life. Both these factors led to an intense morphological and functional differentiation. The *Phycomycetes–Asco-*

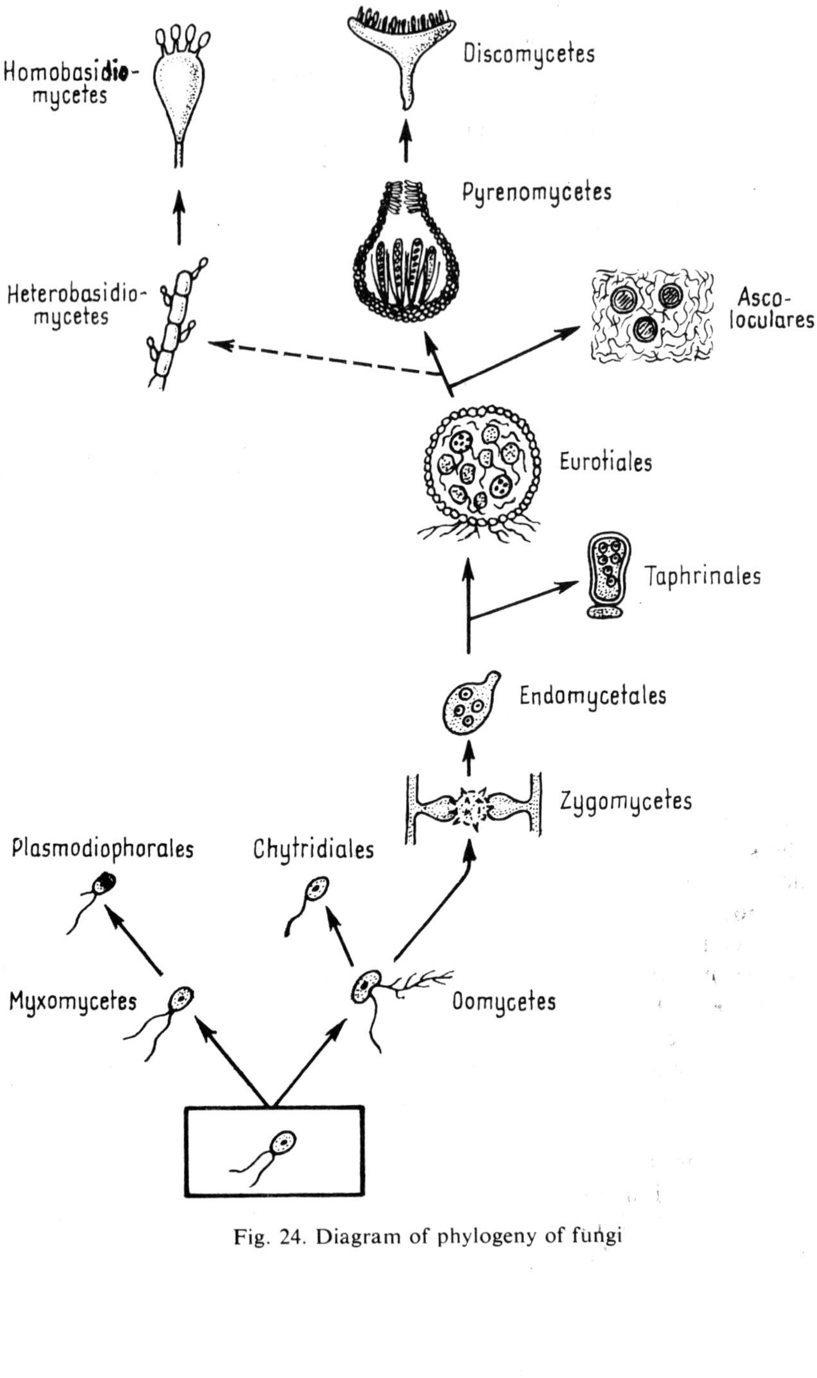

Fig. 24. Diagram of phylogeny of fungi

mycetes–Basidiomycetes line includes an overwhelming majority of the fungi that are known today.

Examples of the phylogenetical relationships are given is a simplified taxonomic chart at the end of this chapter. In the case of the *Myxomycetes–Plasmodiophorales* line, as well as in the aquatic *Phycomycetes*, flagellation serves as the best basis for derivation and classification, while in the case of the *Ascomyctes* and *Basidiomycetes*, sexual reproduction and the formation of fruiting bodies are the best characters to use for a basis of classification.

The *Phycomycetes–Ascomycetes–Basidiomycetes* line is a uniform phylogenetic one, which is also supported by much indirect evidence besides the morphological characteristics mentioned above. These are, for example, the appearance of a dikaryophase during the life cycle or between the developmental sequences of formation. Plasmogamy and karyogamy follow each other immediately without a dikaryophase in the course of ontogenesis in the *Phycomycetes*. A short dikaryophase appears at the end of the life cycle in the cause of the *Ascomycetes*, but in the Basidiomycetes, the whole thallus is derived from dikaryotic hyphae. A similar situation exists in the case of the differentiation process of the formation of the *Ascomycetes* in the septa. The pores functionally simulate the coenocytic hyphae of the *Phycomycetes*. The next step of evolutionary development of the septum pore is the functionally closed dolipore type of the *Basidiomycetes*, supplied with parenthesomes.

The phylogenetic relationships outlined here represent the limits of present knowledge. As mentioned above, increased study of ontogenesis and cell structure of the fungi could result in new phylogenetic considerations in the future (Fig. 24).

The following simplified taxonomic key helps to fix a general characterization of the most important taxonomic groups. This, of course, does not even approximately include all of the known taxa and because of the key's simplicity, arbitrary contractions were made. The main points of the dichotomic key refer in some cases to classes, in others to collective orders, or to even series, but the phytopathologically or systematically more important genera have been listed in brackets in each case.

1 a) Thallus without a cell wall, the motile forms with two whiplash flagella:
Myxomycetes (*Plasmodiophora*)
b) Thallus with cell wall, or if it is naked, then the motile form does not have two whiplash flagella 2
2 a) Thallus is unicellular, rudimentary, no true mycelium, the motile form with one wiplash flagellum:
Chytridiales (*Olpidium*, *Synchytrium*)
b) Thallus more or less well developed mycelium, the motile form—if it occurs—has two flagella, one whiplash, the other tinsel 3
3 a) Thallus aseptate, coenocytic mycelium 4
b) Thallus is septated, multicellular mycelium 5
4 a) Sexual reproduction is oogamy, motile forms occur:
Oomycetes (*Pythium*, *Phytophthora*, *Peronospora*)

b) Sexual reproduction is zygogamy, motile form does not occur:
Zygomycetes (*Mucor*, *Rhizopus*)

5 a) Sexual reproduction known 6

6 a) Sexual reproduction unknown (*Deuteromycetes* or *Fungi imperfecti*) . . 16

b) Sexual reproduction results in ascospores:
(*Ascomycetes*) 7

c) Sexual reproduction results in basidiospores:
(*Basidiomycetes*) 13

7 a) Asci formed freely, but no fruiting body occurs:
(*Hemiascomycetidae*) 8

b) Asci originating in, or on fruiting bodies:
(*Euascomycetidae*) 9

8 a) Non-phytopathogenic species, mainly yeasts:
Endomycetales (*Saccharomyces*)

b) Plant pathogenic species, asci form a layer on the surface of the host:
Taphrinales (*Taphrina*)

9 a) Asci formed in an opened fruiting body:
Discomycetes (*Rhytisma*, *Sclerotinia*)

10 a) Asci produced in a thin-walled cleistothecium:
Eurotiales (*Eurotium*, *Carpenteles*)

b) Asci not produced in cleistothecium 11

11 a) Asci arising in stroma cavities:
Ascoloculares (*Elsinoe*, *Ophiobolus*, *Leptosphaeria*, *Didymella*, *Pyrenophora*, *Pleospora*, *Venturia*, *Mycosphaerella*)

b) Asci arising in a true perithecium 12

12 a) Thallus develops on the surface, mildew fungi:
Erysiphales (*Erysiphe*, *Sphaerotheca*, *Uncinula*, *Microsphaera*, *Podosphaera*, etc.)

b) Thallus immersed mainly into the substratum:
Pyrenomycetes (*Nectria*, *Claviceps*, *Glomerella*)

13 a) Basidium unicellular:
Homobasidiomycetidae (*Exobasidiomycetes*, *Hymenomycetes*, *Gasteromycetes*)

b) Basidium four-celled:
(*Heterobasidiomycetidae*) 14

14 a) Fruiting body is generally gelatinous:
Tremellales

b) Non-fruiting bodies, but probasidium (teliospore) sori arise:
Teliosporae 15

15 a) Basidiospores arising on sterigmata, probasidium is a rust fungus teleutospore:
Uredinales (*Puccinia*, *Uromyces*, *Phragmidium*, *Melampsora*, *Cronartium*, etc.)

b) Basidiospore is sessile, the probasidium is a smut spore:
Ustilaginales (*Ustilago*, *Sorosporium*, *Tilletia*, etc.)

16 a) Conidia or blastospores are formed 17

b) Asexual reproductive cells do not form, the thallus is sterile mycelium, sometimes sclerotia, or pseudosclerotia occur:
Myceliales or *Mycelia sterilia* (*Rhizoctonia*, *Sclerotium*)

17 a) Conidia developed on surface or on conidiophores:
Moniliales or *Hyphomycetales* (*Monilia*, *Botrytis*, *Verticillium*, *Pyricularia*, *Cercospora*, *Helminthosporium*, *Alternaria*, *Fusarium*)

b) Conidia formed in a more or less closed pycnidium:
Sphaeropsidales or *Phomales* (*Phoma*, *Ascochyta*, *Septoria*, *Diplodia*)

c) Conidia formed in acervuli:
Melanconiales (*Gloeosporium*, *Colletotrichum*)

CHAPTER III

ISOLATION OF PHYTOPATHOGENIC FUNGI

Microscopic fungi that are important in agriculture, seldom occur alone in their natural habitat without other microorganisms. If we wish to identify them, or to work with them they have to be separated from other contaminating microorganisms. It is sometimes possible to obtain a pure culture by directly inoculating a suitable medium. However, in the majority of cases, special isolation methods are needed, the specific kind depending on the nature of the fungus in question and the substratum or host plant involved.

Most obligate parasites (the rust, powdery mildew, and the *Peronospora* downy mildew fungi) cannot be cultivated on artificial media. Descriptions of methods connected with these fungi may be found in the chapters on "Artificial inoculation methods" and "Variation and specialization of pathogens". These fungi can be maintained only on living plants so the separation of mixtures of the species or biotypes is only possible on a suitable host plant species or cultivar.*

It is not always necessary to cultivate facultative saprophytic or facultative parasitic fungi in pure culture for identification purposes. For example, for the fungi that form pycnidia or perithecia, it is advisable to examine them on their natural substrate, the host plant. Identification is based on the morphological characters of the fruiting bodies. The majority of the fungi producing pycnidia, do not form their characteristic fruiting bodies on artificial media, possibly because of the abundance of nutrients that are present. They may produce colonies of sterile mycelium or produce conidia on free conidiophores. Many *Ascomycetes* when grown on an artificial medium produce a sterile form also, or only its conidial form appears in the culture. It is recommended, therefore, to wait for the ripening of the fruiting structure in the natural habitat. The plant part excised prematurely should be kept under conditions similar to the ones in the natural state until the fruiting bodies ripen. If stored under too high humidity or temperature, for example, the plant parts may become covered by mould fungi which often make it impossible to identify the pathogenic fungi.

* Williams and co-workers (Williams et al. 1966) published a method for the successful *in vitro* cultivation of *Puccinia graminis* f. sp. *tritici*. In fact the fungus, regarded as an obligate parasite, formed normal and infective uredospores and teleutospores on artificial media (Williams et al. 1967). But the problems of artificially cultivating the majority of the obligate parasitic fungi have not been solved yet.

Isolation of fungi producing aerial spores from living plants, often succeeds with the direct transfer of the mycelium, or spore mass formed on the surface of the plant, onto a new medium. If the fungus growth or spore formation is not perceivable on the surface of the plant, the plant part must be put in a moisture chamber to obtain them. Water-soaked filter paper in Petri dishes serves this purpose well. The fungus grows from the inner portion to the surface of the leaf, stem, or root part when placed on the moistened filter paper, and humid atmosphere. The transfer inoculation from the colony formed in this way can be accomplished easily. Special attention should be given to insure that the moisture chamber is kept at an optimal temperature needed by the fungi for colonization. It often happens that secondary, saprophytic microorganisms present on the surface of plants, grow quickly in the moisture chamber and may inhibit or overgrow the phytopathogenic fungi sought in the isolation. In such cases the surface of the plant part must be surface disinfected with corrosive sublimate or hypochlorite solution before putting them into the moisture chamber. The incubating plant parts will be less mouldy if the plant stem or leaf parts are pushed into a 2 cm thick layer of moist, sterile sand that can be substituted for the filter paper in the glass moisture chamber.

The isolation of living phytopathogenic fungi from inside infected host tissues may be obtained without waiting for fruiting body formation or without other identifiable structures. In this case, after surface disinfection, the infected plant structure will be cut into small pieces with a sterile knife. The excised portions are placed on the surface of a suitable medium and incubated until colonies appear. The hydrogen-ion concentration of the medium may be adjusted to a low pH with 50 per cent lactic acid. Special differential media that can favour the growth of one organism over another are available also.

For the isolation of phytopathogenic fungi from rotting parts of plants which are largely contaminated by other microorganisms or soil, the most frequently used methods are by streaking and diluting. The streak method is used entirely for qualitative purposes. A water suspension is made from the material being investigated. This will be diluted according to need and afterwards, one drop will be picked up and streaked with a wire loop to the surface of a solidified agar medium. At the point where the loop first touches the medium and for a certain distance away from it, continuous microbial growth is observable after a suitable incubation period. But towards the end of the streaking stroke if a suitable initial dilution was made, discrete, isolated colony microorganism will be obtained. By transferring the isolated colonies to fresh media, pure cultures may be obtained. The streaking can be done in two ways: either by making a zigzag line across the surface of the agar, or after two or three short parallel streaks have been completed, by making two or three new parallel ones starting from the leading edge of the first streaks and at right angles to them. These are made with a flamed loop and this pattern repeated twice. The latter procedure causes a much greater decrease in the number of the microorganisms than the zigzag stroke.

The plate dilution technique is simply the dilution of the given material into a medium. This method is suitable for quantitative as well as qualitative purposes, if a known quantity of the test material, such as soil, is used. Soil suspension of 1 : 10, 1 : 100, etc., dilutions are made, and 1 ml aliquots of the various dilutions with water are mixed in Petri dishes with a known quantity of an agar medium, held in a liquid state at 45 °C. Only a few isolated colonies will be obtained for some dilution rates. The microorganism content of the initial material may be determined on the basis of the rate of the dilution, and the number of colonies obtained, while at the same time, isolated colonies in pure culture are acquired.

Warcup's soil plate method is a modified dilution technique. The essence of this method is not to make a soil suspension but to add a small amount of soil (5–15 mg) directly to a liquid agar medium (Warcup, 1950). A great advantage of this method over the plate dilution one is that those fungi grow on the plate which occur only in soil aggregates or adhere to rapidly settling particles and which may not be cultured by plate dilutions.

With especially large populations of microorganisms, even the very small amount of soil suggested by the Warcup method still resulted in uniform, continuous microorganism growth on the plates. Therefore this suggested that the number of colonies could be reduced by using sterile sand to dilute the sampled soil before plating (Johnson and Manka, 1961).

With the Rossi-Cholodny "contact slide" method it is possible to isolate the microorganisms, especially fungi, from the soil in a pattern corresponding to their original arrangement in the soil (Cholodny, 1930). Sterile, microscopic slides are pressed on split-open surfaces on fresh soil. After a few days incubation each slide is freed from the larger soil particles by shaking, then placed on the surface of an agar medium plate with the side down that was in contact with the soil. Fungus growth is directly observable with low magnification and after a suitable incubation period, the colonies are transferred to fresh media.

In the laboratory in which only qualitative data is required, finely powdered soil particles or soil suspensions are sprayed routinely onto the surface of solidified agar media instead of using a dilution method. For the isolation of some soil-inhabiting phytopathogenic fungi (e.g., *Phytophthora* species), various bait methods are used. The essence of these methods is to plant a cultivar of host plants that is extremely susceptible to the pathogen, or to place sterile pieces of the host into the soil. The isolation of the pathogen from an infected plant or plant part, is accomplished easier than from the soil directly.

Agnihothrudu (1962), Cooke (1963) as well as Johnson et al. (1959) have reviewed the isolation methods for fungi occurring in the soil and in the rhizosphere of plants.

CHAPTER IV

MAINTENANCE OF FUNGI

Microscopic fungi must be maintained in a viable form and in a pure culture throughout the course of an experiment or an investigation. The simplest procedure is to transfer the culture at certain intervals to culture tubes of freshly slanted agar media. Definitely uncontaminated cultures are chosen for maintenance after having developed well and having covered the whole surface of an agar slant.

These are stored at room temperature or in refrigerators at a 2–5 °C range. Using the latter procedure, one has to transfer the fungi to fresh media less often because the cultures remain viable for longer periods at lower temperatures. By storage in the refrigerator, drying of the culture is reduced to a considerable extent, too. To further prevent and reduce drying of stored cultures, the tube is flamed and the cotton plug is pushed in below the mouth of the tube and the mouth covered with metal caps, aluminium foil, plastic, or paraffin sheets.

The intervals between transfers of the agar slant cultures to fresh media varies according to the survival time of the fungus. *Aspergillus*, *Penicillium* etc. species generally can survive 8–9 months without being transferred, while other fungi generally need to be transferred every 3 weeks. It is always advisable to keep stock cultures stored in two different places and in duplicate samples to be certain of maintaining the culture. Cultures may easily become lost, because of accidental microbial or arachnid mite contamination or because of mechanical failures in refrigeration.

The longer a fungus is cultivated under artificial conditions, the greater is the possibility of small or even large changes occurring not only in the nutritional and physiological characteristics but also in morphological ones. This is obvious because by means of sexual recombination, parasexuality, mutation or by adaptation, new characteristics can arise or some of the original characteristics can disappear in the course of their cultivation. During the course of an experiment, it is most often the aim of the experimenter to preserve the original characteristics of the fungus in culture on artificial medium as much as possible.

A common phenomenon that occurs with plant pathogenic fungi cultivated on artificial media for long periods of time is that their pathogenicity is lost, or at least their virulence is decreased. In the same way the intensity or the production of certain metabolites being studied may be reduced. Many fungi sporulate well even after their initial isolation, but after many transfers on various media, the culture consists more and more of vegetative mycelium and less and less of spores or at least the intensity of the sporulation decreases. One part of this phenomenon, besides the genetical changes that may occur during cultivation is due just to the abundance of nutrients in the medium that differ from the nutrients supplied by the original host or substrate. Such modification or "degeneration" of the stock culture can often be corrected by cultivating the fungus under conditions that are poor in nutrients (e.g. in soil) or on the original substratum, or in the case of the phytopathogenic fungus, by passage through a suitable host plant several times. Using such corrective measures, the original characteristics of many fungi are often regenerated.

The danger of contamination by other microorganisms or arachnid mites cannot be ignored in the cultures maintained by periodic transfers.

The disadvantages of maintenance in agar slant cultures are hardly compensated by the fact that such a culture does not require special effort for maintenance and that it is available for transfer and for direct examination at any time.

Hence a number of procedures have been worked out which do not require frequent serial transfers or cultivation on artificial media over long periods of time. The most commonly used methods among these procedures are covering agar slant cultures with oil, preserving cultures by the soil or sand culture method, and by lyophilization.

The preservation of the agar slant culture using oil is accomplished by pouring sterile, refined paraffin oil over the actively growing culture until the entire culture is covered by the oil. It is generally accepted that this method increases the survival time of the agar slant culture greatly.

This method is important especially in preserving colonies consisting of sterile mycelia. Such fungi cannot be lyophilized or only inadequately so due to the lack of spores. Covering cultures with sterile water to preserve them was applied for the first time in phytobacteriology only recently (Kelman, 1956). Later water was shown to preserve the viability and the sporulation ability of pycnidious fungi. *Ascochyta* species can be maintained for 20–25 months on agar slants with sterile water covering the culture and storage at 10–12 °C (Person, 1961).

Two methods are used for preserving cultures in sterile soil or sand.

a) The spore suspension in a small amount of water is added to dry sterile soil and after mixing in the tube, it is air dried quickly at room temperature. There is not enough moisture supplied for much, if any, growth.

b) The spore suspension is added to wet sterile soil and the fungi are incubated in the soil. Drying occurs slowly. When the seeded tubes have returned to dryness, the cultures are stored at a low temperature.

The first is actually a spore conservation. Many fungi maintained in this way keep their viability for up to 27 years (Raper and Fennell, 1965). The advantage of maintenance in a soil culture is not only that the fungi are preserved unchanged but that the dry soil culture serves to provide homogeneous inoculum for very many transfers, and over a very long period of time. This method has had support mainly in the field of industrial microbiology (antibiotic production, etc.).

The maintenance of the majority of fungi, especially those that are good spore formers, is most successfully done by lyophilization or freeze drying. Descriptions of this procedure can be found on pages 167–168. Fennell (1960) has published a survey in connection with the maintenance and conservation of fungi.

CHAPTER V

MICROSCOPIC AND OTHER LABORATORY INVESTIGATION METHODS

Most fungi are directly suitable for microscopic examination without previous preparation or sectioning. An appropriate portion of a fungal preparation is chosen using the naked eye or under a preparative, stereoscopic microscope and placed into a drop of water on the slide. Some investigators use 0·1 *M* NaCl in distilled water. The cover slip is placed carefully over the specimen and pressure applied according to need. Under low magnification, the cover slip can be tapped to separate the fungus elements.

Disturbing air bubbles in the preparation can be avoided if the following solution is used instead of water:

glycerol	10 ml
distilled water	20 ml
90 per cent ethanol	30 ml

For such native preparations, it is sometimes better to use Amann's solution or lactophenol, instead of water. The composition of Amann's solution is as follows:

lactic acid	20 g
glycerol	40 g
phenol	20 g
distilled water	20 ml

If the preparation is made from dry material, a 2–10 per cent potassium hydroxide solution is used for swelling or rehydrating the material. If the material being investigated is too dark and not transparent enough, then boiling of the material in 10 per cent potassium hydroxide solution is advised.

Spore germination, growing mycelium and bud-cell formation can be examined most conveniently in a hanging-drop preparation.

Preparations made with water or Amann's solution can be preserved for months if the edges of the cover slips are sealed hermetically. For this purpose a mixture of paraffin and vaselin in a 1 : 1 proportion, colourless nitrolac or other varnish, which dries quickly, can be used.

Glycerol-gelatin mounting medium is the most widely used preparation for fixing fungus preparations on slides and has the following composition:

gelatin	7 g
distilled water	42 ml
glycerol	50 ml
phenol	1 g

The boiled mixture is a dense, gelatinous substance but clear in its cooled state and a small cube is cut out for each mount and placed on a microscopic slide and melted above a low flame. The fungus part being investigated is put into the melted drop and it is covered by a cover slip while still in a warm state. By pressing gently on the cover slip, the surplus glycerol-gelatin medium is eliminated. Such fixed preparations obtained in this way are preserved for a long time.

Microscopic preparations made from fungus culture or fructifications on the surface of the substrate generally do not need staining, in fact staining can be disturbing if the colour of hyphae or spores is the one means of separation used in their identification. Fungus hyphae on the inside of the substrate (e.g., inside plant tissues) can be observed and investigated directly but this is accomplished only with difficulty. Sectioning and staining is required. For staining fungus structures in the inner part of plant tissues, the most appropriate dye is cotton blue (0·05–0·1 per cent) dissolved in lactophenol. Cotton blue preparations can be cleared in lactophenol by heating the slide gently over a low flame.

The Schiff's periodic staining procedure may be used for the detection of fungus elements that are found in animal tissue or in other substrates as well. The essence of the method is as follows:

Fixation: immerse specimen in methanol for 20 min, then rinse first in ethanol, and then in tap water.

Staining: immerse in 1 per cent hydrochloric acid for 20 min, then rinse in tap water. This is followed by a period in Schiff's stain for 15 min. Prepare this staining solution by dissolving 0·5 g basic fuchsin in 100 ml hot distilled water and after the solution has cooled to 50 °C, add 5 ml 1 *N* hydrochloric acid and 1 g potassium metabisulphite. The staining solution is kept in the dark for 1 day. If it is not a light yellow colour, it should be filtered through activated charcoal. After staining the preparation with Schiff's stain, destaining twice for 5 min with a weak hydrochloric acid and potassium metabisulphite solution (0·5 g potassium bisulphite is dissolved in 0·05 *N* hydrochloric acid). After this treatment, the preparation is rinsed in tap water for 8 min.

Contrast staining: stain for 1 min with light green dye solution, followed by washing in tap water (1 g methyl green is dissolved in 100 ml hot 1 per cent acetic acid). As a result of Schiff's staining regime the fungus elements will be stained dark red and the surrounding tissues a light green.

In phytopathological course work, very often a demonstration is required to show the presence of seed-borne pathogens. In the simplest case, the signs of the pathogen are visible to the naked eye on the surface of the seeds, e.g., extensive *Colletotrichum lindemuthianum* infection in bean seeds. The pathogen on the surface of the seed or penetrating the seed is detectable with the help of a moisture chamber when the pathogen grows out and produces spores. Pathogens penetrating the seed can be cultivated from infested seeds that have been surface disinfected and ground up or homogenized and inoculated onto a suitable medium. In the case seed-borne fungi that cause rot or damping off, simply sowing the seed sample into a sterile soil placed at an appropriate temperature, the appearance of symptoms or signs can convince us about the presence of a certain pathogen. These methods are not suitable for the demonstration of all seed-borne pathogens, partly because some pathogens cannot be cultivated on artificial media and partly because their development requires a long time. Such an example is the demonstration of the presence of loose smut infection in cereal seeds. There are many fundamentally similar methods, for this purpose (see chapter on loose smut). For the detection of *Ustilago nuda* infection of barley seeds, e.g., the seeds first are boiled in a mixture of sodium hydroxide and sodium silicate solution (1 molar) for 1 hr and during this time the seed becomes a gelatinous mass. In the course of this treatment,the white embryos containing the hyphae of the pathogen can be pressed from the swollen endosperms. The embryos are collected by washing and sieving through suitable sized sieves. Clearing is accomplished by boiling the embryos in lactophenol. The fungus in the embryos can be stained by using tripan blue (0·05–0·8 per cent) dissolved in lactophenol. The fungus hyphae will stain a dark blue color and are easily observable in the cleared embryo, mainly in the scutellum.

The usual staining procedures not involving infiltration and embedding in paraffin are not adaptive or hardly suitable for studying the infection process and development of ectoparasitic fungi (powdery mildew) due to the danger of separating the fungus from the substrate by washing. Drandarevski and Weltzien (1966) successfully stained such fungi using fumes of iodine.

Staining procedures are used occasionally for very special purposes, e.g., for the identification of species that can be determined or identified only with difficulty. For example, the rust fungi *Cronartium ribicola* and *C. occidentale*, can hardly be distinguished from each other. However, the teleutospores of *C. ribicola* stained with acidic fuchsin solutions buffered to pH 4·8, retain their stain, while the stain from the teleutospores of *C. occidentale* can be easily eluted (Ford and Rawlins, 1956).

In the examination of microscopic fungi and especially in the determination of species, it is of the utmost importance to determine the size of the fungal elements, spores, fruiting bodies, conidiophores, etc. The ocular micrometer serves this purpose well. Micrometer discs are engraved most often with divisions on a scale from 0–50. The glass disc is positioned in the ocular on a flat ring built into the focal plane of the eyepiece. The specimen that is to be measured is placed in such a way that its widely separated points still fall along the divisions of the ocular micrometer. In this way we can easily establish that the diameter of a spore under a certain magnification is for example equivalent to exactly 1 division. A condition of measurement is that we know how many microns each division of the ocular micrometer means. This can be done by calibrating the ocular micrometer with a stage micrometer. A stage micrometer serves this calibration purpose because it has an engraved scale marked in tenths and hundredths of a millimeter. The smallest divisions are then exactly 10 micron divisions. If one ocular micrometer division at a certain magnification coincides with one division of the stage micrometer set at a sharp focus then at this magnification our ocular micrometer's one division would be equivalent to 10 microns. The spores therefore in the example above are 10 microns in diameter. Separate calibrations must be made for any other optical combination. Every change in magnification causes a change in the standardization. We should know the corresponding size of each ocular division exactly for each lens combination. Essentially the procedure is for every combination of lens, two arbitrary divisions of the ocular micrometer are matched with two divisions on the stage micrometer. They must be observed to coincide exactly. The number of divisions on the stage micrometer are multiplied by ten and divided by the exact number of divisions of the ocular micrometer that coincided with the lines on the stage micrometer. In this way we can obtain the number of microns that are equivalent to one division of the ocular micrometer when used at a specified magnification.

The calibration of the Filar crossed-thread eyepiece micrometer differs somewhat from the procedure described above. In this case the cross thread in the eyepiece is movable by means of a screw with a calibrated scale. The extent of movement, the distance that the thread is moved across the field can be read directly from the scale divisions. In making a measurement, the point at which the cross thread coincides with one end of a spore is recorded. After registration of the position on the scale the screw is turned again until the cross thread just reaches the opposite end of the spore. After reading this position on the scale, we may calculate from the differences of the two values, the distance between the two points or the length of the spore in terms of the eyepiece micrometer. The calibration of the eyepiece cross thread micrometer is made with the aid of the stage micrometer described earlier. By this means we may ascertain the number of microns that are equivalent to each scale division made by moving the screw of the cross thread at a certain magnification. A new standardization must be obtained for each new optical combination used.

The determination of the density of a cell or spore suspension as to the number of individual microorganisms is usually counting cells or by making a plate count from a suitable and known dilution of a spore suspension. In the latter case of counting colonies we may calculate the number of viable spores present after a certain incubation time from the number of colonies that develop and a knowledge of the degree of dilution. This procedure requires a lot of work and time. Therefore a direct cell count often is necessary. The Bürker or haemacytometer blood counting chamber is frequently used. This is a slide with a 9 mm^2 ruled surface prepared from a thick glass, having a chamber, 0·100 mm deep, cut into the slide so that a cover slip pressed down by two springs, provides a 0·1 mm thick water layer in the slide. Over each square mm on the ruled surface of the slide below the cover slip is 0·1 mm^3 of water. The boundary lines of the slide square regions, limited by double lines, are 0·2 mm apart. The central square mm is ruled into 25 large and 16 small squares. The basic area between the double lines is 0·04 mm^2. The small square area formed from the crossing of the double lines 0·05 mm apart is 0·0025 mm^2. If the layer of the liquid is 0·1 mm thick, the volume of the liquid columns over each of the 25 large squares is 0·004 mm^3, and the volume over each of the 16 squares is 0·00025 mm^3. In counting the number of spores or cells, one drop of the suspension should be placed at the edge of the cover slip to fill the chamber, as far as possible bubble free, and without overflow. The suspension should be diluted to such an extent that 2–6 cells should be visible in each of the large 0·2 × 0·2 mm squares. The number of the cells is counted diagonally in 12 of the 25 squares in touching pairs and the average number of cells determined. Considering that the quantity of liquid above each large square is 0·004 mm^3, the number of cells in the diluted suspension per mililitre can be obtained with the aid of the formula below: average number of cells above 1 large square × $\frac{1\ \text{ml}}{0{\cdot}004\ \text{mm}^3}$ or simply, number of cells counted × 250,000. This number, of course, has to be multiplied by the original dilution of the suspension to ascertain the density of the original suspension in number of cells per ml. The average number of spores above the small squares, should the counting be made above the 0·05 × 0·05 mm squares as for example in the case of dense bacterial suspensions, has to be multiplied by $\frac{1\ \text{ml}}{0{\cdot}00025\ \text{mm}^3}$ or simply 4,000,000 × average number of cells per small square × the dilution in order to obtain the number of cells per ml in the original suspension.

CHAPTER VI

MEDIA USED FOR THE CULTIVATION OF FUNGI

Fungi can be cultivated on many different kinds of media (see chapter concerned with general laboratory procedures on p. 137). These can be liquid or solid. Solid media are either such because of their very nature, e.g., potato plug and carrot, etc., or they are made so with the addition of a solidifying or jellifying material. Fungi are almost exclusively cultivated on media containing agar-agar in 1·5–2 per cent by weight when a solid medium is required. Due to the great variety of the nutrient requirement of fungi, the composition of the media used for their cultivation can also be varied. Many common moulds grow well on almost any organic material, other fungi, e.g. some *Phytophthora* species, require the presence of certain amino acids found only in certain media such as lima beans, whereas the obligate parasitic powdery mildew fungi and others cannot be cultivated on any artificial media, either natural or synthetic.

According to its composition, natural, semisynthetic, and synthetic media can be differentiated. The majority of fungi grow best on media consisting of natural, basic materials like plant or yeast extracts. But the disadvantage of these is that the same results cannot always be reproduced or only unsatisfactorily. The quality of a medium containing oat or corn meal depends on the variety of the plant the degree of ripening of the seed, the degree of finesses of the ground flour, etc. The second disadvantage of natural media is that their exact composition is not precisely known so they are not suitable for precise metabolic investigations. Synthetic media whose chemical composition is known precisely are used for such investigations. Semisynthetic media are those in which all but one or two ingredients can be duplicated exactly.

Regardless of composition most media generally is or becomes contaminated by different microorganisms. For this reason it is necessary to sterilize media before use. This may be accomplished by physical or chemical means. In most cases this is carried out by heat or, to be more precise, with steam. When one sterilizes with streaming steam as for example in a Koch's pot, the temperature of any liquid media reaches only 100 °C. This only destroys the vegetative forms of life, but heat-resistant spores survive. Therefore for complete sterilization in streaming steam, it must be done fractionally for 3 successive days for 30 min each time. The spores incubated in between the steamings will germinate and the vegetative forms or germ tubes that developed will be destroyed. The higher the temperature of the steam, the greater is the sterilizing efficiency of steaming.

Knowing this, sterilization usually is carried out in an autoclave with steam under pressure unless the materials are heat-sensitive or thermolabile. With each increase in pressure, the temperature of the steam rises and the duration of sterilization time decreases proportionally. The correlation of pressure and temperature in the autoclave may be seen from the following data:

Pressure	*Temperature*
0·5 atm.	112 °C
1·0 atm.	121 °C
1·5 atm.	128 °C
2·0 atm.	135 °C

These data show the temperature for live steam under pressure. However, the temperature of air and steam mixtures is much lower, so their sterilization effect is weaker. This is why it is important that the autoclave should not contain air during sterilization and this is accomplished by seeing that before closing, only live steam streams through the autoclave and all air has been expelled before closing the stopcock.

The sterilization of thermolabile substances can be done by filtration. For this purpose a Seitz fritted-glass filter or a No. 5 glass filter are equally suitable. One has to sterilize the filter already in place, ready for use in a filter flask, and with a suitable, non-absorbent cotton filter placed over the tube opening that goes to the vacuum pump. This is to prevent foreign microorganisms or other materials from penetrating the sterile filtered medium upon completion of filtering.

Other physical methods also may be used to sterilize heat-sensitive materials. Short exposure to the radiation from radioactive cobalt destroys microorganisms occurring in natural and synthetic media.

Exposure to the fumes of propylene oxide was shown in many successful experiments to be a useful chemical method for the sterilization of media without heat (Klarman and Craig, 1960). Thompson and Gardemann (1962) reported that 1 ml of propylene oxide added to 20 ml of medium in a Petri dish 10 cm in diameter, sterilized both the medium and all exposed surfaces within the Petri dish. So they concluded it was not necessary to sterilize by heat either the glass dish or the media before using propylene oxide. After a short but required aeration period, the medium, which had been disinfected by propylene oxide, was ready for use in the cultivation of different microorganisms, with no harmful effects. Later Goss and Marr (1963) found that the above method does not insure the necessary sterility. They proved by experiments carried out with potato-dextrose-agar that the fungicide activity of the propylene oxide was better than its bactericide effect. With the direct addition of propylene oxide to the Petri dish, bacterial contamination occurred very often on the bottom of the dish. To eliminate this, the sterilization was conducted in a closed system and with much better results. Only 1·25 ml propylene oxide was needed per litre of volume to insure proper sterility in the closed system. Propylene oxide can react with many plastic containers and its fumes are inflammable.

The following list contains the formulae of the most commonly used natural, synthetic and semisynthetic media.

NATURAL MEDIA

Potato plug

Healthy, sound potato tubers are disinfected with corrosive sublimate or hypochlorite. After washing and peeling the tubers, 3–4 cm long and 1 cm thick prismatic pieces are cut. These are cut diagonally and the wedge-shaped pieces obtained in this way are put into culture-tubes with the slant surface up. The tubes are closed with cotton plugs and sterilized in streaming steam, fractionally for three successive days.

Rice grain

Two g rice and 6 ml water are put into a tube and sterilized fractionally in streaming steam, similar to that of the potato plug.

Lupine stem

Six cm pieces or piece of lupine stem, or if necessary from other plant materials, are put into tubes or the tube is filled with small cut stem pieces until it is one-third full, then 3–6 ml water are added per tube so that after sterilization, 1–2 ml water will be left standing in the tube. The sterilization is in an autoclave at 1 atm. pressure for 30 min so that the sporeforming bacteria will be destroyed.

Hemp seed

Hemp seeds cut into two halves are sterilized in a surplus amount of water in streaming steam three successive days so that after sterilization, the seeds will still be covered with water. This medium is most suitable for the cultivation of certain aquatic fungi, mainly *Saprolegnia* species.

Prune-juice agar

Forty g dried prunes are boiled in 1 l of water in a waterbath for 1 hr and then filtered. After the addition of 20 g agar, it is sterilized in the autoclave at 0·5 atm. for 15 min. Many fungi grow well on this medium, but bacterial growth is inhibited.

Corn-meal agar

Twenty g corn meal is mixed in 1 l water and boiled in a water-bath for 1 hr and then filtered or centrifuged. After the addition of 20 g agar to the filtrate, it is sterilized at 0·5 atm. pressure for 15 min. The medium can be amended with the addition of 20 g glucose and 20 g peptone. Fungus growth will be more abundant at the expense of spore formation, while fungus growth on corn meal without amendments, spore formation is increased.

Apple-juice agar

Grated apple (200 g) is boiled in 1 l water and filtered and after 2–3 per cent agar is added, it is then sterilized in streaming steam. The apple infusion can be naturalized by sodium hydroxide, if it is desired.

Oatmeal-agar

One hundred g oatmeal or oat flakes are boiled in 1 l of water in a water-bath for 1 hr at 60 °C. After filtering through coarse linen and the addition of 3 per cent agar, it is sterilized in an autoclave at 1 atm.

Oatmeal agar (Johnson et al., 1959)

Oatmeal (17 g) is boiled in 500 ml water in a water-bath for 1 hr and filtered through gauze and 17 g agar is melted in 500 ml water and mixed with the oatmeal filtrate. One g yeast extract is added to the medium, if it is desired. It is sterilized at 0·5 atm. pressure for 20 min. This medium is recommended for the maintenance of fungi and *Actinomycetes* cultures.

Potato-dextrose-agar

Two hundred g scrapings made from surface disinfected, washed and peeled potato tubers are boiled in 1 l of distilled water in a water-bath for 1 hr. After filtration, 20 g glucose (in the case of some fungi, sucrose) and 15–20 g agar is added to the filtrate and sterilized at 0·5 atm. pressure for 15 min. This medium is convenient mainly for the cultivation of phytopathogenic fungi, but many bacteria grow well on this medium too. Cotton discs, soaked in potato dextrose broth and placed into Petri dishes are especially suitable as a medium for the comparison of the stromata formation of the members of *Sclerotiniaceae* (Noviello and Korf, 1961).

Malt extract agar

Twenty to 50 g commercial malt extract concentrate and 20 g agar are melted in 1 l water. The medium is sterilized at 0·5 atm. pressure for 15 min. It is recommended for the cultivation of many fungi e.g. *Mucor* spp. and wood destroying fungi.

Czapek's solution

sodium nitrate	3·0 g
potassium monohydrogen phosphate	1·0 g
magnesium sulphate	0·5 g
potassium chloride	0·5 g
ferrous sulphate	0·01g
sucrose	30·0 g
distilled water	1000·0 ml

Sterilize at 0·5 atm. pressure for 15 min. Instead of sucrose, glucose can be supplied as the carbon source, if it is desired. Without the addition of the agar it can be used as Czapek's nutrient solution. If partial caramelization of the sugar is not desirable, the sugar can be dissolved in a small amount of water and sterilized by filtering through a Seitz filter and added aseptically after autoclaving is completed. The Czapek's medium suitable for the cultivation of many fungi and for the determination of species of *Penicillium* and *Aspergillus*. The species belonging to the *Aspergillus glaucus* group grow better if the sugar concentration is raised (15–50 per cent) in the medium. The Czapek's medium is supplemented often with additional materials like vitamins, amino acids, or other natural substances. The latter amendments make this a semi-synthetic media.

Synthetic acid medium

magnesium sulphate	0·5 g
potassium dihydrogen phosphate	1·0 g
peptone	5·0 g
glucose	10·0 g
agar	25·0 g
distilled water	1000·0 ml

The medium is autoclaved and cooled to 40–50 °C before aspectially adjusting the medium to pH 4·0 with sulphuric acid. If the pH is adjusted before sterilization, the agar will not solidify. This medium is especially suited for the isolation of fungi from the soil.

Peptone-glucose agar with rose bengal and streptomycin or aureomycin (Johnson et al., 1959)

potassium dihydrogen phosphate	1·0 g
magnesium sulphate	0·5 g
peptone	5·0 g
glucose	10·0 g
agar	20·0 g
distilled water	1000·0 ml

"rose bengal"	3·3 mg
streptomycin	30·0 mg
or aureomycin	2·0 mg

The components of the medium, except the rose bengal dye and the antibiotics, are brought to a boil slowly while stirring. Then 10 ml of a 1 : 300 dilution of the "rose bengal" solution is added to each litre of medium. After sterilization at 1 atm. for 30 min the medium is cooled to 45–60 °C and the antibiotic is added. This is the most useful medium for the isolation of soil fungi, especially when used with the plate dilution technique, because this medium inhibits not only the bacteria and *Actinomycetes* found in most soils, but prevents spread of quick-growing fungi that quickly obscure other colonies on the plate. Also the addition of 0·4 per cent sodium propionate into various media may be used successfully (Crook et al., 1950) for the elimination of bacteria and certain fast-growing fungi.

Rose-bengal-streptomycin medium and media amended with other antibiotics are used successfully, especially for the isolation of *Pythium* species from soil (Singh and Mitchell, 1961, Schmitthenner, 1962) as well as fungi from the air in the mycological analysis of air samples (Rogerson, 1958).

Synthetic nutrient solution for the cultivation of Phytophthora infestans (Henniger, 1959)

potassium dihydrogen phosphate	40 mg
sodium nitrate	20 mg
calcium chloride	15 mg
magnesium carbonate	15 mg
sodium sulphate	10 mg
ferrous sulphate	2 mg
succinic acid	20 mg
D- or *DL*-aspartic acid	60 mg
L-glutamic acid	40 mg
alanine	20 mg
glycine	20 mg
cystein hydrochloride	10 mg
thiamine	5–10 gamma
sucrose	1·5 g
distilled water	100 ml

Sterilize at 0·25 atm pressure for 30 min. The unadjusted pH of the medium is 5–5·2. A solid medium can be prepared with the addition of 1·5–2 per cent agar. If it is desirable to complement the substrate further with natural materials in the interest of good growth, powdered potato tuber or rye seed extract can be added. Rye seed extract is prepared by autoclaving 100 g rye seed in 1 l of water at 2 atm pressure for 30–40 min.

Water agar

20 g agar is melted in 1 l water and then autoclaved. For many fungi, such a medium, poor in nutrient supply, increases spore formation. Good sporulation can be obtained on simple water agar, but growth is usually sparse.

SEMI-SYNTHETIC MEDIA

Malt-Czapek's agar

Czapek's agar is complemented by adding 2 per cent malt extract. The majority of moulds (*Aspergillus*, *Penicillium*, *Mucor* species) and plant pathogenic fungi grow well on this medium.

Yeast extract-Czapek's agar

Czapek's agar is complemented by adding 1 per cent yeast extract. This is prepared by suspending 100 g commercial baker's yeast in 400 ml water and autoclaving the mixture at 2 atm. pressure for 30 min. After sedimentation or centrifuging, the supernatant is added to Czapek's medium. *Pythium*, *Phytophthora* and other phytopathogenic fungus species develop well and *Colletotrichum lindemuthianum* produces numerous conidia on this medium.

Leonian's solution agar

potassium dihydrogen phosphate	1·25	g
magnesium sulphate	0·625	g
peptone	0·625	g
maltose	6·25	g
malt extract	6·25	g
agar	20·0	g
distilled water	1000·0	ml

This medium is most suitable for the study of fungal fructification and the formation of fruiting bodies. The mode of spore formation of the fungi cultivated on Leonian's agar can be followed very well. Generally this medium is recommended for the cultivation of fungi producing perithecia or pycnidia.

CHAPTER VII

APPARATUS FOR COLLECTING AND SAMPLING FUNGUS SPORES

In phytopathological work, it is often necessary to collect spores of fungi sporulating on the surface of infected plants or other substrates. A small amount of spores can be obtained easily by touching the spores with a wet inoculation loop, brush, etc., or by shaking the spores from the infected plant over a culture plate or other container. But this method of collecting spores requires much time and it is not suitable for collecting masses of spores although large amounts of spore materials often are required for artificial inoculations, or for the investigation of their chemical composition.

Woodbury and co-workers (1967) invented a simple device for collecting dry fungus spores. This was originally designed for the collection and separation of thin-layer chromatogram zones (Ritter and Meyer, 1962). The equipment actually is a small vacuum flask in which the ground-glass neck is finished in an L-shaped wide tube. Through this opening, the spores from the surface of the plant (e.g. uredospores of the rust fungi) will be pulled into the flask by the vacuum. The spores are collected in a Soxhlet paper tube inside the vacuum flask that is attached below the L-shaped tube. The collected fungus spores are removed from the paper tube by tapping the tube gently over a suitable receptacle.

Both qualitative and quantitative determinations of aerial fungal spores and hypha fragments are important from the point of view of epidemiology and in forecasting as a disease-control measure. A number of spore traps and air sampling equipment have been devised. Among them is Pady's (1959) continuous spore sampler that is capable of determining spore density through a 24 hr period on a silicone-coated microscope slide (26 × 75 mm). The vacuum pump pulls air through an opening 14 × 2 mm aperture less than 1 mm from the slide. The amount of air streaming in is regulated by the size of the blower fan. Two small electric timing motors raise the glass slide up about 3 mm once every hr behind the opening so the daily rhythm of the spore density of the air and its hourly changes can be recorded on the slide in bands. The quality and quantity of the spores can be determined by microscopic examination of the bands on the glass slides. This slide is changed every 24 hr.

PART IV

PRINCIPLES FOR BREEDING PLANTS RESISTANT TO PHYTOPATHOGENIC FUNGI

CHAPTER I

VARIABILITY AND SPECIALIZATION OF PATHOGENS

Species of plant pathogenic microorganisms, like all other living organisms, are composed of different individuals and lines (groups of similar individuals). In other words, species of living organisms are *populations:* they are not uniform, in spite of the apparent morphological uniformity observed among the individuals of the species. Originally it was supposed that species, classified on the basis of morphological characters, are composed of individuals which are also uniform from a *biological* point of view. Eriksson (1894) was the first to show with plant pathogenic rust fungi (*Pucciniaceae*), that apparently uniform species are composed of special forms. Pathogenicity (a biological character) of these special forms ("formae speciales"), is different. For example, *Puccinia graminis* (black stem rust pathogen) comprises a number of formae speciales different in their host range: f. sp. *tritici*, whose host plants are wheat, barley, and many wild grasses, f. sp. *secalis*, whose most important host is rye, f. sp. *avenae*, whose hosts are oats and wild grasses different from those infected by f. sp. *tritici* and so on. There are at least six distinct formae speciales that differ in their pathogenicity. This type of differentiation is called specialization for different hosts. A further differentiation of plant pathogenic microorganisms within f. sp. was shown by Barrus (1911) and Stakman and Piemeisel (1917). These investigators recognized different groups of individuals of pathogenic rust fungi even within formae speciales. These exhibited differences in the degree of pathogenicity on certain *cultivars* within their host plants. These are called *physiologic races.* They are able to infect only some cultivars, in other words rather uniform lines of the host. Cultivars are approximately homogeneous from a genetic point of view. As an example *Puccinia graminis* f. sp. *tritici* was further subdivided into physiologic races which were pathogenic to certain cultivars of wheat. However, some other cultivars of wheat were not hosts for these particular races. Identification of physiologic races is based on their pathogenicity on certain host cultivars.

About that time it was recognized that uniform taxonomic groups of higher plants such as species, which are uniform from a morphological point of view, comprise lines different in their biological characteristics. By inbreeding of self-pollinated plants, pure lines were produced which were different in their physiological properties, in spite of their morphological uniformity. Pure lines are not populations, they are pure from a genetic point of view.

After these findings in plant biology, a new and important method was introduced in microbiology, namely the single spore culture method or single cell culture method. With this procedure it was possible to subdivide microbial species, which were uniform from a morphological point of view, into different monospore cultures. These single spore cultures are similar to the pure lines of higher plants. Monospore cultures of microorganisms are genetically homogeneous and may be regarded as clones. A clone is the vegetative progeny of a single individual, therefore, it is genetically uniform. This genetic uniformity is similar to that of the pure lines of higher plants. Both single cell clones of microorganisms and pure lines of plants are called *biotypes*. In this group of individuals all the members are uniform in their characteristics. In host-pathogen relations in nature, interaction takes place between the biotypes of hosts and biotypes of pathogens. However, in agricultural or plant pathological practice there is a more complicated situation, since plant breeders or pathologists seldom work with plant biotypes, and only rarely with biotypes of pathogens derived from single cell cultures. Identification of biotypes of pathogens is usually carried out on isolations from spore samples or spores of a pustule, and not on the progeny of a single spore isolated by the single cell culture technique. Furthermore, in single spore isolations the question whether or not the spore is composed of one, two or several cells remains. If the originally isolated spore is not single celled, the progeny may be a mixture of two or more biotypes, according to the genotypes originally contained in the isolated spore. If the initial isolate is a single cell spore, the cell may still contain one, two or more nuclei. If it is a multinucleated cell, the question arises whether the nuclei are equivalent or are different from a genetic point of view (see the terms homokaryosis and heterokaryosis in a later paragraph).

In some fungi which cannot grow on artificial media, single spore isolations are rather complicated. Therefore, in the practical work of identifying biotypes the original isolation is usually made from single pustules of spores and not from a single spore. Still the spores in a pustule are relatively homogeneous from a genetic point of view. Consequently, the progeny of this isolate will also be "pure". The increased isolate will comprise one or several biotypes, and the term physiologic race is used to designate it. In fortunate cases the increased isolate will contain only a single biotype.

The genetics of the host should be similarly considered. In the ideal case the plant "cultivar" (formerly "variety") represents only a single biotype, however, in other cases, it is a mixture of different biotypes. Cultivars of cross-pollinated plants are always mixtures of a few biotypes and the composition of this mixture is variable. Still, cultivars of this type are relatively homogeneous if we compare them to species which contain many more biotypes.

It is evident from all these considerations that the species of plant pathogenic microorganisms are populations which consist of several biotypes with different pathogenicity. Similarly, species of host plants also represent populations with biotypes different in their succeptibility or resistance. The next table compares the specialization of both host and pathogen within the species.

Intraspecific subdivision of host and pathogen

Host plant	*Pathogen*
population ↗ **Species** ↘ important morphological differences	population ↗ **Species** ↘ important morphological differences
population ↗ **Varietas** ↘ little differences in morphology	population ↗ **Varietas** ↘ little differences in morphology
	population ↗ **Forma specialis** ↘ no morphological differences; infecting different host *species*
group of biotypes ↗ **Cultivar** ↘ little if any differences in morphology but important differences in physiological characters	group of biotypes ↗ **Physiologic race** ↘ no morphological differences; infecting different *cultivars* of one host species
genetically pure line ↗ **Biotype** ↘ physiological differences exist	genetically pure line ↗ **Biotype** ↘ physiological differences exist

What is the reason for the variability of species? Why are the species composed of biotypes different in their biological characters? The explanation for the cause and mechanism of variability comes from the findings of modern genetics.

Sexual crossing results in new combinations of characters and the appearance of *new genotypes*. Thus, new physiologic races (or biotypes) of pathogens can be produced by sexual recombination especially in the sexually reproducing fungi. However, the finding of Tatum and Lederberg (1946) has demonstrated that fusion and recombination may also occur in bacteria, though at very low frequency (Fig. 1). Thus, one of the factors causing genetic variation in "asexual" bacteria turned out to be a phenomenon which is similar to "sexuality".

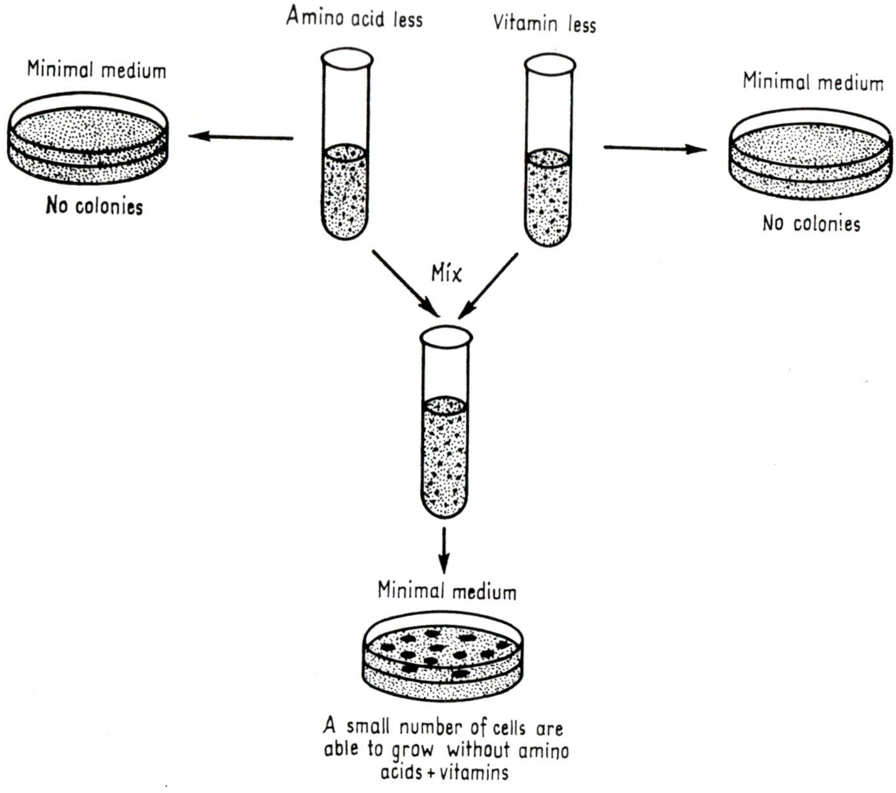

Fig. 1. Genetic recombination

It is also noteworthy that both plant viruses and bacterial viruses can combine characters. Growth of two plant viruses in a plant or two bacteriophages in the same cell results in the formation of recombinant viruses. These carry in their genome genetic characters from both viral parents (Delbrück and Bailey, 1946; Hershey, 1946; Best, 1954; and Watson, 1960b).

MUTATION

Some genetic characters may permanently change and produce hereditary variants or mutants. This mechanism of production of new variations is of great importance in asexual fungi, bacteria and viruses. The role of mutation in producing hereditary variants in fungi was demonstrated by Beadle and Tatum (1941) and later in the case of bacteria by Tatum (1946). Several workers have discovered mutations in plant pathogenic viruses, bacteria and fungi; Kunkel (1940) in tobacco mosaic virus, Lincoln and Gowen (1942) in *Xanthomonas stewartii*, Ark (1951) in *Corynebacterium michiganense*, *Erwinia carotovora* and *Xanthomonas juglandis*, Black (1952) in *Phytophthora infestans*, and Stakman and co-workers (1930) in the rust and smut fungi. In some cases mutations reduced virulence, but in other instances there were stimulations in the virulence of the pathogen (Schwinghammer, 1959).

Special care should be taken with regard to mutations in pathogenicity. Several investigators employed biochemical mutants (auxotrophs) in studying changes in pathogenicity or virulence. However, in these mutants the capacity to exhibit pathogenesis is obscured by their induced nutritional disfunction. In other words, nutritional disfunction of a mutant (a less-strain) does not show a positive correlation with avirulence. Avirulence, in these cases, is due to the inability of the less-mutants to grow in host plants rather than to mutation in virulence. This was convincingly demonstrated by MacNeill and Barron (1966) with avirulent prototrophs (wild types with respect to their nutritional requirements) of *Penicillium expansum*.

The chemical mechanism of induced mutations caused by the action of irradiation, chemicals, heat etc., was not known until recently. The first, and one of the most important results, came from experiments with a plant pathogen, tobacco mosaic virus. The genetic characters of tobacco mosaic virus — a nucleoprotein with protein and nucleic acid components — is determined solely by the nucleic acid moiety (Gierer and Schramm, 1956; Fraenkel-Conrat and Williams, 1957). Viral nucleic acid devoid of the protein moiety is able to induce the same disease symptoms as the intact virus. In other words, the genetic material of the virus is the viral nucleic acid. It is possible to change the viral nucleic acid chemically without loosing its infectious character. However, a hereditary change may occur in the pathogenicity of the virus, i.e., a mutation may be induced (Gierer and Mundry, 1958). When nitrous acid (HNO_2) or $NaNO_2$ in 1 *M* concentration acts on viral nucleic acid at pH 4·2 for 20 min, some of the purine and pyrimidine bases of particular nucleotides will be converted to other bases by de-amination. For example, the pyrimidine base cytosine will be converted to uracil by de-amination. Consequently, some nucleotides will be changed chemically.

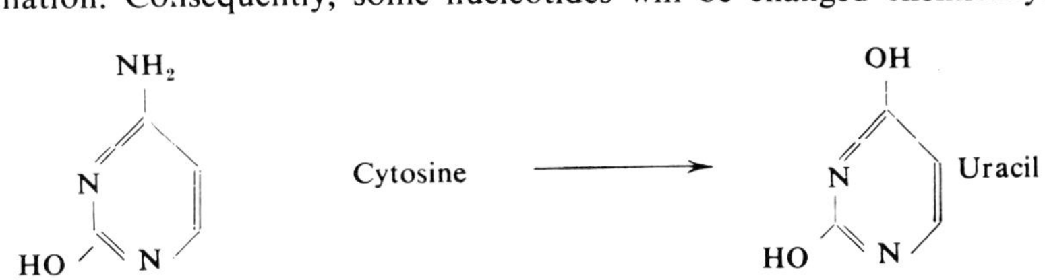

It is known that nucleic acids are composed of linear chains of nucleotides. Schuster and Schramm (1958) have demonstrated that deamination of 2 nucleotides out of the 6000 nucleotides composing the tobacco mosaic virus nucleic acid chain, causes loss of the infectivity of the molecules. This is a lethal change of the virus. However, if only one nucleotide is de-aminated (only one amino group is replaced by an hydroxyl group), the infectivity is not destroyed and a heritable change will take place in infectivity of the virus: a mutation will occur. Indeed a very high incidence of mutants can be observed among the surviving viral RNA molecules. The original *vulgare* strain of tobacco mosaic virus causes a systemic infection on Java tobacco but the chemically (nitrous acid) produced strain, the mutant, induces necrotic lesions on the leaves. Most of the mutations induced by de-amination are different. Deamination of one of the nucleotides of the chain may result in various mutations which produce different symptoms on the host. The type of mutation depends on which particular nucleotide is de-aminated in the chain.

ADAPTATION

An old and seriously discussed question of plant pathology is, which mechanism is responsible for the change of virulence of the pathogen on host plants? Can the host plant influence the virulence of its pathogen directly, or are the changes of virulence the result of the selection of already existing variants of the pathogen by the host plant? How can plant pathogens adapt themselves to resistant cultivars of the host? Can genetically pure races or even biotypes adapt themselves to resistant hosts in successive generations and produce more virulent generations successively until they acquire the ability to grow well on that "resistant" cultivar? Or is the phenomenon of apparent adaptation the result of the selective effect of resistant crop cultivars on a mixed population of biotypes of the pathogen? Christensen and DeVay (1955) discussed the most important plant pathological experiments dealing with adaptation, and came to the conclusion that in all of the experiments the initial inoculum was a population of spores, multicelled hyphae or both, therefore, it was not genetically "pure". This means that the inoculum or the progeny of the inoculum represented a population on which the host exerted a selective action. Increase in virulence was the consequence of selection of the preexisting virulent biotypes in the population, and not a successive adaptation. Christensen and DeVay (1955) were of the opinion that the genetic homogeneity of an isolate of a pathogen is a concept rather than a reality. In many cases, stable lines which were considered as stable biotypes are converted into populations by mutation.

The exact demonstration of the occurrence of adaptation is almost impossible. The selective action of the environment, in our case the host plant, can always change the population of the pathogen, and this gives the impression of changing pathogenicity as a result of adaptation. The analytical work of Stakman and

Christensen (1953) reaches the same conclusion. The phenomenon of the resistance of bacteria to antibiotics raises a similar problem for microbiologists and geneticists. The question is as follows: which method of investigation can demonstrate convincingly that mutation or adaptation is the reason for the resistance to antibiotics observed in bacteria? Resistant mutants originally existing in the mixed population of bacterial inoculum can be observed only as a result of the selective effect of the antibiotic. Similarly, one can experience apparent adaptation in a bacterial population after exposing the bacterial culture to the action of the same agent, the antibiotic. This problem was solved by the application of a replica plate technique proposed by Lederberg and Lederberg (1952). These investigators have demonstrated that a strain of *Escherichia coli* could produce streptomycin resistant cells by mutation even if the cells of the strain did not come into prior contact with streptomycin. Streptomycin thus only selects the resistant cells in the population, however, the progeny of these bacteria will be uniformly resistant to streptomycin, and this gives the impression of converting the bacteria by adaptation.

The replica plating method devised by Lederberg is as follows: The population of the bacterial inoculum is introduced onto plates which do not contain streptomycin. The bacteria will form colonies on the plates. A piece of velvet is then stretched over a block of wood and pressed on the plate. Some bacteria from the discrete colonies are caught in the nap of the velvet and are transmitted to another plate containing streptomycin. By this technique one can make a replica of the discrete colonies of the first plate on the second plate. However, the second plate, containing the antibiotic, will permit growth only of the streptomycin resistant cells preexisting in the original inoculum on the first plate. The position of the colonies on the first plate and on its replica (containing streptomycin) are then compared. Samples can be taken from colonies on the first plate, which have never been exposed to streptomycin and are supposedly resistant to this antibiotic. It will be found that sister colonies of the first and second plates are uniformly resistant to streptomycin. In other words, resistance may arise in the absence of the antibiotic and streptomycin merely selects the preexsisting resistant cells from the mixed population.

These results produced convincing evidence against the theory of adaptation. On the other hand, the experiments of Dean and Hinshelwood (1953) and others call attention to the possibility of stepwise adaptation of bacteria to certain drugs. In experiments with *Bacterium lactis aerogenes* the degree of resistance of bacteria to proflavin was in strong correlation with the concentration of the drug to which the cells were exposed. It seems that this result can be explained by adaptation better than with mutation. A similar explanation was given by Buxton (1958) for the adaptation of *Fusarium oxysporum* f. sp. *pisi* to the toxic root exudate of a resistant cultivar. A 14-day treatment of this fungus with the toxic exudate of the root, induced an acquisition of the ability to tolerate this injurious effect and infect the formerly resistant cultivar. With an increase in the concentration of toxic extract, the ability for adaptation was more pronounced, and a uniform change of the culture treated with exudate took place. The strain of *Fusarium oxysporum* f. sp. *pisi* used in this experiment was genetically "pure". It was derived from parental material which contained a single nucleus. Adaptation in this case was not due to the selective effect of the toxic root exudate, but rather a uniform adaptive change of the whole culture with an increased pathogenicity. Therefore, it seems that one cannot at present, exclude the possibility of the role of adaptation in producing new pathogenic variations of microorganisms, which are independent of mutation and selection. This is an unsolved problem and requires more research.

TRANSFORMATION

In addition to the sexual recombination, there are a number of possibilities for the production of new variants in asexual microorganisms. One of these asexual types of combination of genetic characters is bacterial transformation. The observations of Avery and co-workers (1944) called attention to this process. They were successful in converting the nonpathogenic, no capsule-forming strain of *Diplococcus pneumoniae* into a pathogenic strain, which forms a carbohydrate capsule. This was done by culturing the former strain in the presence of the biologically active fraction of the capsule forming, pathogenic strain. In later experiments this transformation was brought about by the introduction of purified deoxyribonucleic acid (DNA) of the donor. Thus, transformation is a genetic modification induced by the incorporation of DNA purified from other cells into a cell. Transformation may be regarded as a specific form of recombination, involving the naked genes or DNA (Fig. 2).

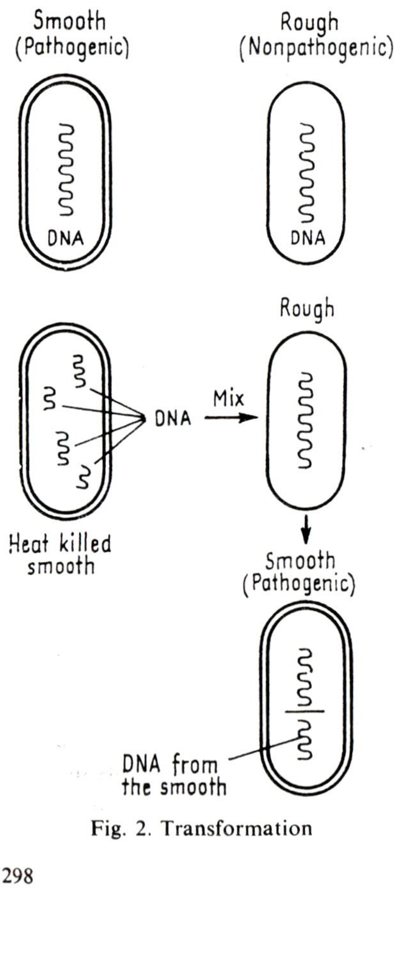

Fig. 2. Transformation

The role of transformation in producing new strains of plant pathogenic bacteria was demonstrated by several workers. Corey and Starr (1957) were able to change the type of colonies and resistance to streptomycin, of *Xanthomonas phaseoli* (a pathogen of beans), by transformation. Similarly, Klein and Klein (1953) have shown that the tumour-inducing ability of *Agrobacterium tumefaciens* can be transferred from virulent donor strains to avirulent strains. Also in this case, the material transforming the avirulent cells was deoxyribonucleic acid (DNA) (Klein and Klein, 1956).

TRANSDUCTION

One of the most recently discovered processes in combining characters genetically, different from sexual recombination, is transduction. Bacteria need not necessarily be destroyed by their infective viruses (bacteriophages). In some cases bacteriophages can be perpetuated during normal growth of the host cell as a part of the bacterial genome. The cells which are symbiotic phage carriers are called lysogenic bacteria (capable of being lysed). Under the influence of different chemical and physical treatments (e.g. UV irradiation) lysogenic bacteria liberate their latent phages by lysis. These are infective viruses having the ability to enter new bacterial cells. During symbiosis, the phage becomes a part of the genetic material of the host. It is known that the deoxyribonucleic acid (DNA) moiety of the phage is responsible for the infection. Phage DNA comes into close contact with host DNA during this symbiosis. This gives the possibility for the DNA molecule carrying genetic factors of the donor bacterium to be transferred, to a recipient bacterial cell, through the vehicle of a phage particle. Thus, bacteriophages in some cases are able to transfer genetic information from one bacterium to another (Fig. 3). Pathogenicity can also be transduced by bacteriophages. Accordingly, transduction is an important mechanism for genetic re-combination in bacteria. This interesting mechanism of genetic exchange was discovered by Zinder and Lederberg (1952). Its importance was demonstrated also in a plant pathogenic bacterium (*Pseudomonas solanacearum*) by Okabe and Goto (1955). One can suppose that transduction has a role in producing new biotypes of plant pathogenic bacteria in nature, however this remains to be shown.

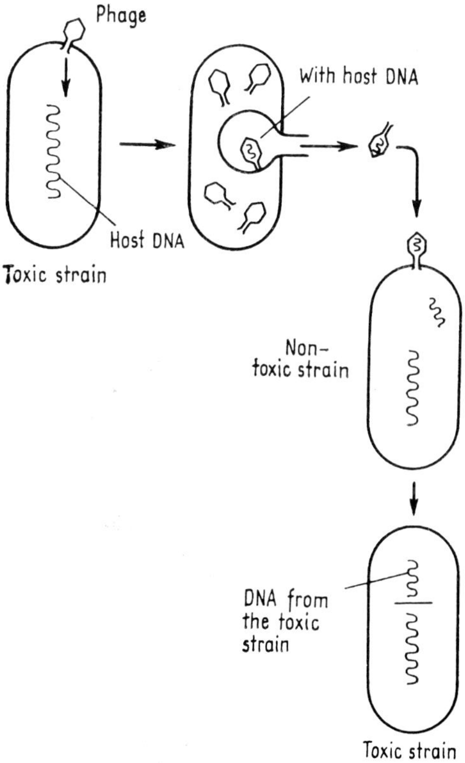

Fig. 3. Transduction

An important possibility of combination of characters in asexual fungi is heterokaryosis as was shown by Pontecorvo (1946) and others. Recent experiments demonstrated that this type of mechanism for producing new variations, also has a role in sexual fungi. A fungus is heterokaryotic when it contains two or more genetically different nuclei, in a single spore or hypha. Filamentous moulds belonging to *Deuteromycetes* (which were the test organism in the original experiments) consist of hyphae containing cells with more than one nucleus. When multinucleate cells contain genetically homogeneous nuclei then the organism is homokaryotic. On the other hand, the cell is heterokaryotic when the nuclei are genetically diverse. Heterokaryotic cells are derived either from homokaryotic ones, if one nucleus in the cell changes its character by mutation, or by the combination of nuclei different in their genetic character. Thus, a new association of nuclei occurs in the case of heterokaryosis. A recombination of the characters of the heterokaryons may occur when the nuclei of the multinucleate cells dissociate and recombine with nuclei of other cells. This system, as was mentioned above, is common among both the sexually and asexually reproducing fungi (cf. Tinline and MacNeill, 1969).

It was shown that the number of nuclei in heterokaryotic cells is variable but small. Each nucleus multiplies independently, in contrast to the multiplication of nuclei in syncytia and multinucleate cells of higher organisms. The rate of multiplication of nuclei in fungi belonging to *Deuteromycetes* is, therefore, different and not synchronized. This is, of course, an additional source of variation in imperfect fungi.

The combination of whole nuclei is possible when hyphae of different strains form anastomoses through which the nuclei migrate from one hypha, to the other. The heterokaryotic hypha may later form branches. Different combinations of nuclei may migrate into different branches (saltation) (Fig. 4). Spores of plant pathogenic fungi are sometimes also heterokaryotic. It was shown by several investigators that heterokaryonts may also originate as a result of anastomoses between germ tubes of spores during or after germination.

The plant pathological role of heterokaryosis in producing new strains (biotypes, races) with altered pathogenicity was shown by numerous investigators in recent years (Buxton, 1954, in experiments with *Fusarium oxysporum;* Nelson, 1956, Wilcoxson et al., 1957, Bridgmon, 1957, with *Puccinia graminis;* Vakili and Caldwell, 1957, with *Puccinia recondita;* and Hrushowetz, 1957, with *Helminthosporium sativum*).

Another finding of the school of (Pontecorvo et al. 1953, and 1956) Pontecorvo has led to the elucidation of the *parasexual process* of fungi. It was demonstrated in species of *Aspergillus* and *Penicillium* that very rarely, two haploid nuclei of a heterokaryotic cell can fuse and later segregate, producing recombination of characters without any sexual process. The heterozygous nucleus formed after nuclear fusion, divides normally (mitosis), however, during the mitotic division crossing over may take place and this produces genetic recombination. Thus the parasexual combination of characters involves "mitotic cross over" in the diploid nuclei within the heterokaryon. Diploid nuclei, by a series of divisions

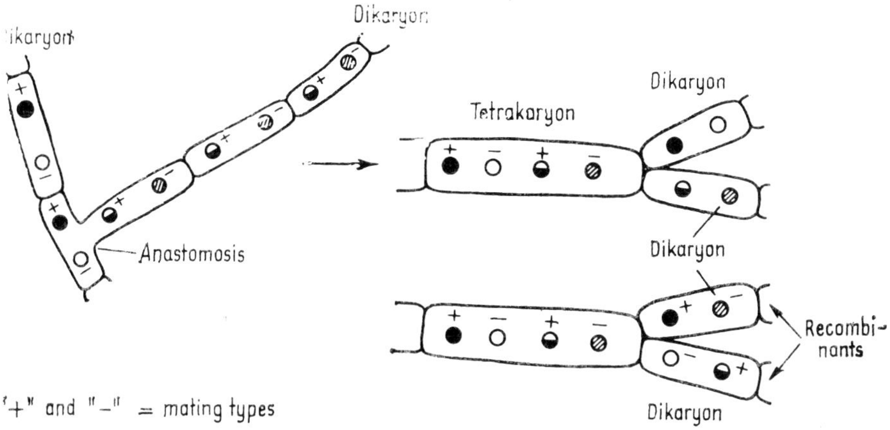

Fig. 4. Heterokaryosis

involving aneuploidy, may eventually produce completely new haploid recombinants.

The parasexual mechanism was demonstrated in *Fusarium oxysporum* f. sp. *pisi* (wilt disease pathogen in peas). New pathogenic types may be produced by this process. Similar phenomena were shown in the case of stem rust (*Puccinia graminis*) by Watson (1957) and Ellingboe (1961), leaf rust (*Puccinia recondita*) by Vakili and Caldwell (1957), *Ustilago zeae* by Holliday (1961), and *Verticillium albo-atrum* by Hastie (1967). For the demonstration of the parasexual cycle in *Fungi imperfecti* a simplified procedure was presented by Barron and MacNeill (1962).

Whether or not there are other mechanisms of microorganisms for the formation of new variations is not known at present. It is clear from the foregoing that possibilities for producing new variations (new biotypes, races) with altered pathogenicity, are very great in plant pathogens as well as in other microorganisms.

CHAPTER II

DETECTION AND IDENTIFICATION OF BIOTYPES AND RACES

In the previous chapter we dealt with variation within the species. From the point of view of plant pathology, those variants which represent new biotypes or physiologic races are important. Pathogenic microorganisms, which grow well on different media, can produce innumerable races on the basis of their behaviour on media. Different races may have different cultural types, colour, growth rate, utilization of components of the media etc. These characteristics, however, have no meaning for plant pathology and plant breeding, since

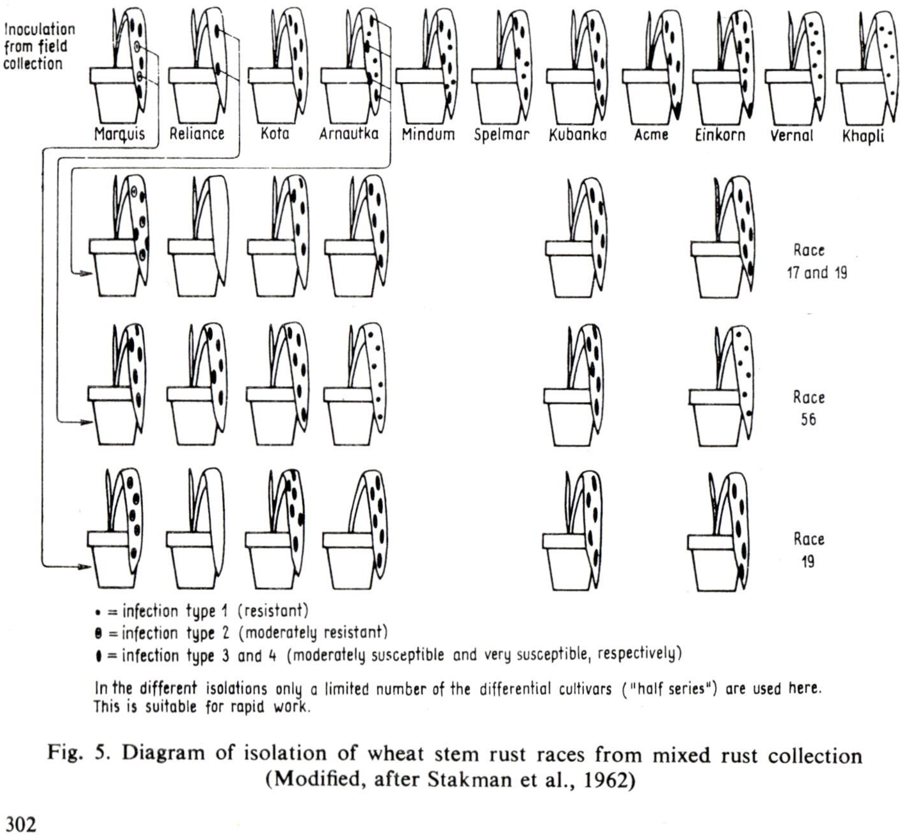

Fig. 5. Diagram of isolation of wheat stem rust races from mixed rust collection (Modified, after Stakman et al., 1962)

they are very rarely correlated with pathogenicity or virulence. Usually, there is no correlation between growth on media and pathogenicity. Therefore, pathogenic races of plant pathogens are determined on the basis of their pathogenicity on certain cultivars of the host plant.

The procedure for identifying rust races is a classic method. This was worked out by the Stakman' school (Stakman et al., 1944, 1962). Samples of uredospores collected from different locations are increased in the greenhouse. Then isolates are made from single rust pustules. Each isolate is tested for pathogenicity on a "differential set" of crop cultivars. This set consists of 12 wheat species and cultivars, the seedlings of which are infected with each isolate. The susceptibility and resistance of the members of the differential set to rust differs. Therefore, they are suitable indicators of pathogenicity or non-pathogenicity of rust isolates. Each member of the differential set will show a reaction type after infection, which is designated as "susceptible", "resistant" or "mesothetic". If the reaction types of members of the differential set are compared, after infection with two or more isolates, then one can determine whether the pathogenicity of the isolates involved in the test is identical or not. On the basis of these reactions it is possible to distinguish between two or more physiologic races, or in other cases determine the identity of two or more isolates which belong to the same physiologic race. In the next Table as an example, the reaction types induced by three races of *Puccinia graminis* f. sp. *tritici* (black stem rust of wheat) on wheats of the differential set are shown:

Race	Twelve members of the differential set											
	1	2	3	4	5	6	7	8	9	10	11	12
17	S	S	R	S	S	S	S	S	S	S	R	R
56	S	S	S	S	R	R	R	S	S	R	R	R
19	S	R	R	S	S	S	S	S	S	S	R	R

S = susceptible, R = resistant

The procedure for the isolation of races is shown in Fig. 5. Using the set of wheat cultivars, which was originally selected by Stakman and co-workers, about 300 races of black stem rust may be identified. One can increase the number of races by using additional differential cultivars (see the section on rust diseases). However, this makes the procedure more complicated though the differentiation of races will be more exact.

Reaction types of the differentials sometimes change according to environmental conditions. Temperature and light are the most important factors which influence the reaction types in some host-pathogen combinations. Therefore, determination of races must be conducted under controlled light and temperature. Reaction types of plants grown in summer or winter in the greenhouse may sometimes be different and this also makes determinations uncertain. In most cases, however, we are not able to control environmental conditions in the greenhouse completely, therefore, a compromise is necessary.

Physiologic races of rusts, mildews and smuts are determined with the differential sets outlined above. If the pathogen can be cultivated on media, one can theoretically distinguish races on different substrates. However, these races are not pathogenic races, since they were not determined on the basis of pathogenicity of the different isolates, or in other words, on the basis of the reaction types on infected plant cultivars.

The methods of the determination of races of particular pathogens are treated under the section on different diseases.

CHAPTER III

INFECTION PROCESS OF FUNGI

Airborne fungal pathogens reach the host passively, but soilborne pathogens arrive or are led usually by active growth through the soil. Before penetration, fungi proceed with various activities on the surface of the plant: spore germination, hyphal growth, formation of various infection structures (appressoria etc.). These active processes are influenced by the host. Various stimuli from the host support spore germination and penetration, but the stimuli are usually not specific. Airborne spores penetrate resistant cultivars of plants as easily as they penetrate susceptible cultivars. Root exudates, as chemical stimuli, usually stimulate spore germination in the rhizosphere. Not only chemical but, sometimes, physical stimuli may have importance in stimulating the infection process. A simple contact with the host surface, in some instances, stimulates the formation of appressoria.

There are, of course, barriers to infection which inhibit or interfere with the penetration of fungi into plants. These are morphological or chemical (phenols, fluorescent glucosides) in nature (Walker et al. 1929, Turner, 1961).

The penetration of the fungus may occur in three ways: (a) directly through the intact surface by a mechanical penetration, (b) through natural openings, and (c) through wounds.

After penetration, fungal pathogens spread from the site of infection into the host organism. This is the colonization of the host. If the pathogen enters into a parasitic relationship with its host, colonization is successful; pathogenesis is initiated. The physiological manifestations of a disease may be diverse. Host cell cytoplasm increases as a rule. The concentratión of ribosomes is also increased. This explains the increased capacity of the infected cell for protein synthesis. Electron-microscope observations show that haustoria of obligate parasites are surrounded by an encapsulation ("zone of apposition") which is neither of the host nor of the fungal cell wall. This separates the haustorium from the host. A specific secretory process is involved in infected cells. Secretory bodies formed in the host cytoplasm, move to the encapsulation, and discharge their contents. They are consumed by the pathogen, as was supposed by Peyton and Bowen (1963).

In plants infected by fungi the rate of photosynthesis is usually reduced and respiration is stimulated. Cell wall degradation is conducted with pectolytic and cellulolytic enzymes. In many instances, growth substances (auxin, cytokinins), incite specific morphogenetic and physiological damages. In other diseases toxins of fungi may induce disease symptoms. One of the most general alterations of the host metabolism is the pathological phenol metabolism. The tissue level of phenols is usually higher in infected plants and oxidation of phenolics induces brown or black discolouration of tissues (cf. Goodman et al., 1967).

CHAPTER IV

DISEASE RESISTANCE OF HOST PLANT

When a pathogen is able to infect some individuals of the host species but not others, then the latter are *resistant*. This definition stresses the importance of behaviour of the host as a *species*. When all individuals of a plant species are uniformly not suitable as hosts for a pathogen, in other words, when the species does not belong to the host spectrum of a particular pathogen, we cannot call this resistance. In this case there is no affinity between a microorganism and the higher plant. Only when some varieties, cultivars, biotypes or individuals of plant species are diseased after infection by a microorganism, can we suppose that there exists some affinity between the species of a higher plant and the pathogen. The chemical or physical basis of this affinity is at present unknown.

A host-parasite pair is a complex of two organisms, therefore, susceptibility or resistance of the host on the one hand, and virulence or pathogenicity of pathogen on the other, are interdependent concepts, which cannot be evaluated independently. It is more accurate to designate the state of a host-parasite relation as *compatible* or *incompatible*. A parasite can be designated as virulent only if the host plant is susceptible or very weakly resistant. A high degree of resistance involves low pathogenicity or virulence. It is more correct to designate the former case as compatibility and the latter as incompatibility. However, for practical purposes, we use the terms susceptibility and resistance or virulence (pathogenicity) and avirulence (apathogenicity) as every one among plant pathologists, plant breeders and growers uses them.

According to Gäumann (1946) two forms of resistance exist: (1) axeny (inhospitableness, incompetence) and (2) defence reactions.

Axeny is equivalent to resistance *sensu stricto*. Axeny is present before infection, prior to the plant becoming a host. It inhibits infection or the disease proper. This is a characteristic of a plant, which exists independently from the attack of the pathogen, e.g. a preformed resistance. Axeny is passive or static based on morphological or chemical barriers.

Defence reaction, on the other hand, is a postinfectional characteristic which begins to operate only after infection. It is active, functional or dynamic; a real "reaction". Defence reaction is an induced capacity, developing in connection with the disease proper.

A strange host-parasite combination develops if the host is susceptible to infection but not to the disease. It is resistant to disease though susceptible to infection. This state of the plant is called *tolerance*.

a) *Axeny* (Resistance)

Morphological resistance hinders the penetration of the pathogen and spread of infection by the thickness of the waxy layer, cutin and cork layer, the structure of the stomata etc.

Chemical resistance may be a consequence of the action of toxic materials in the tissue, or in other cases, of the lack of important compounds in the tissues which cannot be synthesized by the pathogen but are required from the host medium.

b) *Defence reaction* (Immune reaction)

The defence reaction incited by the infecting agent may be a *spontaneous* or *induced* immune reaction, according to the mode of the reaction. On the other hand, according to the degree of the reaction one can distinguish between *normergic* and *hyperergic* (normosensitive and hypersensitive) types of defence reaction.

Most of the plant "immune" reactions belong to the *spontaneous* (autonomous) defence reactions. The host responds to the microorganism directly infecting the plant, and not to a secondary infection which follows the primary one. When the influence of the pathogen on the plant is finished, the ability for a defence reaction to a second infection does not exist. Thus, immunity does not exist against a second attack, and there is no immune state of the plant. This type of defence reaction is very rare in the animal kingdom, but generally distributed among the plants.

The *induced* defence reaction, which is initiated by a primary infection, transforms the host so that it is immune to a second infection. This reaction, therefore, is activated by a pre-infecting agent (induction). The plant will be in an immune state: acquired immunity. This type of reaction is very rare in the plant kingdom. Very recently it was, however, possible to show that a real acquired immunity occurs in a virus disease (Bozarth and Ross, 1964). An interesting form of induced defence reaction is premunity, which is different from acquired immunity. In acquired immunity, recovery from infection protects the host against re-infection, however, in premunity, an *existing* infection protects against super-infection. In the case of premunity the alteration (activation) of the host lasts only as long as the pre-infection continues. There is no state of long-lasting immunity. This is a very important and general phenomenon in the plant kingdom. One can also find it among animals, but its role is limited here, as compared to plants. The practical importance of premunity is that infection by a non-virulent parasite is able to induce a defence against a secondary infecting virulent pathogen.

On the basis of the principles outlined above, one can summarize the different forms of disease resistance as follows:

1. Axeny (preformed resistance)
2. Defence reaction (immune reaction)

- spontaneous defence reaction (spontaneously activated)
- induced defence reaction (activated by a primary infection against a secondary attack)
  - induced immune reaction (acquired immunity is continuous)
  - premunity (acquired immunity during the existence of primary infection)

Detection of postinfectionally produced fungitoxic substances: Phytoalexins

Ipomeamarone, orchinol, isocoumarin derivatives, pisatin, phaseollin, etc. are examples of fungitoxic substances which are synthesized by plants in response to infection with an incompatible fungal pathogen. Müller (1956) called this class of compound *phytoalexins.* Most of the phytoalexins appear to be phenolic compounds which are stable terminal products of disturbed host metabolism.

The synthesis or accumulation of phytoalexins is not specifically induced only by pathogens, however, chemical and mechanical injuries are also able to induce their appearance in some cases (Schwochau and Hadwiger, 1968).

Out of several methods of detection of phytoalexins the "drop-diffusate" technique is the most useful one, which was originally proposed by Müller (1956) and adapted on large-scale operation by Cruickshank and Perrin (1961).

Müller's "drop-diffusate" technique (Fig. 6)

Bean or pea pods are cut longitudinally and the seeds removed. The endocarp lining the seed cavities of pods are inoculated with a drop (0·2 ml) of spore suspension of *Sclerotinia fructicola* (*Monilinia fructicola*). The concentration of spore suspension is 4×10^5 spores/ml. At the place of inoculation, tissue browning (necrosis) appears indicating a hypersensitive type of host reaction. After 24 hr incubation in a moist chamber the inoculation drops are collected with a capillary pipette. The yield of 100 drops is usually about 15 ml. The collected drops are then centrifuged to remove mycelial fragments and spores. With this, the diffusate is ready for examination of its antifungal activity. The liquid will probably contain fungitoxic material(s) as a consequence of the diffusion of chemical substances of small molecular weight from the host cells.

For the bioassay of the diffusate the method recommended by Cruickshank and Perrin (1961) is the most appropriate:

Spore suspension of *Sclerotinia fructicola* (4×10^5 spores/ml) are prepared in the cold. Then tubes of double-autoclaved Difco prune agar (6 ml) are dissolved and cooled to 40 °C in a water-bath. 1·5 ml of spore suspension is used to seed each agar tube, which are then held in the bath for 30 sec and the contents are poured into sterile Petri dishes. Petri dishes are stored at 2 °C prior to use.

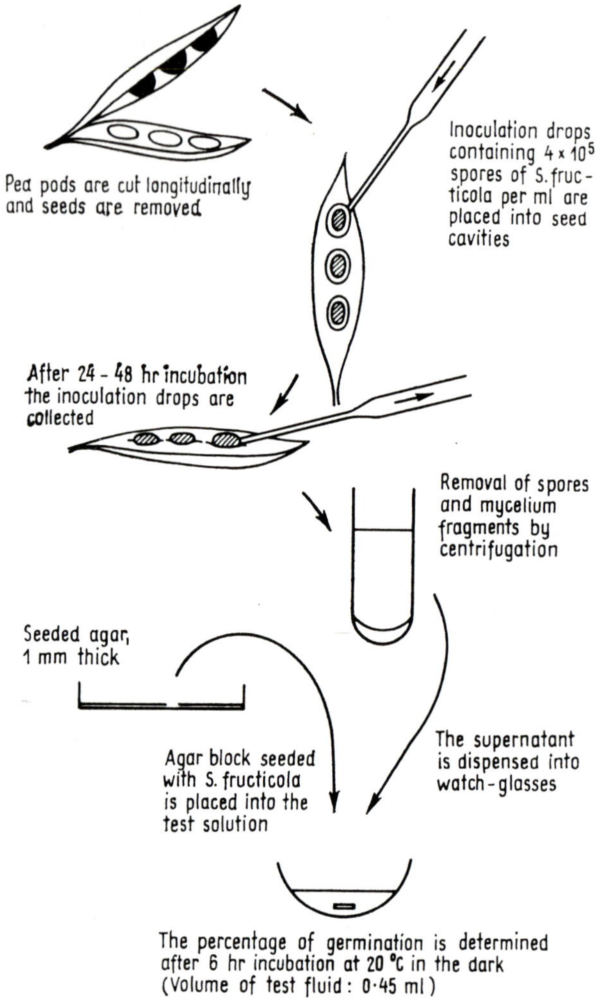

Fig. 6. Detection of phytoalexins

Dilution series with an 0·75 dilution interval is prepared from the diffusates to be tested. They are then dispensed (0·45 ml per watch glass) into open watch-glasses set out on trays lined with moist filter paper, and covered. The seeded agar in the Petri dishes is cut into fives by 1 mm blocks with a sterile stainless steel disc-cutter. One agar block from each of the Petri dishes is placed into each test solution. The assays are carried out at 20 °C in the dark. After 6 hr the spores will be killed by addition of a drop of 1 per cent formalin and cotton blue in lactophenol (see page 276) as a stain. Percentage germination is determined by classifying the first 100 spores in each agar block into germinated and non-germinated. When the spore tube length is greater than spore diameter the germination is taken as positive.

CHAPTER V

BASIC PRINCIPLES OF BREEDING FOR DISEASE RESISTANCE

THE GENE-FOR-GENE THEORY OF HOST-PARASITE RELATIONS

In his studies on rust disease of flax (*Melampsora lini*) Flor (1956) came to the conclusion that for each gene in the host governing resistance, there is a specific and related gene in the pathogen, that governs its virulence (pathogenicity). In other words there is a genetic interrelationship between host and pathogen; this is the complementary gene-for-gene theory. The mechanism for the evolution of this relationship in a host-parasite system is an automatic consequence of the distribution of new resistance genes by mutation and selection in the host, and virulence genes in the parasite population. Genes for resistance are in a complementary interrelationship with genes for virulence in the host-parasite system. The entry and spread of a new gene for resistance in the host population changes the population of the parasite. Members of the population of the parasite having genes for virulence will be selected. In other words genes for resistance favour only those parasites that have the related genes for virulence. Genes for resistance have a high selection value. On the other hand, members of the population of the parasite possessing the genes for virulence are able to survive and reproduce. With their spread, the value of a gene for resistance in the host population will be progressively reduced. At the end of this evolution, all parasites possess the gene for virulence and are able to infect the previously "resistant" host. Thus, genes for virulence in the parasite also have a high selective value. It can be seen that the selective values of genes for resistance and virulence in their respective populations are interdependent. An increase in the frequency of either of them will ultimately lead to a breakdown in the selective mechanism that initiated the increase. This is a self regulating mechanism, which is a necessary condition for the continued co-evolution of the host and its parasite. By this mechanism a pathogen can exist without eliminating its host and the host is able to maintain resistance to its pathogen.

It is seen from the above-mentioned events that the relation between the host and parasite (parasitism itself) induces a genetic polymorphism in both partners (Person, 1959, Person et al., 1962). This genetic variability, exhibiting a self-regulating automatic mechanism, is the reason for the vicious circle which takes place in the alternating prevalence of the resistant host and the virulent pathogen.

PROBLEMS OF BREEDING FOR DISEASE RESISTANCE

It is generally found that resistance of plant cultivars to diseases is not a permanent character. The reason for this phenomenon is to be found in the self-regulating nature of genetic variability in host-parasite relations outlined in the previous paragraph. The evolution of new genes for virulence in the pathogen is the result of the selective action of resistant cultivars on populations of the pathogen which are genetically plastic. Similarly, the genes for virulence in parasites promote selection of genes for resistance in the host population. This automatic regulation of populations of the host and the parasite explains the fact that in many cases the utilization of resistant cultivars is very limited. For example, rust resistance is lost as a rule, in 3–7 years. This is valid, first of all, for pure-line cultivars of self pollinated plant species, e.g. for wheat. Pure line cultivars of wheat suffer under the attack of a new rust race much more heavily than the cross pollinated plants, for example, cultivars of corn. Cultivars of cross pollinated plants are composed of a mixture of different lines (biotypes). This means that some biotypes will be resistant in the host population to the selection effect (attack) of the virulent gene of the parasite. However, there are very few if any resistant biotypes in the susceptible population of pure-line cultivars, since they will have been largely eliminated during the breeding process which produced a high degree of homozygosity. Thus, pure lines will be highly vulnerable to attack by the parasite containing genes for virulence. On the other hand, in cross-pollinated plants the buffering capacity of the genetic pool is relatively high, as compared to the self-pollinated ones.

In recent years plant breeders have tried to incorporate greater variability into self-pollinated plants, such as wheat (Borlaug, 1958, 1965) by developing *multiline cultivars*. The various lines in the multi-line cultivars have different genes or gene systems against the virulent genes of rust fungi. The resulting heterogeneity of the plant population in a multiline cultivar will delay build up inocula in epidemics, and the cultivar as a population, will escape severe damage (Browning and Frey, 1969).

The resistant cultivars of cross-pollinated plants remain resistant for a considerable time. One such example, according to Borlaug (1965), is the relative stability of resistance in corn cultivars against the rusts *Puccinia sorghi* and *P. polysora*. These two rusts of corn are much less important in limiting corn production than the rusts of wheat. The cross-pollinated corn usually possess sufficient variability, that is, sufficient resistant lines are present in the population of the cultivar to resist the selective action of virulence genes of new rust races. This means that development of great rust epidemics in cross-pollinated corn is almost impossible.

Genetic variability of the double-cross corn hybrids is less than that of the best cross-pollinated conventional cultivars. These hybrids are attacked much more severely by rusts, smuts, leaf parasites, and viruses than the ordinary cross-pollinated cultivars. Enhancement of the heterogeneity in these hybrids is thus most necessary and can be carried out by developing synthetic cultivars. These principles are outlined below.

Variability in cross-pollinated and self-pollinated plant cultivars and relative importance of variability for the control of epidemic diseases (After Borlaug, 1965)

Cross-pollinated cultivars (e.g. corn)			Self-pollinated cultivars (e.g. wheat)		
Degree of variability in cultivars			Degree of variability in cultivars		
Less variable than standard	The standard variability is *great*	*More* variable than standard	The standard variability is *zero*	*Considerably more* variability than in standard	*Maximum* variability
Double cross hybrids	Open pollinated cultivars	Synthetic cultivars	Pure-line cultivars	Multiline cultivars	Multiline hybrid cultivars

Browning (1966) also developed multiline cultivars in the breeding program for oats resistant to stem rust. According to his experiments, lines susceptible to rust, yield more under epiphytotic conditions, if these lines are *blended* to give different proportions of stem rust resistance genes in the population. There is less rust on susceptible plants in a blend than in pure stand. When the resistant and susceptible populations yield 100 and 60 bu/a, respectively, in pure stands, a 50 : 50 blend of resistant and susceptible yield 85–90 bu/a. This is above the 80 bushels expected. The reason for this result is that resistant plants in the blend act as spore traps and reduce the spore buildup. Browning stresses the fact that increase of rust in a pure susceptible stand is at the rate of 50 per cent per day, while increase on the susceptible component of a 50 : 50 blend is only 42 per cent per day. This means that rust buildup, necessary for the epidemics, is slower in a blend than in pure stand. In Browning's experiment, rust reached epiphytotic level in only 23 days in pure susceptible stand, while 27 days were required in the blend composed of 50 : 50 per cent resistant and susceptible. The yield increased by 20 per cent in the blend. The commercially grown multiline cultivars will offer a more advantageous situation than the 50 : 50 blend because the resistant lines will be present in a higher proportion. Thus, the yield increase probably will be higher than the above-mentioned 20 per cent. Reduction of the rate of rust reproduction and spread is significant early in the season when the limited rate of spore buildup reduces the damage from rust. In other words, the reduction of inoculum is an important factor in limiting the development of the epidemics. If the crop matures before the mass-spore buildup, which is necessary for heavy epidemics, it will escape severe damage. The resistant lines in the blend do not influence the yield of a 50 : 50 blend of the resistant and the susceptible, mentioned above, since a resistant plant in a blend of isogenic lines would yield the same as in a pure stand (Browning, 1966).

MULTILINE HYBRID CULTIVARS FOR STEM RUST RESISTANCE

Borlaug (1965) proposed a method for the development of multiline hybrid wheat cultivars for a broad base of stem rust resistance. Since stable cytoplasmic sterility and fertility restorer systems have been found, the development of commercial hybrid wheat cultivars became a hopeful endeavour. As regards resistance to infectious diseases, single-cross hybrids will not improve the permanency of rust resistance in hybrid cultivars. However, according to Borlaug (1965), the best possible means of "incorporating" long-lasting rust resistance into wheat is to develop multiline hybrid cultivars. It is possible to convert the single-cross hybrids to multiline hybrids.

The first step in a breeding program is the incorporation of cytoplasmic male sterility into the commercial cultivar. In the next step different lines are produced in a backcross program, which are genotypically different for resistance to stem rust. These lines are phenotypically similar to the male sterile cultivar but none possess the genes for restoration of fertility. Female individuals of a male sterile

Development of multiline hybrid wheat cultivar with provision for resistance to different stem rust races (After Borlaug, 1965)

Cultivar "A" | Cultivar "B"

Cross 1 (For rust resistance) | *Cross 2* (For heterosis)

Cultivar ♀ "A^s"
- X ♂ Cultivar A^n (resistance "x") → Sterile ♀ F_1 X ♂ Cultivar B′
- X ♂ Cultivar A^n (resistance "y") → Sterile ♀ F_1 X ♂ Cultivar B′
- X ♂ Cultivar A^n (resistance "z") → Sterile ♀ F_1 X ♂ Cultivar B′

→ Multiline hybrid cultivar

Cultivar "A": A commercial cultivar into which the cytoplasmic sterile gene will be incorporated

Cultivar "A^s": Cytoplasmic male sterile modification of cultivar A.

Cultivar "A^n": Different lines phenotypically similar to cultivar A^s possessing normal pollination but without genes for restoration of fertility to the cytoplasmic sterile line. (Each line is genotypically different for resistance to stem rust races for example, x, y, z etc.).

Sterile F_1: Lines derived from the first cross. These will be used as females in making the second cross which will produce the hybrid seed.

Cultivar "B": A commercial cultivar into which the genes for restoration of fertility will be incorporated.

Cultivar "B′": A modification of cultivar B with genes for restoration of fertility. Male individuals of B′ are used as pollen parent in the second cross. This cultivar provides the heterotic yield effect when crossed with cultivar A and also possesses the gene for restoring normal fertility to the cytoplasmic sterile derivatives of A′.

cultivar will be crossed with male individuals of rust resistant lines. F_1 plants of each cross will be sterile and will be resistant to stem rust. F_1 plants of different lines will possess resistance to different rust races. Female individuals of these F_1 male sterile but resistant lines will be crossed in a second cross with male individuals of a cultivar which possess the genes for restoring normal fertility. With this cross the heterotic yield effect will be provided. By this method a multiline hybrid cultivar will be developed in which hybrid vigour and a broader resistance to stem rust is incorporated. Rust epidemics will not likely develop in this population since the different lines will possess resistance to a broad spectrum of rust races. The method is given schematically in the Table (page 315).

BREEDING FOR DISEASE RESISTANCE IF PATHOGENS ARE STABLE IN PATHOGENICITY

There is no reason to believe that populations of every pathogen are variable in the manner described in the previous paragraphs. A great number of cases have been recorded in which physiologic races are stable for a long period and only few races have been defined within the population of the pathogen. A good example is the wilt of cotton caused by *Fusarium oxysporum* f. sp. *vasinfectum*. Only four distinct races of this pathogen have been found, two in the United States, one in India and one in Egypt (Armstrong and Armstrong, 1960). These races are very stable, and cotton cultivars with resistance to *Fusarium* remained invariable over many decades. *Fusarium* wilt of tomatoes (*F. oxysporum* f. sp. *lycopersici*), peas (*F. oxysporum* f. sp. *pisi*), and cabbage (*F. oxysporum* f. sp. *conglutinans*) are further good examples of stable pathogenicity. According to Walker (1966) new races of pathogens of this type, which are able to parasitize resistant commercial cultivars, may have poor survival capacity and/or a low level of virulence.

A monogenic inheritance of resistance is characteristic of plants with stable resistance. Cultivars resistant to these pathogens can be developed relatively easily via the backcross method. On the basis of the investigations of Walker (1941, 1953, 1965) the following diseases belong to this group: many wilts caused by *Fusarium* and *Verticillium*, the white rust of radishes caused by *Albugo candida*, bacterial wilt of cucumber, viral mosaic of cucumber and pink rot of onion (*Pyrenochaeta terrestris*). The inheritance of seedling resistance of wheat and other cereals to the rusts is also monogenic and usually due to a single dominant gene. This gives resistance to one or relatively few races. However, the pathogenicity of rusts is not stable, in other words, rusts produce a great number of races. Therefore, too many host resistance genes would be required to give resistance to the complete spectrum of races. It is almost impossible for the breeders to assemble all the genes in one cultivar. A similar problem arises in the case of breeding for smut- and mildew resistance.

ADULT-PLANT RESISTANCE

Mature plant resistance or field resistance is inherited *multigenically*. This resistance is effective over a much wider race spectrum of rusts than is monogenic seedling resistance. Adult-plant resistance is not necessarily expressed in the seedling stage but it begins to operate in the weeks preceding maturity. These plants often are susceptible in the seedling stage but resistant in the adult stage. Recently, increasing attention has been paid to this type of resistance by plant breeders. As a rule, environmental factors can change multigenic resistance more thoroughly than the monogenic type of resistance. Therefore, the former is more labile than monogenic resistance. In addition to adult-plant resistance, some disease resistances, are inherited di-, tri- or multigenically.

The pathological, genetical etc. questions of breeding for disease resistance are treated in the following works (in particular): Vavilov (in original Russian 1935 and 1957, in English 1951), the excellent compilations of Roemer, et al. (1938), Hayes, et al. (1955) and "Handbuch der Pflanzenzüchtung" edited by Kappert and Rudorf (continuously since 1955) which contains several chapters on disease resistance. The books of Stakman and Harrar (1957) and Walker (1966) are also useful introductions to the plant pathological aspects of breeding for disease resistance.

CHAPTER VI

METHODS OF ARTIFICIAL INOCULATION

According to the characteristics of the pathogens and hosts, as well as the nature of the process of infection, methods of artificial inoculation can be grouped as follows:

1. *Inoculation through the soil.* This method is used in the case of the soil-borne diseases, such as club root of cabbage (*Plasmodiophora brassicae*), wart disease of potato (*Synchytrium endobioticum*), wilt diseases (*Fusarium, Verticillium*), take-all, foot rot and eyespot diseases of cereals (*Ophiobolus, Cercosporella*, etc.). It is possible to infest soil naturally when plants, very susceptible to the pathogen, remain in the soil for a long time. In other cases, cultures of pathogens causing soil-borne diseases are mixed with the soil of experimental plots in the field or plots in the greenhouse, or soil of hot and cold beds.

2. *Inoculation of the seeds.* Seeds of the host are powdered with dry spores or treated with other propagules of the pathogen. The surface of the seeds is, in some cases, contaminated with a water suspension of spores in which the seeds are immersed for a short time. It is necessary to use an optimal quantity of spores in seed inoculations, because a very heavy infection may take place as a result of overdosage of inoculum.

Seed inoculation is, in fact, only "contamination" and not at all "infection". With the inoculation of seeds we bring the spores into close contact with the host and this usually helps successful infection. Methods of seed inoculation are applied mainly in experiments with smuts, such as wheat bunt (*Tilletia foetida* and *T. caries*, as well as *T. controversa*) covered smut of barley (*Ustilago hordei*), black loose smut of barley (*Ustilago nigra*), covered smut of oats (*Ustilago kolleri*), head smut of corn (*Sphacelotheca reiliana*), etc.

3. *Inoculation of the leaves.* This is carried out by spraying, brushing, powdering, and rubbing, spore suspensions, dry spores, bacterial suspensions, or extracts of virus infected plant parts on or into the leaves of host plants to be investigated. Injection is also a useful method of leaf inoculation, especially in the case of bacterial pathogens or rusts and smuts. Again, spraying, powdering, or brushing the spores into the leaves does not mean that we have infected the leaves. By these procedures we artificially bring the pathogen into contact with the host. Infection itself will take place later. Leaf inoculation is the most important procedure in plant pathological practice; this is used in the case of rust and powdery mildew diseases, corn smuts and other smuts, in several virus diseases

and bacterial diseases. Plants inoculated with fungal pathogens are usually placed in a moist chamber (100 per cent relative humidity) to induce spore germination and infection. In field experiments, inoculated plants are covered with bell jars or other little boxes to secure high humidity during the night and induce, thereby, spore germination and infection. From the infection centres the inoculum will spread over the whole field plot to be investigated by the local epidemic developed artificially.

4. *Flower inoculation.* This method is applied mainly in the case of loose smut diseases and ergot of rye. Spores are sprayed, powdered or injected into the individual flowers. After infection the pathogen enters the sexual organs of the host and establishes itself in the embryo developed after fertilization. In the loose smut infection process the young embryo is infected by the mother plant without the latter becoming diseased.

CHAPTER VII

METHODS OF DETERMINING DISEASE RESISTANCE

Resistance or susceptibility may, in many instances, be characterized by *crop loss*. In comparing the yields of plots in the field control plots must be protected by chemical control from the damages caused by disease. Yield of protected control plots will be compared to the yields of resistant and susceptible cultivars. The degree of resistance can be expressed on the basis of the amount of crop loss. However, this method of estimation needs complicated field experiments.

In determining the percentage of the diseased or attacked plants, the frequency of the disease is expressed within a plant population. *Disease frequency* may be a very useful characteristic in determining resistance of cultivars, strains or selected lines of host plants. However, results are usually variable, like the estimation of crop loss.

A similar method is to determine disease resistance on the basis of the number of infections on a particular area of leaf or other organ. Actually this signifies the "*incidence of attack*" (see Fig. 14. in the paragraph on rust diseases).

The methods for determining *disease destructiveness* (e.g. the size of the necrotic spot, the destruction of some plant organs etc.) are also used in several cases, but the results are even more variable.

It is also possible in several diseases to determine the rate of the development and distribution of the pathogen in or on the host plant. On the basis of the estimation obtained, one can determine the "*incubation resistance*" of the cultivar to be investigated. In plants possessing this type of resistance, multiplication of the pathogen is delayed to such an extent, that the buildup of the inoculum is not sufficiently high in a given period to incite epidemic distribution.

In addition to the methods described above, it is possible in several cases, to express the *intensity of fructification* of the pathogen in the diseased plants and relate it to the susceptibility (and resistance) of the host cultivar.

One of the most appropriate methods for determining disease resistance is based on the *infection types* (characteristic symptoms). This applies to obligate fungal parasites (rusts mildews etc.) and viruses, where the induced infection types are in close correlation with the defence reaction or the hypersensitive reaction of the hosts. The reaction types caused by black stem rust infection were initially used by Stakman and co-workers in identifying stem rust races. Similar scales of infection types have been worked out by others for the identification of physiologic races of mildews, smuts etc.

PART V

TYPICAL FUNGUS DISEASES

CHAPTER I

PLASMODIOPHOROMYCETES

CLUBROOT
(*PLASMODIOPHORA BRASSICAE* WOR.)

Taxonomic position of fungus

The taxonomy of lower fungi is primarily based upon the number and type of zoospore flagella. The taxonomic position of *Plasmodiophora* has been a question of debate for a long time. At first, members of this genus were considered to be related to the *Myxomycetes* on the basis of a naked thallus. However, it was demonstrated by Couch (1938) that the zoospores of *Plasmodiophora* were produced in thin-walled sporangia and thus not naked. On this basis it was transferred to the *Phycomycetes*, but due to the flagellation it should be placed in the *Myxomycetes* (Martin, 1960, Ubrizsy and Vörös, 1966). The swarm spore of *Plasmodiophora* has two unequal whiplash flagella, but one often is very rudimentary. Elliott (1949) showed that two flagella of the whiplash types also were characteristic of the *Myxomycetes*. Considering that the zoospores of the *Phycomycetes* are uniflagellate or that if two are present, they have two different types of flagella (whiplash and tinsel), thus placing *Plasmodiophora* into the *Phycomycetes* is not justified.

Life cycle

The life cycle of the pathogen, which still has not been completely clarified, is outlined in Fig. 1. The haploid resting spore germinates, producing a biflagellated zoospore. Germination is a process requiring high soil temperatures (27–30 °C) and high moisture (40–50 per cent). Roots of the host plant in an acid soil has a stimulatory effect on germination. The pathogen does not infect root hairs in alkaline soils. The swarm spore moves on the surface of soil particles or in the water contained in the pores of the soil and reaches the surface of the host' root. Watson's (1967) measurements proved that the swarm spores could generally move from 1·2–2·5 cm and sometimes to 5 cm away from the infection source. These measurements refer to sterilized soils that are not watered excessively. In natural soils, however, the movement of swarm spore of *Plasmodiophora* is limited largely by the effects of the antagonistic soil microflora (Watson, 1967).

The zoospore loses its flagella on the host surface and penetrates the host cell, usually a root hair, in the form of an amoeba. Here a haploid plasmodium develops by multiple divisions, spreading from cell to cell. In this developmental stage the cell nuclei divide in a particular way. A chromatin ring surrounds the dumb-bell-shaped nucleolus which appears like a cross from a side view (Alexopoulos, 1962). This so-called "cruciform" cell division is unique among fungi.

When the plasmodium fills the host cell, its content is transformed into zoospores, that is, it becomes a zoosporangium. Williams and Yukawa (1967)

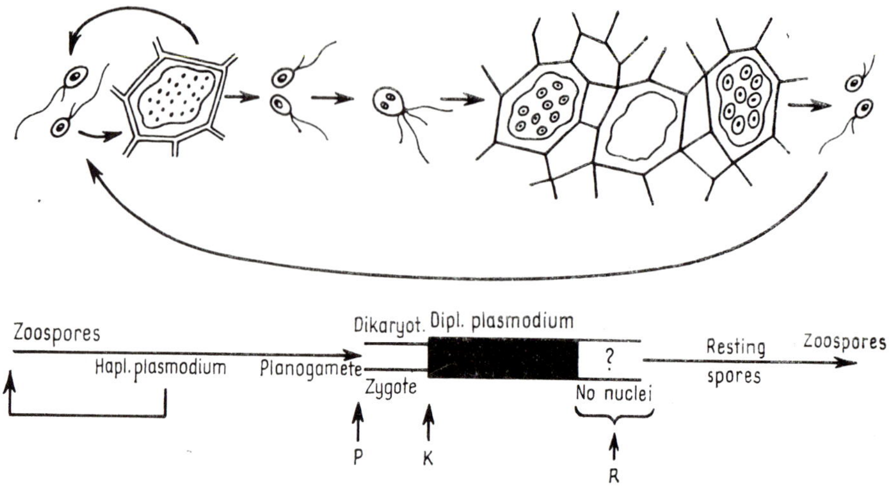

Fig. 1. Life cycle of *Plasmodiophora brassicae* (P = plasmogamy, K = karyogamy, R = reduction)

established that the membrane of the plasmodia were rather thin (250 Å) consisting of two plasmodial membranes. Investigations of the ultrastructure revealed that the cell nuclei contain a single nucleolus. There is no starch in the plasmodium but rather large lipid grains occur. The zoospores are released into the surrounding soil.

The liberated swarm spores may cause new infections. The haploid plasmodium under certain conditions does not give rise to asexual zoospores alone, but at least some of them are differentiated into gametangia. After a few divisions, planogametes of different sexes are formed. The planogametes emptied in the open soil, however, do not infect the host directly but first copulate in pairs. The copulation of planogametes does not mean instant karyogamy. Infection actually occurs by dikaryotic zoospores with a pair of associated nuclei, and the nuclei only fuse sometime after penetration of the host. The division of the nuclei containing the $2n$ chromosomes in the host cell results in a diploid plasmodium. The diploid cycle takes place in the root, while the haploid phase occurs mainly in the root hairs. After the development of the diploid plasmodium is completed, another phenomenon that is also unique among fungi follows, the nuclei of the plasmodia disappear. It is not clear whether the nuclei in this state are really disorganized or whether only the staining characteristics of the nuclei have changed (Miller, 1958).

In the next visible stage of development, the nuclei have gone through meiosis. Cleavage occurs around each haploid nuclei and a plasma membrane separates each resting spore. The spores are surrounded by a cell wall. This is how the 3–4 μ diameter resting spores are formed. Each spore is capable of surviving in the soil for years, but under favourable conditions, germination occurs with haploid zoospores, initiating the life cycle over again.

Symptoms of clubroot

The infection of the fungus causes a tumour of a few mm to several cm size on the main and side roots of the host that have been attacked resulting in clubs of various sizes and shapes. The plasmodia of the pathogen can be found in the cells of the clubbed roots. As a consequence of the infection the development of the root is abnormal and the root structure is deformed even beyond the clubbed-root formation. This is due to the inadequate development of the root that impairs its function, the plant decreases in its above-the-soil development, turns yellow and wilts despite adequate moisture.

Artificial inoculation methods

In the investigations of the races of *Plasmodiophora brassicae*, Williams (1966) successfully used the following inoculation methods:

1. *Storage of clubroot material for inoculum*

If the material, the clubbed root, is not used up immediately for inoculum, then the washed roots can be frozen and stored at −10 °C.

2. *Preparation of suspension of resting spores*

a) Select 100 g frozen root, homogenize for 3 min in 400 ml water in a Waring Blendor.

b) Separate host root pieces from spores by filtering through 8 layers of cheesecloth.

c) Collect the spores by centrifuging (2000 g for 7 min).

d) Wash the spores in 100 ml water, centrifuging and resuspending in 100 ml water for three consecutive times.

e) After the final washing, the number of spores per ml is determined in a blood-counting chamber and then used or placed in storage at 2 °C.

3. *Inoculation*

a) Soil preparation: a mixture of peat moss and black organic soil manure appropriately screened and sterilized (at 90 °C for 2 hr each time on two consecutive days).

b) Soil infestation: homogeneous mixing of the suspension of spores in the soil so that 1 cm^3 of soil will contain 10^8 spores.

c) Planting: 7 day-old seedlings are transplanted into the infested soil in pots. The pots are placed on aluminium trays filled with vermiculite. Uniform watering of the plants in the trays with 1 *N* Hoagland solution should be such that the vermiculite should be kept moist but not soaked with the solution. By holding the plants at 22 °C and with 14 hr of lighting (1,000 ft–c.), the results of the inoculation can be evaluated after 35 days.

4. *Evaluation of the results of artificial inoculation*

For the evaluation of plant inoculations, Williams (1966) suggests the following scheme to calculate a disease index:

Classes established:

0. No clubroot formation or only the resistant type of "sphaeroid" clubroots described by Seaman et al. (1963) occur.
1. A few clubroots on secondary roots.
2. Significant clubroot formation on the side branches.
3. Serious clubroot formation on primary and secondary roots.

Value of classes:

To determine the disease index the number of plants of class 0 are multiplied by 0, class 1 by 10, class 2 by 60, class 3 by 100, and the sum divided by the total number of plants.

Biological specialization of Plasmodiophora brassicae

A system for defining pathogenic races of *Plasmodiophora brassicae* on the basis of differential hosts was described by Ayers (1957) in Canada. In North America, Williams and Walker (1963) found race 6 was the predominant one for the United States. Other systems for classifying pathogenic races were given by Lammering (1965) in New Zealand and by Williams (1966) in Wisconsin.

Williams (1966) worked out a system for the determination of the races on the basis of four differential host plants: *Brassica oleracea* var. *capitata*, "Jersey Queen" and "Badger Shipper", and the "Laurentian" and "Wilhelmsburger" cultivars of *Brassica napobrassica*.

On the basis of the disease index described above for evaulating the artificial inoculation of the four differential host plants, theoretically 16 biotypes can be differentiated. These can be marked in series from 1–16. All of the possible combinations are summarized in the Table below.

This scheme was used by Williams (1966) in investigating 124 *Plasmodiophora brassicae* isolates originating from 16 countries, and 9 races could be distinguished. From among these, four proved to be new races.

According to Williams (1966) with the general acceptance of this system, 16 *Plasmodiophora brassicae* races could be judged on a uniform basis and a series

All possible reactions of four differential host plants for 16 *Plasmodiophora brassicae* races (After Williams, 1966)

Differential host plants	Series of races															
	1	2	3	4	5	6	7	8	9	10	11	12	13	14	15	16
Brassica oleracea																
"Jersey Queen"	+	+	+	+	–	+	+	–	–	+	–	+	–			–
"Badger Shipper"	–	+	–	+	–	–	+		–	+	+	–	+	+	+	
Brassica napobrassica																
"Laurentian"	+	+	+	+	–	–	–	+	+	–	+	–	+	–		
"Wilhelmsburger"	+	–	–	+		–	–	–	+	+	+	+	–	+		+

of races delineated with common hosts. With this procedure we could avoid describing and marking the same *Plasmodiophora brassicae* races with a diversity of different race designations by various authors. If other races are to be delineated using additional hosts, seed should be readily available and genetically pure.

Prior descriptions of the races of *Plasmodiophora brassicae* in the past have been on the basis of arbitrary and spontaneous differential host characteristics. Strandberg and Williams (1967) in their latest publication show that there is a possibility of a genetic basis for classifying and differentiating races.

Clubroot resistant cultivars

Due to the relatively large number of *Plasmodiophora brassicae* races, it is still a rather difficult goal to produce cultivars resistant to the pathogen. Walker (1952) emphasized the relativity of resistant species prepared in England, Sweden and Denmark: *Brassica rapa* — "Bruce", "May", and "Dale's Hybrid"; *Brassica napobrassica*, "Wilhelmsburger", "Resistant Bamgholm" and "Immuna II." The behaviour of resistant species is not the same for all cultivated areas or parts of the world, because of different pathogenic races occurring there. However, in the breeding of clubroot resistant strains of *Brassica pekinensis*, Strandberg and Williams (1967) attained unusually promising results.

Control of clubroot

The most important measures in controlling *Plasmodiophora brassicae* infections are the use of resistant cultivars, healthy seedling for transplanting and a 4–6 years crop rotation. No susceptible host plant should be planted in infested soil, so proper weed control is very important. *Plasmodiophora brassicae* infects not only cabbage and other cultivated Cruciferae, but also a number of weeds in the Cruciferae. As mentioned before, acid soils are favourable for infection. Thus damage can be reduced by the application of lime or alkaline fertilizers. Soil sterilization by chemicals (Vapam, Brassican, etc.) also can be applied to control *Plasmodiophora brassicae* with some success.

CHAPTER II

PHYCOMYCETES

POTATO WART
(*SYNCHYTRIUM ENDOBIOTICUM* [SCHILB.] PERC.)

Taxonomic position of the fungus

From a taxonomical viewpoint, the most important characteristic of the parasitic fungus, *Synchytrium endobioticum*, is the single, posterior whiplash flagellum of the zoospores. The inoperculate sporangium, containing the zoospores, ruptures without any determined or previously formed opening. Reproduction in *Synchytrium* is holocarpic, but the fruiting body forms not one but several sporangia (or gametangia) comprising a sorus. On the basis of these characteristics, the organism causing potato wart belongs to the class of *Phycomycetes*, order *Chytridiales*.

Life cycle

The life cycle of *Synchytrium endobioticum* is outlined in Fig. 2. Uniflagellate, haploid swarm spores stream out from the infected plant parts in spring. The swarming of zoospores takes place only in suitably moist soil. The optimal temperature for the process is 12–19 °C.

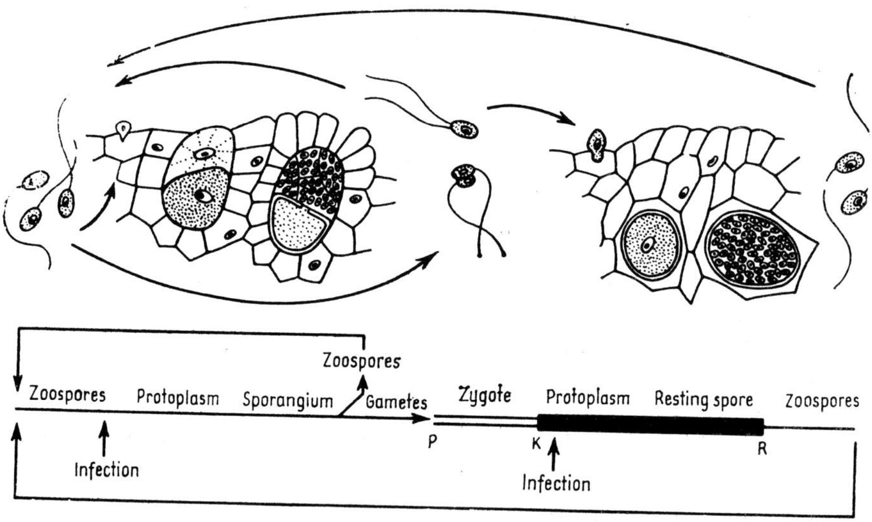

Fig. 2. Life cycle of *Synchytrium endobioticum* (P = plasmogamy, K = karyogamy, R = reduction) (After Roemer et al., 1938)

The swarm spore adheres to the surface of a susceptible potato tuber and lyses a tiny opening into the cell wall of the epidermis. In this way its content reaches the host cell. In the host cell, the amoeba grows in volume; however, its nucleus does not divide but increases in size. After reaching a certain size this structure is surrounded by a yellow-brown cell wall and in this stage it is called a prosorus. In the course of the development of the prosorus, the infected host cell also grows abnormally and as a consequence this and the repeated divisions of the surrounding healthy epidermal and cortical cells that are not invaded, a cancerous tumour appears (rosette formation). Meanwhile, the host cell containing the prosorus dies. Following this, the thick wall of the prosorus opens up and its contents, surrounded by a thin membrane, reaches the upper part of host cell that was empty earlier. After repeated mitotic divisions, generally 32 nuclei develop and the prosorus becomes separated into several segments, each containing a few nuclei. Nuclear division continues in each segment so that finally each prosorus segment contains about 200–300 nuclei. With this, the characteristic sporangium sorus of *Synchytrium* is completed. Now each segment functions as an independent sporangium. The haploid swarm spores stream from them to cause new infections, and this way completes the asexual cycle of the fungus.

In certain cases, for example at the end of the vegetation period or in dry periods, the prosorus segments do not form zoosporangia but gametangia. These do not differ morphologically from the zoosporangia but the planogametes liberated from them are unable to infect the host plant directly. After the fusion of different sex planogametes, accompanied by nuclear fusion, the haploid zygote infects an epidermal cell of the potato tuber or stolon. A hypertrophying host cell encircles the pathogen, and as a result of repeated divisions it is buried deeply in the bud tissue. However, the infection of the zygote does not follow repeated divisions of the surrounding cells and rosette formation cannot be seen after the infection of the zygote. The pathogen as a mature resting sporangium meanwhile becomes slightly swollen and surrounds itself with a thick wall. This formation is the diploid resting sporangium which serves to maintain the pathogen during the winter.

In spring, the resting sporangium is separated from the rotten, cancerous tumour into the soil, and is transformed into a single zoosporangium. Its contents becomes granular and several large chromatin separations can be observed (Curtis, 1921). It is possible that this is not the same as meiosis occurring here but it is a process that is equivalent with it. This results in any case in resting sporangia that upon maturation liberate swarm spores with haploid nuclei at the beginning of the vegetation period. These zoospores are larger than those arising from zoosporangia; however, their function is exactly the same: they serve to initiate the asexual cycle of the fungus.

Symptoms of potato wart

The most important host for the pathogen is the potato, although the fungus is capable of infecting a few other cultivated plants and weeds belonging to the *Solanaceae*. The cancerous tumour-like growths appear mainly on the potato tubers and stolons. The size and shape of the tumours is extremely variable mainly because of many re-infections. The deformations originating in this way are at first yellowish-pink and finally become black. With suitable moisture, the parts above the ground, the shoots and leaves, may also become infected. The symptoms here too show leaf deformation and tumourous hypertrophy.

Artificial inoculation methods

a) Maintenance of the pathogen

It is equally important that the fungus should be available at any time and in a virulent form for artificial inoculations and other investigation of the pathogen. *Synchytrium* cannot be cultivated on artificial media, so it must be maintained in tissue cultures. Mirzabekyan (1965) established that the cancerous tissues can be well maintained and cultivated on a Heller type nutrient medium supplemented with trace elements, vitamins, iron nitrate, glutathion and kinetin. Kinetin may also be substituted by the expressed sap from cancerous tissues. With this method the cancerous tissues can be maintained to retain their original colour and structure, which is also an important condition to be met in respect to histological examinations.

b) Determination of the viability of the pathogen by staining

It is always important to determine whether the cancerous tissues maintained in artificial culture really contain a viable infective pathogen. Nelson and Olsen (1967) worked out a suitable procedure for this. They demonstrated that viable *Synchytrium* sporangia can be stained with triphenyl tetrazolium chloride (TTC) but dead sporangia could not. The staining must be done with 0·5 per cent TTC solution at 35 °C. With the application of too much heat or a lethal fungicide treatment, sporangia lose their staining ability as a sign that they have died. The TTC staining of viable sporangia is inhibited by cyanide, but succinate and 2,4–dinitrophenol does not influence it. In rotting tissues or in materials contaminated by other microorganisms the colour reaction cannot be expected to give reliable results.

c) Inoculation method in the laboratory

According to Zakopal and Spitzova (1964) the procedure is as follows:

1. The tubers to be inoculated are placed in a vessel containing water at a level covering them by about 0·5 cm of water.

2. About 6 mm^2 slices of inoculum (cancerous tumour with viable pathogen) are used to inoculate the tubers.

3. The tubers are kept immersed in water for 4–24 hr at 8–15 °C.

4. Following this treatment, they are planted in a sand-compost mixture at pH 7·0.

5. The development of the disease is observed weekly for three months.

d) Inoculation method in field

Malec (1965) worked out a simple and reliable method for determining which cultivars and lines have proved resistant in laboratory investigations that are also resistant in the field. From among the cultivars to be investigated, each was planted in such a way that after every third row of 25 tubers each, one row of a susceptible cultivar was planted. They considered those cultivars which remained symptomless after 2–3 years of such cultivation to be resistant in the field.

To determine the extent of field infections and to determine the degree of the resistance of certain cultivars in the field, Hille, (1965) in West Germany at three experimental stations (Braunschweig, Lubeck and Munster) applied a scale with six grades. The extent of the disease was determined under a stereo microscope on the basis of the development of the cancerous tumour, its size and structure.

Biological specialization of the pathogen

A number of *Synchytrium endobioticum* races have been described in West Germany. From among these, race 1 (D_1, Dahlem or normal strain), race 2 (G_1, Gieshübel), race 3 (K_1, Koppatz), race 4 (P_1, Pappenheim), race 5 (Czechoslovakian strain), race 6 (E_1, Eulendorf), races 7 and 8 (West German strains), as well as the SB (Südböhmische) races deserve to be mentioned. In addition, a rather aggressive strain of *Synchytrium endobioticum* occurs in an area of the western part of the Russian Carpathians (Ubrizsy, 1965). Blattný (1942) revealed that there were three physiologic races in Central Europe.

A procedure for selecting new biotypes or races from an existing, naturally occurring one was suggested by Malec (1964). The "Bem" potato was used in his investigations, because it was resistant tó the *Synchytrium endobioticum* biotype that occurred in this experimental area. The shoots of this cultivar were inoculated three times at intervals of 14 days by summer sporangia. About 65 days after the first inoculation had been made, typical symptoms of wart were noticed on a single plant. The earlier resistant cultivar Bem proved to be susceptible to the infectious material originating from this tumour. Therefore, the *Synchytrium endobioticum* strain prepared (or selected) in this way may be regarded as a new biotype.

The most successful control against the disease is the cultivation of resistant cultivars. This is especially justified where the weather conditions (yearly average temperature of 8 °C and rainfall above 700–800 mm) make it possible for the epidemic spread of the pathogen (Bojňanský, 1959, 1960). The basis of resistance of cultivars to wart is in the first place related to hypersensitivity. On such cultivars in response to the rapid death of tissues following infection, the pathogen will be destroyed before the formation of the resting spores (Walker, 1950). Kameraz (1964) in the course of breeding virus and *Phytophthora* resistant potatoes, produced strains resistant to *Phytophthora* and *Synchytrium* (combined resistance) from Mexican strains of *Solanum demissum* and *S. longipedicellatum* as the sources of resistance.

Walker (1950) pointed out that resistant and susceptible potato cultivars are often closely related genetically. Thus for example, the resistant Katahdin and the susceptible Chippewa cultivars were selected from exactly the same cross. Walker also showed that the resistant Red Warba was selected from the susceptible Warba cultivar. In the U.S.A. the cultivars listed below are wart resistant: Green Mountain, Burbank, Russet Burbank, Irish Cobbler, McCormick, Spaulding Rose, Katahdin, Mohawk, Sequoia, Mesaba, Norkota and Houma (Walker, 1950). Walker adds to this list: Calrose, Chisago, Kennebec, Ontario, Pawnee and Red Warba.

The most important control is enforcing of quarantine measures. In areas free of infection not even apparently healthy potatoes should be allowed to be brought in from any area where *Synchytrium* has been found. Potatoes in these areas must be consumed locally and not be exported. Potatoes should not be cultivated in infested areas for at least 5 years (this applies to *solanaceous* weed plants, too). The most successful control measure for disease control in the potato wart zones is the cultivation of resistant cultivars. Potato wart cannot be controlled effectively or economically by chemical control measures.

LATE BLIGHT OF POTATO AND TOMATO (*PHYTOPHTHORA INFESTANS* [MONT.] DE BARY)

Symptoms caused by *Phytophthora* may appear on the tops of potato and tomato as brownish-black lesions, which are not restricted in size. Under a favourable environment, rapid blighting of the foliage occurs. Tubers of potatoes are also infected. The tuber symptoms are quite distinctive. On tomato fruits, grey-green water soaked spots appear. The causal organism belongs to the *Phycomycetes*, and is a member of the family *Peronosporaceae*.

Asexual reproduction is by zoospores. These are produced in zoosporangia. From a morphological point of view, sporangia are also called conidia. However, according to their origin and function, they are zoosporangia, the contents of which cleave to form about eight bacillate zoospores. The thin-walled conidium is borne at the tip of a branch of a conidiophore, which arises through stomata on the leaves or through lenticels on the tubers. The conidia are 20–38 μ by 12–23 μ in diameter. Zoospores germinate by means of germ tubes, which occasionally penetrate through the stomata, but usually an appressorium is formed from which a penetration hypha grows directly through the cuticle. Haustoria are produced in the cells.

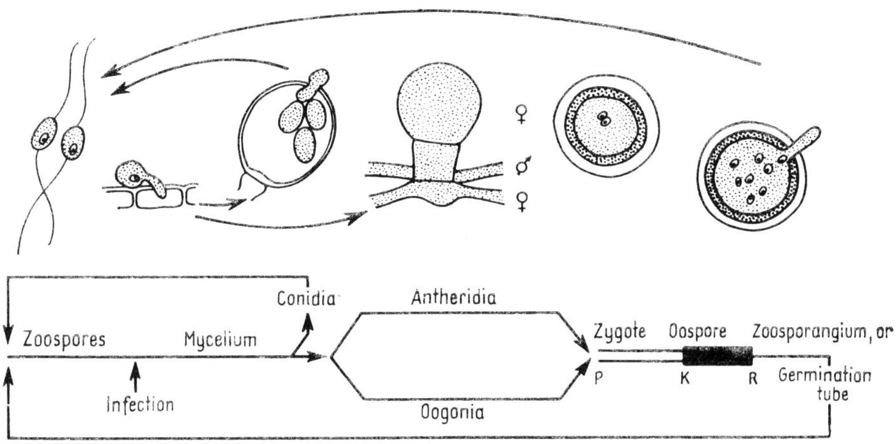

Fig. 3. Life cycle of *Phytophthora infestans* (P = plasmogamy, K = karyogamy, R = reduction) (After Roemer et al., 1938)

Sexual reproduction is carried out by oospore formation. The oogonium is fertilized by an antheridium and a thick-walled oospore is formed. This represents the overwintering stage of the fungus. The germination of oospores is probably accompanied by meiosis (see the scheme in Fig. 3). However, the germination of oospores is very seldom observed in experiments. The reason for this is that the maturation period is unpredictable.

For a long time oospore formation was not observed in nature. Recently, the discovery of two mating types in *Phytophthora infestans* has made possible the formation of oospores under artificial conditions (cf. Gallegly, 1968). But the role of sexual reproduction in the formation of new pathogenic races, as well as in overwintering, remains unknown. Recent investigations in the Toluca Valley in Mexico have referred to the possibility of formation of new pathogenic variations by sexual recombination. In this area the sexual oospore has been observed in nature and the environmental conditions are ideal for *Phytophthora*. All the pathogenic races occur in nature in this restricted valley and the two compatibility types exist also. Therefore, it is supposed that the possibility of the development of new races also exists in Toluca. Interestingly enough, in the rest of the world all the races of *Phytophthora infestans* belong to one compatibility type. These races are not able to hybridize and produce new recombinations. New discoveries of "mating types" (Smoot et al., 1957, 1958) as well as "compatibility types" by Galindo and Gallegly (1960) show that sexual hybridization may occur even in artificial cultures, at least theoretically. The compatibility type is independent of sex. Hybridization, and consequently pathogenic variation, in this fungus is governed by bisexuality, and control of mating behaviour, by compatibility factors. On the basis of these findings *Phytophthora infestans* can be referred to as a heterothallic fungus. However, more research and new and exact results are needed in this field to understand the genetics and variation in *Phytophthora*.

For the testing of resistant cultivars the most appropriate "world testing ground" is in the Toluca Valley in Mexico (Graham et al., 1959, Niederhauser and Mills, 1953, Niederhauser, 1962). Here the pathogen is able to infect potato under the most favourable environmental conditions and all of the known races (perhaps new races too) are present, as well as the two compatibility types. Consequently hybridization, genetic segregation, and recombination take place.

Physiologic races

Pathological specialization of races of *Phytophthora* was discovered as early as 1932 by Schick and by Müller (1933) on different cultivars of potato obtained from *Solanum demissum* × *S. tuberosum*. In these hybridizations, resistance genes from *S. demissum* were transferred to the commercial cultivars. Later, additional new races were identified by different research workers. On the basis of these findings Black et al. (1953) determined four dominant resistance genes (R_1, R_2, R_3, and R_4) in resistant cultivars. At the same time, Black and co-workers suggested an international system of classifying races and genes for resistance. According to this system a particular race receives the number of the resistant gene contained by the plant it is able to infect. For example, the "common field race", race 0 can infect only cultivars of potato with no R genes. Race 1 can infect plants with the gene R_1 in addition to the cultivars with no R genes. Race 2 can infect plants with the gene R_2 plus the plants with no R genes, and so on. Races which infect plants with dominant genes R_1 and R_2 as well as those with no R genes, are designated as race 1.2. The four R genes produce 16 possible combinations. 15 of the total 16 races were identified experimentally.

The nomenclature and identification of races is based on the key as follows:

	Physiologic races															
	0	1	2	3	4	1.2	1.3	1.4	2.3*	2.4	3.4	1.2.3	1.2.4	1.3.4	2.3.4	1.2.3.4
R genes																
R_1	R	S	R	R	R	S	S	S	R	R	R	S	S	S	R	S
R_2	R	R	S	R	R	S	R	R	S	S	R	S	S	R	S	S
R_3	R	R	R	S	R	R	S	R	S	R	S	S	R	S	S	S
R_4	R	R	R	R	S	R	R	S	R	S	S	R	S	S	S	S

R = resistant to the race indicated above
S = susceptible to the race indicated above
* = race not found in nature as yet

Later R genes were found not only in *Solanum demissum* but also in other species of *Solanum* and additional R genes were also identified. Still, the international system outlined above is workable and widely applied by plant pathologists as well as plant breeders.

The resistance of cultivars of potato which have R genes is based on a hypersensitive reaction. This defence reaction is characterized by tissue necrosis around the infection site, which localizes the pathogen. Races identified on *demissum* genotypes are not necessarily the same as those identified on tomatoes. This shows that potato races may be different from tomato races. The R_1–R_4 *demissum* genes are usually referred to as major genes. The cultivars containing major genes have "demissum resistance" (Kennebec is a typical example).

The differential set of cultivars used for the identification of races of *Phytophthora infestans* vary in different laboratories. Schick and Schick (1959) in Germany suggested a clone set of hybrids of *Solanum demissum* × *Solanum tuberosum* as follows:

Flava (r)
Aquila (R_1)
Black 1512 c (R_2)
Lindenhof 1762 (R_3)
Black 1563 c (R_4)
Virginia (R_1, R_4)
Krasnoufinskij (R_2, R_4)

These investigators recently suggested the appropriateness of a differential set composed of seedlings. Seedlings of *Solanum demissum* and *Solanum stoloniferum* carrying different R genes are the members of this differential set. One can maintain this set virus-free more easily and successfully than the original clone set. Virus infections may disturb the evaluation of resistance to *Phytophthora* and in some cases even make it impossible.

Identification of races is carried out as follows: Leaves of plants grown in the greenhouse are put on wet filter paper in Petri dishes. They are inoculated by putting two drops of a suspension containing zoospores of the fungus on the under surface of leaves, usually in the afternoon. Each drop of the suspension should contain about 30 zoospores. The next morning the leaves are turned to the upper surface and the Petri dishes are placed at a temperature of 18 °C in a light chamber. Results are evaluated on the 4th to 6th days. First the degree of sporulation is evaluated under a binocular microscope at 15 × magnification. During the incubation period, the daily temperature should not exceed 24 °C and the night temperature should not be less than 12 °C.

Another type of resistance, the so-called *minor gene resistance* or *field resistance* was found in numerous cultivars of *Solanum tuberosum*. Although these cultivars have marked resistance to *Phytophthora*, generally they need some fungicidal protection under severe epidemic conditions. This type of resistance is relatively weaker than the *demissum* resistance and is not race-specific. It is a recessive character and probably depends on the interaction of several genes. Resistance of the tops is not necessarily correlated with that of the tubers. Cultivars such as Sebago, Minominee, Calrose, and Saco are typical examples of minor gene resistance.

According to Niederhauser (1968) high levels of field resistance have been found in a number of Mexican wild potato species and among clones of some species. The present sources of field resistance are

Solanum andigenum (no R genes)
Solanum phureja (no R genes)
Solanum demissum (with R genes)
Solanum stoloniferum (with R genes)

As a rule, a slower rate of penetration of the leaf, a lower incubation period before penetration, a slower rate of lesion growth and a reduced sporulation contribute to field resistance.

A comparison of the two types of resistance is outlined below:

	Demissum resistance	Field resistance
Mechanism of resistance	Hypersensitivity (incompatibility)	Slower rate of penetration as well as lesion growth and reduced sporulation (compatibility)
Inheritance of resistance	R genes	Minor genes
Races attacking the plants	0, 1, 2, 3 . . . 1. 2 etc. according to the international system	Only restricted race specificity
Source of resistance	*S. demissum* *S. stoloniferum*	*S. tuberosum* *S. andigenum* *S. phureja* *S. demissum* *S. stoloniferum*

Cultivation of the fungus

The most appropriate medium for the cultivation of *Phytophthora infestans* has been given on page 286. It is rather difficult to isolate this fungus from plant material because of the simultaneous occurrence of numerous bacteria as well as fungi on the surface of diseased parts of plants. It is easier to produce pure cultures on sterile tuber tissues of potato than on artificial or synthetic media. Vitality of the strains is often lost if they are cultivated on artificial media for a long time. On the other hand, vitality and race characters of cultures under paraffin oil (see on page 274) remain unchanged for a considerable time (Thurston, 1957, 1959, Henniger, 1959). It is sufficient to make subcultures once a year from strains under paraffin oil. Strains, however, grown on synthetic media without preservation by paraffin oil must be subcultured every 15–18 days. The fungus is very sensitive to metal ions, therefore, artificial inoculation must be carried out with suspensions of zoospores in ion-free distilled water. The water should be entirely free of copper or iron ions and also free from chloride.

Evaluation of demissum resistance based on hypersensitivity

It is advantageous to start with infected tuber tissues. Scrape sporangia from the surface of the infected tuber slices with a scalpel and transfer to water. Then incubate this suspension at 8–12 °C for 1–2 hr in a small Petri dish. After 2 hr examine for zoospores; mount the Petri dish on the microscope stage for examination. At 20 °C the zoospores lose their motility in about 30 min. If a longer period of motility is desired, the Petri dish can be placed in an ice bath.

To inoculate the foliage of potted potato and tomato plants, spray the zoospores on the leaves and incubate the potted plants at 18 °C in a humidity chamber for 24 hr. The suspension should contain about 30 zoospores per drop. To inoculate detached leaves, remove leaflets from plants, place in Petri dishes, and put one drop of zoospore suspension on the under surface of each leaflet. Incubate in a moist chamber at 18 °C for 12 hr. If the material to be tested is from the tip of the foliage, a plant from a bud cutting, or a seedling grown directly from seed, then the evaluation of resistance must be carried out in a greenhouse test, on plants growing in a sand-soil mixture of 3 : 1. In this case a saturated atmosphere is necessary in the greenhouse for 3–4 days after inoculation to stimulate abundant sporulation. Necrotic spots appear usually on the 3rd day after inoculation on infected plant parts. Sporulation begins in 4–5 days as a rule. Abundant sporulation indicates susceptibility.

Inoculation of potato tubers: 5 cm thick slices are inoculated on the surface with a suspension of zoospores. Then two slices are put together, placed in a Petri dish, and kept at 20 °C. The degree of sporulation and the size of the necrosis at the surface of the tuber is evaluated. Tubers should not be kept too moist or too warm since bacterial soft rot may develop. The occurrence of sporulation is the sign of susceptibility.

Assessment of "field resistance"

This type of resistance is sometimes called polygenic, multigenic, or partial resistance. Unlike hypersensitivity, it is effective against all known races of *Phytophthora*. Cultivars possessing field resistance are destroyed more slowly in an epidemic than the more susceptible cultivars. Field resistance received attention because "immunity" to late blight disease was not achieved by transferring R genes from *Solanum demissum* to *S. tuberosum*. Each R gene is able to control only a few races of *Phytophthora*. When such cultivars are produced on a large scale, they are soon attacked by new races of the fungus. Cultivars with field resistance to late blight sometimes show susceptibility if *young* parts of the plant are tested in the greenhouse or in the laboratory. To asses field resistance *adult* plants must be tested in the open field or in the laboratory. Field resistance is mostly, but not necessarily, correlated with late ripening.

A scale of the degree of susceptibility is generally used for the assessment of field resistance as follows:

0 = no lesions
1 = few lesions (1–10 per cent)
2 = slight attack (11–25 per cent)
3 = moderate attack (26–50 per cent)
4 = severe attack (more than 50 per cent)
5 = death of the leaf

Laboratory assessment of "field resistance"

Lapwood (1961a, 1963) compared fourteen different cultivars in laboratory and field tests. He found a good correlation between results in the field and laboratory. Leaves were excised from plants, then inoculated with a zoospore suspension and incubated at 15 °C for 24 hr in a moist chamber. Leaves were then immersed in water by their petioles, and kept cool. The degree of spread and sporulation of the fungus and the rate of lesion formation indicated susceptibility or resistance.

Hodgson (1961, 1962) detached leaves from two-month-old plants and collected leaf material from one-third of the distance from top to base. Discs were punched out of the leaves, and inoculated with a drop of zoospore suspension. Inoculated discs were kept on moist filter paper at 15 °C and assayed after a week or so. This method was further improved by Gallegly (1962) by growing plants in sand in the greenhouse and watering them with a weak Hoagland solution before inoculation (0·1–1·0 Hoagland).

Field assessment of "field resistance"

The method of Lapwood (1961b) is as follows: Plots in field trials are arranged in a 4×4 Latin square design. 10 stems per plot are chosen and assessment of resistance to *Phytophthora* is carried out on 5 leaves per stem. Usually 5 leaflets are used per leaf, the terminal and the two distal pairs of leaflets. Observations are made once a week, after appearance of the first symptoms of the disease. If 4 replicate plots are arranged then 1000 leaflets per cultivar will be assessed to determine their susceptibility to late blight: 5 leaflets per leaf $\times$ 5 leaves per stem $\times$ 10 stems per plot $\times$ 4 replicate plots per cultivar. The number of leaflets destroyed on each occasion is expressed as percentages of the total. With weekly observations one can follow the development of the disease and the expression of disease resistance at different times after inoculation.

The peroxidase activity test for assessment of "field resistance"

On the basis of the original observation of Gretchushnikov (1939), Kammermann (1951), Kedar (Kammermann) (1959) and Umaerus (1959) came to the conclusion that peroxidase activity of leaves is related to field resistance of potato to late blight. If this rapid biochemical test works satisfactorily, this makes artificial inoculations and open field observations unnecessary. The cultivars with a high rate of peroxidase activity usually showed strong resistance to *Phytophthora* in the adult stage. However, Henniger and Bartel (1963) found only a weak correlation between the activity of peroxidase and field resistance. Recently, it has been shown by Sakai and Tomiyama (1964) that the correlation between these two characteristics is quite good in the adult stage but is uncertain or does not exist in young stages. All these investigators stress the fact that the peroxidase test is not suitable for evaluation of *demissum*-resistance. The correlation between resistance and peroxidase activity occurs only in the case of field resistance. Most recently Fehrmann and Dimond (1967) found a positive and striking correlation between the peroxidase activity in different organs of the potato plant and resistance to late blight. According to these investigators there is also some correlation between resistance and the activity of polyphenoloxidase.

The fifth or sixth leaves from the top are used for determining the activity of peroxidase. Ten grams of leaves from each cultivar are stored at —20 °C. Umaerus (1959) claims that leaves stored at this temperature retain their original peroxidase activity.

Leaf samples of 10 g are ground in 10 ml of ice-cold distilled water. The homogenized material is filtered or centrifuged and kept at 1–2 °C. This is the "enzyme extract".

The method of Fehrmann and Dimond (1967) is as follows: 0·1 ml of 0·2 *M* pyrogallol solution as a substrate is added to a cuvette containing 1 ml of 0·1 *M* phosphate buffer (pH 6·1), 0·5 ml of 0·01 M H_2O_2 solution, enzyme extract with appropriate concentration, and water to make a volume of 3 ml. The change in absorption at 430 mμ is measured. The hydrogen peroxide is omitted from the control. The difference between the absorption of reaction mixtures and that of the control is taken as a measure of the amount of the reaction. Enzyme activities are expressed as increase in absorption between 15 and 45 sec after the substrate is added.

The catechol method. 3 ml of 0·04 M catechol, 3 ml of 0·1 M phosphate buffer (pH 6), and 3 ml of 0·1 per cent H_2O_2 is added to a 1 cm cuvette and mixed with 3 ml enzyme extract (leaf extract). The increase in extinction is measured in a photometer. The first measurement is made 30 sec after the reaction starts. Further readings are made every 15 sec for 3 min. Enzyme activity is expressed as the increase in absorption at 470 mμ between 60 and 120 sec.

If a large number of samples are tested routinely Lyr's (1957) method is recommended by Henniger and Bartel (1963): this is the *benzidine-ascorbic acid test.* The following reagents are mixed in a tube of 2·2 cm in diameter: 4 ml of 0·02 per cent benzidine solution in 0·1 N acetate buffer (pH 5·3), 1 ml of 0·1 per cent ascorbic acid, and 1 ml leaf extract (enzyme). This mixture is placed in a 25 °C water bath, stirred and 2 ml of 0·3 per cent hydrogen peroxide (H_2O_2) is added. Ascorbic acid will be oxidized to dehydroascorbic acid. This point will be indicated by the discolouration of benzidine, and the time needed to reach this point is measured from the beginning of the reaction. A short reaction time means a high level of enzyme activity. As a rule, the reaction time is between 10 and 200 sec. On the basis of these values, enzyme activity is expressed in mg ascorbic acid /min/ gram fresh weight.

Late blight disease in relation to breeding for resistance is treated in the following works: Roemer, Fuchs and Isenbeck (1938), Frandsen (1958), Gallegly and Niederhauser (1959), Schick and Hopfe (1962) and Erwin et al. (1963).

DOWNY MILDEW OR BLUE MOULD OF TOBACCO (*PERONOSPORA TABACINA* ADAM)

Taxonomic position of the fungus

The thallus of the obligate parasite, *Peronospora tabacina* consists of coenocytic mycelium without septation and a cell wall made up of cellulose. Sexual reproduction is eucarpic and typically oogamous. On the basis of these characteristics the pathogen belongs to the *Phycomycetes* class and to the collective order of *Oomycetes*.

Life cycle

The life cycle of *Peronospora tabacina* can be seen in Figure 4. The asexual reproductive cell or conidium reaches the susceptible leaf of the tobacco host plant primarily by means of air currents. Here it adheres to the leaf hairs (Deppe and Seehofer, 1962; Tuboly, 1966) and germinates with the production of a germ tube (Fig. 5). The most important environmental factors influencing spore germination have been shown by Cruickshank (1963), Klinkowski and Schmidekneckt (1960) and others, to be mainly moderate temperature and high moisture

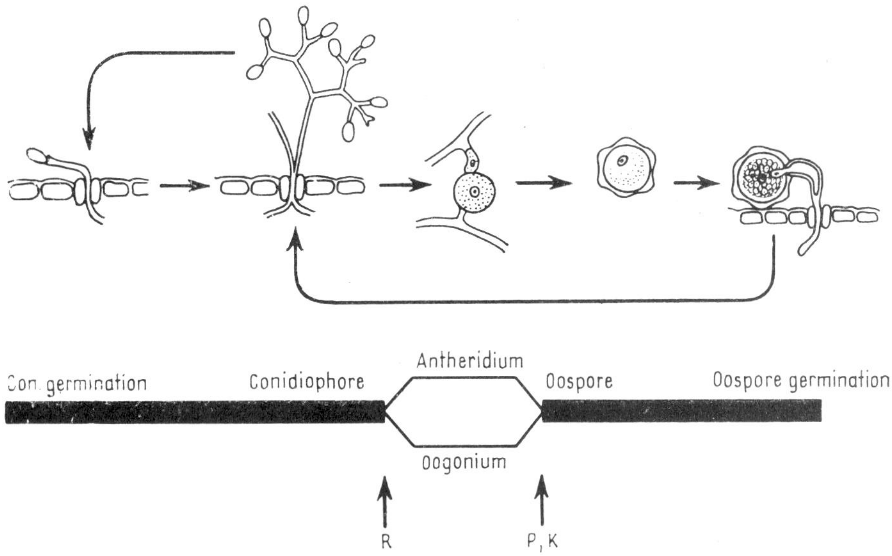

Fig. 4. Life cycle of *Peronospora tabacina* (P = plasmogamy, K = karyogamy, R = reduction

or humidity. Thus germination may take place between 8–27 °C. For the growth of the germ tubes the most favourable temperature initiating germination is 24–27 °C and 15–17·8 °C for subsequent growth. Leaf disc experiments by Shepherd and Mandryk (1962) showed the optimum temperature for the germination of conidia to be between 15–25 °C. The effect of the autoinhibitor and riboflavin content of conidia on the germination process will be discussed under methods of artificial inoculation.

The germ tube penetrates the host either directly through the epidermis or through stomata. The thallus formed inside the leaf tissues in a few days produces conidiophores. These break out in groups usually on the under surface of the leaf. With suitable rainfall and cool weather, the asexual cells may increase rapidly to epidemic proportions. Conidium formation and distribution are largely influenced by environmental conditions (Cruickshank, 1963; Hill, 1961; etc.). The asexual reproductive cells may remain viable for five and a half months (Kröber, 1965), insuring the overwintering of the pathogen.

Especially at the end of the vegetative period, typical sexual organs, oogonia and antheridia, are formed in the tissues of the leaf. The thick-walled oospore originates from a fertilized oogonia having a single oosphere. The oospore is a resting structure that is generally 38–44 μ in diameter. The process of oospore formation has been studied and described in detail by Kröber and Weinmann (1964). The role of oospores in overwintering is debatable. In a very few cases,

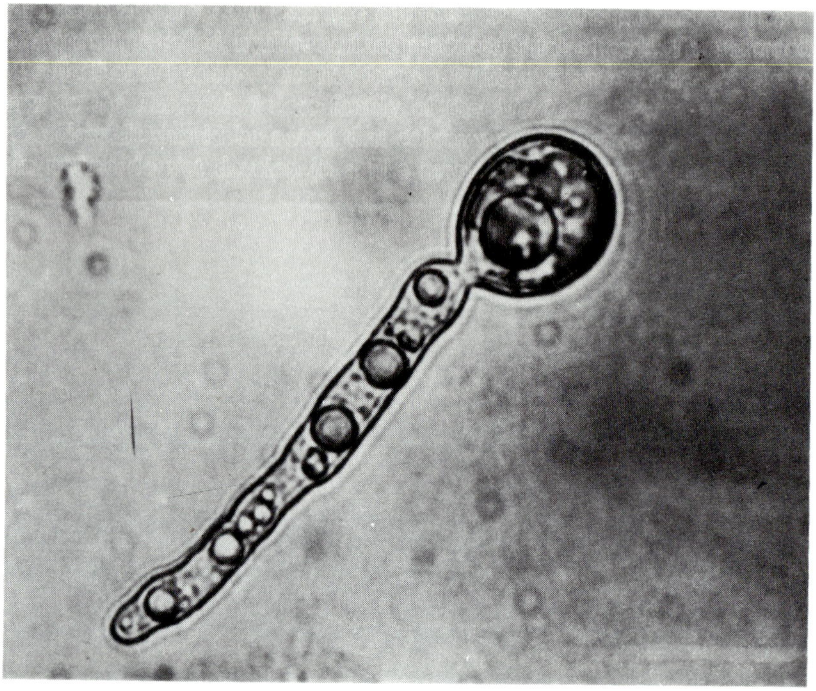

Fig. 5. Germinating conidia of *Peronospora tabacina*

oospore germination and infection caused by oospores were observed. The formation of oospores occurs most frequently in the lower leaves touching the soil (Pawlik, 1961). Covering these lower leaves with a thin layer of soil increases oospore formation.

In discussing the life cycle of the pathogen we have purposely not mentioned the haploid-diploid succession of generations. Sansome (1963a, 1963b) observed meiosis in the oogonia and antheridia of several species of the *Peronosporales* (*Pythium*, *Phytophthora*, *Albugo*). If this is a general phenomenon among the *Peronosporales*, then this means — in contrast with an earlier theory — that the vegetative thallus of these fungi is actually diploid and the *n* chromosome number only occurs in the reproductive organs. *Peronospora tabacina*, so far, has not been investigated in this respect cytologically.

Symptoms of tobacco Peronospora

The pathogen can infect tobacco plants in every stage of development from the seed bed to maturation. The pathogen occurs primarily in the leaves, but in plants in which the stalk is infected, the leaves remain symptom-free (Cruickshank

Fig. 6. Damage on tobacco leaf caused by *Peronospora tabacina*

and Mandryk, 1960). Indefinite yellowish-green spots form on the leaves that later coalesce and envelop the whole surface of the leaf blade (Fig. 6). On the under surface of the leaf, a pale grey, later bluish-grey, cover consisting of a downy mass of conidiophores and conidia develops.

Artificial inoculation methods

According to Hill (1966), the most important principles of *Peronospora tabacina* artificial inoculations are the standardization of methods and conditions. Spore concentration, temperature and moisture, etc. largely influence the severity of the infection. This is true whether we are concerned with comparing different strains of the pathogen or assessing the reactions of different host species. Thus, standard methods and conditions are essential.

a) *Maintenance of the inoculation material*

Since *Peronospora tabacina* is an obligate parasite it must be maintained generally on tobacco plants in a properly isolated growth chamber or glass house and kept in culture according to quarantine regulations. This procedure, however, requires a great deal of work and even when the rules are strictly observed, some inoculum or inoculation material may still become a source of early spring infection in the field. To avoid this danger, Jankowski (1965) suggested that spores obtained from inoculated tobacco and *Nicotiana glutinosa* plants could be dried and stored at 0 °C in Petri dishes. The inoculation material stored in this way could be successfully used for cotyledon pathogenicity tests from autumn until the next spring.

b) *Storage of the inoculation material*

Bromfield and Schmitt (1967) established that conidia can be stored for at least 25 months and still be viable and infective.

Their procedure is essentially:

1. Suspending conidia in dimethyl sulphoxide (DMSO)
2. Cooling the suspension for 1–2 hr to −20 °C
3. Immediately following this pre-cooling, the suspension is transferred to −180 °C for storage at the low temperature provided by a liquid nitrogen freezer.
4. Before the inoculum can be used again as inoculation material, it is most important that the induced dormancy be broken, that is the application of a heat shock just above 40 °C to each cell. For this treatment the most suitable is to place the frozen vial directly into a +40 °C water-bath from the liquid nitrogen freezer. After the suspension is melted and the temperature of the suspension rises to +40 °C for a few min (an hour is not harmful to the germination of spores) the dormancy of the conidia is broken and their germination is assured when used as inoculum. (This procedure probablv can be applied successfully for storing species other than *Peronospora*, too)

c) *Isolation and maintenance of monospore cultures*

In investigations assessing the resistance of cultivars to a particular biotype, it is of the utmost importance to use genetically uniform, homogeneous inocula. In the case of the obligate parasite *Peronospora tabacina* two such methods are available to provide single spore cultures:

A) Peresypkin and Arkhipova's method (1964) consists of the following steps:

1. Water drops are examined from a suitably diluted suspension of conidia under the microscope and those containing single conidia are selected.
2. A water drop containing a single conidia is deposited on the lower surface of a tobacco leaf that was grown under infection-free conditions.

3. The plants are kept in a moisture chamber at 16–25 °C for 20 to 24 hr to accomplish infections of the leaf.

4. After removal from the moist chamber the plants are kept individually isolated for a 10 day incubation period. Isolation of the plant is accomplished by covering the plant with cellophane that is well sealed and keeping the plant in the shade in a greenhouse at 18–25 °C.

5. Ten days after inoculation the plants are placed in the moist chamber for 24 hours without removing their cellophane covering to bring about sporulation of the fungus.

6. By washing the conidia from the infected leaves into water, a suspension of a monospore culture is obtained.

B) Shepherd, Stuart and Mandryk's (1963) method for obtaining monospore cultures is essentially the artificial inoculation of small leaf discs. We shall refer later to this method.

1. An agar plate streaked with a conidial suspension is examined for isolated single conidia and small blocks are cut from the plate.

2. The agar block is placed in a drop of liquid medium on the reverse side of a 15 mm leaf disc floating in 1 ppm concentration of kinetin in the solution.

3. The leaf discs are incubated at 20 °C and provided with a daily photoperiod of 16 hr of light at 800 ft–c (8600 lux).

4. After 10 days conidia are formed on the leaf discs (monospore culture) and may be transferred to newer leaf discs or to potted plants for further multiplication.

d) *The cotyledon test for determining resistant tobacco lines*
(Schiltz and Izard test (1962), modified by Pawlik et al. (1963)

1. Surface sterilize tobacco seeds in 0·5 per cent Perhydrol (hydrogen peroxide) solution and rinse in sterile water.

2. Place 100 seeds into each 10 cm Petri dish on two layers of 9 cm filter paper moistened by 5 ml sterile water.

3. The Petri dishes are placed in a growth chamber for 6 days at 27 °C at 4000 lux (372 ft–c) of continuous light.

4. The seedlings are artificially inoculated on the evening of the sixth day with 1 ml of suspension containing 50,000 conidia/ml density that was prepared from 12 hr-old conidia.

5. The inoculation is repeated after 24 hr.

6. Following the second inoculation, the seedlings are subjected to light between 8 a.m. and 5 p.m., and kept at 24 °C during the daytime and 16 °C at night.

7. The results of the inoculations are checked by observing the plants for the formation of conidiophores on the cotyledons every second day beginning with the 7th day and through the 17th under a stereomicroscope. The cotyledon method in growth chambers does not always give the same results when testing for plant resistance in the field. Fantechi and Lorito (1965) for example, with

this method showed that a hybrid marked C 101 that was moderately resistant in a growth chamber test did not prove to be so in the field.

e) *The leaf disc method for determining resistant tobacco lines*

This method has many variations that are based on the same principles. From among these we shall describe Shepherd, Stuart and Mandryk's (1963) procedure in detail. It is essentially the same as that described for isolating singles-pore cultures of the fungus.

1. Tobacco discs cut from uniform leaves of each line, 15 mm in diameter, are floated in Petri dishes containing 1 ppm kinetin in solution with the lower leaf surface up. In this way the inoculated discs remain green for up to 3 weeks, but without kinetin they turn yellow in 5–7 days.

2. A drop of water suspension containing 2×10^3 conidia is placed in the middle of each leaf disc.

3. The inoculated discs are incubated at 20 °C and given a daily photoperiod of 16 hr of light at 800 ft–c.

4. In successful infections, traces of necrosis can be observed after two days, and conidial formation, after 6–7 days. The conidiophores arise from the still green discs and conidial formation is highest 9–10 days after the inoculation.

5. Twice as many conidiophores are formed per leaf disc, if the discs are incubated in a medium containing 1 ppm kinetin, 2 ppm *p*-amidobenzoic acid, and 2 ppm nicotinic acid.

f) *Method for investigating seed transmission*

Egerer (1962) examined the possibility of seed transmission of the pathogen by the following method:

1. Slants were prepared using 80 ml of a nutrient medium in wide mouth, 300 ml capacity Erlenmayer flasks. (Composition of the nutrient medium: 14 g agar and 2·5 g nutrient salts suitable for growing plants in water culture, in 1 l of water. The flasks were plugged with cotton, autoclaved and slanted.)

2. After solidification of the sterilized agar medium in a slanted form, 0·04 g of surface sterilized seed are sown per flask on the surface of the agar.

3. The resulting seedlings are observed for signs of possible *Peronospora tabacina* infection macroscopically.

Two other factors influencing artificial inoculation require mentioning. One has already been referred to briefly, the autoinhibitory effect. Shepherd and Mandryk (1962; 1963) demonstrated that *Peronospora tabacina* conidia contain substances that inhibit their own germination. The inhibiting material could be separated into two active components by paper chromatography. They also demonstrated that the inoculated tobacco (e.g. in the case of stalk infection) contains inhibitors affecting conidial germination. Another factor for successful spore germination and plant inoculation was pointed out by Shepherd and Tosic (1966). They noticed that the conidia shaken from sporulating lesions of infected leaves germinated well but the washed ones did not. In the latter case, conidial germination could be improved by the addition of riboflavin. In the field, the poor germination of spores, at least in some cases, was due to the suboptimal riboflavin content of the spores.

Biological specialization of the pathogen

With regard to *Peronospora tabacina* no well-defined and internationally accepted biotypes are recognized, although many researchers have described strains with deviating infectivity. In some cases, morphologically different strains were observed, too.

In New Zealand, Wark and co-workers (1960) demonstrated the presence of at least four different biotypes on the basis of the reactions obtained on the following resistant species: *Nicotiana debneyi*, *N. knightiana*, and *N. goodspedii*.

In Hungary, Tuboly (1966 a, b) described a new race which could infect resistant plants of *N. debneyi*, Hicks cultivar of tobacco, etc. The new biotype with its characteristic spiral germ tube is easily differentiated from the type normally encountered.

Iren (1965) established that the races of *Peronospora tabacina* in Turkey originated from Australia. However, the infectivity of isolates coming from various parts of the country deviated on *Capsicum*.

Cultivars resistant to tobacco Peronospora

Reviewing the earlier results in connection with the breeding of the blue mould resistant cultivars, Lea (1961) referred to the crossings of *Nicotiana tabacum* and resistant wild species. The applied sources of resistance were *Nicotiana debneyi*, *N. megalosiphon*, *N. plumbaginifolia*, the amphidiploid *N. debneyi-tabacum*, *N. exigua*, *N. knightiana N. suaveolens* and *N. velutina*. From the point of view of resistance and the internal characteristics of the cultivar, the crossings of the amphidiploid *N. debneyi-tabacum* hybrid with the Hicks tobacco proved to be the best hybrid. This line could be fixed by selection. Apart from the resistance to *Peronospora* its economic value surpassed that of any other species cultivated up to that time.

One year later Kröber and Massfeller (1962) reported that of 13 commercial tobacco and 2 *N. rustica* cultivars, all proved to be susceptible in field experiments, although differences could be noticed in the degrees of susceptibility.

On the basis of results obtained from the investigations in a large number of species and lines of different origin, Endemann and co-workers (1962) suggested further directives in breeding for resistance. They showed that all the examined cultivars were susceptible to the pathogen and that resistant lines could not be obtained from these by selection. The most susceptible were the "aurea-forms" and the "Gendestheimer" cultivars. The greatest resistance was found in the above mentioned "Resistant Hicks" cultivar. The heredity of resistance or susceptibility was generally uniform in all progeny. Segregation of a resistant strain to a susceptible line and a resistant line was noticed only in one case. The F_1 generation of resistant × susceptible hybrids proved to be resistant in every case.

The results of Babayan and co-workers (1966) is especially noteworthy. Every local line and species examined proved to be susceptible to *Peronospora tabacina*. The Bel 16–12 cultivar proved to be immune whereas the Hicks cultivar was largely resistant. In additional crosses, if one parent was resistant, resistance was inherited as a dominant characteristic in the F_1 generation, but in the segregating F_2 generation, a rapid decline in resistance was observed.

Ternovskii and co-workers (1965) found 8 species to be immune from among 17 different *Nicotiana* species. From the breeding point of view, the following species are the most important sources of resistance: *N. debneyi*, *N. exigua*, *N. goodspedii* and *N. ingulba*. From among the different cultivars Bel 1 61–9, Bel 1 61–10 and SO 1 were the most promising. According to Fantechi and Lorito's (1965) data, the A_2-2–27–2 line in the cotydelon tests and in the field proved to be equally resistant.

Control of tobacco Peronospora

Apart from the breeding for resistance, the control of *Peronospora* in tobacco requires cultural and chemical measures. The latter is especially justified in the case of rainy, cold weather causing conditions favourable for blue mould epidemics. For chemical control, the dithiocarbamate type of chemicals are used the world over (Zineb, Maneb, Antracol etc.). The destruction of infected crop residues and providing new locations each year for seed beds and a suitable location for growing tobacco are the most important cultural practices.

CHAPTER III

ASCOMYCETES

APPLE SCAB
(*VENTURIA INAEQUALIS* [COOKE] WINT.)

Taxonomic position of the fungus

The sexual reproduction of the heterothallic *Venturia inaequalis* exhibits typical ascogamy, the asci forming not freely, but inside a fruiting body. For this reason, the fungus belongs to the subclass *Euascomycetidae* of the *Ascomycetes* class. The asci (Fig. 7), however, develop not in a true perithecium but in a stroma-cavity. Fungi having these characteristics, such as the genus, *Venturia*, are grouped together in the collective order *Ascoloculares*. The ascostroma or pseudothecium has a single cavity with the asci originating in a hymenium. The higher developed *Ascomycetes* have a true perithecium with separate walls. Among the asci one can find sterile filaments, pseudoparaphyses, that originate

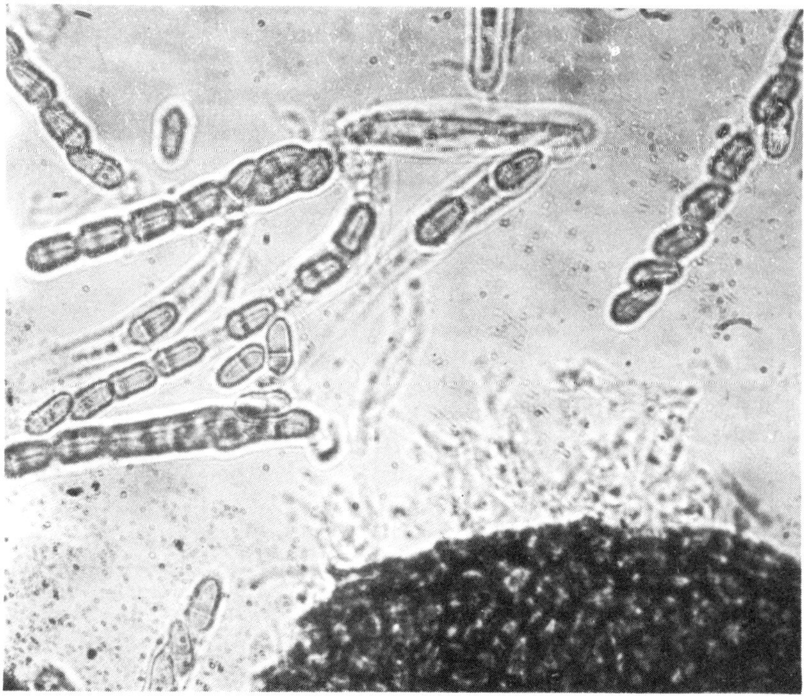

Fig. 7. Asci of *Venturia inaequalis*

not from the hymenial layer but from the upper part of the fertile cavity. The two-celled ascospores are brown. On the basis of these characteristics, *Venturia inaequalis* belongs to the order *Pleosporales*, family *Venturiaceae*.

Life cycle

The life cycle and the alternation of generations of *Venturia inaequalis* may be seen in Fig. 8. The mature, two-celled ascospore germinates and produces a stroma under the cuticle of the host leaf. Soon conidiosphores, 10–65 × 4·5–9 μ, are formed. These break to the surface of the leaf and produce conidia of 12–35 × 5–10 μ. The young conidia are single celled but the mature ones are two celled. The conidial form *Spilocea pomi* Fr. ex. Fr. or an older name, [*Fusicladium dendriticum* (Wallr.) Fuckel] ensures the distribution of the fungus and causes scab epidemics. The fungus actually becomes parasitic in this asexual phase. At the end of the vegetative phase in the host, the fungus penetrates between the tissues from beneath the cuticle in a saprophytic way with the onset of necrosis of the leaf tissues and leaf drop. Here the ascostromata, 90–145 μ in diameter, are formed. Sexual fusion and plasmogamy take place in the ascostroma. This occurs between the trichogyne arising on the apex of the spiral hypha that initiated the ascogonium and the antheridium arising from a hypha of a male individual on a different thallus. The fungus is hermaphroditic, but self-sterile, with ascospores falling into two groups and being intergroup fertile and intragroup sterile (Keitt and Palmiter, 1938). Maturation of ascocarpia with the ascogenous hyphae and the asci 80–110 × 10–15 μ and pseudoparaphysis lasts from autumn

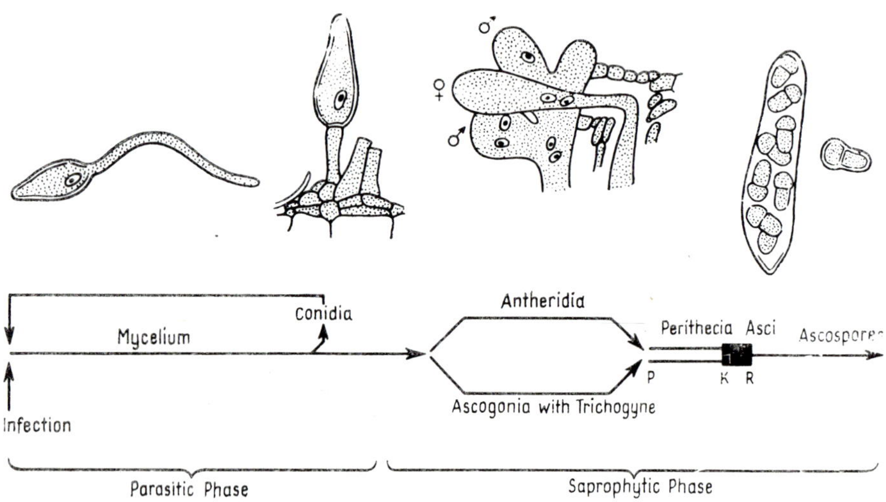

Fig. 8. Life cycle of *Venturia inaequalis* (P = plasmogamy, K = karyogamy, R = reduction) (After Roemer et al., 1938)

to spring. The two-celled ascospores are forcibly ejected from the ascocarp in spring, 11–16 × 4·5–8 μ and initiate the asexual cycle of the fungus, and subsequently provide for its epidemic distribution.

The maturation of the ascospores as well as the formation of the conidia and thereby the epidemic spread of the fungus are equally influenced by temperature and moisture. For the asci to mature in the ascocarpium from the end of winter to the time of discharge in the spring, a determined amount of effective temperature is required. In this way for example counting the total temperature from March 1st 108–150 °C and from February 7th 285–298 °C total temperature is required for the beginning of the discharge of the mature ascospores. The wilted leaf and ascostroma must be saturated with water for a long period of time before spores are ejected (Anderson, 1956). From the same ascostroma infective ascospores may be ejected over a period of several weeks. Water droplets are needed on the leaf for the germination of ascospores and an optimum temperature of between 15–24 °C. The formation of the asexual reproductive cells also is largely dependent upon temperature and moisture. The conidia generally develop in 36–60 hr at the germinal end of the conidiophores. The temperature requirements for the germination of conidia is similar to that for ascospores. The germinating conidium produces appressoria in 24 hr. The most important condition for the epidemic spread of the fungus is a suitable quantity of rain (yearly average of 700–900 mm).

The optimal temperature in artificial cultures for the fruiting bodies to develop is 13 °C. In connection with the nutrient-physiological factors of ascostroma formation, Ross and Hamlin (1965) established that among the nitrogen sources, nitrate, urea (carbamide), and amino acids stimulate the formation on fruiting bodies while ammonium sulphate inhibits them. The disaccharides generally have a stimulatory effect, but xylose, ribose, fructose, mannose, and galactose inhibit ascostroma formation. Also, it was established that in order for ascospores to form in the ascostromata, it is essential that certain vitamins be present in the medium.

Symptoms of apple scab

Symptoms of scab appear most prominently on the leaves. On the young leaves of the susceptible cultivars, grey and then black, olivaceous spots of 2–6 mm in diameter arise. The severely infected leaves soon fall and with this the development of the tree and the formation of buds etc. is hindered. In the second half of the vegetation period the so-called "late scab" symptoms appear. These are more conspicuous because of the deeper penetration of the fungus into the tissues. On the spots, the leaf tissue dies and cracks. Infections of the fruit cause serious economic losses and infections usually start at the same time on the young fruit as the expansion of the core. The spots formed here initially are similar to the leaf spots but become larger in size (6–12 mm). Later, due to the formation of cork and as a consequence of fruit growth, the spots crack radially and cause fruit distortions. The fungus may infect the flower and leaf buds as well as the shoots.

A detailed description of symptoms on those *Malus* species or apple cultivars that are susceptible and on those that are resistant to varying degrees, will be given later in the rating scale for the evaluation of artificial inoculations.

Artificial inoculation method

The greenhouse procedure below according to Shay and Hough (1952) can be successfully used in evaluating *Venturia inaequalis* resistance of *Malus* species and cultivars.

Preparation of inoculum: The mycelium from agar cultures of *V. inaequalis* is planted on cheesecloth wicks that are adhered to the sides of 250 ml Erlenmeyer flasks with an excess of 1 per cent Difco malt extract to keep the wicks wet. After about 10 days, the fungus will have covered the wicks with mycelium and conidia. The excess medium is decanted and replaced with sterile, distilled water. The conidia are freed from the cheesecloth wicks by shaking. The final conidial suspension used for inoculations is diluted in such a way that 100 conidia can be seen in the low power field of the microscope (100 × magnification).

Preparation of plant material: Prepare a single shoot from the plant to be tested either by:

a) using dormant bud shoot growth or

b) the growth resulting from the cutting back of 1–2 year-old nursery saplings.

Inoculation

a) Thoroughly spray the young leaves with the prepared conidial suspension.

b) Cover the plants with a five-layer muslin tent supported by a wooden frame.

c) Maintain the plants under constant high moisture and under the muslin tent for 40–44 hr.

d) To ensure infection, a temperature of 15–18 °C in the greenhouse must be maintained during and after the inoculation.

e) Use one of these susceptible cultivars for comparison: Jonathan, Golden Delicious or Red Rome.

Rating scale for inoculations

For the evaluation of the host symptoms, the following scheme is advised:

0 = no macroscopically observable symptoms (immune reaction),

1 = few or many rather tiny (about 0·5 mm in diameter), very quick to appear (about two and a half days), pinpoint-like spots sometimes coalesce to form a common but irregular 1–2 mm diameter hollow spot. No spore formation (resistant hypersensitive reaction),

2 = regular or irregular chlorotic or necrotic spots, 2–5 mm in diameter. No spore formation (moderately resistant reaction),

3 = limited spreading of necroses, weak or abundant conidial sporulation. Leaves do not drop from the tree prematurely (susceptible reaction),

4 = spreading spots, with abundant sporulation. Infected leaves fall from tree prematurely (highly susceptible reaction).

Biological specialization of the pathogen

Shay and Williams (1956) demonstrated three races of *Venturia inaequalis* using the "Dolgo" (*Malus baccata*), and the "Geneva" (*Malus pumila* var. *niedzwetzkyana*) species and a Russian cultivar marked R 12740–7A. The separation of the races on differential host plants is given in the Table below.

Reaction of *Venturia inaequalis* races on differential host plants (Shay and Williams, 1956)

Malus species or cultivar	Rating or inoculation class		
	Race 1	Race 2	Race 3
Dolgo	2	4	2
Re-crossed R 12470–7A progeny	2	4	2
Geneva	2	2–4	4

2 = chlorotic or wilted spots but without spore formation
4 = well distributed spots with aboundant spore formation
2–4 = necrotic spots with abundant spore formation in a humid environment

Cultivars resistant to apple scab

Anderson (1956) as well as Shay and Williams (1956) established that there are no apple cultivars resistant to scab among those grown in America and Europe. The so-called "relatively resistant" cultivars are only of local significance. Shay and Williams (1956) at the same time emphasized that the most important sources of resistance were *Malus floribunda*, *M. astrosanguinea* as well as *Malus* species introduced from Japan and China such as *M. prunifolia*.

According to Karamisheva's (1965) determinations, the European species (*Malus sargenti*, *M. sieboldii*, *M. kaido*, *M. sylvestris*) are much more resistant to scab than the species from East Asia. However, the cultivation of the cultivars below is suggested by her in the area around Leningrad: Vinnoe, Bessemyanka, Welsey and Novogodnee.

The most important species as regards resistance to scab are *Malus floribunda*, *M. prunifolia*, *M. atrosanguinea*, *M. baccata*, *M. prunifolia-microcarpa*, and

Malus sp. MA–1255. The resistance to scab in these species is governed by the same pair of allele genes (Williams et al., 1966). This gene was marked as *Vf* by Williams and co-workers. A recent review on resistance in *Malus* to apple scab was published in *Ann. Rev. Phytopath.* **7**, 223–246, 1969 by Williams and Kuć.

Control of apple scab

The initial infections in spring can be reduced to a considerable degree with the elimination of inoculum from the infected leaves and shoots.

The basis of the spraying program carried out during the cultivation period is local prognosis. The necessary rainfall and cool temperatures are the most important prerequisites for the initial distribution of ascospores and the inducing of infection in the period following this dispersal of ascospores.

At the beginning of the cultivation period, before blossoming, is the most suitable time for eliminating the primary infections by spraying with organic mercuric compounds. These have an eradicating effect. In later stages of cultivation, Bordeaux mixture or copper-oxychloride preparations are suggested to spray cultivars that are not sensitive to copper.

Captan, Dithion and Dodine are among the organic fungicides that may be used successfully against the scab organism in controlling the disease.

POWDERY MILDEW OF BARLEY
(*ERYSIPHE GRAMINIS* DC. F. SP. *HORDEI* EM. MARCHAL)

Powdery mildew is a widely distributed and very important disease of barley. Especially spring barley is susceptible to the attack of this pathogen.

Biology and aetiology

The fungus is an *ascomycete*. Ascospores or conidia cause the primary infection. Perithecial formation and development of ascospores occurs on infected leaves in late summer. However, the formation and role of the perfect stage of this fungus is uncertain. Ascospores do not have an important role in the propagation of the pathogen or in the distribution of the disease. On the other hand, because of the heterothallic nature of this fungus, sexual processes in the asci and the formation of ascospores are important in producing new physiologic races (Powers and Moseman, 1957). The mycelium is able to persist from season to season in the temperate zone, therefore, asexual conidia have an eminent role in the distribution of this fungus. In the spring grain area, however, ascospores also may have some role as sources of primary inoculum.

For conidial formation and germination, a humid and cool atmosphere is important in the absence of free water.

Infection takes place by direct cuticular penetration into the epidermal cells. The fungus forms appressoria on the surface of the plant and haustoria penetrate into the cells. The mycelium spreads over the surface of the infected plant parts and only the haustoria are found within the cells. The mycelium produce conidia (asexual spores) on special conidiophores. In the perithecium plasmogamy, karyogamy and meiosis take place, which result in the formation of ascospores. However, there has been considerable disagreement regarding the sexual processes of powdery mildew of barley. The possible life cycle is shown on Fig. 9.

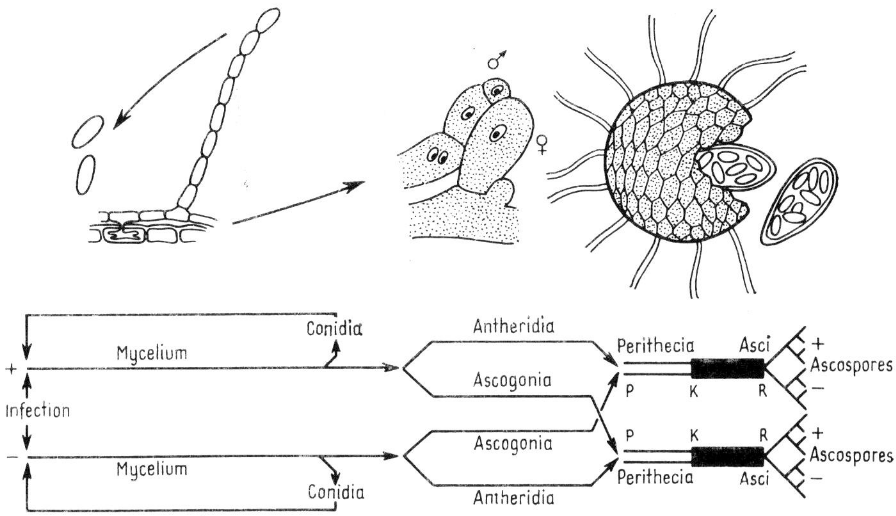

Fig. 9. Life cycle of *Erysiphe graminis* (P = plasmogamy, K = karyogamy, R = reduction) (After Roemer et al., 1938)

Artificial inoculation

Erysiphe graminis can be cultivated easily on seedlings of barley in the greenhouse. The success of the inoculation can be checked in 10 days at 10 °C or in 4–5 days at 20 °C. Conidia should be simply powdered over the seedlings to be inoculated, by shaking the leaves of infected plants over them. Conidia are very light and spread easily, therefore, it is hard to operate with different physiologic races within one greenhouse. For the cultivation of different races without the danger of mixing the different lines, well separated chambers or perhaps cabinets with separate entrances are necessary.

Different cultures of powdery mildew can also be separated by keeping the seedlings in pots under a lamp chimney. The lamp chimney is usually plugged with cotton or with other suitable material.

Separation of different cultures can be carried out if the cultures in pots are in plastic cages of 60 × 60 × 60 cm (2 × 2 × 2 ft). The cages are covered with four layers of cheesecloth. The most appropriate temperature for the cultures is about 20 °C. Seedlings of Heil's Hanna (C. I. 682) and Manchuria (C. I. 2330) barley cultivars are the most suitable for the maintenance and increasing of different cultures. Of course, other susceptible cultivars are also used for this purpose.

For mycological and genetic studies it may be necessary to produce single-spore cultures from ascospores. A method was outlined by Moseman and Powers (1957). The perithecia (cleistothecia) collected from different locations are matured by removing the individual fruiting bodies from the mycelial mats. The perithecia are kept continuously moist on filter paper suspended over water in a Petri plate. Then filter papers are place over 2 per cent water agar in Petri plates. The spores are discharged on the agar and can be transferred with a fine-pointed needle to seedlings under a lamp chimney. Here the spores germinate and develop into a mycelial mat. About 10 per cent of the spores transferred will infect the host plant. This culture can be increased on a susceptible host to develop a pure-line culture.

It is difficult to preserve conidia of powdery mildews. Conidia usually loose their ability to germinate in about 6 days. Keeping the conidia in a cool atmosphere does not change the viability of this fungus very much. Consequently, the collection and preservation of samples for a longer time is rather difficult.

Cultures in test tubes

The cultures or physiologic races can be maintained on seedlings of susceptible cultivars grown in 300 × 25 mm test tubes. The tubes are stored in a refrigerator at a temperature of 12 °C, and equipped with a daylight fluorescent tube. This way the individual cultures can be stored for 4–6 weeks (Moseman, 1956). If we would like to maintain the cultures for a longer time, conidia can be transferred every 4–6 weeks to new seedlings in the test tubes and stored in a refrigerator. Heil's Hanna and Manchuria cultivars are appropriate host plants for cultivation of different physiologic races.

Detached leaf culture method using benzimidazole

On the basis of the method of Mosaman and Greeley (1966) there is a possibility of testing detached barley leaves with two or more physiologic races without danger of mixing the races. It is also possible to obtain seed from the individual plants tested for their resistance to more than one race. This is an advantage for plant breeders.

The primary or secondary leaf is cut into 3–4–5 segments (1–2 cm in length). These are floated with the upper side of the leaf up, on benzimidazole solution (50 ppm) in 2 × 2 cm plastic chambers. Fifty of these chambers fit into a box (12 × 24 cm). By placing each segment from an individual leaf in different chambers and placing the chambers in different boxes, one can inoculate the leaf segments in each box with a different physiologic race. Inoculation is carried out by blowing on infected plants, so that the conidia of powdery mildew float down on the segments in the chambers. Following inoculation each chamber is covered and the boxes are kept in a room at a temperature of about 20 °C. Light of 1,300 ft–c is maintained for 16 hr per day. Reactions of resistance or susceptibility on the leaf segments are read after about 7 days. According to these authors the reactions of the leaf segments are similar to those on attached leaves. The leaf segments remain green for 3 weeks on the benzimidazole solution and permit abundant development of the fungus. Less time and equipment are required to test detached leaves than to test attached leaves for resistance to powdery mildew. Similar experiments were published by Caldwell (1960), Benada and Brückner (1963), Hermansen (1966), and Wolfe (1965) who were successful in culturing *Erysiphe* on benzimidazole solutions.

The following *infection types* can be distinguished on barley (Moseman, 1956):

0. immune, without visible symptoms,
1. very resistant, with limited mycelial development,
2. moderately resistant, the development of the mycelium is moderate, with some conidia, slight necrosis and a little chlorosis,
3. moderately susceptible, the development of the mycelium is moderate or abundant with limited sporulation, some necrosis and chlorosis,
4. very susceptible, great pustules, abundant sporulation without necrosis.

Direct microscopic observation of the fungus on the host plant

On the basis of the staining method of Drandarevski and Weltzien (1966) it is possible to investigate directly the different stages of the life cycle of powdery mildews, such as the deposition of the conidia on the surfaces, germination of conidia, formation of appressoria etc. Powdery mildew exhibits a goldish red or brown discolouration in iodine vapour.

Physiologic races

There are more than 20 races indentified in America and more than 50 races in Europe, employing several sets of differentials. Unfortunately, the composition of the differential sets was always changed by the research workers. Even so, the key for determination of physiologic races has begun to be standardized in America and more or less, also in Europe. The first key for the identification of races was published by Honecker (1937), who differentiated among 9 physiologic races applying a differential set of 8 cultivars. The Honecker-races can be differentiated with the set of cultivars as follows:

Cultivars	Races of Honecker								
	A	H	I	D	B	G	E	C	F
Weihenstephan CP 127422	0	0–1	0	0	4	4	4	4	4
Ragusa b	0	2–3	3	4–3	0	0–1	1–2	4–3	0
Hordeum spontaneum nigrum	0	0	0	0	0	0	0	0	4
Schweden 860	4	2	4	4	4	4	4	4	2–3
Samaria	1–2	3	1–2	2	2	3	1–2	3	2
Peruvian	1–2	4	4	4	4	2	4	1–2	1–2
Hohenfinower	4	4	4	4	4	4	4	4	4
Gopal	0	0	0	0	0	0	0	0	0

0 = very resistant; 4 = very susceptible

Nover (1957) used a new differential set, although he used a few cultivars in his set from that of the Honecker's. Nover differentiated 16 physiologic races by the use of the new set. The following cultivars are the members of the set:

Haisa (spring barley, *distichon*)

Weihenstephan CP 127422 (spring barley, *distichon*)

Ragusa b (semi-winter type, *hexastichon*)

Weihenstephan Stamm 37/136 (winter barley, *hexastichon*)

Weihenstephan Stamm 41/145 (winter barley, *hexastichon*)

Rinn 4 (spring barley, *hexastichon*)

Hordeum spontaneum nigrum, Voldagsen Stamm 8141/44 (spring barley, *distichon*)

Gaterslebener Mutante 511 (winter barley, *hexastichon*)

In North America about 21 distinct physiologic races have been identified. Research workers (Cherewick, 1944; Newton and Cherewick, 1947; Moseman, 1956; Powers and Moseman, 1957) used six cultivars in different tests which composed the American differential set:

Black Hulless (C.I. 666)
Chevron (C.I. 1111)
Goldfoil (C.I. 928)
Heil's Hanna 3 (C.I. 682)
Nepal (C.I. 595)
Peruvian (C.I. 935)

The reaction types of cultivars of the set and physiologic races are indicated in the following Table:

Physiologic races of *Erysiphe graminis* f. sp. *hordei* in Germany (After Nover, 1957)

Physiologic race	Haisa	Weihen-stephan CP 127422	Ragusa b	Weihen-stephan St. 37/136	Weihen-stephan St. 41/145	Rinn 4	H. spon-taneum nigrum Voldag-sen St. 8141/44	Gaters-lebener Mutante 511
1	4	0	0	0	0	0	0	4
2	4	0	0	4	0	0	0	0–1n
3	4	0	0	0	4	0	0	4
4	4	0	0	4	0	0	0	4
5	4	0	0	0	0	0	3–4	0–1n
6	4	0	4	4	4	0	0	0–1n
7	4	0	4	4	4	0	0	4
8	4	4	0	4	4	0	0	0–1n
9	4	4	0	0	4	0	0	0–1n
10	4	4	0	0	0	0	0	—
11	4	4	0	4	0	0	0	4
12	4	4	0	4	0	4	4	4
13	4	4	0	4	0	4	0	0–1n
14	4	4	4	4	4	4	0	0–1n
15	4	4	4	4	4	4	0	4
16	4	4	4	4	4	4	3	0–1n

n = necrosis, 0 = very resistant, 4 = very susceptible.

Physiologic races of *Erysiphe graminis* f. sp. *hordei* in America

Physiologic race	Black Hulless	Chevron	Goldfoil	Heil's Hanna 3	Nepal	Peruvian
1	0–2	0	0	1–2	1–2	0–1
2	1–2	0–1	0	4	4	1
3	3–4	0–1	0	4	4	3–4
4	4	0	0	3–4	4	1
5	3	–	4	4	4	4
6	0–2	0–1	0	3–4	0–1	0–1
7	0–1	0	3–4	3–4	0–1	0–1
8	3	3–4	0	3–4	3–4	0–1
9	1–2	4	0	4	1–2	0–1
10	1–2	0–1	0	4	1–2	3
11	4	4	0	4	4	4
12	1	3	4	4	1–2	1
13	2	1	4	4	4	1–2
14	1–2	4	0	4	3–4	1–2
15	1–2	1	0	4	3–4	4
16	1	3–4	0	4	2	3–4
17	1	4	0	4	4	4
18	1–2	0–1	4	4	0	3–4
19	3–4	4	0	4	2	2
20	4	4	0	4	2	4
21	4	4	4	4	4	2

Another differential set was composed by Hiura and Heta (1955) in Japan. These investigators have differentiated 11 physiologic races on the basis of the reaction types of the cultivars in the set as follows:

Physiologic races of *Erysiphe graminis* f. sp. *hordei* in Japan

Physiologic race	Goldfoil C. I. 928	Hanna C. I. 906	Heil's Hanna 3 C. I. 682	Kairyo-bozu-mugi	Nakaizumi-zairai	Nigrate C 2444
1	0	4	0	1	1	0
2	0	4	4	3–4	1	0
3	2	3–4	0	1	1	0
4	0	4	4	1	1	0
5	0	3–4	0	1	3–4	0
6	0	4	0	4	1	0
7	2	4	3–4	1	1	0
8	0	4	3–4	1	3–4	0
9	0	0–1	4	1–2	1–3	0–1
10	0	4	0	1	1	4
11	2	4	0	1	1	4

Resistant cultivars

The sources of resistance to powdery mildew of barley are as follows: Duplex (C.I. 2433), Hanna (C.I. 906), Goldfoil (C.I. 928), Arlington Awnless (C.I. 702), Chinerme (C.I. 1079), Algerian (C.I. 1179), Kwan (C.I. 1016), Psaknon (C.I 6305), Monte Christo (C.I. 1017), West China (C.I. 7556). In addition, several commercial cultivars such as Atlas 46, Erie, etc. also contain resistance factor(s) against powdery mildew.

The resistance of seedlings may be different from that in the adult stage. Cultivars which are susceptible in the seedling stage sometimes are resistant in the adult stage (mature plant resistance). This latter form of resistance is usually effective against several physiologic races, while the resistance in the seedling stage is race-specific.

Chemical control of powdery mildew

Sulphur, dithiocarbamates and recently antibiotics and derivatives of benzoxazolinone are effective fungicides against powdery mildews. With powdery mildews of cereals, chemical control is not applied in crop production, only for experimental purposes.

POWDERY MILDEW OF WHEAT
(*ERYSIPHE GRAMINIS* DC. F. SP. *TRITICI* MARCHAL)

Methods used with powdery mildew of barley may be applied in the case of wheat, too. Identification of physiologic races of *Erysiphe graminis* f. sp. *tritici* in Europe may be carried out by applying the key of Wolfe (1965). According to Wolfe races may be determined by the use of wheat seedlings or by the detached leaf culture method using benzimidazole. This latter method is quick, simple, and the reaction types are not significantly different from that of the leaves of intact seedlings.

Physiologic races of *Erysiphe graminis* f. sp. *tritici* in Europe
(After Wolfe, 1965)

Physiologic races	Huron C. I. 12663	H 14/44	Ulka	Ax-minster	Nor-mandie	Halle Stamm 13471	Weihen-stepha-ner M. I	Hope	Chul
0	S	–	–	–	–	–	–	–	–
1	S	–	–	–	–	–	–	S	–
2	S	S	–	–	–	–	–	–	–
3	S	–	–	S	–	–	–	–	–
4	S	S	–	S	–	–	–	–	–
5	S	–	S	–	–	–	–	–	–
7	S	–	S	S	S	–	–	–	–
8	S	–	S	S	S	S	–	–	–
9	S	S	S	S	S	–	–	–	–
10	S	S	S	S	S	S	–	–	–
11	S	S	S	–	–	S	–	–	–
12	S	–	–	–	–	–	S	S	–
13	S	–	–	S	–	–	–	–	S
14	S	–	–	–	–	–	–	–	S
15	S	–	–	S	–	–	–	S	–
16	S	–	–	S	–	–	–	S	S
17	S	–	S	S	S	–	–	S	–
18	S	–	S	S	S	–	–	–	S
19	S	–	–	–	–	–	–	S	S
20	S	–	–	–	–	–	S	–	–
21	S	–	S	–	–	–	–	–	S
22	S	–	S	–	–	–	–	S	S
23	S	–	–	S	–	–	S	–	–
24	S	–	S	–	–	S	–	–	–
25	S	–	S	–	–	S	–	–	S
26	S	S	–	S	–	–	–	–	S
27	S	S	–	S	–	–	–	S	–
28	S	–	S	–	–	–	–	S	–
29	S	S	–	–	–	S	–	–	–
30	S	–	–	S	–	S	–	–	–

S = susceptible; — = resistant

It has been found that the adaxial leaf surface supports a greater development of powdery mildew than the abaxial surface (Wolfe, 1963).

In addition to the basic differentials, "supplementary" and "additional" wheat cultivars were used recently by Wolfe (1967) to determine new races in England. The most common races in Europe are 0 and 3.

APPLE MILDEW
(*PODOSPHAERA LEUCOTRICHA* [ELL. ET EV.] SALM.)

This is a widely distributed disease of apple. Severe damage occurs primarily under relatively dry conditions when the summer wheather is warm. The disease is conspicuous in the early spring. The leaves of infected shoots are small with a white cover („powdery mildew"). Infected leaves later become rolled and dried out, especially at the edge of of leaves. Young infected shoots usually die back. In addition to leaves, the fungus attacks flowers and buds.

Life cycle and aetiology

The white cover on the infected parts consists of mycelium and asexual spores (conidia). Conidia are 21×12 μ in diameter. For the germination of the conidia, 18–25 °C is the most suitable temperature. Above 33 °C conidia are not able to germinate. Perithecia are produced only under optimal conditions on the infected parts of the plant. There is a single ascus in the perithecium, and 8 or fewer ascospores in each ascus. The ascospores are 24×13 μ in diameter.

This fungus overwinters as mycelium, established within buds during the previous summer. After winter dormancy the infected buds (first of all fruit buds) provide a source of conidia in the early spring. The mildew present at bud break on leaves, flowers, and shoots constitutes the "primary infection phase". From these primary sources succeeding leaves are infected. The resultant infections represent the "secondary infection phase". These latter infections provide the inoculum for further spreading of the disease. Wet conditions usually restrict the secondary infections.

Artificial inoculation method for greenhouse evaluation of fungicides

The method of Kirby and Frick (1963) makes it possible to inoculate a large number of potted apple rootstocks with conidia of *Podosphaera*, using a settling tower. Conidia blow, via a central pipe, into the upper part of the settling tower by a fan, and later settle on the foliage of the host. The scheme of the apple mildew inoculation tower is seen in Fig. 10. The source of inoculum is infected foliage from the field or from a greenhouse. The central pipe reaches about two-thirds of the way above the floor of the tower. Below, it ends in a metal cone leading to a fan set horizontally. In the inoculation tower of Kirby and Frick (1963) this is a 22·8 cm axial-flow fan. According to the description of Kirby and Frick "a small port, cut in the pipe between the fan and the tower floor, is normally closed by a flap of metal that can be replaced by another carrying a central clip in which infected apple shoots are placed and so held in the pipe. The fan is then switched on for a short time, sufficient to carry the spores to the top of the tower, where they form a cloud and later settle on the plants below. The height of the tower was the maximum that could be accommodated in the glasshouse, and the diameter chosen was such as to house one hundred 12 cm pots". Plants are put in the inoculation tower 24 hr before artificial infection and left enclosed for another 24 hr. Then the pots are placed on the bench in a greenhouse. The results are recorded about 9 days after inoculation. The most successful infections are seen on the third and fourth leaves down the shoots, therefore, the records are taken on the 3–4–5 leaves. The lower (older) leaves are not susceptible. The temperature must be kept below 27 °C because of the sensitivity of powdery mildews to high temperatures. Plants kept at 20 °C and a relative humidity not exceeding 70 per cent grow well and are susceptible to powdery mildew. High humidity is unnecessary for infection and development of this fungus.

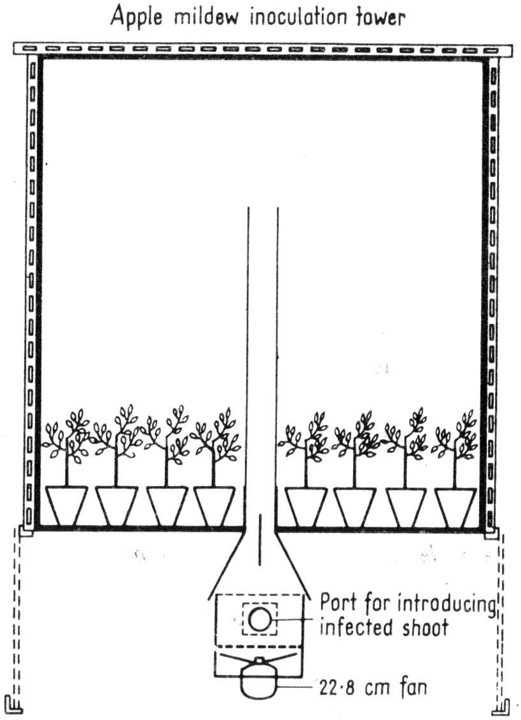

Fig. 10. Artificial inoculation with apple mildew in an "inoculation tower" (After Kirby and Frick, 1963)

Susceptible cultivars serve as host plants. The potted stocks are stored at 1–2 °C through the winter. Rootstocks should be placed in the greenhouse weekly from April to June, so as to produce small plants continuously. In about 4 weeks the shoots produce six leaves. At this stage the plants are suitable for experiments. After use, the stocks can be cut back to three or four dormant buds and allowed to break again. These plants usually produce shoots and leaves for another experiment in about six weeks. In September, day length and temperature become too unfavourable for the growth of the plants, therefore, experiments are seldom carried out, after September. The infection method of Kirby and Frick (1963) may also be used for barley. A simpler procedure for the production of apple and pear seedlings has been developed by Groves (1965).

Other authors also deal with artificial inoculation methods, like Müller (1957), Siebs (1958), Torgeson and Lindberg (1961).

The question of whether different *physiologic races* of *Podosphaera leucotricha* exist or not, is hard to answer. Yarwood (1957b) and Siebs (1958) are convinced that a few races do exist. This was shown by Siebs experimentally. However, there is no generally accepted method for the identification of physiologic races of this fungus.

Chemical control of apple mildew is carried out by the application of sulphur containing fungicides or dinocap type chemicals (like Karathane, Crotothanex) or with chinomethionates (Morestan).

A slide spore-germination method of evaluating fungicides against mildews

In vitro screening tests on fungi other than powdery mildews usually are of no value in evaluating fungicides against this group of fungi. Growth tests on nutrients *in vitro* cannot be used for these fungi because they are obligate parasites. Methods for spore germination tests are of limited value since the conidia of powdery mildews do not germinate normally in water. On glass they show erratic germination. Therefore, the conventional *in vitro* tests are unsatisfactory for screening effective fungicides against powdery mildews (Yarwood, 1951a).

Recently, a new spore-germination method has been published by Zaracovitis (1964). The germination of conidia of different powdery mildews on glass slides coated with cellulose acetate, was over 85 per cent. A 96 per cent germination was obtained if a polyvinyl acetate derivative, Gelva 7, was the coating material. One can study the effectiveness of fungicides towards conidia of powdery mildews by incorporating chemicals into a cellulose acetate layer used as the coating. Zaracovitis (1964) incorporated an analogue of Captan and members of a homologous series of *n*-alkyl fatty acids into the cellulose acetate coating. Satisfactory fungicidal action was demonstrated against *Erysiphe betae*, *E. horridula*, *E. martii* and *Oidium begoniae*. The germination of conidia was suppressed and the *in vitro* action of these fungicides correlated with the *in vivo* action in the greenhouse.

Glass slides are dipped in the solution of cellulose acetate and left to dry. The conidia are detached by contact with the slides or by shaking the source plant over the slides. Incubation of conidia is carried out in a saturated atmosphere in the dark at 21 °C for about 6–24 hr. The chemicals to be investigated are dissolved in a small amount of acetone and then added to the solution of cellulose acetate (2·5 per cent, w/v, in a mixture of 3 parts of acetone and 1 part ethyl lactate). The percentage germination of conidia is recorded on each slide. Germination is completed if the conidia form appressoria at the end of the incubation period. Therefore, the percentage of formation of appressoria is also assessed.

Questions dealing with the action of fungicides are treated in the works of Niemann (1964), Dekker and Oort (1964). There is an extensive literature on the general problems of powdery mildews, however the following works must be mentioned: Cherewick (1944), Yarwood (1957b), Newton and Cherewick (1947), Nover (1957), Nover (1963), Brückner (1963), Moseman (1964, 1966), Moseman et al. (1965), Wolfe (1965).

CHAPTER IV

BASIDIOMYCETES

RUST DISEASES

Fungal pathogens causing rust diseases attack several species of plants. Cereal rust diseases cause the most spectacular damage among diseases of plants. Rust fungi belong to the *Basidiomycetes*. The life cycle of some of these fungi is rather complicated: that of black stem rust (*Puccinia graminis*) is schematically indicated on Fig. 11. The thick-walled, dark teliospores are overwintering spores. These spores appear in old uredia, or they develop in independent sori as the host approaches maturity. Teliospores are diploid. Following a period of rest, they germinate to form a four-celled basidium (promycelium) and sporidia (basidiospores) are borne laterally on the basidium. Sporidia are haploid. They are carried by wind to barberry (*Berberis* spp.) which is the intermediate host of the fungus. Some species of *Mahonia* are also intermediate hosts for *Puccinia graminis*. The true diploid phase of this fungus occurs during teliospore maturation or prior to germination. Meiosis occurs upon germination and the haploid nuclei move into the sporidia. The haploid mycelium from sporidial infection forms the pycnium (spermogonium) beneath the upper epidermis of the barberry leaf and this forms another type of fructification, the aecial primordium.

In the flask-shaped pycnia, pycnospores (spermatia) are produced which are exuded at the ostiole in a "honey dew". The mating type of the pycnia derived from sporidia, which infect the intermediate host are different, being of opposite "sign" (+ or —). These mating types are intercompatible. Each pycnospore is either + or — (heterothallism). The pycnospores in the exudate are transmitted to other exudate drops by insects. A pycnospore belonging to one mating type fertilizes a "receptive hypha" of the opposite mating type, which emerges onto the leaf surface and is connected to the aecial primordium. Thus cells of some hyphae of the aecial primordium will contain a + and — nucleus (binucleate or dikaryotic phase). If conditions do not permit this type of fertilization, pycnospores may grow into the leaf and fuse with the hyphae of the opposite mating type. The honey dew, containing pycnospores, is distributed by insects thereby mixing the two mating types. It also has been found that pycnospores may copulate by means of their germ tubes, but the fate of these fusions is not known. However, after the fusion of the pycnospores and the receptive hyphae, maturation of the aecial primordia begins and binucleate (dikaryotic) aeciospores are formed. For complete development of the primordia, a dikaryotic condition must be established within the mycelia. The dikaryotic phase is a consequence of *plasmogamy* but not of

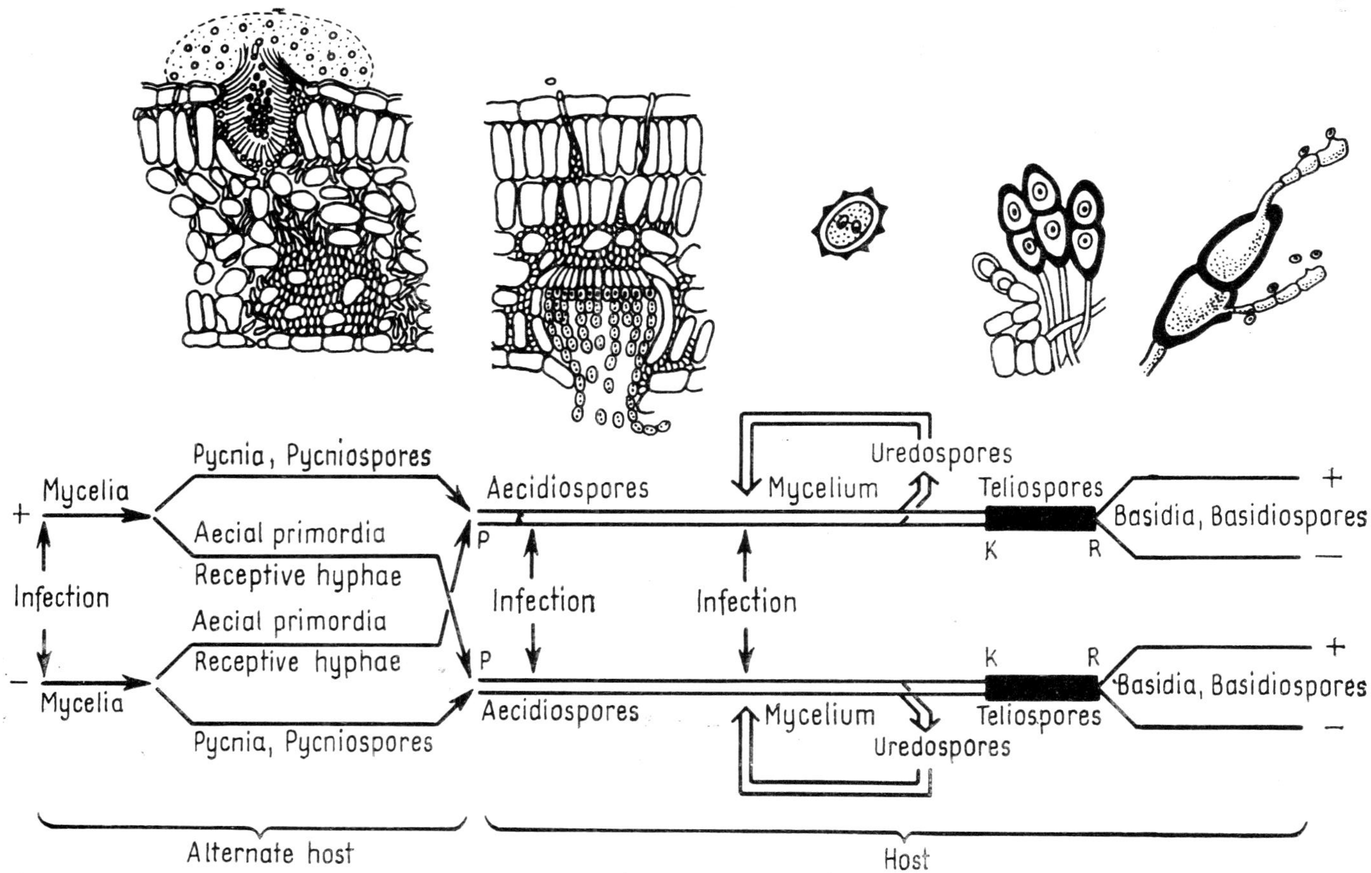

Fig. 11. Life cycle of *Puccinia graminis* f. sp. *tritici* (P = plasmogamy, K = karyogamy, R = reduction) (After Roemer et al., 1938)

karyogamy. The mycelia and the aeciospores contain two types of nuclei but the nuclei are still haploid. The aeciospores will infect a cereal host but not the intermediate host (barberry).

Accordingly the sexual phase of the life cycle of stem rust is confined to growth on the intermediate host. Genetic recombination, accompanying the sexual events, is the most important cause of the production of new variants in rust fungi. Intermediate hosts also have a role in the production and liberation of inoculum for epidemics. A single *Berberis* bush is able to produce about a billion aeciospores of *Puccinia graminis* during spring and early summer. The role of intermediate hosts in the sexual recombination and heterothallism of rust fungi was clarified in the studies of Craigie (1927, 1931).

Damage to crop plants is caused by the uredial generation of cereal rusts. Following aecial infections, pustules of the uredial generation are produced on the crop plants. Uredospores are rust-brown in colour ("rust"). They lose their ability to germinate in several weeks as a rule. Uredospores are dikaryotic and infect only the cereal host plants, producing new generations. The number of uredial generations depends on weather conditions.

At the end of the vegetative period, teliospores are produced on the crop plants. These are two-celled spores, black or dark brown in colour. They germinate after a rest period to form the promycelium (basidium). Each cell of the teliospore contains two nuclei. The haploid nuclei in each cell fuse during the maturation of the teliospore: *karyogamy*. The teliospore is the only diploid stage of the fungus. Upon germination of the teliospore the diploid nuclei become haploid by meiosis (*reduction division*) and the haploid nuclei move into the sporidia (basidiospores) as was mentioned above. With this, the life cycle of the rust fungus starts again.

STEM RUST OF WHEAT
(*PUCCINIA GRAMINIS* PERS. F. SP. *TRITICI* ERIKSS. ET HENN.)

Uredospores are ellipsoid, 17–20 × 25–30 μ in diameter. On the basis of microscopic observation of the shape of uredospores, one can distinguish between stem rust and leaf rust (*Puccinia recondita*) of wheat. The uredospores of the latter are spherical (not ellipsoid!) and 18–26 μ in diameter (Figs 12 and 13).

Artificial inoculation of wheat seedlings in the greenhouse

Seedlings for inoculation can easily be grown under ordinary greenhouse conditions, except during the winter. They should be grown in good soil, under good light. The seedlings can be inoculated when 5–10 cm tall; e.g. in the 2 or 3 leaf stage. Primary leaves of wheat are used for inoculation both to increase the amount of the uredospores and for the identification of stem rust races.

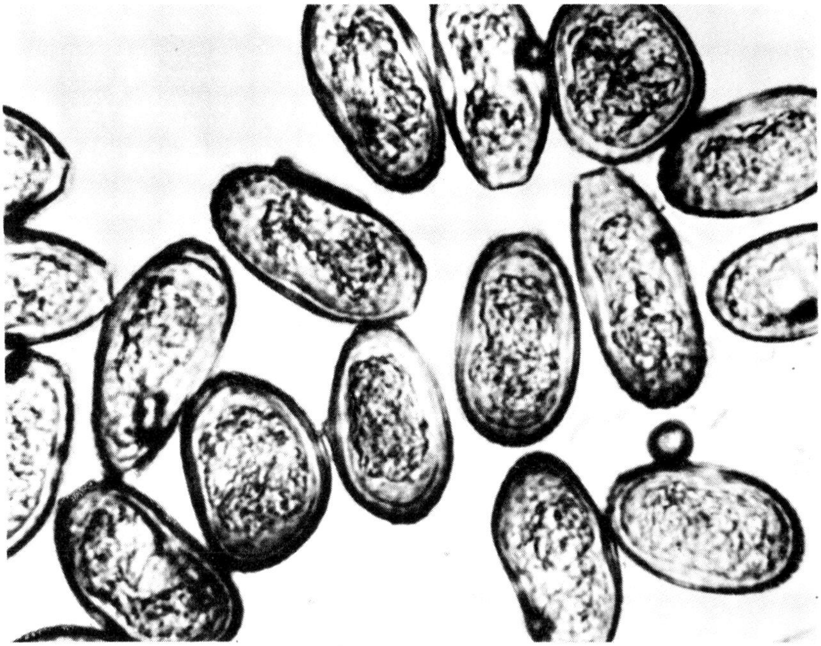

Fig. 12. Uredospores of *Puccinia graminis* f. sp. *tritici*

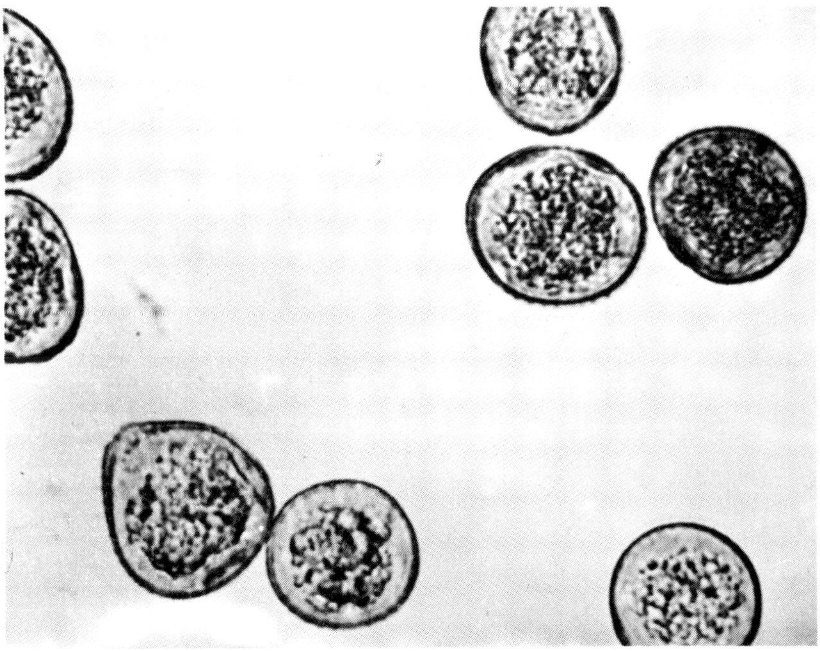

Fig. 13. Uredospores of *Puccinia recondita* f. sp. *triciti*

Leaves are rubbed gently between moistened fingers. This way the waxy layer is removed from the surface of the leaf. This layer hinders the penetration of the fungus. Following the removal of the waxy layer, leaves are atomized with distilled water and inoculated. Then immediately atomized again to produce artificial dew on the leaves. Inoculum may be applied by several methods. The most appropriate is the *spatula method.* Uredospores are taken from rust-infected leaves or stems with a wet spatula and applied to leaves by running the spatula over the surface to distribute the inoculum. The inoculated leaves are sprayed once more with water and incubated in a saturated atmosphere in a moist chamber overnight. On the following day the plants can be transferred to a greenhouse or a controlled environment chamber. This method is particularly useful when inoculum in a collection is scarce or when inoculum is composed of spores of individual uredopustules of different types. When inoculum is abundant, spores can be dusted with a "baby cyclone" with a rubber bulb. In this case a spore-talc mixture is usually dusted on the leaves previously atomized with water. By reversing the valve in the rubber bulb, spores can be collected by suction. Spores can also be brushed on leaves in the form of a suspension. When the uredial or aecial material is too dry or old it should be put into a moist chamber for several hours prior to inoculation. This usually results in more successful infections.

For collection of spores, different cyclone separators have been described. Woodbury et al. (1967) recommended a simple small glass vacuum collector for collecting uredospores from wheat leaves. This was mentioned in a previous chapter on page 288. In a more direct way, spores can also be shaken from leaves or scraped from pustules. However, spores that are shaken off germinate better than those that are scraped off. The former inoculum contains only matured uredospores but the latter one probably contains immature spores too, with no germination ability.

Injection of wheat seedlings with a spore suspension

When inoculum is scarce or when the viability of uredospores is very low, the injection of a suspension of uredospores into seedlings by means of a hypodermic syringe is very useful. With aecial material the inoculum is usually so scarce that this method seems to be the most appropriate. Inoculated plants need not be kept in a moist chamber as in the case of superficial inoculation, and this makes this method extremely simple. The injection method is also suitable in rust nurseries for inoculating plants in susceptible border rows around the plots to be investigated for rust resistance. Of course, this method is not suitable for race determinations by inoculating differential sets of cultivars because the characteristic symptoms (reaction types) do not develop. However, it is especially useful in establishing new cultures of stem rust or other cereal rusts in the greenhouse.

Field assessment of resistance of different wheat cultivars to stem rust is carried out in "rust nurseries". Cultivars to be investigated are sown in little plots of 1 m^2 in an open field. Usually a series of these plots are planted in a row between border rows or "spreader rows" of a susceptible wheat cultivar. Here and there in the border rows, plants are inoculated in the boot stage by hypodermic injection method. As a consequence, hundreds of infection centres will be created. These provide abundant inoculum for the adjacent plots between the spreader rows. Some workers prefer to spray the spreader rows at different locations with a suspension of uredospores in a 0·1 per cent agar. In this case the spray must be applied in the evening to use natural dew formation. Generally, better infection can be attained by inoculating plants in the spreader row hypodermically. As many plants as possible should be inoculated by injecting the uredospore suspension into the stems or leaf sheaths. The young leaves in the boot are very susceptible to rust infection. The emerging leaves and heads are usually heavily rusted. From these plants in the infection centre of the border row, inoculum is distributed to the plots of different cultivars. A local epidemic is developed with this method. A scale has been developed in the United States to designate rust intensity at the heading stage, on the basis of the amount of pustules on the leaf surface (Fig. 14). The size of the pustules and chlorosis or necrosis accompanying the hypersensitive reaction should also be noted in the field test, usually from the heading stage onwards.

In field tests of rust resistance the inoculum applied to infect the border rows has to contain all the prevalent races which exist in the country or in a certain area. In some precise investigations however, the inoculum may be composed of only a single race. In field experiments where the soil was fertilized heavily with nitrogen, the cover of rust uredia on leaf surfaces is increased, in other words, plants exhibit a greater susceptibility to rust. This effect of nitrogen fertilization,

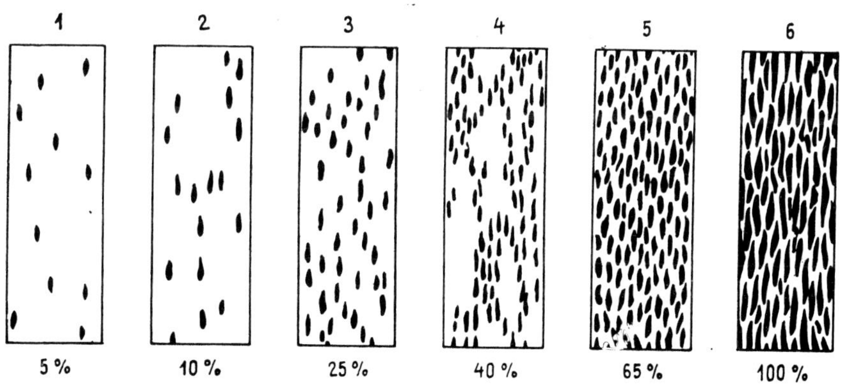

Fig. 14. Scale for determining susceptibility to rust in the field

however, does not change the reaction type. The hypersensitive reaction (incompatibility) will not change to a susceptible (compatibility) type of reaction, although the number of infections (the incidence of attack) will be increased.

Collection of rust samples and preservation of inoculum

In collecting uredospores avoid contaminating hands with spores. Glassine end-opening envelopes are useful for collecting rusted plant material. In collecting samples of stem rust, the tops of culms can be cut off and inserted into the envelope. In other cases rust-infected leaves are the samples which are collected in envelopes. Mixing should be avoided by holding rusted plant materials from the outside when folding the open end of the evenlope.

Samples of rusted plant organs or uredospores should usually be preserved for a short- or long-time prior to inoculation on living plants. Cool and dry conditions are necessary for the preservation of rusted plant material. Uredospores also require low humidity for storage. The relative humidity should be below 40–50 per cent. The most suitable temperature is about +4–5 °C. One can preserve the viability of uredospores for several months in a refrigerator under the conditions mentioned above. However, spores kept in this environment (50 per cent humidity and 5 °C) can survive no longer than a year, which seems to be the maximum. Low humidity can be maintained in a desiccator. Uredospores can be kept above $CaCl_2$ in the desiccator, which is placed in a refrigerator. For short-time storage, uredospores should be mixed with talc of the same weight and kept in small vials in the refrigerator. For long-time preservation, when storage of the spores for more than 1 year is necessary lyophilization is the most suitable method for maintaining viability. (A method of lyophilization was already mentioned on page 167.) The small glass vials which are closed after lyophilization should be held at +4–5 °C in a refrigerator. Uredospores remain viable for several years in this case. Lyophilized spores should be placed in a moist environment with 100 per cent relative humidity for at least 12 hr prior to using them as inoculum. Loegering et al. (1961) has published a new method for long-time storage. According to this procedure uredospores are frozen in liquid nitrogen at —195 °C. Spores will germinate normally if thawed at +40 °C prior to use as inoculum. Spores can be preserved by this method for a very long time.

Identification of physiologic races of stem rust

For the determination of races a standard differential set of cultivars is used as follows:

Little Club (C.I. 4066)	—	*Triticum compactum*
Marquis (C.I. 3641)	—	*Triticum aestivum*
Reliance (C.I. 7370)	—	*Triticum aestivum*
Kota (C.I. 5878)	—	*Triticum aestivum*
Arnautka (C.I. 1493)	—	*Triticum durum*

Mindum (C.I. 5296) — *Triticum durum*
Spelmar (C.I. 6236) — *Triticum durum*
Kubanka (C.I. 2094) — *Triticum durum*
Acme (C.I. 5284) — *Triticum durum*
Einkorn (C.I. 2433) — *Triticum monococcum*
Vernal (C.I. 3686) — *Triticum dicoccum*
Khapli (C.I. 4013) — *Triticum dicoccum*

A limited number of differential cultivars of wheat, called a "half series" can be used for many purposes with satisfactory results. Little Club can be omitted from this series because of the uniform susceptibility of this cultivar to most of the races. Arnautka, Mindum and Spelmar react alike to many races, therefore, it is sufficient to use only Arnautka in the half series. Kubanka and Acme react similarly to most of the races. Therefore, only Kubanka is included in the half series. Vernal and Khapli are resistant to many races, thus they are also omitted. The half series includes the following cultivars: Marquis, Reliance, Kota, Arnautka, Kubanka and Einkorn. (See the figure on page 302)

However, if there are two or more infection types on either cultivar, it must be included in the series. If there is any doubt regarding determinations on the half series, the isolate of rust to be identified must be tested on the complete set of differential cultivars. In spite of these precautions race identifications on the basis of a half series can simplify the procedure in many instances (Stakman et al., 1962).

Infection types (reaction of cultivars)

The reaction classes of cultivars of the differential set by Stakman and co-workers is regarded as a classic key for identifying races of stem rust. The following Table shows infection types and reaction classes of host plants (see also Fig. 15)

Infection type		Reactions of cultivars and reaction classes
Resistant	0	Immune: No uredia nor other symptoms
	0;	Nearly immune: No uredia, but hypersensitive spots present
	1	Very resistant: Uredia minute, surrounded by distinct necrotic areas
	2	Moderately resistant: Uredia small to medium, usually in green islands surrounded by a chlorotic or necrotic border
Susceptible	3	Moderately susceptible: Uredia medium in size, no necrosis but chlorotic areas may be present. Coalescence of uredia is infrequent
	4	Very susceptible: Uredia large, and often coalescing. No necrosis, but chlorosis may be present
Mesothetic	x	Heterogeneous: Uredia variable, sometimes including all infection types and intergradations between them on the same leaf. No mechanical separation possible, on reinoculation small uredia may produce large ones, and vice versa

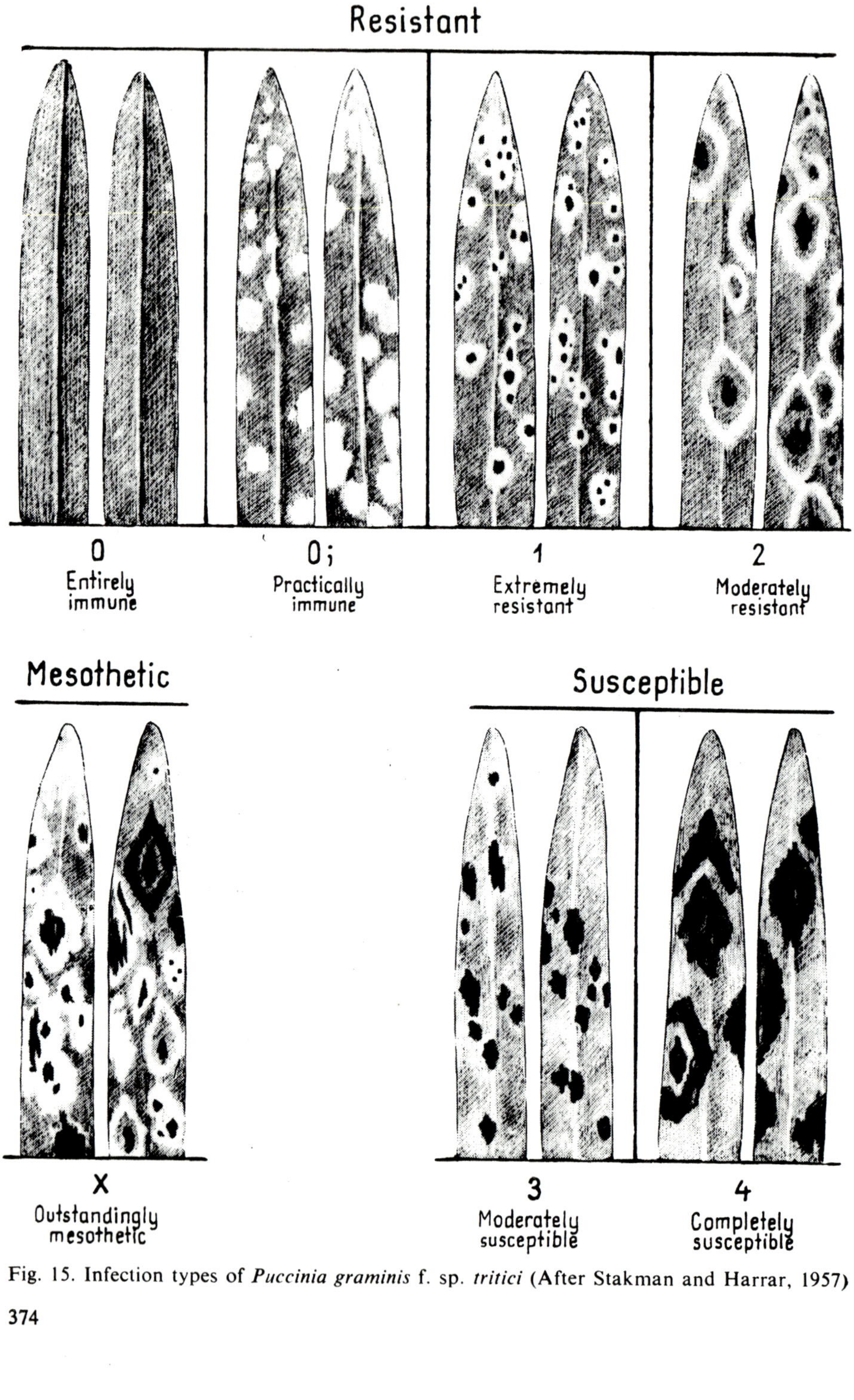

Fig. 15. Infection types of *Puccinia graminis* f. sp. *tritici* (After Stakman and Harrar, 1957)

Variations within a certain infection type are indicated with plus and minus signs, + + and = indicate the upper and lower limits of a certain type. The sign ± indicates a variation between + and —. The letter "c" indicates exceptionally pronounced chlorosis, "b" indicates browning with a tendency towards necrosis, and "n" indicates a tendency towards necrosis.

For the identification of reaction types a temperature of 21 °C and at least 1000 foot-candles of light are desirable in the greenhouse. At a lower light intensity or at very high temperatures infection types are variable. Therefore, environmental conditions must be kept under some control in the greenhouse.

Identification of races in mixtures

Rust samples collected in the field should be increased on a very susceptible cultivar of wheat. Uredospores harvested from these plants can be used in inoculating the differential set of cultivars. In some of the samples one can find only a single race, but most of the field collections contain a mixture of different races, which produce different infection types on a particular cultivar of the differential set. Figure 5. in Part IV (page 302) schematically shows the method of isolating races from a mixed field collection. The mixtures of races are most apparent on the following members of the differential set: Marquis, Arnautka, Mindum, Spelmar, Einkorn, Vernal. On Marquis mainly two infection types occur, type 2 and type 4. On the other cultivars mentioned above there are two distinct types: type 1 and type 4 (or type 3+).

Rust from pustules of different reaction types are inoculated on a limited number of differential cultivars (the half series). On the basis of the infection types it is possible to determine whether or not the isolates from the original field collection represent different races. In many instances it is also possible to identify the races with some certainty. If there is any doubt concerning the physiologic race identified this way, the isolate must be tested on the complete set of differential cultivars.

Physiologic races of Puccinia graminis f. sp. *tritici*

Physiologic races and cultivar reactions of the differential set (after Stakman et al., 1962) are shown in the Table below.

Race	Little Club	Marquis	Reliance	Kota	Arnautka	Mindum	Spelmar	Kubanka	Acme	Einkorn	Vernal	Khapli
1	4	4−	0	3+	1=	1	1=	3+	3++	3	0;	1=
2	4	2=	2=	2=	1−	1	1=	1+	3++	3+	1−	0;
3	4	4−	4=	3+	1=	1=	1−	1+	3++	3+	1=	0;
4	4+	2−	1−	2=	4=	3+	3++	2	3++	3++	1=	1=
5	4	4−	0;	3	4=	3++	3++	1++	3+	3	0;	0;
6	4	2	1=	0;	3+	2=	2=	1	3+	3	0;	0;
7	4	2=	3+	1=	1=	1++	1−	1	3++	3−	1	1−
8	4	4	0;	4−	4=	3++	4=	0;	3	3	4	0;
9	4	4−	0	3++	4−	4=	4=	4=	3++	3+	4±	1−
10	4+	2−	3++	2	4	4	4	3++	4−	3+	1=	1=
11	4−	4=	3++	3+	4=	4=	4=	3++	3++	3	1=	1=
12	4+	4−	4=	3+	4=	1	1++	1++	3++	3++	1=	0;
13	4	4−	3++	3++	4=	3++	3++	2−	3++	3	1	1=
14	4+	2−	1−	1++	3++	3++	3++	3++	3++	3	1=	0;
15	4	4−	4=	3++	4=	4=	4=	3++	3++	3++	4±	1=
16	4−	2=	0	1	3++	3+	3++	3+	4=	1=	1=	1
17	4	4−	0;	3+	4=	4=	4=	3++	3++	3	1=	1=
18	4	4−	4=	3++	1	1=	1−	3++	3++	3+	1−	1±
19	4	2−	0;	3−	4=	4=	4=	3++	3++	3	0;	1=
20	4	4=	4=	4=	1++	1−	1++	3++	1++	3+	1=	1−
21	4	4	0	3++	4−	4−	4−	4=	3++	1=	0;	1=
22	4+	4+	4	3	1	4	4−	0;	3+	3	1−	0;
23	4	2	1−	1=	1	1=	1−	3+	3++	3	0;	0;
24	4	4=	0;	2=	4=	4=	4=	3++	3+	3+	1=	0;
25	4	4	3+	3	1−	3	1=	3	3++	3	1=	1−
26	4	4	0;	3	4−	1=	3++	1=	4=	3	1−	1+
27	4=	2	0	0;	1=	1	1−	4=	3++	1=	4±	1++
28	4	2	0;	3	4−	1	4=	3	3	3	1=	0;
29	4	4−	0	3	X++	X±	X+	X	X+	3	1−	1−
30	4	4	0;	3++	X++	X±	X+	X±	X++	3+	4=	1

31	4+	4	3++	2−	X−	X+	X±	X	X±	3+	1	1
32	4	4=	4=	3+	X+	X±	X±	X	X+	3	1=	1
33	4+	2	4	1+	1=	1−	1	4=	3++	3	1−	1
34	4+	4	4−	4=	4	4=	4=	4±	3++	1=	0;	1±
35	4	4=	3+	0;	1=	1−	1=	3+	3++	3	0;	1
36	4	4	4−	3++	1=	1=	0;	X	3++	3+	0;	1−
37	4	4−	0	3++	4=	4=	4=	X±	3+	3	1=	1−
38	4	2=	4−	3−	X+	X±	X+	X+	X++	4−	1=	1+
39	4−	2=	4=	3+	4+	3++	4−	4=	3++	4=	1=	1−
40	4+	4+	4	4+	4+	4+	4	4=	4	0;	4=	1=
41	2++	4	0	0;	4=	4	4+	4+	4−	4−	1−	4c
42	4	4	1±	0;	4+	4	4	4	4	4=	2=	4c
43	4	3++	0	0;	0;	0;	0;	X	1	3	1	0;
44	4	3++	0	0;	0;	0;	0;	3+	3+	3	1	0;
45	4	2	0	2−	4	4	4	X	X	3	3	1
46	4	3++	0	2−	4	4	4	1	1	3	3	1
47	4	2−	0	1±	X−	X;	X++	X;	X±	1=	0;	1=
48	4+	1	0;	1+	X+	X=	X±	X+	4−	4=	1=	1+
49	4	4−	0	4=	1=	1−	0;	X=	3+	1−	0;	1=
50	4+	2+	0;	2++	1−	0;	0;	X=	X±	0;	0;	0;
51	4	2=	3+	0;	0;	0;	0;	4	3++	3+	4−	0;
52	4	4	4−	4	1=	1=	1=	X±	4	4−	4+	1−
53	4	2±	0	1	4	4	4	4	4	3±	3±	1
54	4	3++	0	0;	0;	0;	0;	1	3+	3	1	0;
55	4	4	0	2−	4	4	4	X	X	3	3	1
56	4	3+	3+	3+	1=	1=	1=	3+	3+	1=	1=	1−
57	4	4−	0;	3+	1	1	1	4	3+	3	3	1
58	X+	X+	0	0;	1+	1=	1±	4+	3+	3++	4+	1++
59	4+	2+	0	0;	1=	1=	1−	X−	3++	3+	1=	0;
60	4++	4++	0;	3++	0;	0;	0;	0;	3+	1±	0;	1=
61	4+	4	0	3+	0;	0;	0;	X	4+	4	0;	0;
62	4	X±	X+	X	4	4	4	4	4	3++	0;	1−
63	4+	X++	4	3++	4+	4++	4+	4+	4n+	1=	1=	0;
64	4++	4+	3++	1±	1=	0;	1=	X±	3c+	1=	0;	1
65	4	2	0	3	0;	0;	0;	X	4	4	4	1

Race	Little Club	Marquis	Reliance	Kota	Arnautka	Mindum	Spelmar	Kubanka	Acme	Einkorn	Vernal	Khapli
66	4	2	4	0;	0;	0;	0;	X+	1=	3	0	0
67	4±	4+	4	4	1±	2=	2=	X=	X+	3−	4±	1−
68	X+	2+	0	0;	1=	1+	1−	X++	4±	4	4++	0;
69	3++	2+	0;	0;	0;	0;	0;	3+	4=	3	3++	1=
70	4+	X+	0;	0;	0;	0;	0;	0;	0	1=	X±	0;
71	4+	X−	0	0;	0;	1=	0;	1=	0;	0;	0;	0;
72	X−	2	0;	0;	1±	1+	1+	4+	4+	3+	3±	X
73	4	X	0	X	X	X	X	X	3±	4−	1	1
74	4	4−	3=	2+	0;	0;	0;	X−	3+	3±	0;	1
75	4	3+	2+	0;	3+	3+	3+	4−	3+	1	0;	1
76	4	4−	0	3+	X	X	X	X	3+	3+	X−	1
77	4	4	3=	3−	3−c	3−c	3−c	4−	3+	1±	X	1
78	4	X	0	3=	3−c	3−c	3−c	3+	3+	3+	1	1
79	4	4−	0	1−	1−	1−	1−	4−	3+	0;	3+	1
80	4	3−c	0	3−	0;	0;	0;	X−	3+	1−	3+	1
81	4	X	0	1+	4	4	4	4	4−	1−	1−	1−
82	4	3+	3+	3+	0;	0;	0;	X	3+	3+	X	1
83	4	1+	3−	1−	3+	3+	3+	3+	3+	3+	3+	1
84	4	X	3−c	3−	0;	0;	0;	X	4	3+	X	1
85	4	4−	0	3±	4−	4	4	4	4	3+	X	1
86	4	3+	3+	1±	3+	X	X	X	3+	3+	3+	1
87	4	4	4	3±	4	X	X	X	3+	3+	4	1
88	4	X	0	1±	4	4	4	4	4	3+	1	1
89	4	2	3+	0;	4	X	X	X	3±	3+	3+	1
90	4	4	0	3±	4	X	X	X	3+	1	1	1
91	4	X	0	X	4	4	4	4−	3+	3+	X	1
92	4	X	0	X	1	1	1	4	4	1	1	1
93	4	3+	0	3=	0;	0;	0;	X	3±	3+	3+	1
94	4	X	0	X	4−	4−	4−	4−	3+	1	1	1
95	4	2	3±	0;	4	4	4	4	3+	1	1	1
96	4	X	4	X	4	4	4	4	3+	3+	1	1
97	4	3+	4	3+	0;	0;	0;	X	3	0;	3+	1

98	4	2=	4+	3+	4−	3++	3	4=	3−	1±	1	1−
99	1+	X	3++	3	0;	2+	2++	4+	2+	2+	3++	3++
100	3±	4	3+	3=	3	3−	3=	4=	1	1±	X±	1
101	4+	4++	4−	4±	1	0;	0;	X+	X−	3	0;	1±
102	4±	0;	1+	0;	0;	0;	1=	1=	0;	3+	0;	1=
103	X++	0	0;	0	0	0;	0;	0;	0;	2−	1=	0;
104	4	X=	0	0;	0;	0;	0;	0;	0;	3+	X+	0;
105	4	X−	0	3±	0;	0;	0;	X−	3+	3	X	1=
106	4	X	3	3−	4	X	X	X	3	3+	4−	1−
107	4	3±c	3−	0;	4	4	4	4	3	1−	0;	1−
108	4	4	4−	0;	1±	0;	0;	X+	4−	3	3+	1−
109	4	2±	4	3−	1±	1±	1+	4	3−	3+	4	1
110	4	4−	3	3−	3+	3+	3+	3+	3+	3	X−	1
111	3±c	1−	0	0;	0;	0;	0;	0;	0;	1−	0;	1−
112	4	X	0	0;	0;	0;	0;	X−	0;	3	0;	1−
113	4	X	3=	3±	3±	X−	X−	X−	4	3+	0;	1
114	4	3+	X	3+	4	4	4	4	4−	3+	1−	1
115	4	2−	3=	3=	4−	4−	4	4	4	3+	3±	1−
116	4	4−	0	3	4	4	4	4	4	1	4	1−
117	4	4	0	0;	4−	4−	4−	4	4	3+	3+	1−
118	4	X	0	1−	1−	1−	1−	4	4	3+	1−	1−
119	4	X	0	0;	4	4	4	4	4	3+	3+	1−
120	4	X	0	3=	4	4	4	4	4−	3+	3	1−
121	4	4−	0	0;	1−	1−	1−	4	3+	3	3+	1
122	4−	3+	4−	0;	4	4	4	4	4−	3+	1−	1
123	4	2−	0	3−	3+	4−	4	4	4	3+	3+	1
124	4	3−c	0	3=c	0;	0;	1=	3+	X	3	X−	1
125	4±	4	4	4	0;	0;	0;	X	4	1=	0;	1−
126	4=	4=	3+	3+	X±	X++	X+	X−	X	1±	1=	1−
127	4−	4−	3++	3+	1=	1±	1=	X	X	0;	0;	1−
128	4	3++	0;	3±	1−	1=	1	X−	X±	3±	1=	1−
129	4	4	4	X	4	4	4	4	4	1−	4	0
130	1−	2	0	0	4	4	4	4	4	3+	0	1−
131	1−	4	0	0	4	4	4	4	4	4+	1	1−
132	4	4	0	2	1	4	4	1	4	1	1	1

Race	Little Club	Marquis	Reliance	Kota	Arnautka	Mindum	Spelmar	Kubanka	Acme	Einkorn	Vernal	Khapli
133	4	4	0	0	1	2	2	4	4	0	1	2
134	4	4	0	0;	0;	0;	0;	1	3	3	3	1
135	4	4	4	0;	0;	3	2	4	3	1	1	1
136	4	X	0	3=	0;	0;	0;	4−	4	4−	1−	1
137	4	4−	4−	0;	1−	1−	1−	3	1−	3−	X−	1
138	2	0;	0	0;	0;	0;	0;	0;	0;	1+	X−	0;
139	4±	2+	1−	2±	1−	1=	1	X	4+	1=	1−	0;
140	4+	2−	1	4=	1±	0;	1±	X+	4±	4−	1±	1±
141	4	4	1+	0	4	1−	4	1+	4	4	1−	1+
142	4	4−	2	0	3++	X	4	X+	4	4	1	1++
143	4	4−	4−	0	4−	4−	4−	4−	3++	2	3++	1=
144	X	4	4	3+	4−	4−	4−	4−	3++	1+	3++	1++
145	4	0	2	4	0;	0;	0;	4	4	0;	4	2
146	4+	4	4±	3++	1−	1±	1=	X±	3++	1=	X±	1+
147	4++	4+	4	4±	1=	1−	1	4+	4	1=	4±	1±
148	4	3+	3+	3+	1−	0;	0;	0;	3±	1−	0;	1−
149	4	X+	0	3−	X+	X	X	X	X	X	X	1−
150	4	3−	0	3−	3+	X	X	4−	X	3+	3+	1−
151	4	1	3—	0;	4−	X	X	X	4—	3+	0;	1—
152	4	1+	3−	0;	1	0;	0;	X−	3+	3+	0;	1−
153	4−	1+	3	0;	0;	0	0	0;	0	0;	X−	0;
154	4	X	0	3=	1	0;	0;	4	4	3±	4−	1
155	4	X	0	3=	1	0;	0;	X	4	3+	0;	1
156	4	X	0	3=	4	X	X+	3+	4−	3±	4−	1
157	4	X	0	3=	1−	0;	0;	4	4	3+	X	1−
158	4	X	0	3=	4	4−	4−	4−	4−	3+	X	1
159	4−	4	0	3++	1	1=	1=	X++	X	4=	X±	1
160	X±	0;	0	0;	0;	0;	0;	1±	0	3+	1	0;
161	X	4	0	3++	4=	4−	4−	4=	3++	1=	0;	0;
162	4	X±	0;	X−	0;	0;	0;	X	3	3	0;	1
163	4−	4=	4−	X	4−	4	4	4−	X	3++	X±	1=
164	4	4−	4=	3+	0;	0;	0;	3++	X=	1−	3+	0;

165	4	X±	2	0;	X−	X−	X=	X±	X=	3	4−	1±
166	4−	1++	2±	0;	1=	0;	0;	4−	0;	3+	1=	1−
167	4	2−	2++	1−	0;	0;	0;	X	0;	3++	X	1
168	4	X	0	X	0;	0;	0;	X	0;	3	3+	1=
169	4	X	0	0	4	4	4	4	4	0;	3+	1
170	4	4	1	X	4	4	4	4	4	0;	4	1
171	4	X±	3−	0;	1	0;	0;	X−	3+	3+	0;	1−
172	4	X	3−	2−	1−	0;	0;	4−	4	3	0;	1−
173	4	2	4−	3−	0;	0;	0;	4	4−	1−	0;	1
174	4	2	4−	3=	4−	X	X	4−	4−	1−	0;	1
175	4	2+	4	1±	1−	0;	0;	X	4−	3±	X	1−
176	4	4	0	4−	1−	1−	1−	4	4−	1	0;	1−
177	4	2	4−	3−	1−	1−	1−	4−	3+	3	0;	1−
178	4	1	0;	0;	4	X−	X	X+	4	3+	0;	1
179	4	X	4	4−	4	4	4	4	4−	3+	1−	1−
180	4+	2=	0;	1=	1±	1=	1−	X±	1=	3+	0;	1±
181	4	X+	0;	0;	0;	0;	0;	X	1−	3±	3+	0;
182	4−	2++	2±	0;	1++	1±	1	X±	1	3	3	1±
183	4	X	0	3=	1−	0;	0;	X	4−	1−	4−	1−
184	4	3	0	0;	3	3	3−	4=	4	1	4	1
185	4	X−	0	0;	1=	0;	1=	1=	4	1=	0;	1
186	4	1	0	0;	1	0;	0;	4=	4=	1	0;	1
187	4	4	3	3	4−	3++	2	1++	1++	0	2	1
188	4	X+	X±	X±	X++	X++	X+	X+	X±	3+	3++	1±
189	4	4	4	3+	4	4	4	4	4	4	4	4c
190	4	4−	0;	0;	4=	X+	X−	X	3±	3	3+	0;
191	4	3++	2−	1+	0;	0;	0;	4=	0;	3+	0;	1
192	4	4−	4−	4−	4	0;	0;	0;	3±	1	X−	1
193	4	X	4−	3	0;	0;	0;	X	4	3+	0;	1
194	4	4−	0	3++	4	4	4	0; 1b	0; 1b	0; 1−	0; 1−	0; 1−
195	4	2=	0	0;	1−	0;	0;	X++	1−	1++	3+	2=
196	3+	3+	3+	3+c	1−	1−	1−	3+	0;	1	1−	1−
197	4	2	0	1−	X	X	X	X++	3+	1−	4	1−
198	4	4	0	3+	0;	0;	0;	4	3+	1−	X	1−
199	4	3+	3+	3+	X−b	X−b	X−b	3+	0;	3	1−	1++

Race	Little Club	Marquis	Reliance	Kota	Arnautka	Mindum	Spelmar	Kubanka	Acme	Einkorn	Vernal	Khapli
200	4	3+c	0	3+	X++b	X++b	X++b	4	1−	3	1−	1−
201	4	3+	0	3++	1−	1−	1−	3+	0;	1−	1−	2−
202	4	3+	2	0;	X	X	X	3+	0;	3	3+	1−
203	4	X	0	X	0;	0;	0;	3+	3+	3	0;	1−
204	4	X	0	X	0;	0;	0;	X−	1−	1−	0;	2−
205	4	X	0	0;	Xb	Xb	Xb	3+	0;	0;	0;	0;
206	4	X	2	1−	0;	0;	0;	X	0;	1−	0;	2−
207	4	2−	0	3−	4	4	4	4	4	1−	1−	1−
208	4	2−	0	1±	3+	3+	3+	3+	3+	1−	3+	1−
209	4	2	0	2	3+	X	X	X−	3+	3	3+	1±
210	4	2	2	0;	X++	X=	X=	X	1−	3	1−	1++
211	4	2−	0	1±	0;	0;	0;	3+	0;	3++	3+	1++
212	4	2−	0	2−	1−	1−	1−	4	1±	1−	1−	1−
213	4	4	3	3	4	4	4	X	X	1	0	0
214	4	3	3	3	0	0	0	0	0	3	0	0
215	4	4	3	2	0	0	0	3	3	0	0	0
216	4	4	4	X	2	2	2	4	4	3	0	0
217	4	2	3	0	0	0	0	4	4	0	3	0
218	4	3	3	0	3	3	3	3	3	3	4	0
219	4	4	3	0	2	2	2	3	3	3	3	0
220	4	2	0	3	4	4	4	1	1	4	4	0
221	4	3	0	3	0	0	0	1	1	1	0	0
222	4	3	3	3	3	3	3	1	1	0	0	0
223	4	2	0	0	3	3	3	0	0	3	3	0
224	4	2	0	0	2	2	2	1	1	3	3	0
225	4	2	0	0	3	3	3	1	1	1	0	0
226	4	3	0	0	3	3	3	0	0	0	0	0
227	4	3	0	0	0	0	0	1	1	3	0	0
228	4	2	3	0	4	4	4	1	1	3	0	0
229	4n	X	0	0	0	0	0	1	1	3	0	0
230	4	2	0	0	0	0	0	0	3	3	3	0
231	4	4	2−	3=	3	0;	0;	1	3	1	0;	1

232	4	2±	0	1–	4±b	4±b	4±b	4±b	0;	3	1–	1
233	1=	4	0;	0;	0;	3	1	2	4	4	0	4
234	0	0	0	0	2–	3	2+	2–	0	1–	0	0
235	1++	1++	1++	1++	3=	1++	1–	1–	1	1+	1++	3=
236	2	2	1+	2++	3–	3	3	1++	3	0	0	0
237	1+	3–	1+	2–	3–	3–	1++	3–	3=	2=	1=	1=
238	1	1+	0	3–	1	1	1	1	1	0	1	3
239	1++	2	1	1	2	1+	1++	3–	3–	3	1	1++
240	2–	2–	0;	2	3=	3	2	2+	0	0;	0	0
241	2+	4	0	2+	2++	2++	2++	4	2++	0;	0;	0
242	1	3=	0	2=	2=	3=	1	3=	3=	1–	0	1
243	1+	0	0	2=	1=	1++	3=	1+	1+	1	0	1+
244	1	4	0	3	4	4	3++	3	3++	2	1	2–
245	4	2	0;	0;	4	4	4	4	4	4	0;	4
246	4	1+	1+	0;	0;	4	4–	4	4–	3++	1	0;
247	3+	0;	0;	3+	4–	4	2+	4–	4	3–	1+	1+
248	4	0;	0;	X	0;	4–	4–	4–	4	4	1	1
249	3+	1	0;	0;	4	4	4	X	4–	4–	1	X–
250	3	1	0	3	2+	3+	3–	3+	3–	1	0;	0;
251	3	2+	0	0	3+	0	1	0	1	1+	1	1
252	4	1+	0	0	4–	3	3+	2+	3	3–	0;	4–
253	3++	1+	2	2+	4	4	4	X	4	2–	0;	0;
254	3–	1++	1=	1	1++	1+	3=	3=	3=	1–	0	0
255	4	0	0	0	2–	4=	4–	0	0	0	0	0
256	4	1–	0	1++	3	4	1	1	3=	3	1	2
257	4	2	3	1	3	3	3	1	0	0	3	3
258	4	2+	1=	4	4	4	4–	4=	1+	1	0;	1
259	3=	1	0	1	1+	3=	3–	3–	1	2–	0	0
260	4	X	0;	0;	0;	4–	3+	4	0;	4–	1	0;
261	4	X	0;	X	4–	4	4–	4	4–	4–	1+	2–
262	4–	X	0;	0;	3+	4–	4	2++	4–	4	3+	3–
263	3+	X	0;	0;	4–	4–	4–	X	X	3++	1	2=
264	4	4	0;	4	0;	4	4	4	4	4	0;	0;
265	4	4–	0;	0;	3++	4	0;	4–	4–	4	1–	1–
266	4	4–	0;	1=	4=	4	4–	4–	X	0;	0;	1–

Race	Little Club	Marquis	Reliance	Kota	Arnautka	Mindum	Spelmar	Kubanka	Acme	Einkorn	Vernal	Khapli
267	4	4	4	4	3	3	2	2	3	0;	4	4
268	4	4	0	2	4	3	4	4	2+	3−	1	0;
269	3+	3	0	4−	4−	4	0	4	0	1−	1−	0
270	4	4	1+	4	3+	3++	4	4	4	4	3+	3+
271	4	4−	3	4−	4	1	0	1	4−	2±	1	1+
272	4−	4−	0	4	4	2	0	4	2++	2+	1	2−
273	3=	3	0	1++	3+	2	3	3+	3++	3=	0;	0
274	4−	3−	0	1++	2	3	1+	3	3−	1	0;	1
275	4	4	0	2	3+	3+	3=	1+	3	1−	0;	0;
276	4−	3=	0	3=	4	3+	3=	1++	3−	1	1	1
277	3+	3−	0	3−	3−	3−	1++	3=	3=	1+	0;	1−
278	3+	3	0	1+	3	3++	3	3+	1++	1−	0;	1−
279	3	3+	0	3	3+	3+	3	3=	1++	1	1−	0;
280	3	3+	3=	1±	3=	3=	1++	1+	1−	1−	0	0
281	3	3−	0	2−	1++	3	3−	3=	3=	0;	0;	0;
282	3=	3=	1−	3−	3=	1+	3=	1−	1	1=	1	1
283	4−	4−	0	0	1−	1−	3	4	4−	2	1	1
284	3+	3=	4	0	0	0	0	0	4	0;	0;	2
285	4	4	0	1	3+	1−	0	4	1	1	0;	0;
286	3=	3	0	3=	1+	3=	1++	1+	1+	1	1	1
287	4	4	0	3+	2	2	4	4	2	1	1	0
288	4	3+	0	1+	1++	4−	3	1++	1++	1	1	0;
289	3++	3	0	1	3+	3+	1	0	3++	1	0	0
290	4	4	0	3+	0	4+	4+	4	3+	0;	0;	0;
291	3−	3	0	2	3=	3	1+	1	1+	2−	1	1
292	3+	3	0	3	3	3	0	1++	1	0	1	1
293	3	3	3−	2	3−	3+	4	4−	2−	0;	0;	0;
294	1++	1	0	1++	2	2	2=	1++	2	1	0	0

Subdivision of races of stem rust and the supplemental differential cultivars

The standard races enumerated above may be subdivided by the use of supplemental cultivars. Thereby, races are divided into *subraces*. The following temporary sets of cultivars are recommended by Stakman et al. (1962) for distinguishing between certain races and subraces:

Lee (C.I. 12488)	Distinguishes race 15B from race 15
Golden Ball (C.I. 5059)	Used in subdividing races 11, 15B and 29
Selkirk (C.I. 13100)	Distinguishes certain isolates of 15B at 18·5 °C (65 °F) and of race 11 at 30 °C (85 °F)
Bowie (C.I. 13146)	Resistant to isolates of 15B, but susceptible to races 11E, and 48A. Helps in subdividing race 29
Kenya Farmer (C.I. 12880)	Resistant to common races in North America except one isolate of race 29
F.K.N. (C.I. 13154)	Resistant to common races in North America except some isolates of races 11 and 29
Yuma (C.I. 13245)	Used in subdividing race 15B

A useful procedure for identifying rust races in relation to practical plant breeding has been worked out by Watson and Luig (1961) in Australia. Rust collections are increased in the greenhouse on a susceptible cultivar and then inoculated onto a set of six cultivars containing the resistance genes used in the local breeding work. So the classification of races is based on the reaction of the six cultivars which represent useful resistance sources in Australia. Naturally, it is not essential to know the genes in each resistance source and the resistance sources used may differ in different countries. In this classification a particular race overcomes the resistance of a particular resistance source, hence, it resembles the scheme which is used in classifying races of *Phytophthora infestans*. This is based on the ability of races (biotypes) to overcome the resistance of the R genes in potatoes (see page 334).

Identification and detection of races in the greenhouse and in the field (rust trial projects)

Identification of races in the greenhouse under rather controlled conditions is a fairly exact method of investigation. Still, many research workers have come to the conclusion that the appearance of new races in the field is easily and quickly detected when "rust trial projects" are conducted throughout several countries. Differential sets (Fangsortiments) are composed of cultivars with known susceptibility or resistance to particular physiologic races. These trials are spread and inspected throughout the whole of Europe and America. The appearance of new races are very conspicuous on the trials. Moreover, the geographical distribution of the various races of rusts and the resistance of some important cultivars of host plants to the various races, can be detected under different climatic conditions (Watson and Luig, 1961).

Naturally, new races which appear in field observations have to be identified in the greenhouse on seedlings of the standard differential set. Thus, greenhouse observations complete the data obtained in field trials.

Different rust trial projects were organized in the last few years in Europe as well as in America and Australia. It is hoped that a good international control of rust epidemics and cooperation in breeding for resistance will develop in the near future.

Resistance to stem rust

The most conspicuous form of resistance is the hypersensitive reaction. In the *seedling stage*, resistance is controlled by single dominant genes. A single gene gives resistance to one or relatively few races of the pathogen. On the other hand, *adult-plant resistance* (mature plant resistance) is inherited multigenically and is effective over a wide race spectrum. This type of resistance becomes expressed in the weeks preceding maturity, and is not necessarily expressed in the seedling stage. Adult-plant resistance is evaluated in field experiments in the rust nursery. With many cultivars of wheat there is no correlation between seedling resistance and mature plant resistance. When a particular cultivar is susceptible to rust in the seedling stage this does not mean that it will be susceptible also in the adult stage, because it may have adult-plant resistance. However, resistance in the seedling stage will not turn to susceptibility in the adult stage (except in a few, rare cases). Thus, selection for rust resistance should be carried out in the open field, in a rust nursery under the effect of artificial epidemics. As to the technical requirements for developing artificial local epidemics see a previous paragraph on rust nurseries.

Sources of resistance to stem rust

A high degree of resistance is exhibited by *Triticum timopheevi* and derivatives of this species of wheat as well as Khapli (*Triticum dicoccum*) and *Triticum carthlicum*. The important sources of stem rust resistance are the *durum* wheats: Golden Ball, Iumillo, Mindum, Beladi, Tremez Rijo etc. Of common wheat the following cultivars are well-known sources of resistance: Hope, Kentana, Kenya 58, Kenya 117A, McMurachy, Selkirk, Red Egyptian, Egypt Na-95, Thatcher and Timstein. These are all spring wheat cultivars. In winter wheat breeding programs Hope-type rust resistance has been utilized in the past by plant breeders. On the basis of their experiments in Europe and America, Stewart et al. (1967) recently recommended new winter wheats as sources of resistance: Oro, Redcoat, and Dimitrovka 5–14-IZR. These cultivars are resistant to leaf rust (*Puccinia recondita*) and stripe rust (*Puccinia striiformis*) in addition to stem rust. It would be very useful to utilize these cultivars in breeding programs in the future to make a marked improvement in rust resistance of winter wheat.

PHYSIOLOGIC RACES OF *PUCCINIA RECONDITA* ROB. EX DESM. F. SP. *TRITICI* ERIKSS. ET HENN.

The following differential set of cultivars of wheat is used for identifying races of leaf rust:

Malakoff (C.I. 4898)
Carina (C.I. 3756)
Brevit (C.I. 3778)
Webster (C.I. 3780)
Loros (C.I. 3779)
Mediterranean (C.I. 3332)
Hussar (C.I. 4843)
Democrat (C.I. 3384)

Later Johnston (1956) omitted cultivars Carina, Brevit and Hussar from this set. The races identified previously have been grouped and given new numbers. These are the "UN races" (unified number of races).

The next Table shows the relation of standard races to the UN races, as well as the reaction of differential cultivars to each race.

Standard race	UN Race	Malakoff	Carina	Brevit	Webster	Loros	Mediterranean	Hussar	Democrat
1	1	R	R	R	R	R	R	R	R
2	2	R	R	R	R	R	S	S	S
3	3	R	R	R	R	I	S	R	S
4	4	R	S	S	R	S	S	R	R
5	5	S	R	R	R	R	S	R	S
6	6	S	R	S	R	S	S	S	S
7	7	S	R	R	S	R	S	R	S
8	8	S	R	S	S	S	S	R	R
9	9	S	R	R	S	S	R	R	R
10	9	S	S	S	S	S	R	R	R
11	10	R	R	S	R	S	R	R	R
12	3	R	S	S	R	S	S	S	S
13	9	S	S	R	S	S	R	S	R
14	10	R	R	S	R	S	R	S	R
15	2	R	R	R	R	R	S	R	S

Standard race	UN Race	Malakoff	Carina	Brevit	Webster	Loros	Mediterranean	Hussar	Democrat
16	1	R	R	R	R	R	R	R	R
17	11	S	R	R	R	R	R	S	R
18	12	R	S	R	S	S	R	S	R
19	9	S	S	R	S	S	R	R	R
20	9	S	S	S	S	S	R	S	R
21	13	S	S	R	S	S	S	S	S
22	14	I	S	S	R	S	R	R	R
23	9	I	S	R	S	S	R	S	R
24	15	S	S	R	S	S	R	S	I
25	2	R	R	S	R	R	S	R	S
26	10	R	S	S	R	S	R	S	R
27	15	S	R	R	S	S	R	I	I
28	6	S	R	R	R	S	S	S	S
29	9	S	I	I	I	S	R	S	R
30	13	S	I	I	S	S	S	R	S
31	9	S	R	R	S	S	R	S	R
32	3	R	R	I	R	S	S	R	S
33	16	R	R	R	R	S	R	R	S
34	2	R	R	R	R	R	S	I	S
35	13	S	R	R	S	S	S	R	S
36	11	I	R	R	R	R	R	R	R
37	14	S	R	R	R	S	R	R	R
38	10	R	R	R	R	S	R	R	R
39	6	S	I	I	R	S	S	R	S
40	6	S	S	R	R	S	S	S	I
41	15	S	I	R	S	S	R	R	S
42	13	S	I	S	S	S	S	S	S
43	14	S	S	S	R	S	R	R	R
44	3	R	R	R	R	S	I	R	S
45	17	R	S	R	S	S	S	R	S
46	16	R	I	R	R	S	R	R	S
47	18	S	S	R	S	R	R	R	R
48	8	S	S	R	S	S	I	R	R
49	14	S	S	S	R	S	R	S	R
50	14	S	R	S	R	S	R	R	R
51	16	R	R	S	R	S	R	S	S
52	5	S	R	R	R	R	S	S	S
53	1	R	R	R	R	R	R	S	R
54	13	S	S	R	S	S	S	R	S
55	19	R	S	S	I	S	S	S	R
56	20	R	S	S	R	R	S	S	R
57	17	R	S	S	S	S	S	S	S
58	3	R	R	S	R	S	S	R	S
59	2	R	R	S	R	R	S	I	S
60	7	S	R	R	S	R	S	I	S

Standard race	UN Race	Malakoff	Carina	Brevit	Webster	Loros	Mediterranean	Hussar	Democrat
61	3	R	R	S	R	S	S	S	S
62	2	R	R	S	R	R	S	S	S
63	1	R	R	I	R	R	R	R	R
64	14	S	R	S	R	S	R	S	R
65	13	S	I	I	I	S	I	R	I
66	12	R	S	S	I	S	R	S	R
67	17	R	S	S	S	S	S	S	I
68	12	R	S	S	S	S	R	R	R
69	14	I	S	S	R	S	R	I	R
70	9	I	S	S	S	S	R	R	R
71	19	R	S	S	I	S	I	S	R
72	12	R	S	S	I	S	R	I	R
73	19	R	S	S	S	S	S	I	R
74	10	R	R	R	R	I	R	R	R
75	19	R	S	S	I	S	I	R	R
76	17	R	I	S	I	S	S	R	S
77	13	S	S	S	S	S	S	S	S
78	3	R	S	S	R	S	I	I	I
79	16	R	R	I	R	I	R	R	I
80	13	S	I	S	I	S	I	S	I
81	17	R	I	I	I	I	S	R	S
82	13	S	S	I	S	S	I	R	S
83	6	S	R	I	R	S	I	R	S
84	3	R	S	S	R	S	S	R	S
85	17	R	S	S	I	S	S	S	S
86	19	R	S	S	S	S	S	R	R
87	17	R	S	S	S	S	I	R	I
88	17	R	I	I	S	S	I	R	S
89	13	S	I	I	I	S	S	S	S
90	17	R	R	R	I	S	S	R	S
91	18	S	S	S	S	R	R	S	R
92	20	R	R	R	R	R	S	R	R
93	11	S	R	R	R	R	R	R	R
94	13	S	S	I	S	S	I	S	S
95	10	R	S	S	R	S	R	S	R
96	12	R	S	S	S	S	R	R	R
97	12	R	S	S	S	S	R	R	R
98	3	R	I	I	R	I	I	I	I
99	15	S	R	R	S	S	R	S	S
100	5	I	R	R	R	R	S	I	S
101	13	S	I	I	I	S	I	S	I
102	3	R	R	I	R	I	S	I	S
103	6	S	R	R	R	I	I	R	I
104	13	S	I	S	I	S	S	I	S
105	6	S	R	S	R	S	S	R	S

Standard race	UN Race	Malakoff	Carina	Brevit	Webster	Loros	Mediterranean	Hussar	Democrat
106	12	R	I	S	I	S	R	R	R
107	12	R	S	S	S	S	R	S	R
108	9	S	S	S	I	S	R	S	R
109	21	S	I	S	R	R	R	S	S
110	14	S	R	I	R	I	R	R	R
111	3	R	R	I	R	I	I	R	S
112	13	S	I	I	S	S	I	I	S
113	13	S	I	I	I	I	S	R	S
114	13	S	S	S	I	S	S	R	S
115	13	S	S	S	S	S	I	I	I
116	13	I	I	I	I	I	I	I	I
117	15	S	S	S	S	S	R	S	S
118	22	R	S	S	S	R	R	R	R
119	18	S	R	S	S	R	R	S	R
120	1	R	R	S	R	R	R	S	R
121	18	S	R	R	S	R	R	S	R
122	13	S	S	S	S	S	S	R	S
123	1	R	R	S	R	R	R	R	R
124	23	R	R	R	R	R	R	R	S
125	18	S	R	R	S	R	R	R	R
126	6	S	R	R	R	S	S	R	S
127	2	R	S	S	R	R	S	R	S
128	14	S	R	R	R	S	R	S	R
129	24	S	S	S	R	S	S	S	R
130	13	S	R	R	S	S	S	S	S
131	10	R	S	S	R	S	R	R	R
132	10	R	I	I	R	I	R	R	R
133	10	R	I	R	R	S	R	S	R
134	12	R	I	S	S	S	R	R	R
135	12	R	I	I	I	I	R	I	R
136	17	R	R	R	I	I	S	I	S
137	19	R	I	I	I	I	I	I	R
138	12	R	I	I	S	I	R	I	R
139	12	R	R	R	S	S	R	R	R
140	17	R	R	R	S	S	S	R	S
141	3	R	R	R	R	S	S	S	S
142	17	R	I	I	S	S	I	R	I
143	17	R	S	S	S	S	S	R	S
144	6	S	S	S	R	S	S	S	S
145	13	S	I	S	I	S	S	S	S
146	9	S	I	S	I	S	R	S	R
147	9	S	R	S	S	S	R	S	R
148	9	S	R	S	I	S	R	S	R
149	13	S	R	S	S	S	S	S	S
150	13	S	S	S	I	S	S	S	S

Standard race	UN Race	Malakoff	Carina	Brevit	Webster	Loros	Mediterranean	Hussar	Democrat
151	2	R	S	S	R	R	S	S	S
152	15	S	R	S	S	S	R	R	S
153	2	R	S	R	R	R	S	S	S
154	10	R	S	R	R	S	R	R	R
155	10	R	S	R	R	S	R	S	R
156	4	R	R	S	R	S	S	R	R
157	19	R	R	S	S	S	S	R	R
158	16	R	S	S	R	S	R	S	S
159	25	R	R	R	S	R	S	R	S
160	21	S	I	S	R	R	R	R	R
161	3	R	R	R	R	S	S	R	S
162	17	R	S	S	S	S	S	I	S
163	4	R	R	R	R	S	S	R	R
164	12	R	R	S	S	S	R	S	R
165	20	R	R	S	R	R	S	S	R
166	12	R	I	S	S	S	R	S	R
167	17	R	R	S	S	S	S	S	S
168	12	R	R	S	S	S	R	R	R
169	23	R	R	S	R	R	R	R	S
170	26	R	R	R	S	R	R	R	S
171	26	R	R	S	S	R	R	R	S
172	8	S	S	S	S	S	S	R	R
173	26	R	R	R	S	R	R	S	S
174	22	R	S	R	S	R	R	S	R
175	27	R	R	R	S	S	R	R	S
176	17	R	R	S	S	S	S	R	S
177	22	R	R	R	S	R	R	R	R
178	16	R	S	S	R	S	R	R	S
179	23	R	R	R	R	R	R	S	S
180	24	S	S	R	R	S	S	R	R
181	27	R	S	S	S	S	R	R	S
182	14	S	S	R	R	S	R	S	R
183	4	R	I	S	R	I	I	I	R
184	8	S	S	S	S	S	S	S	R
185	17	R	S	S	I	S	S	R	I
186	13	S	S	S	S	S	I	S	I
187	8	S	S	S	S	S	I	S	R
188	17	R	R	S	S	I	I	S	S
189	17	R	S	S	S	I	I	S	S
190	12	R	S	S	I	S	R	R	R
191	27	R	S	S	S	S	R	I	I
192	7	S	S	S	S	R	S	S	S
193	16	R	S	S	R	S	R	R	I
194	27	R	S	S	I	S	R	I	I
195	12	R	S	R	S	S	R	R	R

Standard race	UN Race	Malakoff	Carina	Brevit	Webster	Loros	Mediterranean	Hussar	Democrat
196	13	S	I	S	I	S	I	I	I
197	27	R	S	S	S	S	R	I	S
198	19	R	S	S	S	S	I	I	R
199	16	R	R	S	R	S	R	I	I
200	3	R	S	S	R	S	S	I	I
201	25	R	S	S	S	R	I	S	I
202	1	R	S	S	R	R	R	S	R
203	17	R	S	S	S	S	I	I	S
204	17	R	S	R	S	S	I	I	S
205	19	R	S	S	S	S	I	R	R
206	27	R	S	S	S	S	R	S	S
207	17	R	S	S	I	S	I	I	I
208	17	R	S	S	S	S	I	I	I
209	17	R	S	S	S	S	S	I	I
210	13	S	S	S	S	S	S	I	I
211	4	R	S	S	R	S	S	S	R
212	17	R	R	S	I	S	S	R	S
213	22	R	S	R	I	R	R	R	R
214	15	S	S	R	S	S	R	R	S
215	12	R	R	S	I	S	R	R	R
216	22	R	S	S	I	R	R	R	R
217	23	R	R	S	R	R	R	S	S
218	19	R	S	S	S	S	S	S	R
219	4	R	R	S	R	S	I	R	R
220	19	R	R	R	S	S	S	S	R
221	23	R	S	S	R	R	R	I	I
222	25	R	S	S	S	R	S	S	S
223	10	R	R	R	R	S	R	S	R
224	20	R	S	S	R	R	I	I	R
225	3	R	R	S	R	S	I	I	I
226	3	R	S	S	R	S	I	R	I
227	10	R	I	S	R	S	R	R	R
228	3	R	R	S	R	S	S	I	I

Supplemental differential cultivars of wheat for identification of races of Puccinia recondita f. sp. *tritici*

There are different possibilities for supplementing the international standard method of identifying races of leaf rust. If it is desirable to further characterize the virulence of a culture of *P. recondita* f. sp. *tritici*, a new group of cultivars is used in addition to the "International Standard Differential". This group is the "Supplemental Differential Set". Young and Browder (1965) placed the following five cultivars into this set:

Dular (C.I. 13373), Lee (C.I. 12488), Waban (C.I. 12992), Sinvalocho (C. I. 12595), Exchange (C.I. 12635)

The races characterized by the aid of this set were called "NA65" races. In preparing the key, only two reaction classes were distinguished: resistant and susceptible. Reactions designated as 0, 0; 1, 2–, 2, and 2+ were placed in the resistant class, and those of 2++, X, 3, and 4 in the susceptible class. Using five cultivars and two reaction classes one can determine 32 possible races. This is shown in the next Table.

The reactions of cultivars of the 1965 Supplemental Differential Set and the NA65 races of *Puccinia recondita* f. sp. *tritici*
(After Young and Browder, 1965)

"Na65" race	Dular	Lee	Waban	Sinvalocho	Exchange
1	R	R	R	R	R
2	R	R	R	R	S
3	R	R	R	S	R
4	R	R	R	S	S
5	R	R	S	R	R
6	R	R	S	R	S
7	R	R	S	S	R
8	R	R	S	S	S
9	R	S	R	R	R
10	R	S	R	R	S
11	R	S	R	S	R
12	R	S	R	S	S
13	R	S	S	R	R
14	R	S	S	R	S
15	R	S	S	S	R
16	R	S	S	S	S
17	S	R	R	R	R
18	S	R	R	R	S
19	S	R	R	S	R
20	S	R	R	S	S
21	S	R	S	R	R
22	S	R	S	R	S
23	S	R	S	S	R
24	S	R	S	S	S
25	S	S	R	R	R
26	S	S	R	R	S
27	S	S	R	S	R
28	S	S	R	S	S
29	S	S	S	R	R
30	S	S	S	R	S
31	S	S	S	S	R
32	S	S	S	S	S

S = susceptible, R = resistant.

PHYSIOLOGIC RACES OF *PUCCINIA STRIIFORMIS* WEST F. SP. *TRITICI* ERIKSS. ET HENN.

The Table on p. 395 shows the races of stripe rust and the reactions of the differential set (after Fuchs, 1965) (IV = very susceptible, 0 = resistant).

Recently, the appearance of a dangerous new race of stripe rust was detected in Northern- and Central-Europe (Macer and Doling, 1966; Fuchs, 1967). This is more aggressive than any other known races, and severely attacks most of the cultivars in the field. On the basis of the reaction types exhibited by the members of a reduced differential set, Fuchs (1967) was able to demonstrate that it represented a new race. She suggested the designation of race 60 for it. The most important European races produce the following infection types on the "reduced" differential set: (IV = very susceptible, 0 = resistant).

Races	Vilmorin 23	Heines Kolben	Chinese 166
3/55	IV	0	0
54	0	IV	0
27/53	0	0	IV
8	0	0	0
60	0+	IV	IV

The recent demonstration by Zadoks (1961) of the existence of "field races" helped clarify the race situation for plant breeding. Field races are distinct from races identified on wheat seedlings in the greenhouse. They are identified by the reaction of specific cultivars in field nurseries. According to Zadoks (1961) a field race is "a unit distinct from other units by so high a compatibility with one or more varieties (cultivars) of a differential set in the mature plant stage that it may produce epidemics on these varieties (cultivars) but not on others". This means that field races appear on specific wheat cultivars after these cultivars have been grown for some time over large areas. The next Table shows the reaction types of the most important cultivars of wheat to different races in the field (adult plant reactions) (Fuchs, 1967) (+ + + = susceptible, — = resistant).

Cultivars	3/55	54	27/53	8	60
Capelle	+++	—	—	—	—
Nord Desprez	+++	—	—	—	—
Heines Kolben	—	+++	—	—	+++
Heines Peko	—	+++	—	—	+++
Chinese (166)	—	—	+++	—	+++
Heine VII	—	+++	—	+++	+++
Merlin	—	+++	—	+++	+++
Rothwell Perdix	—	—	—	—	+++
Ibis	—	—	—	—	+++
Harvest Queen	+++	+++	+++	+++	+++
Michigan Amber	+++	+++	+++	+++	+++

Race	Michigan Amber	Blé rouge d'Ecosse	Strubes Dick-kopf	Webster	Vilmorin 23	Heines Kolben	Carsten V	Spaldings prolific	Chinese 166	Rouge prolifique barbu	Lee	Reichersberg 42	Nord Desprez
1	IV	IV–	IV	IV–	IV–	IV	0–IV	i	i–0	i–0	0	0	IV
2	IV	IV	IV	0–IV	IV	0	0–IV	IV–	i–0	IV–	0	0	IV
2A	IV	IV–	IV–	0–IV	IV	0	0–IV	II–IV	i–0	IV–	IV	IV	IV
3/55	IV	IV	IV	0–IV	IV	0	0–IV	i–0+	i–0	i–0+	0	0	IV
7	IV	IV	IV	0+	0	0	IV	0+	i–0	0+	0	0	0
8	IV	IV	IV	0+	0	0	0–IV	0+	i–0	0+	0	0	0
8x	IV	IV–	IV–	0+	0+	0	0+	i–0	i–0	i–0	0	0	0+
11/29	IV	IV	0+	0–IV	0	0	0+	i–0	i–0	i–0	0	0	0
11/29A	IV	IV–	i	IV–	0	0	i	i	i–0	i	IV	0+	0
13	IV	i–0	i–0	i–III	i–0+	i–0	i	i	IV–	i	0	i–0	i–0+
15/43	IV	IV–	0+	0–IV	0	0	0+	IV	i–0	IV–	0	0	0
18	IV–	i–0	i–0	0+	i–0	0	i–0	i–0	i–0	i–0	0	0	0
20	IV	0	0	0–IV	0	0–IV	i–0	i–0	i–0	i–0	0+	0+	0
$20A_1$	IV	i–0	i–0	IV–	0	0–II+	i–0	i–0	i–0	i–0	IV	IV	0
$20A_2$	IV	i–0	i–0	IV–	0	II–IV	i–0	i–0	i–0	i–0	IV	0–IV	0
25	IV	i–0	i–0	IV–	i–0	i–0	i	i	i–0	i	0+	0	i–0
26	IV	IV	IV	0–IV	0	0	IV	IV	i–0	0–IV	0	0	0
27/53	IV	IV	IV	0–IV	0	0	IV	0+	II–IV	0	0	0	0
32	IV	IV	0+	0–IV	0+	0	0–IV	0+	i–0	i–0	0	0	
32A	IV	IV	0+	0–IV	II+	0	0+	i–0	i–0	i–0	IV	O–IV	0+
32/43A	IV	IV	0+	0–IV	II+	0	0+	0–IV	i–0	0–IV	IV	IV–	0+
$42A_1$	IV	0+	0+	IV	0	II±	i–0	i	IV	i	IV	IV	0
$42A_2$	IV	0+	0+	IV	0	II±	i–0	i–0	IV	i	IV	0+	0
52	IV	0–IV	i–0	IV	IV	0	0	i–0	i–0	i–0	0	0	IV
54	IV	IV	IV	0–IV	0	IV	0–IV	0+	i–0	0+	0	0	0
58	IV	IV	IV	0–IV	IV	0	0–IV	0+	II–IV	0+	0	0	IV
59	IV	0–IV	i–0	0–IV	II	IV	i	i–0	i–0	i	0+	0	IV

THE USE OF BENZIMIDAZOLE AND KINETIN FOR THE DETACHED LEAF CULTURE OF LEAF- AND STRIPE-RUST

Using compounds like kinetin or benzimidazole, which exert a cytokinin effect, it is possible to culture detached leaves on cytokinin-solutions for even a month (Person et al., 1957). Rusts grow well in leaf cultures of this type. On the basis of these findings, several investigators tried to determine races of rusts and mildews, as well as maintain obligate parasites on detached leaves floated on cytokinins (see also the paragraph on powdery mildews on page 356).

Infection types incited by different leaf- and stripe-rust races are nearly identical, in detached leaf cultures, with those incited on attached leaves of wheat seedlings. Reactions of cultivars of the differential sets in leaf cultures are also very similar to those which are exhibited by intact seedlings (Doling, 1966: Hermansen, 1966; Wolfe and Macer, 1966).

Two cm segments of the youngest expanding leaves of wheat seedlings are floated on a solution of kinetin (10 ppm) or benzimidazole (40 ppm). Leaf segments are placed on the solution in a Petri dish or in a polystyrene box of 5 × 4 × 2 cm. Leaf segments are inoculated by blowing a uredospore-talc mixture on the leaf surface. Detached leaf cultures are then placed in a room or chamber at 10–12 °C under 18 hr of daily light. Sporulation occurs in 10–14 days as a rule.

The experiments of Lovrekovich and Király (cf. Goodman et al., 1967) have shown that detached leaves treated with kinetin and benzimidazole resembled the attached leaves in protein composition. On the other hand, leaves detached and floated on water have lost at least two proteins, as was determined with serological analysis using rabbit antisera. In addition, these leaves changed their reaction type to rust. This may explain the fact that detached leaves floated on cytokinins (kinetin and benzimidazole) exhibit almost identical rust reactions as those of attached, intact leaves.

Experiments conducted in detached leaf cultures have several advantages. One can cultivate different races in closed Petri dishes or boxes without contamination. Experiments can be repeated quickly and easily, and require only limited room in the laboratory or in the refrigerator. A further, and most important, advantage of this method is that it becomes possible to inoculate different leaf segments from a plant with different races and even different leaf pathogens without infecting the whole plant. One can test individual plants against several diseases or several races of the same pathogen and, at the same time, grow them and use them in cross breeding programs as partners. On the other hand, there are usually no distinct chlorotic and necrotic spots around rust infections on detached leaf segments. But for all these uncertainties, susceptible and resistant type reactions are clearly distinguishable from each other on detached leaf segments.

BEAN RUST
(*UROMYCES PHASEOLI* [PERS.] WINT.)

Uredospores (Fig. 16) lose their germination ability in a very short time. When the spores are stored in a refrigerator they remain viable for 1–2 months, at −18 °C for more than one year (Davison and Vaughan, 1963a).

Artificial inoculation in the greenhouse

Plants are inoculated with a spore suspension containing about 20,000 uredospores/ml. When large quantities of inoculum are used, ivory soap (125 ppm) serves as a spore carrier and 0·8 ml of a 0·1 per cent Triton B1956 solution is added to each 100 ml of soap solution as a wetting agent.

Inoculum is sprayed on hand-rotated plants held 25 cm from the sprayer. 10 ml of spore suspension/5 bean plants produces successful infection. The inoculum is applied to the upper and lower surfaces of the primary leaves. Pots are placed in a moist chamber for 16–18 hr, then moved to a greenhouse.

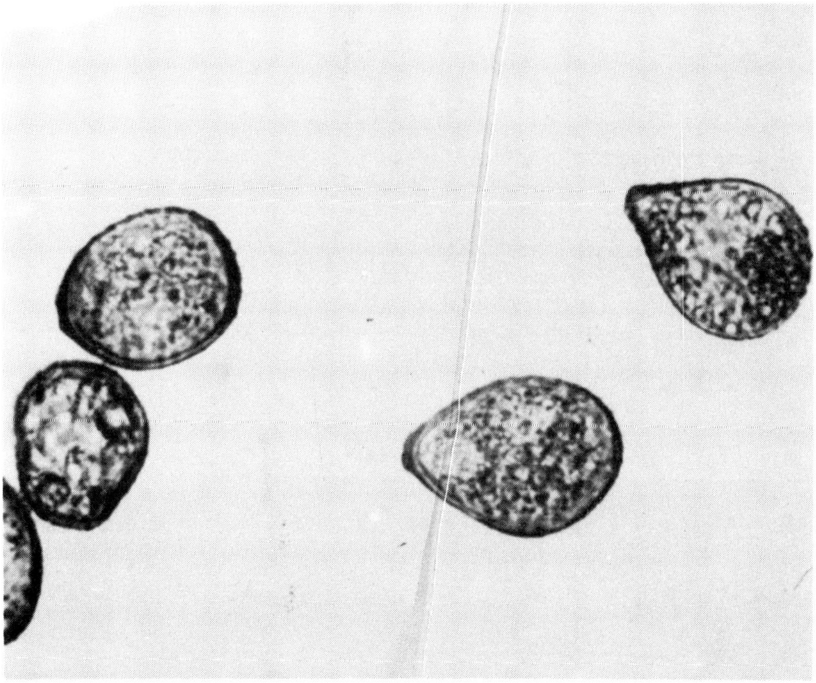

Fig. 16. Uredospores of *Uromyces phaseoli*

Artificial inoculation in the field

Cultivars to be investigated for rust resistance are sown in little plots. The plot rows are surrounded with a strip planting of susceptible cultivars. The border rows are inoculated with a spore suspension at the time of the emergence of the second and third trifoliate leaves. Following the spraying, inoculated plants are covered with polyethylene bags. Inoculations are carried out in the late afternoon or evening. The bags are removed in the morning before sunrise to avoid higher temperatures because the fungus is more sensitive to heat than the host.

Physiologic races

Seven differential cultivars were used by Harter and Zaumeyer (1941) for distinguishing races of bean rust. These are as follows:

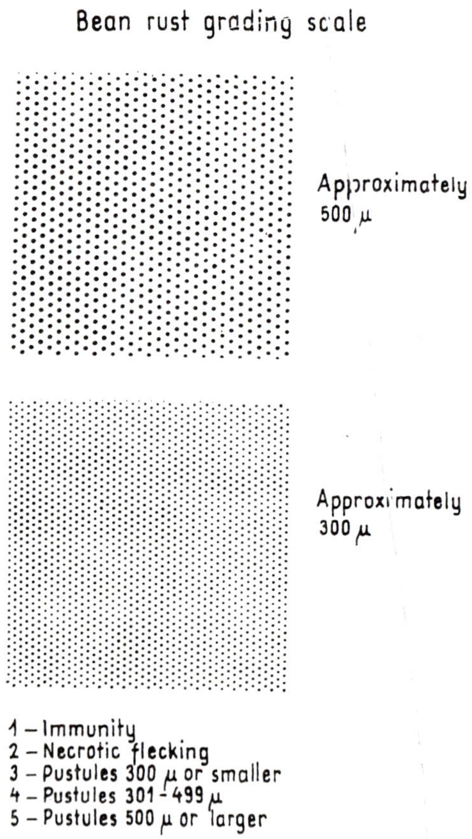

Fig. 17. *Scale* for determining susceptibility and resistance to bean rust on the basis of the size of the pustules (After Davison and Vaughan, 1963b)

No. 643 (Calif. Small White)
U.S. no. 3. (White Kentucky Wonder)
No. 650 (U.I. Pinto 111)
No. 765
No. 814
No. 780
Golden Gate Wax

The reaction scale was composed of 11 reaction grades. This scale of reaction types was based on the relative size of pustules, but the distinct size of pustules for each grade has not been defined. Therefore, a new, simple, and more exact scale was proposed by Davison and Vaughan (1963b). The reaction grades are as follows (see also Fig. 17):

1. Immune. No evidence of infection.
2. Necrotic flecking without uredial-sori.
3. Pustule diameter 300 μ or smaller.
4. Pustule diameter 301–499 μ.
5. Pustule diameter 500 μ or larger.

Thirty-four races of bean rust have been identified. It is noteworthy that a new differential set was proposed by Crispin and Dongo (1962) for identifying Mexican races of bean rust because all the American cultivars of the differential set were susceptible to Mexican races.

SMUT DISEASES

Smut fungi belong to the *Basidiomycetes*. They infect several crop plants and cause serious losses. The flowers, leaves and stems of the host plants are attacked by smut fungi. In the case of *Ustilago zeae* (corn smut) an abnormal proliferation of cells of attacked tissues takes place and tumorous tissues develop.

Biology and aetiology

Three distinct phases of development can be distinguished, which are closely related to nuclear condition (see Fig. 23 and page 416).

(1) *Haplophase* begins with the germination of the teliospore, and is represented by the basidium (promycelium) and sporidia produced on it.

(2) *Dikaryophase* is initiated by conjugation of sporidia or by the development of fusion tubes between cells of the basidium. The two sporidia usually represent

two opposite "sexes". In the dikaryotic cells there are two nuclei with opposite "sex", but fusion of the nuclei does not take place (there is no diploid phase).

(3) *Diploid phase* begins with the fusion of nuclei (karyogamy). This takes place in the teliospore, as a rule.

The sporidia of the haplophase can function as specialized sex cells or as propagative cells. The ability of the haplophase to grow parasitically is very limited.

The pathogenic phase is the dikaryophase. The two different nuclei in the dikaryotic cell may represent two pathogenic types, in addition to the two different sexes. The pathogenic dikaryophase cannot be propagated on artificial media nor on living plants as a clone. The dikaryophase completes the cycle with the formation of the diploid teliospores. However, the sporidia of the haplophase can be propagated saprophytically on artificial media. Monosporidial diploid lines can be found in smut fungi which are able to parasitize their host plant.

Physiologic races of smut fungi cannot be purified to pure lines or biotypes (genetically uniform lines) because the diploid teliospores are necessary for artificial inoculation and because the dikaryophase cannot be propagated clonally (cf. Holton, et al., 1968; Day, 1960, and Halisky, 1965).

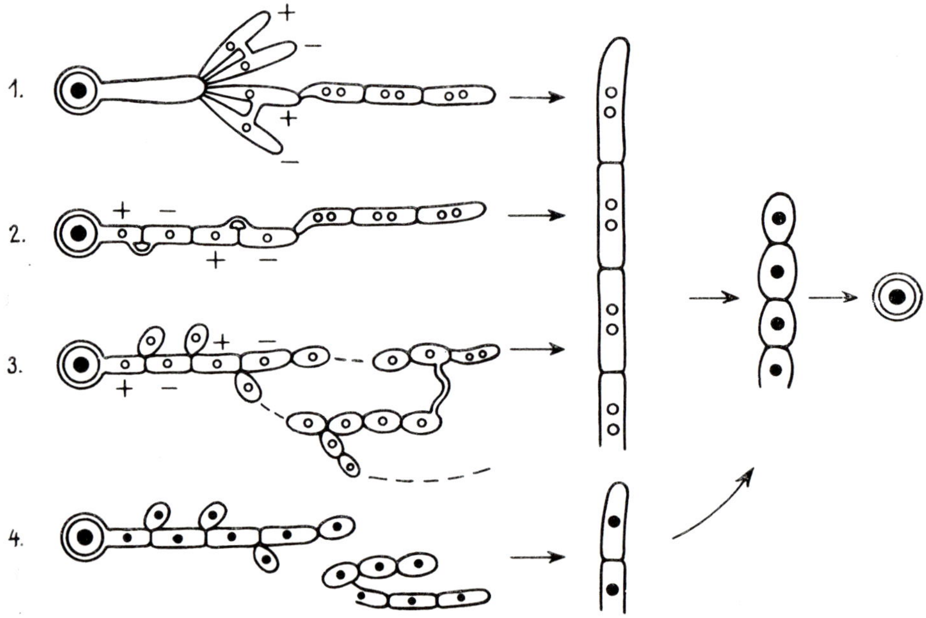

Fig. 18. Life cycle of smut fungi. Haploid-, dikaryo- and diploid-phase in the case of the different smut species. 1. *Tilletia caries*, *T. foetida*, *T. controversa*. 2. *Ustilago tritici*, *U. nuda*. 3. *Ustilago hordei*, *U. nigra*, *U. avenae*, *U. kolleri*, *U. zeae*. 4. Particular lines of *Ustilago zeae* remain diploid during the whole life cycle. These monosporidial lines are able to infect corn (After Stakman and Harrar, 1957)

The mode of spore germination of smut fungi may be grouped as follows:

Type 1. A non-septate tube of promycelium develops from the diploid teliospore with eight terminal haploid sporidia. Pairs of sporidia copulate with the aid of a connecting bridge between the two sporidia. A dikaryotic infection hypha grows out from the fusion of the sporidia (Fig. 18). This type of spore germination is characteristic for *Tilletia* species.

Type 2. A non-sporidial type of germination is characteristic for *Ustilago nuda*. The promycelium (basidium) forms either fusions between adjacent cells of different "sexes" or connecting bridges between branches of the same promycelium or promycelia of near-by spores. After fusion a dikaryotic infection hypha grows out (Fig. 18).

Type 3. Teliospores germinate with a four-celled basidium (promycelium) bearing four lateral sporidia. Sporidia represent different "sexes". They copulate and infection hyphae grow out. These represent the parasitic phase. This type of spore germination is characteristic for *Ustilago zeae*, *U. hordei*, *U. nigra*, *U. kolleri* and *Sphacelotheca* species (Fig. 18).

Sporidia may be separated by a micromanipulator and cultured on artificial media; producing thereby, haploid lines. Spore germination is influenced by the environment. By changing the medium or temperature one can easily induce abnormal types of germination. Optimal temperature for spore germination is about 20–22 °C in most cases. However, *Tilletia caries* requires a lower (18 °C) and *Ustilago zeae* a higher (25–30 °C) temperature as an optimum. The optimal temperature for spore germination is not necessarily advantageous for infection itself.

Spore germination begins on distilled water, but the growth of the germ tube is restricted. For the isolation of sporidia from *Ustilago* spp. the most suitable medium on which the teliospores germinate is a one per cent malt agar, or a weak Knop's solution, or a weak agar made with tap water.

Artificial inoculation

Methods of artificial inoculation depend on the type of spore germination and natural infection.

Seed infection or seedling infection is necessary in the cases of *Tilletia* spp., *Ustilago hordei* and *Sphacelotheca* spp. Dry grain is dusted with dry sieved teliospores or the spores are introduced into the soil. Seedlings will be infected in the soil. The dikaryotic mycelium reaches the growing point and destroys the inflorescence.

Flower infection. Teliospores are introduced into the flowers. After germination the infection mycelium grows into the ovary and reaches the embryo where it lies dormant. This method of infection is applied in the case of loose smut (*Ustilago nuda*).

A somewhat different technique is used on grain in the husk, in the case of oat smuts (*Ustilago avenae* and *Ustilago kolleri*). In natural infections, spores reach the flowers and produce the mycelium on the inner, parenchymatic surface of the pales. This will be the source of infection when the seeds germinate. In the case of artificial inoculation, grains are placed in a suspension of teliospores and evacuated in a desiccator attached to a vacuum pump. The inoculated seed is sown in the greenhouse or in the field. Shelled grain may also be inoculated by dry spore dusting.

A further method of infection ("*local infection*") is used in the case of *Ustilago zeae*. Spore- or sporidial suspensions are injected into tissues of the corn plant. Almost all organs of the host can be infected. It is also possible to inoculate injured tissues by spraying suspensions of sporidia onto the mechanically injured surface.

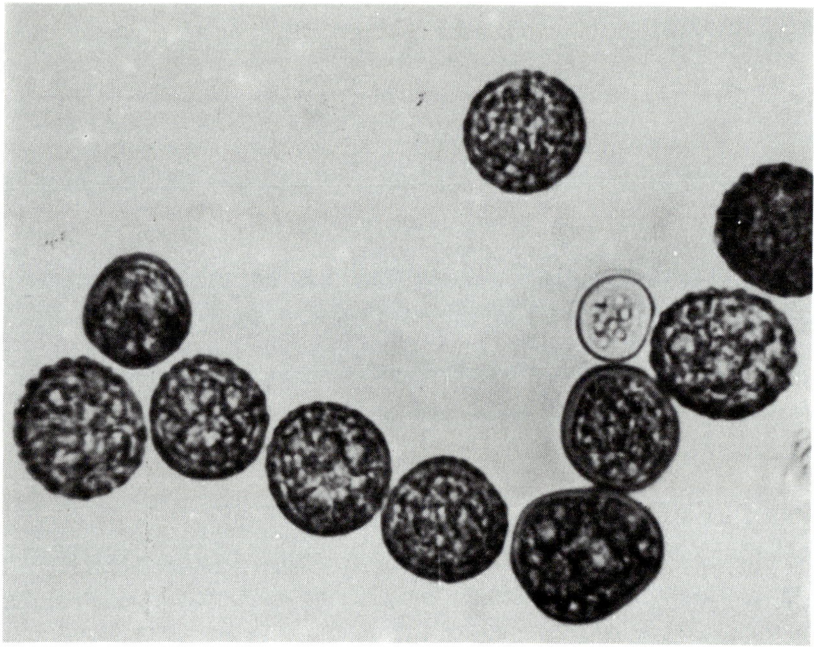

Fig. 19. Teliospores of *Tilletia caries*

WHEAT BUNT

(*TILLETIA CARIES* [DC.] TUL. AND *TILLETIA FOETIDA* [WALLR.] LIRO)

Teliospores of *T. caries* are reticulate, and 15–23 μ in diameter. The spores of *T. foetida* are, on the other hand, smooth, and 17–20 μ in diameter (Figs. 19 and 20).

Artificial inoculation is carried out by powdering 100 grams of wheat seeds with 0·5 g of dry spores. In this case about 35,000–150,000 spores contaminate a single wheat seed. The powdered seeds are sown 4·5 cm deep, in a soil at 50 per cent of its moisture capacity. The pH value is 5·5–7·5. Successful infection will take place in the soil if germination of the seeds occurs at 5–10 °C.

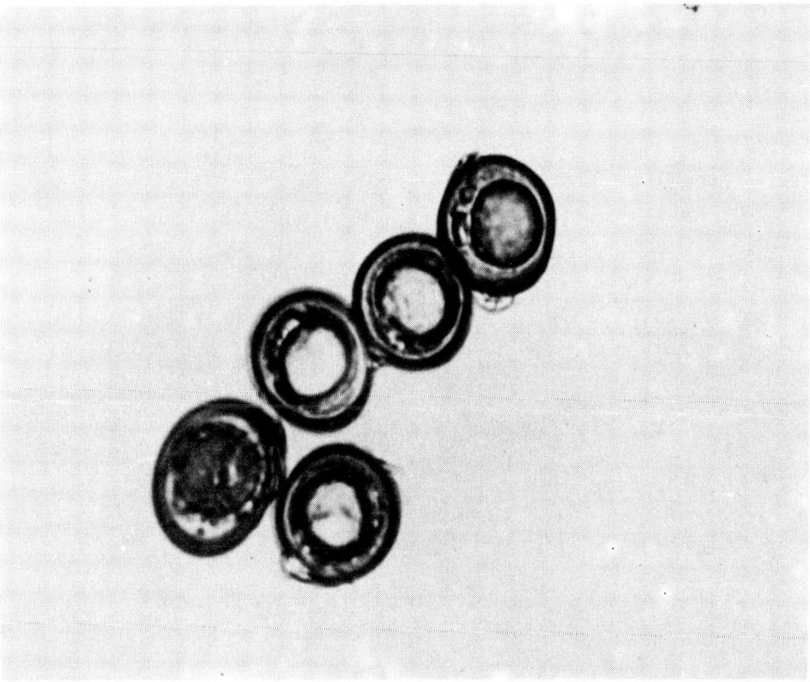

Fig. 20. Teliospores of *Tilletia foetida*

Physiologic races

After the original work of Rodenhiser and Holton (1937) the classification of physiologic races was modified by several workers. At present 18 races of *T. caries* ("T" races) and 10 races of *T. foetida* ("L" races) are known. Due to the simplification of resistance test programs, Kendrick (1961) worked out a new classification. On the basis of pathogenicity on 7 wheat cultivars, the 28 known races

Races and race groups of *Tilletia caries* and *T. foetida* and reactions of cultivars of the differential set
(After Kendrick, 1961)

	Differential set							
	Hybrid 128	Albit	Selection 50077	Selection 1403	Orin	Hohen-heimer	Ridit	Omar
	Resistance genes							
	–	M	M_2	H	T or R	Ho	rd	TM
Race groups and races:								
Martin								
T–14	S	S	R	R	R	R	R	R
T–1, L–1, L–2	S	R	S	R	R	R	R	R
T–2	S	I	S	R	R	R	R	R
T–4, T–6, L–4	S	S	S	R	R	R	R	R
Hussar								
T–3, L–3	S	R	S	S	R	R	R	R
T–5, T–7, T–8, L–5, L–6, L–7	S	S	S	S	R	R	R	R
Turkey								
T–17	S	R	S	R	S	R	R	R
L–8	S	R	S	S	S	R	R	R
T–16	S	R	S	S	S	S	R	R
Hohenheimer								
T–10	S	R	R	R	R	S	R	R
T–9	S	R	S	R	R	S	R	R
T–12	S	S	S	R	R	S	R	R
T–15	S	S	S	S	R	S	R	R
Ridit								
T–11	S	R	R	S	R	R	S	R
L–10	S	R	S	S	R	R	I	R
T–13, L–9	S	S	R	S	R	R	S	R
Omar								
T–18	S	S	S	R	S	R	R	I

were re-classified into 17 race-groups. These groups contain both "T" and "L" races. It is possible to re-group these 17 race groups into 6 groups, on the basis of their pathogenicity on the 6 generally recognized resistance types. This is a further simplification of the work of breeding for disease resistance. The 6 resistance types are represented by the cultivars Martin, Hussar, Turkey, Hohenheimer, Ridit and Omar. Each of the 6 race groups is only pathogenic on a single resistance type. Thus, related pathogenic races may be grouped in a few classes. For testing bunt resistance of cultivars, it is sufficient to use only a single race for inoculation from each race group. This means that testing of cultivars for disease resistance will be carried out against a reduced number of races. In other words, many cultivars can be tested in a season. The classification of the simplified race-groups is shown on page 404. Cultivars of the differential set possess the following resistance genes:

Albit (C.I. 8275): M gene
Selection 50077 (C.I. 13561): M_2 gene
Selection 1403 (C.I. 13562): H gene
Orin (C.I. 12687): T or R gene
Omar (C.I. 13072): M, and T genes combined
Hohenheimer (C.I. 11458): not investigated
Ridit (C.I. 6703): not investigated

Hybrid 128 (C.I. 4512) is also a member of the set as a susceptible check cultivar. (Reactions of the cultivars of the differential set to the races or race groups: R = resistant, 0–10 per cent infection. I = intermediate, 11–40 per cent infection. S = susceptible, 41–100 per cent infection.)

Many resistant cultivars were derived from crosses with the six resistant types (Kendrick, 1965). In addition to the production of resistant crops, bunt can be controlled with copper compounds and organic mercuric fungicides.

Controlled environment during infection

Wheat seeds, powdered with teliospores, are sown in the greenhouse after planting, to obtain environmental conditions that favour infection in the plant. The seedlings are kept in the greenhouse under controlled temperature and soil moisture. The water content of the soil is an important factor in influencing successful infection (Roemer and Bartholly, 1933). Seedlings, which were supposedly infected in the greenhouse, are planted in 2–3 leaf-stage in field plots. Bunt resistance or susceptibility is noted at the ripening stage. If winter wheats are infected with bunt, seeds powdered with spores are sown in big pots or little boxes, and the seedlings are put in the cold for vernalization. Treated seedlings are planted in field plots the next spring.

Testing the effectiveness of fungicides with spore germination-tests

The percentage of spore germination is determined most conveniently on washed clay. Teliospores of bunt are distributed on the surface of the clay in Petri dishes by spreading (screening). Cultures are kept at 18–20 °C. Spore germination is checked on the 4–6th day under a microscope with low magnification. To test the seed dressing materials, spores previously immersed in the fungicides, are spread over the surface of the washed clay. If the effectiveness of the dressing material on the spores located on the surface of the seed has to be tested, clean seeds should be infested artificially and then treated with the fungicide. These treated seeds are pressed on the surface of clay or washed soil and germination of the adhering spores is checked on the 4–6th day.

Experiments with bunt and other smuts are treated in particular, by Fischer and Holton (1957) and Schuhmann (1966).

DWARF BUNT OF WHEAT (*TILLETIA CONTROVERSA* KÜHN)

The infection by this smut causes excessive dwarfing and tillering of the host. Spores have prominent reticulations and they are 17–22 μ in diameter (Fig. 21). Optimal temperature for spore germination is 5 °C.

Methods of artificial inoculation

Powdering the seeds with smut spores does not give successful infection. Germination occurs only on the surface of the soil. Therefore, a simple method of artificial inoculation is to distribute teliospores on the soil, 1·5 g/m^2, after sowing the grain. This is a very simple and successful method which is used in testing resistance of cultivars (Baylis, 1955).

A more accurate method is to inoculate seeds or coleoptiles with germinated spores under controlled conditions (Niemann, 1956; Meiners, 1959). The spores are plated on washed soil or on a 2·5 per cent soil-extract agar in Petri dishes. The dishes are incubated at 5 °C in continuous light. Germination occurs in 3–6 weeks. At the time of inoculation, Petri dishes will contain spores in all stages of germination. Then 2 ml of sterile distilled water and about 100 seeds of the wheat cultivar to be inoculated, are placed in a Petri dish containing the germinating teliospores. The seeds are stirred with a glass rod until well covered with a spore suspension. Then the seeds are placed in another dish containing a 2 cm layer of moist vermiculite. The seeds are covered with another 2 cm layer of vermiculite. The dish is covered and kept in the greenhouse at 10–15 °C for 10–14 days. The seeds will germinate and seedlings then are trans-

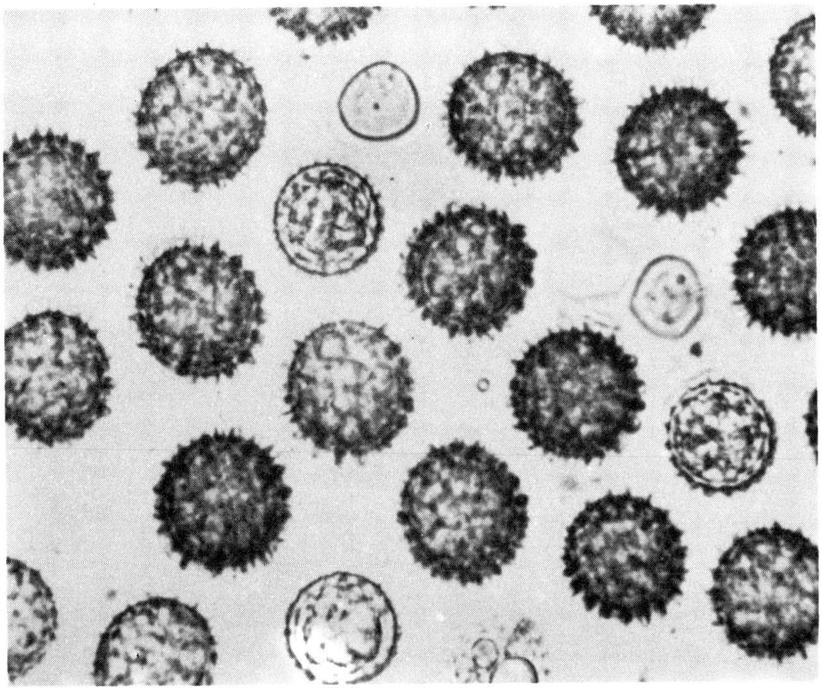

Fig. 21. Teliospores of *Tilletia controversa*

planted to plant bands containing a mixture of soil and sand. When the seedlings become well established, the plants can be transplanted to the field. In the case of winter wheat, the seedlings will be placed out of doors for vernalization, and only in the following spring are they transplanted to the field. This method is complicated and time consuming, but suitable for race determinations.

According to Novacky (1964), 2—4 mm long coleoptiles are slightly injured by touching them to sand-paper and then passing them through germinated teliospores. The inoculated seedlings are kept at 5 °C for two weeks or so under a 12 hr photoperiod. After the appearance of the first leaf the seedlings are transplanted to sand-culture in the greenhouse under 16 hr photoperiod where possible until maturity.

Physiologic races

Dwarf bunt races were classified by Hoffmann et al. (1962, 1967) using six differential cultivars. Ten races with different pathogenicity were identified. Cultivars of the differential set are as follows:

Albit (C.I. 8275): M resistance gene
Selection 50077 (C.I. 12561): M_2 gene
Selection 1075: H gene
Turkey (C.I. 1558): T gene
Selection 3432: Ho (Hohenheimer) resistance
Ridit (C.I. 6703): rd resistance

In addition to these cultivars, a check-cultivar namely Hybrid 128 (C.I. 4512), is also included in the differential set. The particular races are designated with a D (dwarf bunt) and numbers. The next table shows reaction types of cultivars of the differential set and classification of races.

Physiologic races of *Tilletia controversa* and the reactions of cultivars of the differential set (After Hoffmann et al., 1967)

	Differential set						
	Hybrid 128	Selection 50077	Selection 3432	Selection 1075	Albit	Turkey	Ridit
	Resistance factors						
Races	Check	M_2	Ho	H	M	T	rd
D–1	S	S	S	R	R	R	R
D–2	S	S	S	R	R	S	R
D–4	S	S	S	S	S	R	R
D–5	S	S	S	R	S	S	R
D–10	S	S	S	S	R	R	S
D–7	S	S	R	S	R	S	S
D–6	S	S	S	S	S	S	R
D–8	S	S	S	S	S	R	S
D–3	S	S	R	S	S	S	S
D–9	S	S	S	S	S	S	S

R = resistant (0–10 per cent infection
S = susceptible (11–100 per cent infection)

A recently produced wheat cultivar (Moro, C.I. 13740) is very resistant to all races of dwarf and common bunt (Kendrick and Hoffmann, 1963).

LOOSE SMUT OF BARLEY AND WHEAT
(*USTILAGO NUDA* [JENS.] ROSTR. AND *U. TRITICI* [PERS.] ROSTR.)

Loose smuts are flower-infecting or embryo-infecting pathogens. The heads of barley and wheat are destroyed as a result of infection. The spore mass is powdery. The spores are minutely echinulate, 5–9 μ in diameter. They germinate with a promycelium without the production of sporidia (Fig. 18). Fusions take place between the haploid adjacent cells of the promycelium. After this the infection hypha becomes dikaryotic. The promycelium of *Ustilago tritici* is twice as long as the promycelium of *U. nuda* (Popp, 1955).

Methods of artificial inoculation

Flower inculation with spore suspension

The most effective concentration is 0·5–1 g teliospores per litre of water, or 1 pc cent glucose, or extract of flower heads made with boiled water. After screening, this suspension is used as inoculum. Using a *hypodermic needle* the technique of inoculation is as follows: Aristae of the heads are removed so that the heads can be easily handled. A 2·54 cm 25-gauge hypodermic needle is inserted into each floret through the lemma or palea, and one drop of spore suspension is injected into the florets. The most convenient time for inoculation is one or two days after the exertion of the inflorescence. The needle is connected with a rubber bulb containing the spore suspension (Poehlman, 1945). This is a fairly quick method and does not damage the embryos. The percentage of infection is very high, therefore, this method is appropriate for testing resistance of cultivars. 30–40 heads can be inoculated in one hour. Physiologic races can be easily cultured by this method without danger of contamination.

With the *partial vacuum method* of Moore (1936) more than one head can be treated at one time. By this technique one can bring the suspension into direct contact with heads of growing barley or wheat. Heads are placed in a glass chamber; the stems are placed in a slit in a soft rubber plug. A partial vacuum is created with the help of a large pump (250 mm Hg or more). The air is withdrawn from the florets and replaced by the spore suspension after normal air pressure is suddenly restored. By this technique the inoculum will be introduced into each floret. The method was modified by several workers (Oort, 1939; Atkins, 1943). Two persons can infect 180–200 heads per hr. This procedure is not simple but it is very quick. The percentage of infection is not always satisfactory. By creating partial vacuums repeatedly, one can increase the percentage of infection. The optimum period for inoculation lasts only for a few days: 1–2 days before and 4–5 days after flowering. The inoculum to be used should be freshly collected teliospores.

The germination ability of the teliospores is diminished after 2–3 months and will be almost zero after one year. However, they remain viable for more than one year at 0 °C in dry air. Lyophilization is also an appropriate method for prolongation of viability. Spore germination is successful on water, and usually takes place in 14–17 hr. If the germination of spores occurs at low temperature (2–4 °C), cells of the promycelium will be separated from each other. Therefore, it is possible to isolate single cells with a micromanipulator, or by other means, to produce haploid cultures of this fungus (Lange de la Camp, 1936; Thren, 1937).

Inoculation of seedlings with a spore suspension

Wheat or barley seeds are surface-disinfected with 1 per cent sodium hypochlorite and germinated on water agar at 20 °C for about 72–96 hr. At this time the coleoptiles are 0·5–1·0 cm long. Then 2 ml of coleoptile tissue is cut from the tip of the seedling shoot with a razor blade. Seedlings are immersed in a suspension of teliospores containing 5–10 g of spores per l in sterile distilled water plus 0·02 per cent Tween 20. The immersed seedlings are then placed in a vacuum at 500 mm mercury for 2 min and the pressure is suddenly restored. Inoculated seedlings are sown in autoclaved soil in a greenhouse.

The most important advantage of this method is that one can produce symptoms very quickly. In the case of flower infection a 6–12 month cycle, involving 2 generations of the host, is necessary for testing cultivar resistance. On the other hand, spores are produced on the infected host plants in 8 weeks, if we apply the seedling inoculation technique outlined above. Thus, one host is sufficient for this test. An additional advantage is that contamination between different races can be avoided, and furthermore, a single head collection, or different races can be increased without danger of contamination. A disadvantage is that the pathogen infects the host in an unnatural way. This method was worked out by Kavanagh (1961) on the basis of the technique proposed by Rowell and DeVay (1953) for the inoculation of seedling corn with *Ustilago zeae*.

Resistance to loose smut

Different types of host resistance have been described (Oort, 1944; Király and Lelley, 1956; Gaskin and Schafer, 1962; Mumford and Rasmusson, 1963).

1. The pathogen grows into the pericarp but mycelia do not reach the scutellum and the embryo. Emmer wheats (Hope, Hussar) have this type of resistance.
2. Mycelia are present in the scutellum but do not reach the growing point of the embryo. This type of resistance is characteristic of the cultivar Kawvale (Fig. 22).
3. Hypersensitive reaction. The mycelium reach the ear primordia, but the host recover from the infection. This is accompanied by reactions leading to dwarfing or killing of the host (hypersensitivity). This type of resistance occurs only in particular host-parasite pairs (e.g. after infection of the barley cultivar Jet).

Quick methods for detecting loose smut mycelium in embryos

The original method of Skvortzoff (1937) was improved by Simmonds (1946), Ohms and Bever (1955) and others. For separating embryos from kernels, the seed sample to be investigated is placed in a 10 per cent NaOH solution overnight.

The separated embryos are then collected and washed repeatedly with boiling water. The embryos are cleared in boiling lactophenol (1 : 1 : 1 : 5 lactic acid, phenol, glycerine, water) for 2–3 hr. Lactophenol clears the embryos so thoroughly,

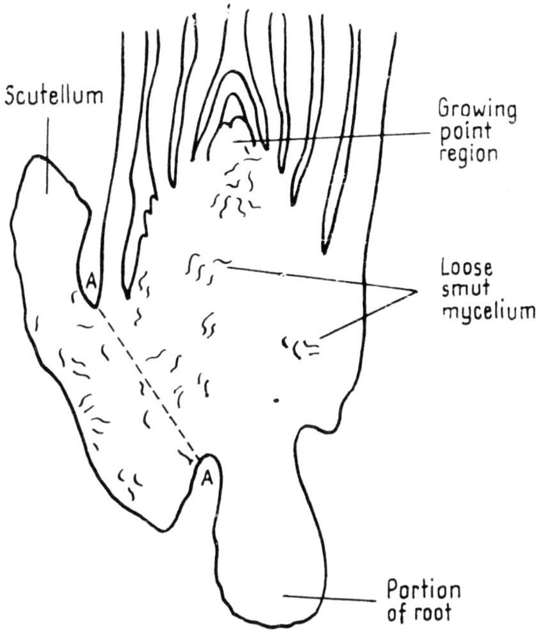

Fig. 22. Longitudinal section of a wheat seedling infected by *Ustilago tritici* 8 days after sowing. The line from A to A indicates the spread of mycelium in the case of embryo-susceptible but field resistant cultivar (After Batts and Jeater, 1958)

that detection of the mycelium under a microscope is carried out very easily. However, it is better to stain the mycelium with cotton blue (50 mg/100 gram lactophenol) by boiling for 3–4 min and then washing out the cotton blue with pure lactophenol. Mycelia will be stained blue. According to another procedure, the embryos are placed in a nearly saturated aqueous solution of chloral hydrate (4·5 parts chloral hydrate to 1 part H_2O) containing 0·01 per cent acid fuchsin. Then embryos are heated in this solution for 5 min until the embryos become clear and the mycelia are stained. Then the embryos are placed in lactophenol for examination under a microscope at low magnification. Cotton blue can also be used as a stain. Morton (1961) uses trypan blue (0·05–0·1 per cent) in 5 per cent NaOH, which separates embryos from the kernels. This effectively stains loose smut mycelium in barley embryos. Popp (1958) described equipment for making cross sections of the embryos.

All these methods are not suitable for detecting mycelia of loose smut in cultivars which are embryo-susceptibe hosts, but are resistant in the field (Batts and Jeater, 1958).

Detection of loose smut mycelia in seedlings

Seedlings of wheat cultivars to be investigated are used for the test in the 3–4–5 leaf stage (14–21 days old). The crown nodes which bear the growing point are excised and the leaves and roots removed. The excised crown nodes are cleared for 15 min in boiled lactophenol or in lactophenol containing 0·001–0·005 per cent trypan blue. Then the growing point is smeared in lactophenol and examined for the presence of loose smut mycelia under a microscope (Ohms and Bever, 1955; Morton, 1961). In some cases there is an important difference between the results of the embryo test and the seedling test. Sometimes the percentage of infection detected by the two methods is identical. This depends on the particular host cultivar-pathogenic race interaction.

Physiologic races of loose smut of barley

Four physiologic races were identified by Tapke (1955) with the following differential set:

Race	Charlotte-town (C. I. 2732)	Compana (C. I. 5438)	Valki (C. I. 5748)	Golden Melon (C. I. 3192)
1	S	R	R	S
2	S	S	S	S
3	S	S	R	S
4	R	R	R	S

S = more than 10 per cent smutted heads
R = 0–10 per cent smutted heads

Moseman and Reid (1958) described a fifth race of barley loose smut. The sources of resistance are in the cultivars Jet and Trebi. Resistance factors of these cultivars were incorporated into several commercial cultivars.

Physiologic races of loose smut of wheat

Bever (1953) reported the occurrence of 19 distinct races. The differential set composed the following cultivars of wheat:

Wabash (C.I. 11384)
American Banner (C.I. 6943)
Purdue No. 1. (C.I. 11380)
Hussar (C.I. 4843)
Early Premium (C.I. 11858)
Nabob (C.I. 8869)
Forward (C.I. 6691)
Trumbull (C.I. 5657)
Kangip (C.I. 13094)
Leap (C.I. 4823)

Physiologic races of *Ustilago tritici* and cultivar reactions of the differential set
(After Bever, 1953)

Physiologic races	Wabash	Amer. Banner	Purdue No. 1	Hussar	E. Premium	Nabob	Forward	Trumbull	Kangip	Leap
1	—	R	R	R	R	R	R	R	R	R
2	—	—	—	—	—	R	R	R	R	R
3	—	—	—	—	—	—	R	R	R	R
4	—	—	—	—	—	—	—	R	R	R
5	—	—	—	—	—	—	—	—	R	R
6	—	—	—	—	—	—	—	—	—	—
7	—	—	—′	—	—	—	—	—	R	—
8	—	R	—	—	R	R	—	—	R	R
9	—	R	—	R	R	R	R	—	R	R
10	—	R	R	R	R	R	R	R	—	R
11	R	—	R	—	R	R	R	R	R	R
12	R	—	R	R	R	R	R	R	R	—
13	—	R	—	R	—	R	—	—	—	—
14	—	R	—	—	—	R	—	—	—	—
15	—	—	R	R	R	R	—	R	R	—
16	—	—	—	R	—	R	—	—	R	—
17	—	—	R	R	—	R	—	—	R	—
18	—	—	—	—	—	—	—	R	R	—
19	—	R	R	R	R	R	—	R	R	R

R = resistant (0–10 per cent infection "—" = susceptible (11–100 per cent infection)

The source of loose smut resistance is as follows: Hope and Hope derivatives, Preston, *Triticum timopheevi* and crossings of *T. timopheevi* × *T. aestivum*. Resistant winter wheats are Kawvale, Leap, Currell, Thorne, Trumbull, Grüne Dame, and Jubile.

The control of loose smut is carried out by the hot water treatment of seeds or, recently, with an anaerobic treatment, which kills the fungus selectively.

BLACK LOOSE SMUT
(*USTILAGO NIGRA* TAPKE)

Symptoms caused by this pathogen are similar to those of loose smut, however the type of spore germination is not the same: sporidia are formed on the promycelium. Teliospores are 6·5–7·0 μ in diameter.

Artificial inoculation is the same as that used with *Ustilago hordei*.

Identification of *physiologic races* is carried out with the same differential set of barley cultivars which is used for the classification of races of covered smut (*U. hordei*). Tapke (1951) differentiated 13 physiologic races of black loose smut, as is shown in the next Table (R = resistant, S = susceptible).

Physiologic races of *Ustilago nigra* and cultivar reactions of the differential set (After Tapke, 1951)

Race	Excelsior	Hannchen	Himalaya	Lion	Nepal	Odessa	Pannier	Trebi
1	R	R	R	R	R	S	R	R
2	R	S	R	R	R	S	R	S
3	R	R	R	S	R	S	R	(S)
4	R	S	R	S	R	S	R	S
5	R	S	S	S	S	S	R	S
6	S	S	S	S	S	S	R	S
7	R	S	S	R	S	S	R	S
8	(S)	R	R	S	S	S	R	S
9	R	R	R	R	S	S	R	R
10	R	S	R	S	R	S	R	R
11	R	S	R	S	R	S	S	(S)
12	S	S	R	S	S	S	S	S
13	S	S	S	S	S	S	S	S

COVERED SMUT OF BARLEY
(*USTILAGO HORDEI* [PERS.] LAGERN.)

Four sporidia are produced on the promycelium (Fig. 18). Teliospores are smooth, 5–9 μ in diameter.

Methods of artificial inoculation

Seeds are powdered with teliospores, and these surface-contaminated seeds are sown in the field. Infection is more successful if the pales are removed from the seeds mechanically, but germination is usually reduced as a result of this procedure.

Another procedure, called the spore-suspension method, is applied to grain in the husk. The technique is as follows: A spore suspension is made by shaking 1 g of teliospores in 1 litre water. Seeds are shaken in this and then allowed to soak for 15 min. The spore suspension is decanted and the seeds are placed on wet filter paper, covered, and incubated for 24 hr at 20 °C. Then the seeds are dried and are sown in soil at a temperature of about 15 °C. This method is very successful in the case of covered smut (Tapke, 1935) and *U. nigra* (Tapke, 1937).

Physiologic races

The following cultivars of barley were used by Tapke (1945). They compose a differential set for the determination of physiologic races of covered smut:

Excelsior (C.I. 1248)
Himalaya (C.I. 1312)
Hannchen (C.I. 531)
Lion (C.I. 923)
Nepal (C.I. 595)
Odessa (C.I. 934)
Pannier (C.I. 1330)
Trebi (C.I. 936)

Physiologic races of *Ustilago hordei* and cultivar reactions of the differential set (After Tapke, 1945)

Race	Excelsior	Himalaya	Hannchen	Lion	Nepal	Odessa	Pannier	Trebi
1	R	R	I	R	R	S	R	I
2	R	R	R	I	S	S	R	R
3	I	R	R	I	S	S	R	R
4	R	I	I	R	S	S	I	I
5	R	R	R	I	R	S	R	I
6	R	R	I–S	I	R	S	R	S
7	R	R	R	R	S	S	R	I
8	R	R	R	R	R	S	R	R
9	I	R	I	I	S	S	R	R
10	R	I	I	I	S	S	I	I
11	R	I	I–S	I	I	S	R	S
12	R	I	I	R	S	S	R	R
13	I	R	I	I	S	S	R	I

R = resistant; I = intermediate; S = susceptible

SMUTS OF OATS
(*USTILAGO AVENAE* [PERS.] ROSTR. AND *USTILAGO KOLLERI* WILLE)

Flowers of the infected plants are destroyed by the mass of teliospores. On the promycelium, which is produced after germination of the teliospore, four sporidia cut off from the apex and the septa (Fig. 18). At very low temperature (about 0–5 °C) sporidia are not produced but cells of the promycelium are separated singly or in pairs. At high temperature (30–35 °C) haploid hyphae grow out from the promycelia, and again, sporidial production is absent. Teliospores remain viable for a long time in a cool and dry environment.

Natural infection takes place at seed germination from teliospores which have drifted between the pales at anthesis. These spores produce resting mycelium on the inner surface of the pales and on the pericarp. Another possibility for infection occurs by means of teliospores, lying within the pales, which germinate at the time of the sowing of seeds. The life cycle of *U. avenae* is seen in Fig. 23.

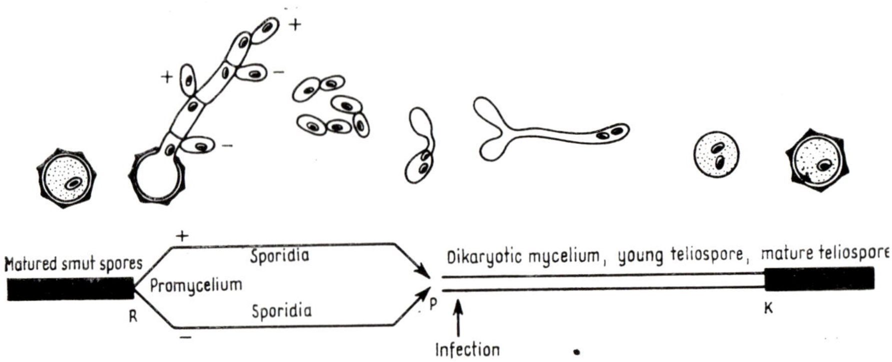

Fig. 23. Life cycle of smut fungi as exemplified by *Ustilago avenae* (P = plasmogamy, K = karyogamy, R = reduction) (After Roemer et al., 1938)

Methods of artificial inoculation

1. Dusting grain with dry spores

First of all it is necessary to remove the pales before dusting the grain with teliospores. This can be done by placing the seed sample in absolute alcohol for 15 min, then in water for 15 min. Remove the pales with a scalpel. The dry seeds are powdered with sieved teliospores, 100 mg per 100 seeds. The contaminated seeds are sown in sand with a moisture content of 20 per cent saturation. The temperature must be maintained at 20–22 °C during the first three days of seed germination. Then the seedlings are transplanted to the field.

2. Partial-vacuum method

100 grains are placed in 10 ml of a spore suspension and evacuated in a desiccator for 20 min. Upon restoring normal pressure, the spores will be forced between the pales and the pericarp. This procedure can be repeated several times. Then the sample is allowed to dry for 12 hr to avoid seed germination. Before sowing, the inoculated seeds are stored at low temperature (2 °C) for 24 hr in a moist chamber to initiate spore germination and the production of resting mycelium (Zade, 1928; Haarring, 1930).

Physiologic races

As a result of the investigations of Holton and Rodenhiser (1948) and Hansing (1954) and others one can distinguish 20 races with distinct pathogenicity. The classification of physiologic races of *Ustilago avenae* was further developed by Luke (1964) and Luke et al. (1961). The next Tables show the names of cultivars used in the differential set for identification of races and reactions of cultivars to different races of oat loose smut as well as covered smut.

Physiologic races of *Ustilago avenae* and cultivar reactions of the differential set (After Holton and Rodenhiser, 1948, and Hansing, 1954)

Race	Anthony	Black Diamond	Victory	Goth-land	Monarch	Fulghum	Black Mesdag	Camas	Nicol	Lelina
A1	S	S	S	R	R	R	R	R	R	R
A2	S	S	S	R	R	R	R	S	R	R
A3	S	S	S	I	S	R	R	R	R	R
A4	S	S	S	R	S	R	S	R	R	R
A5	S	S	S	S	R	R	R	R	R	R
A6	S	S	S	S	S	R	R	R	R	R
A7	S	S	S	S	S	R	R	S	I	R
A8	S	S	R	R	R	R	R	R	R	R
A9	S	S	R	R	R	S	R	R	R	R
A10	S	R	S	S	R	R	R	R	R	R
A11	S	R	S	S	R	R	R	R	S	R
A12	R	S	R	I	R	S	R	R	R	R
A13	R	R	R	R	S	R	R	R	R	R
A14	S	S	I	R	R	S	R	R	R	S
A15	R	S	R	S	S	S	R	R	R	S
A16	S	S	S	S	R	S	R	R	R	R
A17	R	S	R	S	S	S	R	R	R	R
A18	R	S	R	S	R	S	R	R	R	R
A19	S	S	S	S	S	R	R	R	I	R
A20	S	S	S	S	S	R	R	R	R	S

R = resistant; I = intermediate; S = susceptible

Physiologic races of *Ustilago kolleri* and cultivar reactions of the differential set (After Holton and Rodenhiser, 1948, and Hansing, 1954)

Race	Anthony	Black Diamond	Victory	Goth-land	Monarch	Fulghum	Black Mesdag	Camas	Nicol	Lelina
K1	S	S	S	R	R	R	R	R	R	R
K2	S	S	S	R	S	R	R	R	R	R
K3	S	S	S	S	R	R	R	R	R	R
K4	S	S	R	R	S	S	S	R	R	R
K5	S	S	S	R	S	R	S	R	R	R
K6	S	S	R	R	R	R	R	R	R	R
K7	S	S	S	S	R	R	R	R	R	S
K8	S	S	S	R	R	S	R	R	R	R

R = resistant: S = susceptible

CORN SMUT
(*USTILAGO ZEAE* [BECKM.] UNG.)

Tumorous galls are produced by this smut on infected corn tissues. The teliospore mass of the pathogen fills the galls. Germination of the teliospore is by formation of a promycelium. On the four-celled promycelium, terminal and lateral sporidia are borne (Fig. 18). Spores are 8–11 μ in diameter. They are very resistant to the influence of environmental conditions. This pathogen induces localized infection of the host. It is a characteristic feature of this smut that haploid mycelium from the haploid sporidium is also able to infect host tissues but induces only a very weak disease. The dikaryotic phase is produced by the fusion of haploid hyphae of different sexes. This dikaryotic condition is necessary for the initiation of the real parasitic stage. Rarely, some monosporidial lines may also be pathogenic ("solopathogenic lines"). These lines most probably develop from diploid sporidia. When meiosis is hindered for some reason in the diploid teliospores, diploid promycelia and sporidia will be produced. Thus, solopathogenic lines do not represent homothallism in this fungus. This is only an exceptional diploid phase of the pathogen.

Methods of artificial inoculation

Teliospores do not germinate in water, only on media rich in nutrients. One per cent potato-glucose-agar is the most appropriate medium for spore germination. To maintain the stock culture for a longer time, subcultures have to be prepared every four weeks.

1. Rowell and DeVay's (1954) method for seedling inoculation

Corn seedlings with 1 cm coleoptile are injected with a suspension of teliospores. This causes localized infections (tumorous galls) on the shoot.

In another procedure, the tip of the coleoptile is removed and the decapitated seedlings are infiltrated with a spore suspension under vacuum. In both cases susceptibility and resistance can be determined in 1–3 weeks, in the greenhouse or in the laboratory. However, seedling resistance or susceptibility may change at the adult plant stage in the field. Thus, results in the seedling stage or in adult plant stage may be quite different.

2. Hypodermic injection with sporidial suspension

Monosporidial lines of corn smut are grown in a liquid medium composed of 2 per cent glucose and 1 per cent malt syrup. The different monosporidial cultures are allowed to grow in the medium for about 2–3 weeks. Then the cultures are strained through cheese-cloth to remove solid materials. The lines to be tested are mixed just before inoculation and then injected into tissues with a hypodermic needle. Inoculations have to be made as near to the growing point as possible (Tisdale and Johnston, 1926; Stakman and Christensen, 1927). At 27–32 °C 3–4 weeks are necessary in the greenhouse, for smut galls to develop on the infected plants. When young one-week-old plants are inoculated, they sometimes, exhibit susceptibility, though these cultivars or strains of corn may be resistant in the field (in the adult plant stage of development).

3. The drop method of Middendorf (1958)

This is the most suitable procedure for testing corn smut resistance in the field. A sporidial suspension is dropped in the rolled top leaf tube on three subsequent days when the young plants are in the 2–5 leaf stage (about 10–18 days after sowing). Inoculum is composed of suspensions of different sporidial lines which represent several mating types.

Physiologic races

The growth of different strains or lines of corn smut on artificial media takes place quite differently. Many races or biotypes have been distinguished on the basis of their characteristics. However, these are not pathogenic races, therefore they do not have any importance for plant pathology or plant breeding.

Resistance

The basic principles of breeding for resistance were already mentioned in a previous chapter (see pages 313–314). Several inbred lines of corn as well as some hybrids are more susceptible to smut than the old commercial cultivars.

HEAD SMUT
(*SPHACELOTHECA REILIANA* [KÜHN] CLINT.)

This fungus infects sorghum, sudangrass and corn. The pathogen, isolated from sorghum, is able to infect sorghum and some cultivars of corn too. However, head smut isolated from corn can only infect corn. On the basis of this finding Al-Sohaily et al. (1963) proposed the name *Sphacelotheca reiliana* (Kühn) Clint. var. *reiliana* for the fungus which infects sorghum and *S. reiliana* (Kühn) Clint. var. *zeae* Pass. for the one on corn. This is a specialization at the generic host level. Further specialization within each variety may lead us to designate them as races.

Methods of artificial inoculation are identical with those of *Ustilago zeae*. Compatible sporidial cultures are mixed and serve for preparation of inoculum. The inoculum is introduced into the tissues by injection.

Physiologic races

Five physiologic races can be identified with the aid of seven differential cultivars of sorghum (Al-Sohaily et al., 1963; Mehta et al., 1967). Individual teliospores are heterozygous for pathogenicity, in other words, one can isolate compatible sets of monosporidial lines from individual teliospores with different pathogenicity. Therefore, all the compatible combinations of sporidial lines from individual spores and different spores of our smut collection have to be tested in the procedure for race determination.

"Physiologic races" of *Sphacelotheca reiliana* and reactions of compatible monosporidial combinations on the differential host cultivars of sorghum (After Mehta et al., 1967)

Differential set of cultivars	"Physiologic races" (compatible monosporidial combinations with different pathogenicity)				
	1	2	3	4	5
Lahoma	R	R	R	R	R
Feterita (Tx 09)	R	R	R	R	R
Hi-hegari	S	R	R	R	R
Sourless	S	S	R	R	R
Caprock	S	S	S	R	R
Sumac	S	S	S	S	R
Combine Kafir (B407)	S	S	S	S	S

R = resistant; S = susceptible

Sources of resistance

Different selections from Feterita and derivatives of Feterita are all resistant to head smut. Additional resistance is in Hegari (P.I. 34911), White Kafir, Wheatland, Dwarf Kaoling, Dwarf Yellow Milo, Ryer Milo, Dwarf White Durra, Tx 414, and Lahoma (Mehta et al., 1967).

CHAPTER V

DEUTEROMYCETES (FUNGI IMPERFECTI)

ANTHRACNOSE OF CUCURBITS (*COLLETOTRICHUM ORBICULARE* [BERK. ET MONT.] ARX [= *COLLETOTRICHUM LAGENARIUM* (PASS.) ELL. ET HALST.])

Taxonomic position of the fungus

Conidia, hyaline, one-celled, 13–19 × 4–6 μ, are formed in acervuli with 90–120 μ long, brown, tapering setae or bristles (Fig. 24). The conidial form of the fungus belongs to the *Melanconiales* order of the *Deuteromycetes* class.

In recent years, although it has not become generally well known, the name and taxonomic position of the fungus causing anthracnose of cucurbits has undergone certain changes. Arx and Velden (1961) established that the description of the pathogen published under the name of *Colletotrichum lagenarium* was invalid as it had already been published as *Gloeosporium orbiculare* Berk. et Mont. Therefore on the basis of priority, its rightful and valid name is *Colletotrichum orbiculare* (Berk. et Mont.) Arx. Further investigations have clarified that the

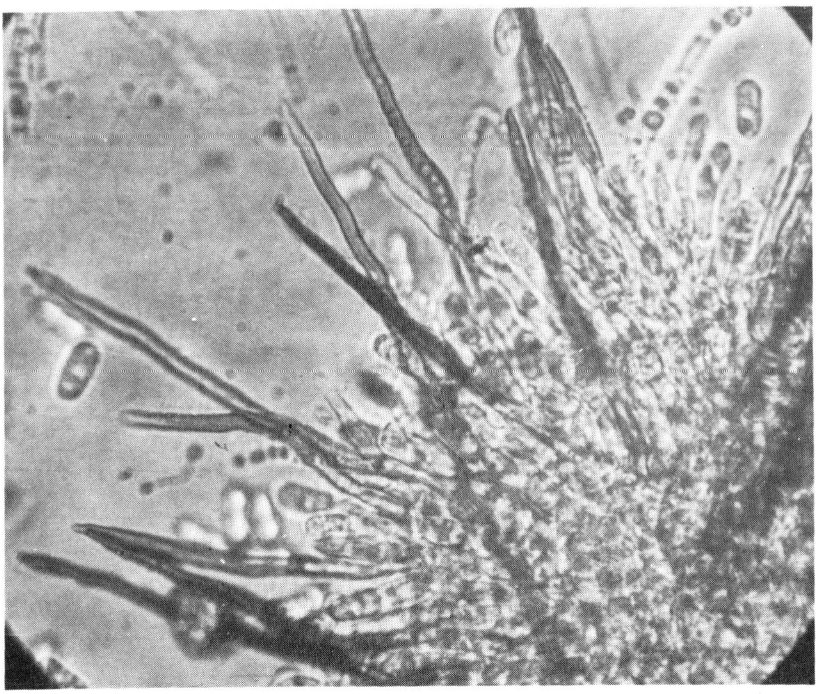

Fig. 24. Acervulus of *Colletotrichum orbiculare*

Colletotrichum species originating from watermelon is identical in every respect with the asexual conidial form of *Glomerella cingulata* (Stonem.) Spauld. et Schrenk in the *Ascomycetes*. Every strain of *Glomerella cingulata* and *Colletotrichum orbiculare* investigated have proven uniformly pathogenic on the following host plants: cucumber, watermelon, apple, tomato and orange. Then according to Arx's investigations, the legitimate name for this fungus is *Colletotrichum orbiculare* and is the imperfect form of *Glomerella cingulata*.

Life cycle

The epidemic spread of the pathogen is insured by the asexual reproductive cells or conidia. Splashing raindrops carry the conidia to the surface of the host. Here the conidia germinate (optimal temperature 22–27 °C) and the fungus enters the host by means of appressoria. Infective hyphae penetrate the host through appressoria. The thallus formed here consists mainly of intracellular hyphae. Towards the end of this disease cycle, acervuli are formed on the necrotic sunken spots induced by the fungus in the host. The overwintering formation has not been detected in nature. It is likely that the fungus is maintained in infected plant residues, but it is also possible that the fungus survives to the next season by adhering to the surface of seeds of infected plants. The development of the disease requires high humidity (more than 87–93 per cent), a temperature of not below 15–18 °C and rainy, windy weather is necessary for epidemic spread.

Symptoms of anthracnose

The pathogen can cause death or damping off of seedlings already in or on the surface of the soil. Typical anthracnose symptoms in the form of sunken, dark spots can develop everywhere on the plant parts (leaf, stalk, fruit) above the ground (Fig. 25). The symptoms on the leaves are interveinal, light-green, water-soaked spots, and when dry, they are reddish-brown. The spots on shoots are darker. Acervuli (Fig. 24) are formed in large numbers on the surface of deeply sunken and often confluent spots of the fruit under wet conditions. For this reason their surface has a yellow-orange colour due to the masses of conidia.

Artificial inoculation methods

Artificial inoculation can be accomplished by spraying susceptible plants with a suspension of conidia and keeping the plants in the moist chamber until infection has taken place. But in investigations to determine plant resistance, a knowledge of biotypes and a standard inoculation method is necessary in order that the results obtained in various places and times may be comparable. The method of Littrell and Epps (1965) serves this purpose.

Fig. 25. Disease symptoms on musk melon seedlings caused by *Colletotrichum orbiculare*

1. Preparation of the inoculum for inoculations.

a) Inoculate sterilized, 2·5–4 cm long, green bean-pod pieces with the mycelia originating from young, vigorous agar cultures of a single physiologic race.

b) Incubate the pods in the dark at 26 °C for 5–7 days. During this time large amounts of conidia are formed on the surface of the pod pieces.

c) Macerate the green bean pods in a Waring Blendor in sterile tap water.

d) Filter the homogenate through cheese cloth.

e) Dilute the conidial suspension obtained with sterile tap water so that the adjusted suspension contains 75,000 conidia per ml.

f) This suspension—if it is not used immediately—can be preserved at 5 °C for 72 hr without decreasing the viability of the conidia.

2. Inoculation of the plants.

a) Spray the selected plants or fruits with the suspension.

b) Keep the plants in a dew chamber at 18–24 °C for 16 hr.

c) Incubate the plants at 23–29 °C until the appearance of symptoms.

Artificial inoculations in the field, usually conducted during the evening hours, will be successful if temperatures do no decline below 15–18 °C and the relative humidity remains high (Littrell and Epps, 1965).

Biological specialization of the pathogen

Goode (1958) differentiated 3 physiologic races of the pathogen in the course of his investigations of the strains (monospore cultures) isolated from different sources and hosts. The differential host plants and the reactions on them caused by the 3 races are summarized according to Goode (1958) in the following Tables:

Relative susceptibility of seedlings of various watermelon and cucumber cultivars to races 1, 2, and 3.

Differential host plant	Race 1	Race 2	Race 3
Watermelon			
"Charleston Gray"	2·1	5·0	1·1
"Congo"	2·3	5·0	1·1
"Fairfax"	2·3	5·0	1·1
"Garrison"	5·0	5·0	5·0
"New Hampshire Midget"	5·0	5·0	5·0
Cucumber			
"Model"	5·0	5·0	4·9
"Palmetto"	4·9	4·9	4·9
Squash			
"Butternut"	2·5	4·0	1·0

1 = immune or highly resistan
2 = resistant
3 = moderately resistant
4 = susceptible
5 = highly susceptible

The reaction of fruits of watermelon and cucumber cultivars to races 1, 2, and 3

Differential host plant	Degree of ripeness	Race 1	Race 2	Race 3
Watermelon				
"Charleston Gray"	not ripe	+	+	—
	ripe	—	+	—
"Congo"	not ripe	+	+	—
	ripe	—	+	—
"Fairfax"	not ripe	+	+	—
	ripe	—	+	—
"Garrison"	not ripe	+	+	+
	ripe	+	+	+
Cucumber				
"Model"	not ripe	+	+	+
	ripe	+	+	+
"Palmetto"	not ripe	+	+	+
	ripe	+	+	+

+ = serious or moderately serious symptoms
— = immune

The race differentiation method outlined in the above Tables was improved by Dutta and co-workers (1960 a) using fewer cultivars. From among a large number of host species and cultivars surveyed, the "Chriss Cross" watermelon cultivar was added to the differential hosts, to distinguish a fourth race. They further established that none of the watermelon cultivars showed much resistance to race 2. The following Table serves to separate races 1, 2, 3 and 4, using 5 differential host plants (Dutta et al., 1960 a).

The reaction of seedlings of various differential hosts to races 1,2, 3 and 4

Differential host	Race 1	Race 2	Race 3	Race 4
Watermelon				
"Black Diamond"	5·00	5·00	5·00	5·00
"Charleston Gray"	1·10	4·90	1·24	1·30
"Chris Cross"	3·71	4·61	2·98	3·87
Cucumber				
"Model"	4·96	2·81	4·71	3·02
"PI 163213"	4·04	1·17	1·76	1·40

0–2·5 = resistant
2·6–3·4 = moderately resistant
3·5–5·0 = susceptible

The nature of resistance to anthracnose and resistant cultivars

During the course of the investigations on 135 cucumber lines (Busch and Walker, 1958) resistance was demonstrated in 3 cases. In agreement with earlier made observations they found that the inheritance of resistance was directed by a single gene. The histological and cytological observations in connection with the host-parasite relationship are especially worthy of note. It turns out that the pathogen penetrated both susceptible and resistant hosts. In the latter case, however, the growth of the hypha is much slower. The cell wall of the resistant host as a consequence of infection thickens and a dark red sedimentation can be seen in the intercellular spaces.

We can conclude from this observation that resistance to anthracnose is dependent on the reaction of the host tissues. In experiments with biochemical mutants of the pathogen Dutta and co-workers (1960 b) were able to discover resistance factors of another nature. With ultra-violet irradiation of race 2, they could prepare mutants requiring different amino acids and pyridoxine. The mutants requiring amino acids (leucine, isoleucine, serine, lysine, histidine, proline, alanine, valine and inositol) or pyrimidine were avirulent on certain muskmelon and watermelon cultivars. Most avirulent strains requiring amino acids regained their pathogenicity if the amino acid in question reached the leaf surface in the form of a water solution. For those requiring pyrimidine, such a regeneration of pathogenicity could not be achieved. The missing amino acids could be substituted with cucumber and watermelon epidermal extracts in the case of some mutants but the entire tissue extracts were more effective in this regard. In the case of mutants requiring pyrimidine, infectivity could not be replaced by even entire tissue extracts (see also page 295 on less–mutans). It may be supposed that in the epidermal cells of resistant plants there are specific inhibitors acting against some races.

This theory seems to have been confirmed by the results of Akai and co-workers (1959). Also, according to them, the reaction responsible for resistance takes place in the epidermis. If the pathogen penetrates deeper than the epidermis, the fungus spreads equally well in the tissues of susceptible and resistant cultivars. However, at the same time, the cell contents collected from the epidermis of resistant cultivars did not inhibit the germination of the pathogen or the formation of appressoria. This indicates resistance is a function of the intact epidermis.

Hadwiger and Hall's (1961) work is outstanding in connection with the chemical nature of resistance to the anthracnose fungus. They observed that in field conditions the anthracnose spots occurred at first on the dark green stripes of ripe "Garrison" watermelon fruits. On the lighter part of the striped fruit, the pathogen did not occur or only to a slight extent. It turned out that the dark green tissues contained greater amounts of citrullin and glutamine and smaller amounts of alanine and aspartic acid than the lighter green ones. They supposed that the higher citrulline and glutamine content of the dark green tissue better supported the growth of the pathogen and in turn, caused greater damage.

Concerning resistance to anthracnose a rather large number of watermelon, cucumber, pumpkin etc. cultivars and lines already have been examined (Akai, et al., 1959; Winstead et al., 1959; Dutta et al., 1960 a). From these investigations, it was unanimously established that the cultivars resistant to race 1 and 3 were relatively abundant. Winstead and co-workers (1959) found simply that every cultivar resistant to race 1 was also resistant to race 3. They demonstrated that the resistance to race 1 and 3 was directed by the same dominant gene. According to them the results of investigations completed up to then, not even one watermelon cultivar had shown resistance to race 2. However Winstead and his colleagues (1959) found that the progenies from the W 695 line of the only "Citron" type watermelon originating from Africa to be resistant to race 2.

Control of anthracnose

Apart from the elimination of the infected plant residues and crop rotation, the most important measure in preventing the disease is the use of pathogen-free seed. The seeds of uncertain origin must be treated with organic mercuric chemicals to eliminate the fungus. Most cultivated *Cucurbitaceae* are copper sensitive. We may prevent the spread of the disease by the application of organic fungicides such as the dithiocarbamates, dinitrorodanebenzene, etc. in the course of cultivation—when climatic conditions make this possible.

Where the pathogenic races 1 and 3 are widely distributed, it is best to cultivate resistant or tolerant watermelon cultivars.

VERTICILLIUM WILT
(*VERTICILLIUM ALBO-ATRUM* REINKE ET BERTH., *V. DAHLIAE* KLEB. ETC.)

Taxonomic position of the fungus

The *Verticillium* species causing wilt has no typical means of sexual reproduction. Their slender verticillate-branched, hyaline conidiophores and one-celled, hyaline conidia are formed freely on the surface of the substrate (Fig. 26). For this reason these species belong to the *Moniliales* (*Hyphomycetes*) order of the *Deuteromycetes* (*Fungi imperfecti*) class.

Life cycle

The life cycle of the *Verticillium* species–if it is at all possible to refer to a life cycle when there is no sexual reproduction–is very simple. New asexual reproductive cells and resting structures arise on the thallus. These structures are microsclerotia

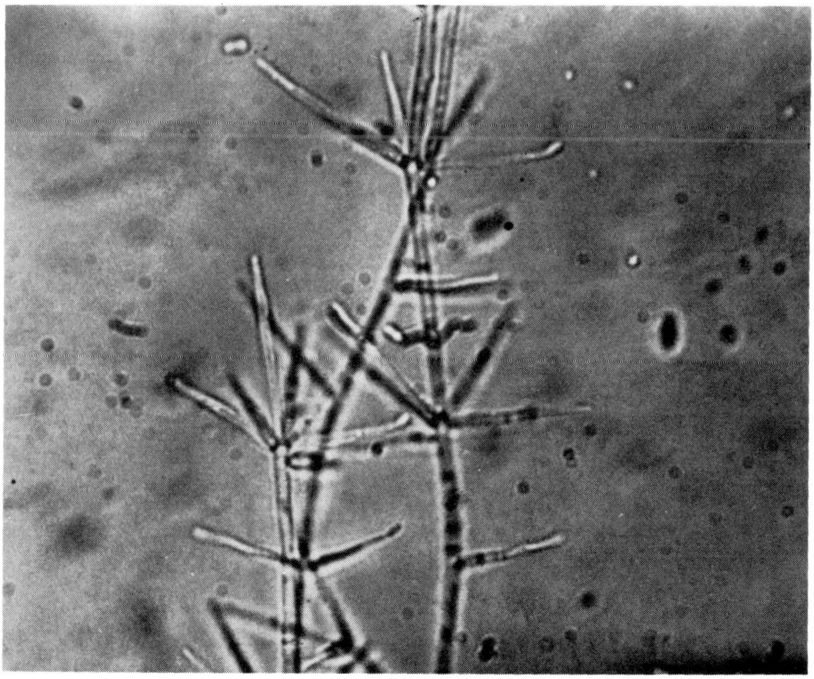

Fig. 26. Conidiophores of *Verticillium*

(*V. dahliae*), chlamydospores (*V. tricorpus*, *V. nigrescens*) and dark thick-walled, torulose hyphae (*V. albo-atrum*). Among the factors influencing germination and growth of these cells and structures, temperature is of great importance and will be discussed in detail later in the differentiation of certain species. Malca and co-workers (1966) established in *in vitro* experiments, that the optimum growth of *V. albo-atrum* is between pH 5·9–6·3 and the optimum temperature is between 24–27 °C. Also they established that although the pathogen utilized all of the carbon-sources tested, the best growth was on *L*-arabinose and on maltose. Among the nitrogen sources ultilized were *L*-arginine, *L*-alanine, *L*-asparagine, *L*-glutamine, *DL*-amino-*n*-butyric acid and *L*-serine that stimulated its growth much better than the inorganic N-compounds, glycine, *L*-histidine or *L*-leucine.

The survival and saprophytic growth of the fungus in the soil is a very important problem, especially from an epidemiologic point of view. Menzites and Griebel (1967) noticed an increase followed by a slow decrease of the fungus population in soil artificially inoculated with the microsclerotia of *V. dahliae*. The population rise can be explained by the fact that conidia are formed from microsclerotia. The extent of conidial formation was not influenced by the addition of organic materials to the soil, indicating that this process goes on with the help of reserve materials of microsclerotia. They also proved that the tolerance to heat as well as the tolerance to dessication is markedly decreased in microsclerotia that have already germinated once. In spite of this, the sclerotia can form a few conidia under suitable conditions even after a long time of unfavourable conditions too, which is enough to initiate infection.

The *Verticillium* species, especially *V. albo-atrum*, are extremely variable organisms. The formation of new strains in respect to morphology, physiology or virulence is the result of the parasexual process (processes), because recombinations cannot be formed sexually in *Verticillium* species. A frequent phenomenon is the appearance of white variants in black microsclerotial stock cultures during periods of artificial cultivation (Brandt, 1965). More new white variants arise at the point of contact between the two types when cultivated simultaneously than could be explained as a result of mutations or by parasexuality. Hastie (1962, 1967) intensively studied the nature of the recombination in *Verticillium albo-atrum*. He established that "mitotic crossovers" are extraordinarily common in the conidiophores. This phenomenon thus accounts for the high degree of variability encountered between the two types. The separation of diploid strains is not rare during the parasexual cycle. The conidia of the diploid *Verticillium* are larger in size than haploid conidia.

Symptoms of the Verticillium wilt and the role of toxins

The pathogen originates from the soil, infects the plants through the roots and invades the xylem which then causes systematic symptoms, yellowing of the leaf and wilting. Browning of the vascular bundles is observable in the roots and in the lower stem crown, but the external appearance of the roots and stem is healthy. The pathogen penetrates the potato through the stolons and causes browning of the vascular bundles here too without causing external symptoms. The pathogen can infect plants belonging to about 70 genera but is the most serious on potato, tomato, cotton, apricot and some ornamental plants. The pathogen does not penetrate very deeply inside the plant from the point of infection. The wilt symptoms, for example, in the apricot tree are induced by the toxins produced in the roots (Berend, 1958). But at the same time in cotton, Garber and Houston (1967) could demonstrate the presence of conidia of *Verticillium albo-atrum* throughout the artificially root-inoculated plants. It is not clear how conidia get into the sap stream of the leaves. Graber and Houston, in any case, suppose that the appearance of the leaf symptoms is connected with the germination of conidia *in situ*.

The wilt symptoms are induced by the toxins produced by *Verticillium* species. This has been confirmed by many researches to show that typical symptoms may be obtained on plants placed in the pathogen-free filtrates of cultures (Zatykó and Garay, 1959).

A detailed analysis of the *Verticillium* toxin is reported by Stoddart and Carr (1966). They separated the culture filtrate of *Verticillium albo-atrum* cultivated on a synthetic nutrient solution into a light and heavy molecular-weight fraction using a "Sephadex G-25" column. The law molecular weight fraction although causing leaf necrosis and yellowing (on alfalfa), has no essential role in producing the typical symptoms. The heavy molecular weight fraction has moderate cellulolytic activity and has proven to be a protein-fructosan mixture. According to Stoddart and Carr the thermostable carbohydrate components of the heavy molecular fraction causes symptoms more quickly than the protein moicty, and is responsible for the wilting of leaves, whereas the water-insoluble protein component gives rise to lesions in the xylem of the stalk. Heat treatments, causing protein inactivation, also caused the wilt-inducing effect to cease.

Patil and Dimond (1967b) pointed to the role of the polygalacturonase enzyme produced by *V. albo-atrum* in causing wilt. They established that glucose was necessary for the *in vitro* growth of the fungus and 2-amino-*D*-glucose is required for enzyme synthesis. With the addition of glucose, saccharose, or ribitol to the medium, a 2 to 4 fold rise in the level of the enzyme is reached, but 2-amino-*D*-glucose induced a 20 times increase. A single C-source insures only a weak-inducing substance and a low enzyme level. Monogalacturonic acid decreased polygalacturonase synthesis. The basic work of Patil and Dimond made it possible to clarify the part played by the polygalacturonase enzyme in the wilt. The authors established in further investigations that the polygalacturonase of *V. albo-atrum* was inhibited by potato phenoloxidase, chlorogenic acid or caffeic acid (Patil and Dimond 1967c). The inhibitory effect of the oxidative product of chlorogenic acid could not be compensated by cysteine or glutathione.

MacMillan and Brandt (1966) concluded on the basis of their detection of the very weak polyphenoloxidase activity of *Verticillium albo-atrum* that the melanin production of this fungus is connected with the peroxidase process.

Data concerned with the aetiology of *Verticillium* wilt, the role of the factors influencing infection, and host specificity of the pathogen are extremely confusing and contradictory in many cases. This is explained by the fact that the wilting can be caused not only by one species, but at least 4 different *Verticillium* species. Their nutrient and temperature requirements and host ranges are often dissimilar. Therefore it becomes especially important to know which pathogenic *Verticillium* species one has under investigation.

According to Litvinov (1964) the following *Verticillium* species can cause wilting of different host plants: *V. albo-atrum*, *V. dahliae*, *V. nigrescens*, *V. nubilum* and *V. tricorpus*. The first two species are the most widespread and important from an economical point of view according to most observers. The species differed little in the shape and size of conidiophores and conidia.

Selective medium for the differentiation of the Verticillium species (*Talboys*, 1960)

Prune extract is made boiling for 30 min 50 g dried prunes, cut into small pieces in 1 l of water and pressing the liquid from the pulp through a gauze and filter. Composition of the medium is 100 ml prune extract: 5 g lactose; 1 g Difco yeast extract; 30 g agar, completed to 1000 ml with distilled water, and the pH is adjusted to 5·8–6·0. After sterilization, the medium is poured into 100 mm Petri dishes (20 ml/dish). On this medium the reverse side of the colony is used to differentiate the species:

V. dahliae: As a result of the black microsclerotia, the colony structure is nodular and granulated.

V. albo-atrum: Dark coloured hyphae form a distinct radial structure. Microsclerotia are not formed.

V. tricorpus: Besides the dark coloured radial hyphae and microsclerotia, chlamydospores are formed too. The reverse side of the colony can, at the same time, be radial as well as granulated in appearance.

Intensive studies by Devaux and Sackston (1966) were concerned with the differentiation of *Verticillium* species causing wilt. They established that the 23 *Verticillium* strains isolated from 10 different horticultural plants, represented 3 species. The *V. albo-atrum* species was characterized by radial, black hyphae and the *V. dahliae* species by microsclerotia, and the third species, *V. nigrescens* by chalmydospores. It is also remarkable that the *V. albo-atrum* could be isolated only from potato and in one case from tomato. The host range of the other two species was varied.

Separation of V. albo-atrum and V. dahliae on the basis of their temperature requirements

Martinson and Englander (1967) worked out a reliable method, based on differences in temperature requirements, for the separation of these two *Verticillium* species, both important from an agricultural point of view. Cultivation of these fungi at 30 °C and 33 °C was used as the differential temperature for this purpose. The diameters of colonies of *V. albo-atrum* on the potato dextrose agar varied between 0–0·5 cm at 30 °C, after 16 days incubation, while the colonies of *V. dahliae* varied between 3–4 cm. The *V. albo-atrum* colonies at 33 °C, after 16 days incubation, did not grow at all or only a trace, while the *V. dahliae* colonies reached 1–2 cm in diameter under the same conditions.

Monospore isolates of *V. albo-atrum* did not grow at 30 °C or produced colony diameters of less than 0·5 cm, while the colonies of *V. dahliae* were at least 1 cm in diameter. At 33 °C for a longer time than 16 days was generally lethal to the conidia and hyphae of *V. albo-atrum* but the vegetative hyphae and asexual reproductive cells of *V. dahliae* were not destroyed at this temperature.

Artificial inoculation methods

a) Seedling-toxin method for the evaluation of resistance

The investigations of Michail and Carr (1966) showed that the degree of wilt of lucerne seedlings from the effect of contact with *Verticillium albo-atrum* culture filtrate agreed with the reaction produced by the pathogen in the field. This seedling method is suitable for a preliminary selection in breeding plants for resistance. The wilt-causing activity of the filtrate generally depended on the composition of the medium and not so much upon the age of the culture. An exception was the Czapek-Dox nutrient solution where the wilt-causing effect was rapidly increased with incubation time.

b) Root infection methods

The simplest and most commonly used method is the mixing of soil with an appropriate density of *Verticillium* spores or mycelial fragments in suspension. The roots of plants grown from seed are in constant contact with the pathogen in the soil and become infected without any prior injury occurring.

The root-dipping method of inoculation can be applied in the field too. The seedlings are immersed into the suspension of propagules just before transplanting them (Wright, 1966). There is some root injury prior to incubation.

Garber and Houston's (1967) procedure for inoculating and infecting potted cotton plants was to remove the plants with the soil-ball intact and spraying the exposed roots on the surface of the soil-ball with the suspension of the pathogen. The injuries to the roots are unavoidable but less in such a procedure than in the root dipping method of inoculation. Artificial inoculation was successful on cotton seedling plants incubated at 28 °C in the daytime and 18 °C at night after inoculation but not on the plants kept after inoculation at 36 °C in daytime and 18 °C at night.

Cultivars resistant to Verticillium wilt

The polyphagous nature of *Verticillium* spp. makes it very difficult in breeding for resistant cultivars. A further difficulty in this problem is that for a long time the various *Verticillium* species considered to be variants of the same species or the species were not recognized and identified as separate species.

Cotton is one of the most seriously damaged crops by *V. albo-atrum*. For this reason it is one of the most studied disease in the field from the point of view of host resistance. The most resistant cultivars have come from selections of the *Gossypium barbadense* types. Other species have less resistance (Dickson, 1956). Garber and Houston (1967) found that the *Gossypium hirsutum* "Acala 4–42" cotton in their experiments was tolerant to *V. albo-atrum*.

Wilhelm and co-workers (1967) obtained in the course of coordinated laboratory and field tests, resistant lines from *Gossypium barbadense* marked B-1, 3, 6, 10, 11, 104 and 106 as well as from cotton lines from California, New Mexico and Alabama. The B-106 line that originated from the Seabrook 12-B-2 cultivar proved to be almost immune. The F_1 generations from the crosses of six susceptible cultivars and B-106 segregated into resistant, moderately resistant and susceptible progenies.

Verticillium resistance is dependent upon one gene in the case of tomato. The moderately resistant cultivars to wilt in California are the Riverside and the Essar cultivars (Walker, 1952).

Busch (1966) conducted artificial inoculation experiments on ten potato cultivars with both "dark mycelial" and "microsclerotial" isolates of *V. albo-atrum*. The former corresponds probably to the *V. albo-atrum* species, the latter to *V. dahliae*. He established that the "dark mycelial" culture was much more pathogenic to the potato than the "microsclerotial" strain, although the degree of their occurrence in nature is roughly the same. Based on yield reductions, the Kennebec, Cobbler and the Cherokee cultivars proved to be the most susceptible ones under artificial inoculation, while the yields of Sebago, Keswick and, in one experiment, Viking, were hardly reduced at all.

Control of Verticillium wilt

Sterilization of the soil with heat or soil disinfectants can only be used economically in the greenhouse and in seed beds. But damage in the field can be greatly reduced by certain cultural measures. The more important ones are crop rotation and using healthy propagating material (e.g. potato seed pieces). The roots of seedlings or saplings for transplanting should be free of injury (e.g. apricot) and planting members of the *Solanaceae* (potato, tomato etc.) should be avoided as interim crops (e.g. in apricot plantations) since they are also susceptible to *Verticillium* wilt. Applications of potassium fertilizers generally decrease the appearance of the disease and lessen the degree of damage to the host. In some cultures (e.g. cotton) the growing of resistant or tolerant cultivars is the best control.

RICE BLAST (BRUSONE) DISEASE (*PYRICULARIA ORYZAE* CAV.)

Taxonomic position of the fungus

The sexual reproduction of *Piricularia oryzae* is not known. The obclavate conidia or asexual reproductive cells are 20–22 × 10–12 μ in size and are three-celled, colourless, hyaline. The conidia are borne terminally on slender, simple, septate conidiophores that arise on the surface of the host plant. The individual conidiophores, 60–120 × 4–5 μ, emerge through stomata in clusters. The fungus therefore belongs to the *Moniliales* order of the *Deuteromycetes* class.

Life cycle

The conidia of the pathogen germinate on the surface of the susceptible host. The infection is primarily the function of the microclimatic conditions. According to Padmanabhan (1965) epidemics may be expected if the lowest temperature is between 20–26 °C at night and the relative humidity above 90 per cent during the susceptible period. Knowing the relation between meteorological data and disease epidemics helps to prevent more extensive damage through timely forecast.

The conidia and conidiophores are formed on the necrotic spots of infected plants. Dense layers are gray in color. In regards to conidial formation, the most important single factor is the relative humidity: under 88 per cent no formation, at 90 per cent sporadically produced, and above 93 per cent asexual cells arise profusely.

The conidia may survive and be virulent for 7 months while the thick-walled chlamydospores (under dry conditions) for 20 months. The pathogen therefore overwinters with the help of these structures and in infected crop residues. The spread of the pathogen by seed-borne inoculum has not been proven (Sokolova, 1964). Thus the apperance of blast in new areas cannot be explained other than by wind dissemination of conidia and the movement of crop residues.

Pyricularia oryzae strains with different infectivity and physiologic characteristics arise by means of asexual recombinations. The presence of parasexuality seems very probable on the basis of Chu and Li's (1965) observations. They established that the cells of *Pyricularia oryzae* hyphae and conidia are multinucleate. The nuclei divide mitotically in the cells and each nucleus contains two large and two small chromosomes. The authors noticed that hyphal anastomoses occurred frequently, which may insure parasexual recombinations.

Symptoms of rice blast

Beginning with the seedling stage, the pathogen may attack rice plants in any stage. The spots that form on the leaves are brown and their centers are gray due to the dense layers of conidiophores. Similar spots arise on the leaf sheaths too. The nodes on the attacked stalks become brown and break over easily. The pathogen may infect any portion of the rice panicle. If the plant is infected in the early stages of its development, the panicles of the infected plant are reduced and grain formation does not occur (the grain blasts).

Artificial inoculation methods

In addition to insuring favorable relative humidity conditions for infection, an important fundamental principle of successful, artificial inoculations is the preparation of a suitable spore suspension. Tseng and co-workers (1965) investigated the extent of conidial formation of 5 different *Pyricularia oryzae* physiologic races on various media. The best sporulation was achieved for every race on the Misata kind of medium (g/l: dissolvable starch 10 g, yeast extract 2 g). For field experiments large amounts of virulent conidia for inoculum can be prepared in fermentors (Minsov and Peruanskii, 1965).

The results of Asai and co-workers (1967) are worthy of note in clarifying the role of factors influencing the success of artificial inoculations. In the field, they investigated the artificial inoculation of susceptible cultivars. They measured daily the occurence of dew, rain, temperature, relative humidity, light intensity, soil temperature and size of the leaf blade. In contrast with earlier investigators, they found that only the extent and duration of dew formation, and the size of the leaf blade significantly influenced the average number of spots per plant.

There are numerous artificial inoculation procedures that are essentially based on the same principles. Two examples may be cited: Südi and Podhradszky's (1959) quick glasshouse method which serves to evaulate the resistance of rice cultivars to *Pyricularia oryzae* and Podhradszky and Südi's (1957) artificial field inoculation method.

Artificial inoculations in the glasshouse

a) Prepare a spore suspension from oatmeal-agar or rice-grain cultures, using 0·25 per cent gelatine and 0·05 per cent sodium oleate solution.

b) Adjust the density of the spore suspension in a Bürker or similar counting chamber to 100,000 conidia/ml.

c) The seedlings of the rice cultivars to be investigated are grown in the greenhouse at 22–28 °C in pots of muddy clay soil placed in water-filled cement tubs or containers.

d) Inoculate the plants by brushing the conidial suspension on the leaves when the second leaf begins to develop, usually 9–12 days after sowing.

e) A saturated atmosphere for 24 hr after inoculation is necessary to secure infection. So the plants are covered with filter paper with the edges in contact with the water in the tub to obtain the required above 93 per cent relative humidity.

f) Evaulate the symptoms after 6–8 days incubation.

Artificial inoculation method in the field

a) Large amcunts of conidia can be prepared by artificially inoculating seedlings of a susceptible rye cultivar (FB 101, FB 104, Coloess, Olli, Blue Nackte, etc.) in the greenhouse.

b) Two weeks after the inoculation of rye plants, wash the conidia from the leaves with a solution consisting of 0·25 per cent gelatin and 0·05 per cent sodium oleate.

c) Adjust the spore suspension to contain 100,000 conidia/ml.

d) Spray rice plots in the field in any stage with 80–150 ml ot the spore suspension/square metre, the amount depending upon the stage of development of the host.

e) Cover the inoculated plots with two-layered canvas tents over reinforced steel frames, with the ends of the canvas in the water flooding the plants.

f) Moistening the tents every 2–3 hr, between 7 A.M. and 7 P.M., will keep the relative humidity high around the plants.

g) Remove the tents after 24 hr and after 6–8 days incubation, evaluate the symptoms.

In determining the resistance of rice cultivars, Südi and Podhradszky (1959) distinguished four main types of symptoms:

Type 1. Pin-point like necrotic spots (highly resistant).

Type 2. Spots of 2 mm in diameter or less with strongly limiting necrotic margins (resistant).

Type 3. Spots larger than 2 mm in size but with locally limiting necrotic margins (susceptible).

Type 4. Large, spreading spots, seemingly limitless and without necrotic margins (highly susceptible).

Soon after the inoculation, the type 1 symptoms appear, but no later than 6–8 days, the type 4 develops in susceptible cultivars. The resistant plants show only type 1 or possibly type 2 symptoms, the susceptible plants only types 3–4.

A differential series of 8 rice cultivars was described by a team of American and Japanense research workers for the separation of 32 *Pyricularia oryzae* races found in the USA and Japan (Atkins et al., 1967). These cultivars are as follows:

Characteristics of *Piricularia oryzae* races on eight differential rice cultivars

Race group	Race No.	Raminad	Zenith	NP 125	Usen	Dular	Kanto 51	Shatiao-tsao (S)	Caloro
IA	1	S	R	R	S	S	S	S	S
	2	S	R	R	S	R	R	S	S
	3	S	R	R	S	R	R	S	R
IB	1	R	S	S	S	S	S	S	S
	2	R	S	R	S	S	S	S	S
	3	R	S	R	S	R	R	R	S
	4	R	S	R	R	S	S	S	S
	5	R	S	R	R	S	R	S	R
	6	R	S	R	R	R	R	S	S
IC	1	R	R	S	S	S	S	S	S
	2	R	R	S	S	S	S	R	S
	3	R	R	S	R	S	S	S	S
	4	R	R	S	R	S	S	R	S
	5	R	R	S	R	R	S	S	S
ID	1	R	R	R	S	S	S	S	S
	2	R	R	R	S	S	S	R	S
	3	R	R	R	S	S	R	S	S
	4	R	R	R	S	S	R	S	R
	5	R	R	R	S	S	R	R	S
	6	R	R	R	S	S	R	R	R
	7	R	R	R	S	R	S	S	S
	8	R	R	R	S	R	R	S	S
	9	R	R	R	S	R	R	S	R
	10	R	R	R	S	R	R	R	S
	11	R	R	R	S	R	R	R	R
IE	1	R	R	R	R	S	S	S	S
	2	R	R	R	R	S	S	R	S
IF	1	R	R	R	R	R	S	S	S
	2	R	R	R	R	R	S	R	S
IG	1	R	R	R	R	R	R	S	S
	2	R	R	R	R	R	R	S	R
IH	1	R	R	R	R	R	R	R	S

R = resistant, S = susceptible

Raminad Str. 3., Zenith, NP 125, Usen, Dular, Kanto 51 Shatiaotsao(S), and Caloro. The races were established on the basis of their behaviour on the differential host plants and could be separated into eight international race groups (IA, IB, IC, ID, IE, IF, IG and IH). The reactions which occurred in the interaction between the host cultivars and the pathogenic races, are summarized in the Table on page 436 (Atkins et al., 1967).

Rice blast resistant cultivars

By artificially inoculating a large number of cultivars and lines in the greenhouse and field, Podhradszky and Südi (1957) and Südi and Podhradszky (1958, 1959) determined that the cultivars Dunghan-Shali and Vross 3716 were the most susceptible. They found that the Vross-213, Ooba, Hokkaido Early and the Precoce Allorio cultivars were moderately susceptible. They established that the following were resistant to blast: Uzrosh-17, Uzrosh-72, Dubovsky-129, Santakhessky, Agostano and Linia-45. The pseudo-resistance of certain cultivars in the field, as for example Hokkaido Early, could be explained on the basis of that they had short vegetative periods. They found that the susceptibility of leaves decreased with their ageing in every cultivar they examined.

According to the observations of Sokolova (1964), the resistance of the Dubovsky-129 cultivar has gradually decreased over the years, and this is probably due to the build-up or the development of new races of *Pyricularia oryzae*.

The International Rice Research Institute (Anonym., 1965) made the statement that a large proportion of the rice cultivars examined since 1962 proved to be more or less resistant to *Pyricularia oryzae*. In the course of artificial inoculation experiments in the greenhouse, 9 cultivars were resistant to all the races found in the Philippine Islands. In this area the "Japonica" types were generally much more resistant than the "Indica" cultivars. This points out the fact that the reaction of a certain cultivar depends primarily on the *Pyricularia oryzae* races present in the area and not on the type of cultivar or its geographical origin. The inheritance of resistance to *Pyricularia oryzae* is dominant.

In the search for resistant sources of rice, Murti and co-workers (1963) have brought attention to the Ninnidhan cultivar. Ninnidhan 6517 and 6522 have proven to be resistant in field experiments. Chang and his colleagues (1965) examined 161 rice cultivars and lines between 1956 and 1964. They established that Chianung-242 was an important source of resistance to *Pyricularia oryzae*.

FUSARIUM WILT
(*FUSARIUM OXYSPORUM* SCHL. EMEND. SNYDER ET HANSEN)

Taxonomic position of the fungus

Sexual reproduction or a perfect stage of *Fusarium oxysporum* is unknown. It reproduces with one-celled microconidia and typical sickle-shaped multicellular macroconidia and also by chlamydospores and sclerotia that are formed freely in culture. Often chlamydospores are formed by the mycelium and by the macroconidia by laying down thick walls. Blue-black sclerotia also are produced on 25 per cent potato, 2 per cent dextrose, 2 per cent agar (Toussoun and Nelson, 1968). The macroconidia (sometimes the microconidia too) arise typically on short, upright, branched, densely-crowded conidiophores on the surface of the substrate. These structures are the sporodochia. In some cases abundant macroconidia form a slimy coat over the entire substrate, the pionnotes. The microconidia are single-celled, hyaline and are produced in false heads. The macroconidia are 4-celled, and hyaline. On the basis of these characteristics *F. oxysporum* belongs to the *Tuberculariaceae* family of the *Moniliales* order of the *Deuteromycetes* class. It should be noted that colonies of the fungus on artificial media and in nature do not develop in a like manner. Sporodochia (Fig. 27) are not

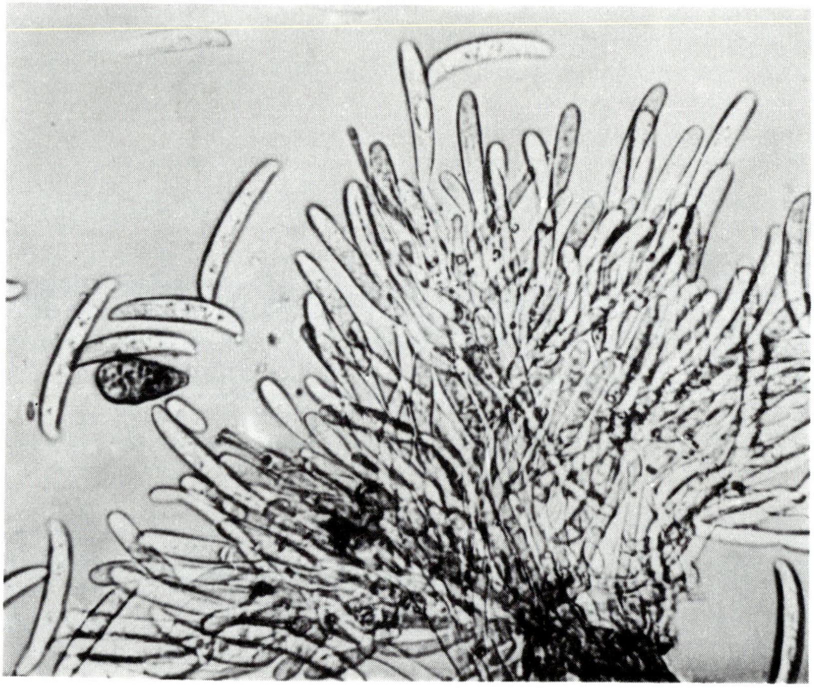

Fig. 27. Sporodochium of *Fusarium*

always found in culture, but conidia are produced on relatively scattered conidiophores on a loosely woven hyphal stroma. In this case, the colony structure resembles that found in the *Moniliaceae* family.

Life cycle

The pathogen is a successful soil inhabitant. It can live without an intervening host and can grow and fructify in a saprophytic way in the soil for long, indefinite periods of time in the presence of suitable moisture and temperature and organic materials. The chlamydospores and sclerotia of the fungus maintain their viability for more than one year in soil cultures. In the presence of appropriate host plants, the fungus changes its way of life to a parasitic one (facultative parasite). It causes damping off of seedlings and tracheomycosis on older plants, often causing the latter to wilt. It is very difficult to uniformly outline the environmental factors influencing the development of *Fusarium oxysporum* because of the large number of *formae speciales* (*f. sp.* — see later under biological specialization). Warm temperatures favor most of the formae speciales. The optimal soil temperature for example is 28 °C for *F. oxysporum* f. sp. *lycopersici*, 22 °C for f. sp. *callistephi*, 23–30 °C for f. sp. *tuberosi* and 28–32 °C for f. sp. *vasinfectum*. Different formae speciales have rather diverse soil moisture and pH requirements.

The *Fusarium* infection is greatly enhanced by any mechanical injury of the root system, especially the damage caused by nematodes (Dickson, 1956).

The microconidia usually are formed on the aerial mycelia of the fungus on the surface of the diseased plant or other substrate. These are egg-or elliptical-shaped, unicellular, 5–12 × 2–3·5 μ conidia. The macroconidia arise in the latter phases of disease development on aerial mycelium or in sporodochia. Frequently the macroconidia are septated, usually with 3 but sometimes 4–5 (7) cross-walls. The size of the four-celled conidia is 20–63 × 2·5–6 μ. Additional reproductive cells are terminal or intercalary chlamydospores as well as sclerotia that are formed quite often too.

The pathogen thus reproduces only asexually and the appearance of any new characteristics in the progenies such as the formation of biologically specialized strains must take place parasexually. Such a phenomenon was reported by Buxton (1958) in the case of *F. oxysporum* f. sp. *pisi* causing pea wilt. Hoffmann (1966), in the course of mixed cultivation in culture of two heterotrophic *F. oxysporum* strains differing in three characteristics from each other, demonstrated four new combinations from among the six possible new combinations that he said arose parasexually.

Symptoms and the role of toxins

The symptoms are about the same that result as a consequence of the interaction between the different host plant species and *F. oxysporum* f. sp. but the degree and development of the symptoms depend largely on the resistance of the host cultivar as well as on environmental factors.

The damping off and death of seedlings caused by *F. oxysporum* is sometimes encountered in the fields. The pathogen does not attack the vascular bundles but mainly the cortex of seedlings. Full-grown plants are infected by the pathogen through the roots and causes irreversible changes in the vascular system. This is partly due to mechanical plugging and partly to the effect of toxins. The first symptoms occur on the lower leaves in the form of yellowing. The yellowing spreads upwards in the plant and at the end of the disease cycle, the whole plant begins to wilt and then dies. Often the yellowing and wilting of the leaves is unilateral with symptoms only on the one side of the host. This depends on whether the pathogen has invaded the entire ring of vascular boundles or only a portion.

Fusaric acid, among the wilt-causing toxins is not host-specific, but one nevertheless that can at least partly be the reason for host specificity in some *F. oxysporum* formae speciales. Davis (1967) proved in the case of *F. oxysporum* f. sp. *niveum* that the pathogenic strains of the fungus produces more than 5 ppm fusaric acid only in the watermelon. Strains of f. sp. *niveum* are more virulent than other formae speciales that produce no fusaric acid, or that produce only trace amounts of fusaric acid in the watermelon. So in this case, the host specificity of the pathogen is explainable by the selective production of fusaric acid.

The investigations of Patil and Dimond (1967a) shed light on the role of the polygalacturonase enzyme of *F. oxysporum* in wilt. They demonstrated that the polygalacturonase enzyme production of *F. oxysporum* f. sp. *lycopersici* is inhibited to a large extent (from 50–80 per cent) by sugars and sugar-like alcohols (glucose, monogalacturonic acid, sorbitol, galactitol, 2-deoxy-*D*-glucose). These materials *in vivo* considerably moderated the wilt symptoms caused by the pathogen. It may be supposed that this is also may account for the inhibition of polygalacturonase enzyme production.

Methods used for isolation and artificial inoculations

Fusarium oxysporum can be isolated on most media both from infected plant parts or from infested soil. The isolation of *F. oxysporum* is very often contaminated by bacteria in the substrate. To eliminate this and to determine quantitatively the *F. oxysporum* content of soil samples, Papavizas' method (1967) is recommended. Soil dilutions are made on peptone-pentachlornitrobenzene or V-8 juice agar media supplemented with the following components per litre:

Oxgall	1·0 g
pentachlornitrobenzene	0·5 g
chlortetracyclin	0·5 g
streptomycin sulphate	0·1 g

The soil bacteria disturbing the determination can be successfully eliminated especially with the use of the above modified pentachlornitrobenzene medium.

A new method for the estimation of *Fusarium oxysporum* population was developed by MacNeill and Sabanayagam (1968). As the first step of the process, dodine (*n*-dodecylguanidine acetate) resistant strains of *Fusarium oxysporum* f. sp. *melonis* were obtained by ultraviolet irradiation. Selective reisolation and at the same time the exact identification of these strains from inoculated soils were easily carried out on potato-sucrose agar containing 250 mg dodine/ml.

Walker (1950) has reviewed the artificial inoculation procedures conducted with *F. oxysporum*, particularly with f. sp. *lycopersici*. The essence of the most satisfactory method is as follows:

a) Plant seedlings of the cultivar under investigation into heat-sterilized soil or sphagnum moss-soil mixtures.

b) Maintain 24–30 °C temperatures in the greenhouse.

c) A furrow is made with a knife running beside the rows to injure the roots of the seedlings when they reach the four-leaf stage.

d) Pour a suspension of conidia or hyphal fragments made from freshly-isolated vigorous *F. oxysporum* cultures into the furrows made by the knife.

e) Repeat the inoculation of the plants after 10 days.

f) Observe the symptoms 30 days after the first inoculation.

A less drastic artificial inoculation can be carried out by simply mixing the fungus suspension into the soil before planting or pouring the suspension over the soil surface. The roots are not injured in this way, but the frequency and intensity of infection is less.

Biological specialization of the pathogen

The species *F. oxysporum* is really a collective species. It includes many different virulent forms. Snyder and Hansen (1940) distinguished 25, Gordon (1965) already has delineated 66 *F. oxysporum* formae speciales. Listing all of them would be too long and it would be impossible to describe these in detail here. Therefore only the economically important formae speciales are listed with their most important host plants and types of symptoms.

Fusarium oxysporum

f. sp. *batatas*	— sweet potato and tobacco wilt
f. sp. *callistephi*	— aster wilt
f. sp. *conglutinans*	— cabbage yellows
f. sp. *cucumerinum*	— cucumber foot rot and wilt
f. sp. *dianthi*	— carnation wilt
f. sp. *lini*	— flax wilt
f. sp. *lycopersici*	— tomato wilt
f. sp. *lupini*	— lupine wilt
f. sp. *niveum*	— watermelon wilt
f. sp. *pisi*	— pea wilt
f. sp. *tracheiphilum*	— soybean and cowpea wilt
f. sp. *tuberosi*	— potato tracheomycosis
f. sp. *vasinfectum*	— cotton wilt

Physiologic races can be differentiated within certain formae speciales on the basis of virulence on different host plant cultivars. So, for example, the f. sp. *batatas* is known to have races 1 and 2; f. sp. *conglutinans* races 1, 2 and 3; the f. sp. *lupini* races 1, 2 and 3; the f. sp. *lycopersici* races 1 and 2; and the f. sp. *vasinfectum* races 1, 2, 3 and 4 (Gordon, 1965).

Fusarium wilt resistant cultivars

Although *F. oxysporum* is a rather highly variable organism nevertheless the breeding for resistance against certain formae speciales is not quite as hopeless as it appears. A good example of this is the case of *F. oxysporum* f. sp. *vasinfectum* causing cotton wilt in which there are numerous, stable, resistant host cultivars that are still in use (Walker, 1950). Control of cotton wilt is based mainly on the cultivation of resistant cultivars (Dickson, 1956; Walker, 1950). In the same way resistant cultivars have all but eliminated the occurrence of *F. oxysporum* f. sp. *conglutinans* causing cabbage yellows as well as *F. oxysporum* f. sp. *lycopersici* which causes tomato wilt. Among the cultivars resistant to cabbage yellows "Wisconsin Hollander", the "Wisconsin All Seasons" and the "Wisconsin Brunswick" are examples worth mentioning. Among the tomato cultivars resistant to *Fusarium*, "Louisiana Pink", "Louisiana Red", "Marglobe", "Norton", "Pritchard" and "Red Currant" are the most important. Breeding for resistance has already been treated on page 316.

Control of Fusarium wilt

In the greenhouse and seed beds, the pathogen can be eliminated or reduced by soil disinfection (with heat or with chemicals). In the field, crop rotation is not recommended since the pathogens are very successful soil inhabitants and after the invasion of a soil, they remain there idenfinitely. The best and most reliable control is the cultivation of resistant cultivars.

LITERATURE

Agnihothrudu, V. (1962): A comparison of some techniques for the isolation of fungi from tea soils. *Mycopath. Mycol. Appl.* **16,** 235–242.

Ainsworth, G. C. and Sampson, K. (1950): The British Smut Fungi. *Commonw. Mycol. Inst. Kew*, Surrey.

Akai, S., Yasumori, H. and Terasawa, H. (1959): On the resistance of cucumber varieties to anthracnose and the behavior of the causal fungus in the invasion of the host tissues. *Shokubutsu Byogai Kenkyu*, Kyoto. **6,** 97–103.

Alexopoulos, C. J. (1962): *Introductory Mycology*. 2nd ed. J. Wiley, New York and London.

Al-Sohaily, I. A., Mankin, C. J. and Semeniuk, G. (1963): Physiologic specialization of Sphacelotheca reiliana to sorghum and corn. *Phytopathology* **53,** 723–726.

Anderer, F. A., Uhling, H., Weber, E. and Schramm, G. (1960): Primary structure of the protein of tobacco mosaic virus. *Nature* **186,** 922–925.

Anderson, H. W. (1956): *Diseases of Fruit Crops*. McGraw-Hill Book Co., Inc. New York.

Anonymous (1950): Common names of virus diseases used in the Review of AppliedMycology. *Rev. Apl. Mycol.* **24,** 513–556.

Anonymous (1965): Plant pathology. *Rep. Int. Rice Res. Inst.* 1964. pp. 127–156.

Ark, P. A. (1951): Phenotypic variations induced by chemicals in Corynebacterium michiganense and Xanthomonas juglandis. *J. Bact.* **61,** 293–298.

Armstrong, G. M. and Armstrong, J. K. (1960): American, Egyptian, and Indian cotton-wilt fusaria: their pathogenicity and relationship to other wilt fusaria. *USDA Tech. Bull. No.* 1219.

Arx, J. A., and van der Velden, F. J. J. A. (1961): Das Colletotrichum der Gurkengewächse. *Phytopath. Z.* **41,** 228–235.

Asai, G. N., Jones, M. W. and Rorie, F. G. (1967): Influence of certain environmental factors in the field infection of rice by Piricularia oryzae. *Phytopathology* **57,** 237–241.

Astier-Manifacier, S. and Cornuet, P. (1965): Isolation of turnip yellow mosaic virus RNA replicase and asymetric synthesis of polynucleotides identical to TYMV-RNA. *Biochim. Biophys. Res. Commun.* **18,** 283–287.

Atkins, I. M. (1943): Reaction of some varieties and strains of winter wheat to artificial inoculation of loose smut. *J. Amer. Soc. Agron.* **35,** 197–204.

Atkins, J. G., Robert, A. L., Adair, C. R., Goto, K., Kozaka, T. et al. (U. S. and Japan Cooperative Study) (1967): An international set of rice varieties for differentiating races of Piricularia oryzae. *Phytopathology* **57,** 297–301.

Averre, C. W. and Kelman, A. (1964): Severity of bacterial wilt as influenced by ratio of virulent to avirulent cells of Pseudomonas solanacearum in inoculum. *Phytopathology* **54,** 779–783.

Avery, O. T., MacLeod, C. M. and McCarty, M. (1944): Studies on the chemical nature of the substance inducing transformation of pneumococcal types. Induction of transformation by a desoxyribonucleic acid fraction isolated from Pneumococcus type III. *J. Exp. Med.* **79,** 137–158.

Ayers, G. W. (1957): Races of Plasmodiophora brassicae. *Canad. J. Bot.* **35,** 923–2.

Babayan, A. A., Barsegyan, S. G., Keclzek, N. A. and Vartanyan, V. S. (1966): A study of the resistance of tobacco varieties and hybrids to Peronospora disease. (In Russian) *Izv. Akad. Nauk Armenian SSR* **18,** 3–7.

Babos, P. (1966): The ribonucleic acid content of tobacco leaves infected with tobacco mosaic virus. *Virology* **28,** 282–289.

Babos, P. and Kassanis, B. (1962): Unstable variants of tobacco necrosis virus. *Virology* **18,** 206–211.

Badami, R. S. (1958): Changes in the transmissibility by aphids of a strain of cucumber mosaic virus. *Ann. Appl. Biol.* **46,** 554–562.

Bald, J. G. (1949): A method for the selective staining of viruses in infected plant tissues. *Phytopathology* 39, 395–402.

Bald, J. G. (1964): Cytological evidence for the production of plant virus ribonucleic acid in the nucleus. *Virology* **22,** 377–387.

Bancroft, J. B. and Keasberg, P. (1960): Macromolecular particles associated with alfalfa mosaic virus. *Biochim. Biophys. Acta* **39,** 519–528.

Barron, G. L. and MacNeill, B. H. (1962): A simplified procedure for demonstrating the parasexual cycle in Aspergillus. *Canad. J. Bot.* **40,** 1321–1327.

Barrus, M. F. (1911): Variation of varieties of beans in their susceptibility to anthracnose. *Phytopathology* **1,** 190–95.

Bartels, R. (1954): Serologische Untersuchungen über das Verhalten des Kartoffel-A-Virus in Tabakpflanzen. *Phytoplath. Z.* **21,** 395–406.

Bartels, R. (1957): Ein Beitrag zum serologischen Nachweis des Y-Virus in der Kartoffel. *Phytopath. Z.* **30,** 1–16.

Bartels, R. (1958): Die Konzentration des Kartoffel-Y-Virus in Kartoffelpflanzen. *Zentrbl. Bakteriol. Paras. Hyg.* II. Abt. **111,** 185–190.

Bartels, W. (1955): Der gegenwärtige Stand der Forschung auf dem Gebiet der Inaktivierung pflanzenpathogener Viren insbesondere des Tabakmosaikvirus. *Phytopath. Z.* **24,** 117–178.

Batts, C. C. V. (1955): Observations on the infection of wheat by loose smut (Ustilago tritici [Pers.] Rostr.). *Trans. Brit. Mycol. Soc.* **38,** 465–475.

Batts, C. C. V. and Jeater, A. (1958): The interaction of wheat varieties to loose smut as determined by embryo, seedling, and adult plant tests. *Ann. Appl. Biol.* **46,** 23–29.

Bawden, F. C. (1950): *Plant Viruses and Virus Diseases.* Chronica Botan. Comp., Waltham, Mass. USA.

Bawden, F. C. (1955): The classification of viruses. *J. Gen. Microbiol.* **12,** 363–366.

Bawden, F. C. (1959): Physiology of virus diseases. *Ann. Rev. Plant. Physiol.* **10,** 239–256.

Bawden, F. C. and Freeman, G. G. (1952): The nature and behaviour of inhibitors of plant viruses produced by Trichothecium roseum Link. *J. Gen Microbiol.* **7,** 154–168.

Bawden, F. C., Hamlyn, M. G. and Watson, M. A. (1954): The distribution of viruses in different leaf tissues and its influence on virus transmission by aphids. *Ann. Appl. Biol.* **41,** 229–239.

Bawden, F. C. and Kassanis, B. (1945): The suppression of one plant virus by another. *Ann. Appl. Biol.* **32,** 52–57.

Bawden, F. C. and Kleczkowski, A. (1945): Protein precipitation and virus inactivation by extracts of strawberry plants. *J. Pomol. Hort. Soc.* **21,** 2–7.

Bawden, F. C. and Nixon, H. L. (1951): The application of electron mycroscopy to the study of plant viruses in unpurified plant extracts. *J. Gen. Microbiol.* **1,** 104–109.

Bawden, F. C. and Pirie, N. W. (1936): Experiments on the chemical behaviour of potato virus X. *Brit. J. Exp. Pathol.* **17,** 64–74.

Bawden, F. C. and Pirie, N. W. (1938): A plant virus preparation in a fully-crystalline state. *Nature* **141,** 513.

Bawden, F. C. and Pirie, N. W. (1940a): The inactivation of some plant viruses by urea. *Biochem J.* **34,** 1258–1277.

Bawden, F. C. and Pirie, N. W. (1940b): The effects of alkali and some simple organic substances on three plant viruses. *Biochem. J.* **34,** 1278–1292.

Bawden, F. C. and Pirie, N. W. (1943): Methods for the purification of tomato bushy stunt and tobacco mosaic viruses. *Biochem. J.* **37**, 66–70.
Bawden, F. C. and Pirie, N. W. (1945): Further studies on the purification and properties of a virus causing tobacco necrosis. *Brit. J. Exp. Pathol.* **26**, 277–285.
Bawden, F. C. and Pirie, N. W. (1952): Physiology of virus diseases. *Ann. Rev. Plant. Physiol.* **3**, 171–188.
Bawden, F. C. and Roberts, F. M. (1948): Photosynthesis and predisposition of plants to infection with certain viruses. *Ann. Appl. Biol.* **35**, 418–425.
Baylis, R. J. (1955): Field infection experiments with dwarf bunt of winter wheat in Ontario. *Plant. Dis. Reptr.* **39**, 159–160.
Beadle, G. W. and Tatum, E. L. (1941): Genetical control of biochemical reactions in Neurospora. *Proc. Nat. Acad. Sci.* Wash. **27**, 499–506.
Beltukova, K. I., Matishevskaya, M. S., Kulikovskaya, M. D. and Sidorenko, S. S. (1968): *Investigation Methods of Plant Diseases Induced by Bacteria.* Nauk. Dumka, Kiev (In Russian).
Benada, J. and Brückner, F. (1963): Powdery mildew, Erysiphe graminis DC. on barley. IV. Resistance tests performed on detached leaves. *Rostlinna Vyroba* **6**, 585–592.
Benda, G. T. A. and Naylor, A. W. (1954): On the tobacco ringspot disease I. Localization of recovery. *Amer. J. Bot.* **41**, 799–803.
Bennett, C. W. (1939): The nomenclature of plant viruses. *Phytopathology* **29**, 422–430.
Bennett, C. W. (1940): Acquisition and transmission of viruses by dodder (Cuscuta subinclusa). *Phytopathology* **30**, 2.
Bennett, C. W. (1944): Studies on dodder transmission of plant viruses. *Phytopathology* **34**, 905–932.
Bennett, C. W. (1953): Interactions between viruses and virus strains. *Adv. Virus Res.* **1**, 39–67.
Bennett, C. W. (1956): Biological relations of plant viruses. *Ann. Rev. Plant Physiol.* **7**, 143–170.
Beraha, L., Varzandeh, M. and Thornberry, H. H. (1955): Mechanism of the action of abbrasives on infection by tobacco mosaic virus. *Virology* **1**, 141–151.
Bercks, R. (1954): Untersuchungen über Änderungen des Virusgehaltes in Tabakpflanzen während der Vegetationsperiode. *Phytopath. Z.* **22**, 215–226.
Berend, St. (1958): Zur Ätiologie der Aprikosenwelke. *Phytopath. Z.* **33**, 399–402.
Bergey's Manual of Determinative Bacteriology. Seventh Edition. The Williams and Wilkins Co., Baltimore, 1957.
Best, R. J. (1954): Cross protection by strains of tomato spotted wilt virus and a new theory to explain it. *Austr. J. Biol. Sci.* **7**, 415–424.
Best, R. J. (1961): On maintaining the infectivity of a labile plant virus by storage at −69°. *Virology* **14**, 440–443.
Bever, W. M. (1953): Further studies on physiologic races of Ustilago tritici. *Phytopathology* **43**, 681–683.
Bils, R. F. and Hall, C. E. (1962): Electron microscopy of wound-tumor virus. *Virology*, **17**, 123–130.
Blattny, C. (1942): Vorläufige Mitteilung über die Rassen der Kartoffelkrebses Synchytrium endobioticum (Schilb.) Perc. *Ann. Akad. Tschechoslov. Agr.* **17**, 40–46.
Black, C. M. (1941): Further evidence for multiplication of the aster yellows virus in the aster leaf hopper. *Phytopathology* **31**, 120–135.
Black, L. M. (1953): Transmission of plant viruses by cicadellids. *Adv. Virus Res.* **1**, 69–89.
Black, L. M. (1955): Concepts and problems concerning purification of labile insect-transmitted plant viruses. *Phytopathology* **45**, 208–216.
Black, L. M. (1958): Transmission of virus by leafhoppers. *Proc. 10th Int. Congr. Entomol.* **3**, 201–204.
Black, L. M. (1959): Biological cycles of plant viruses in insect vectors. In: *The Viruses*, (Ed.) F. M. Burnett and W. M. Stanley. Vol. 2. Acad. Press, New York and London.

Black, L. M. and Markham, R. (1963): Base-pairing in the ribonucleic acid of wound-tumor virus. *Netherl. J. Plant Pathol.* **69,** 215.

Black, W. (1952): A genetical basis for the classification of strains of Phytophthora infestans. *Proc. Royal Soc.* Edinburgh. **65,** 36–51.

Black, W., Mastenbroek, C., Mills, W. R. and Peterson, L. C. (1953): A proposal for an international nomenclature of races of Phytophthora infestans and of genes controlling immunity in Solanum demissum derivatives. *Euphytica* **2,** 173–179.

Blendon, D. C. and Goldberg, H. S. (1965): Silver impregnation stain for Leptospira and flagella. *J. Bact.* **89,** 899–900.

Blumer, S., Stalder, L. und Harder, A. (1955): Über die gegenseitigen Beziehungen zwischen Gurkenmosaik und Gurkenmehltau. *Phytopath. Z.* **25,** 39–45.

Boardman, N. K., Francki, R. I. B. and Wildman, S. G. (1965): Protein synthesis by cell free extracts from tobacco leaves. II. Association of activity with chloroplast ribosomes. *Biochemistry* **4,** 8 2–876.

Boardman, N. K., Francki, R. I. B. and Wildman, S. G. (1966): Protein synthesis by cell free extracts from tobacco leaves. III. Comparison of the physical properties and protein synthesizing activities of 70 S and 80 S cytoplasmic ribosomes. *J. Mol. Biol.* **17,** 470–489.

Boardman, N. K. and Zaitlin, M. (1958): The association of tobacco mosaic virus with plastids. II. Studies on the biological significance of virus as isolated from a chlorplast fraction. *Virology* **6,** 758–768.

Bojnansky, V. (1959): A burgonyarák és jelentősége szárazabb éghajlatú vidékeken. (The importance of potato wart under dry climates) *Növénytermelés* **8,** 281–286.

Bojnansky, V. (1960): *Ekologia a prognoza rakoviny zemiakov.* SAV, Bratislava.

Bollard, E. G. and Matthews, R. E. F. (1966): The physiology of parasitic disease. In: *Plant Physiology. A Treatise.* (Ed.) F. C. Steward. Vol. IV. B. Pp. 417–550. Academic Press, New York and London.

Böning, K. und Diercks, R. (1955): Versuche über den Einfluss der Mineralsalzernährung auf die Empfindlichkeit der Kartoffelpflanze für Blattroll- und Strichelkrankheit. *Bayer. Landw. Jahrb.* **32,** 276–323.

Borlaug, N. E. (1958): The impact of agricultural research on Mexican wheat production. *Trans. New York Acad. Sci.* **20,** 278–295.

Borlaug, N. E. (1958): The use of multilineal or composite varieties to control airborne epidemic diseases of self-pollinated crop plants. Pp. 12–27. *Proc. First. Int. Wheat Genetics Symp.*, Univ. of Manitoba, Winnipeg, Canada, August, 1958.

Borlaug, N. E. (1965): Wheat, rust, and people. *Phytopathology* **55,** 1088–1098.

Bos, L., Hagedorn, D. J. and Quantz, L. (1960): Suggested procedures for international identification of legume viruses. *Tijd. Pl. ziekten* **66,** 328–343.

Bozarth, R. R. and Ross, A. F. (1964): Systemic resistance induced by localized virus infections: Extent of changes in uninfected plant parts. *Virology* **24,** 446–455.

Bradley, R. H. E. (1952): Studies on the aphid transmission of a strain of henbane mosaic virus. *Ann. Appl. Biol.* **39,** 78–97.

Bradley, R. H. E. (1954): Studies of the mechanism of transmission of potato virus Y by the green peach aphid Myzus persicae (Sulz.) (Homoptera: Aphidae). *Canad. J. Zool.* **32,** 64–73.

Bradley, R. H. E. (1956): Effects of depth of stylet penetration on aphid transmission of potato virus Y. *Canad. J. Microbiol.* **2,** 539–547.

Bradley, R. H. E. (1962): Different areas of tobacco leaves as sources of potato virus Y for aphids. *Virology* **16,** 366–370.

Bradley, R. H. E. and Ganong, R. J. (1955a): Evidence that potato virus Y is carried near the tip of the stylets of the aphid vector Myzus persicae (Sulz.). *Canad. J. Microbiol.* **1,** 755–782.

Bradley, R. H. E. and Ganong, R. J. (1955b): Some effects of formaldehyde on potato virus Y in vitro, and ability of aphids to transmit the virus when their stylets are treated with formaldehyde. *Canad. J. Microbiol.* **1,** 783–793.

Brakke, M. K. (1960): Density gradient centrifugation and its application to plant viruses. *Adv. Virus Res.* **7,** 193–224.
Brandes, J. (1956): Über das Aussehen und die Verteilung des Tabakmosaikvirus im Blattgewebe. Elektronenmikroskopische Untersuchungen an ultradünnen Mikroschnitten. *Phytopath. Z.* **26,** 93–106.
Brandes, J. (1957): Eine elektronenmikroskopische Schnellmethode zum Nachweis faden- und stäbchenförmiger Viren, insbesondere in Kartoffeldunkelkeimen. *Nachrbl. Deut. Pfl. schutzdienst,* Braunschweig, **9,** 151–152.
Brandes, J. and Wetter, C. (1959): Classification of elongated plant viruses on the basis of particle morphology. *Virology* **8,** 99–115.
Brandt, W. H. (1965): Loss of melanin-containing structures by Verticillium cultures. *Phytopathology* **55,** 1200–1202.
Braun, A. C. (1955): A study on the mode of action of the wildfire toxin. *Phytopathology* **45,** 659–664.
Braun, A. C. (1962): Tumor inception and development in the crown gall disease. *Ann. Rev. Plant Physiol.* **13,** 533–558.
Brčák, J. (1959): Transmission of beet mosaic virus by the green peach aphid starved before infection, feeding under different conditions of air humidity. *Biol. Plant.* **1,** 330–332.
Bridgmon, G. H. (1957): The production of new races of Puccinia graminis var. tritici by hyphal fusion on wheat. *Phytopathology* (Abtr.) **47,** 517.
Brinkerhoff, L. A., Samayoa, E. and Murray, J. C. (1967): Verticillium wilt reactions in cotton as affected by post-inoculation temperatures. *Phytopathology* (Abstr.) **57,** 803.
Bromfield, K. R. and Schmitt, C. G. (1967): Cryogenic storage of conidia of Peronospora tabacina. *Phytopathology* (Abtsr.) **57,** 644.
Browning, J. A. (1966): Discussion In: *Plant Breeding.* A Symposium Held at Iowa State Univ. (Ed.) Kenneth J. Frey. Iowa State Univ. Press, Ames, Iowa. Pp. 233–236.
Browning, J. A. and Frey, K. J. (1969): Multiline cultivars as a means of disease control. *Ann. Rev. Phytopath.* **7,** 355—382.
Brückner, F. (1963): Getreidemehltau (Erysiphe graminis DC.) an Gerste. III. Erforschung des Auftretens physiologischen Rassen auf tschechoslowakischem Gebiet in den Jahren 1960–1961. *Rostlinna Vyroba* **9,** 1–8.
Buddenhagen, I. and Kelman, A. (1964): Biological and physiological aspects of bacterial wilt caused by Pseudomonas solanacearum. *Ann. Rev. Phytopath.* **2,** 203–230.
Buddenhagen, J., Sequeira, L. and Kelman, A. (1962): Designation of races in Pseudomonas solanacearum. *Phytopathology* **52,** 726.
Burdon, R. H., Billeter, M. A., Weissmann, C., Warner, C., Ochoa, R. O., and Knight, C. A. (1964): Replication of viral RNA. V. Presence of a virus specific double stranded RNA in leaves infected with tobacco mosaic virus. *Proc. Natl. Acad. Sci.* U. S. **52,** 768–775.
Burill, T. J. (1880): Pear and apple tree blight. *Trans. Illionis Hort. Soc.* **14,** 157–167.
Burkholder, W. H. (1945): The longevity of the pathogen causing the wilt of the common bean. *Phytopathology* **35,** 743–744.
Burkholder, W. H., McFadden, L. A. and Dimock, A. W. (1953): A bacterial blight of chrysanthemums. *Phytopathology* **43,** 522-526.
Busch, L. V. (1966): Susceptibility of potato varieties to Ontario isolates of Verticillium albo-atrum. *Amer. Potato J.* **43,** 439–442.
Busch, L. V. and Walker, J. C. (1958): Studies of cucumber anthracnose. *Phytopathology* **48,** 302–304.
Buxton, E. W. (1954): Heterokaryosis and variability in Fusarium oxysporum f. gladioli Snyder et Hansen. *J. Gen. Microbiol.* **10,** 71–84.
Buxton, E. W. (1958): A change of pathogenic race in Fusarium oxysporum f. pisi induced by root exudate from a resistant host. *Nature* **181,** 1222–1224.
Cadman, C. H. (1960): Inhibition of plant virus infection by tannins. In: *Phenolics in Plants in Health and Disease.* (Ed.) J. B. Pridham. Pergamon Press, Oxford. Pp. 101–105.

Cadman, C. H. (1962): Evidence for association of tobacco rattle virus nucleic acid with a cell component. *Nature* **193**, 49–52.
Cadman, C. H. and Harrison, B. D. (1960): Studies on the behaviour in soils of tomato black ring, raspberry ringspot, and arabis mosaic viruses. *Virology* **10**, 1–20.
Caldwell, J. (1932): Aucuba or yellow mosaic of tomato in Nicotiana glutinosa and other hosts. *Ann. Appl. Biol.* **19**, 144–152.
Caldwell, J. (1936): Factors affecting the formation of local lesions by tobacco mosaic virus. *Proc. Royal Soc.* London Ser. B. **119**, 493–507.
Caldwell, J. (1962): Seed-transmission of viruses. *Nature* **193**, 457–459.
Caldwell, R. M. (1960): Culture of powdery mildew (Erysiphe graminis DC.) on excised wheat leaves in solutions of benzimidazole. *Indiana Acad. Sci.* **69**, 109. (Abstr.)
Canova, A. (1957): Richerche intorno ad una virusi del pomodoro (mal della striscia). Presenze del virus sui semi e nel terrano. *Phytopath. Z.* **4**, 415–422.
Capoor, S. P. (1949): The movement of tobacco mosaic viruses and potato virus X through tomato plants. *Ann. Appl. Biol.* **36**, 307–319.
Čech, M., Pozdena, J. and Blattny, C. (1957): Mechanical transmission of infectious sterility of hop. *Folia Biol.* **3**, 314–318.
Chalcroft, J. and Matthews, R. E. F. (1966): Cytologigal changes induced by turnip yellow mosaic in Chinese cabbage leaves. *Virology* **28**, 555–562.
Chang, W. L., Wang, M. K. and Yang, S. C. (1965): Reaction of rice varieties and strains to blast in the uniform blast nursery. *J. Taiwan Agric. Res.* **14**, 1–10.
Chapman, G. B. (1959): Electron microscope observations on the behaviour of the bacterial cytoplasmic membrane during cellular division. *J. Biophys. Biochem. Cytol.* **6**, 221–224.
Cherewick, W. J. (1944): Studies on the biology of Erysiphe graminis DC. *Canad. J. Res.* Sect. C. **22**, 52–86.
Cherewick, W. J. and Cunningham, R. H. (1956): Further improvements in the partial-vacuum method of inoculation with loose smut fungi. *Phytopathology* **46**, 355–358.
Chester, K. S. (1937): Serological studies of plant viruses. *Phytopathology* **27**, 903–912.
Cholodny, N. G. (1930): Über eine neue Methode zur Untersuchung der Boden-mikroflora. *Arch. Mikrobiol.* **1**, 620–652.
Christensen, J. J. and DeVay, J. E. (1955): Adaptation of plant pathogens to host. *Ann. Rev. Plant Physiol.* **6**, 367–392.
Chu, O. M. Y. and Li, H. W. (1965): Cytological studies of Piricularia oryzae Cav. *Bot. Bull. Acad. Sin. Taipei*, N. S. **6**, 116–130.
Clark, M. F., Matthews, R. E. F. and Ralph, R. K. (1964): Ribosomes and polyribosomes in Brassica pekingensis. *Biochim. Biophys. Acta* **91**, 289–304.
Clements, F. E. and Shear, C. L. (1954): *The Genera of Fungi.* Hafner, New York.
Commoner, B. (1959): The biochemical basic of tobacco mosaic virus infectivity. *IVth Int. Congr. Biochem.* Vol. 8. *Biochemistry of Viruses.* Pergamon Press, London. Pp. 17–30.
Commoner, B. and Mercer, F. L. (1952): The effect of thiouracil on the rate of tobacco mosaic virus biosynthesis. *Arch. Biochem. Biophys.* **35**, 278–289.
Commoner, B. and Nehari, V. (1953): The effects of tobacco mosaic virus synthesis on the free amino acid and amide composition of the host. *J. Gen. Physiol.* **36**, 791–805.
Commoner, B., Yamada, M., Rodenberg, S. D., Wang, T. Y. and Basler, E. (1953): The proteins synthesized in tissue infected with tobacco mosaic virus. *Science* **118**, 529–534.
Cook, A. A. and Stall, R. E. (1963): Inheritance of resistance in pepper to bacterial spot. *Phytopathology* **53**, 1060–1062.
Cook, M. T. (1947): *Viruses and Virus Diseases of Plants.* Burgess Publ. Co., Minneapolis.
Cooke, W. M. B. (1963): A laboratory guide to fungi in polluted waters, sewage, and sewage treatment systems. Their identification and culture. *Public Health Publ.*, Cincinnati.
Copeland, H. F. (1956): *The Classification of Lower Organisms.* Pacific Books, Palo Alto.
Corey, R. R. and Starr, M. P. (1957): Genetic transformation of colony type in Xanthomonas phaseoli. *J. Bact.* **74**, 141–145.

Costa, A. S. and Carvalho, A. M. (1960): Comparative studies between Abutilon and Euphorbia mosaic viruses. *Phytopath. Z.* **38,** 129–152.
Couch, J. N. (1938): Observations on cilia of aquatic Phycomycetes. *Science* **88,** 476.
Craigie, J. H. (1927): Discovery of the function of the pycnid in the rust fungi. *Nature* **120,** 765–767.
Craigie, J. H. (1931): An experimental investigation of sex in rust fungi. *Phytopathology* **21,** 1001–1040.
Cramer, R. and Svensson, H. (1961): Density gradient electrophoresis as a new tool in virology. *Experientia* **17,** 49–57.
Crispin, A. and Dongo, S. (1962): New physiologic races of bean rust, Uromyces phaseoli typica, from Mexico. *Plant Dis. Reptr.* **46,** 411–413.
Crook, P., Carpenter, C. C. and Klens, P. F. (1950): The use of sodium propionate in isolating Actinomycetes from soil. *Science* **112,** 565.
Crosse, J. E. (1956): Bacterial canker of stone-fruits. II. Leaf scar infection of cherry. *J. Hort. Sci.* **31,** 212–224.
Crosse, J. E. (1963): Bacterial canker of stone-fruits. V. A comparison of leaf surface populations of Pseudomonas mors-prunorum in autumn on two cherry varieties. *Ann. Appl. Biol.* **52,** 97–104.
Crosse, J. E. (1965): Bacterial canker of stone-fruits. VI. Inhibition of leaf-scar infection of cherry by a saprophytic bacterium from the leaf surfaces. *Ann. Appl. Biol.* **56,** 149–160.
Crosse, J. E. (1966): Epidemiological relations of the pseudomonad pathogens of deciduous fruit trees. *Ann. Rev. Phytopath.* **4,** 291–310.
Crosse, J. E. and Garrett, C. M. E. (1966): Bacterial canker of stone-fruits. VII. Infection experiments with Pseudomonas mors-prunorum and Ps. syringae. *Ann. Appl. Biol.* **58,** 31–41.
Crosse, J. E. and Shanmuganathan, N. (1961): Pathogenicity test with isolates of Pseudomonas mors-prunorum from leaf surface of sprayed and unsprayed cherry trees. *Ann. Report of the East Malling Res. Station.* 1960. Pp. 87–89.
Crowley, N. C. (1955): The effect of seed extracts on the infectivity of plant viruses and its bearing on seed transmission. *Austr. J. Biol. Sci.* **8,** 56 67.
Crowley, N. C. (1957): Studies on the seed transmission of plant virus diseases. *Austr. J. Biol. Sci.* **10,** 449–462.
Crowley, N. C. and Hanson, J. B. (1960): The infection of apical meristems of tomato with tobacco mosaic virus after treatment with ethylenediaminetetraacetic acid. *Virology* **12** 603–606.
Cruickshank, I. A. M. (1963): Environment and sporulation in phytopathogenic fungi. IV. The effect of light on the formation of conidia of Peronospora tabacina Adam. *Austr. J. Biol. Sci.* **16,** 88–98.
Cruickshank, I. A. M. and Mandryk, M. (1960): The effect of stem infestation of tobacco with Peronospora tabacina Adam on foliage reaction to blue mold. *J. Austr. Inst. Agric. Sci.* **26,** 369–372.
Cruickshank, I. A. M. and Perrin, D. R. (1961): Studies on phytoalexins. III. The isolation assay, and general properties of phytoalexin from Pisum sativum L. *Austr. J. Biol. Sci.* **14,** 336–348.
Curtis, K. M. (1921): The life history and cytology of Synchytrium endobioticum (Schilb.) Pers., the cause of the wart disease in potato. *Trans. Royal Soc.* London, **210,** 409–478.
Davis, D. (1967): Fusaric acid in selective pathogenicity of Fusarium oxysporum. *Phytopathology* (Abstr.) **57,** 808.
Davison, A. D. and Vaughan, E. K. (1963a): Longevity of urediospores of race 33 of Uromyces phaseoli var. phaseoli in storage. *Phytopathology* **53,** 736–737.
Davison, A. D. and Vaughan, E. K. (1963b): A simplified method for identification of races of Uromyces phaseoli var. phaseoli. *Phytopathology* **53,** 456–459.
Day, P. R. (1960): Variation in phytopathogenic fungi. *Ann. Rev. Microbiol.* **14,** 1–16.
Dean, A. C. R. and Hinshelwood, C. N. (1953): Observations on bacterial adaptation. *Symp. Soc. Gen. Microbiol.* **3,** 21–45.

Dekker, J. and Oort, A. J. P. (1964): Mode of action of 6-azauracil against powdery mildew. *Phytopathology* **54,** 815–818.
Delbrück, M. and Bailey, W. T. (1946): Induced mutations in bacterial viruses. *Cold Spring Harbor Symp. Quant. Biol.* **11,** 33.
De Ley, J. (1968): DNA base composition and hybridization in the taxonomy of phytopathogenic bacteria. *Ann. Rev. Phytopath.* **6,** 63–90.
Deppe, K. und Seehofer, F. (1962): Der Haftmechanismus der Konidiospora von Peronospora tabacina Adam. *Beitr. Tabakforsch.* **5.**
Devaux, A. L. and Sackston, W. E. (1966): Taxonomy of Verticillium species causing wilt of horticultural crops in Quebec. *Canad. J. Bot.* **44,** 803–811.
Dickson, J. G. (1956): *Diseases of Field Crops.* 2nd edition. McGraw-Hill Book Co., Inc., New York, Toronto, London.
Diener, T. O. (1963): Physiology of virus-infected plants. *Ann. Rev. Phytopath.* **1,** 197–213.
Diener, T. O. and Weaver, M. L. (1959): Inefectivity of phenol-extracted preparations from cucumber leaves infected with necrotic ringspot virus. *Virology* **8,** 531–532.
Dimond, A. E., Stoddard, E. M. and Chapman, R. A. (1952): Chemotherapeutic investigations on the common bacterial blight of beans. *Phytopathology* **42,** 72–76.
Doling, D. A. (1966): The use of benzimidazole for the detached leaf culture of Puccinia striiformis. *Cereal Rust Conf.*, 1964. Plant Breeding Inst., Cambridge, Pp. 55–57.
Donk, M. A. (1954): A note on sterigmata in general. *Barthalia* **6,** 301–302.
Dowson, W. Y. (1949): *Manual of Bacterial Plant Diseases.* Adam and Charles Black, London.
Dowson, W. J. (1957): *Plant Diseases Due to Bacteria.* Univ. Press, Cambridge.
Drandarevski, Ch. and Weltzien, H. C. (1966): Über die Anfärbung von Erysiphaceen durch Joddämpfe. *Naturwissenschaften* **53,** 616–617.
Dutta, S. K. and Garber, E. D. (1962): Genetics of phytopathogenic fungi. VII. Fungicide resistance in Colletotrichum lagenarium. *Phytopathology* **52,** 35–38.
Dutta, S. K., Hall, C. V. and Heyne, E. G. (1960 a): Observations on the physiological races of Colletotrichum lagenarium. *Bot. Gaz.* **121,** 163–166.
Dutta, S. K., Hall, C. V. and Heyne, E. G. (1960 b): Pathogenicity of biochemical mutants of Colletotrichum lagenarium. *Bot. Gaz.* **121,** 167–170.
Dye, D. W., Starr, M. P. and Stolp, H. (1964): Taxonomic classification of Xanthomonas vesicatoria based upon host specificity, bacteriophage sensitivity and cultural characteristics. *Phytopath. Z.* **51,** 394–407.
Dyson, J. G. and Chessin, M. (1961): Effect of auxins on virus-induced leaf abscission. *Phytopathology* **51,** 195.
Egerer, A. (1962): Zur Frage der Übertragung von Peronospora tabacina Adam durch Saatgut. *Ber. Inst. Tabakforsch., Dresden* **9,** 50–57.
Ehrlich, H. G. and Ehrlich, M. A. (1963): Electron microscopy of the host-parasite relationships in stem rust on wheat. *Amer. J. Bot.* **50,** 123–130.
Ellingboe, A. H. (1961): Somatic recombination in Puccinia graminis var. tritici. *Phytopathology* **51,** 13–15.
Elliott, E. W. (1949): The swarm cells of Myxomycetes. *Mycologia* **41,** 141–170.
Elrod, R. P. and Braun, A. C. (1947): Serological studies of the genus Xanthomonas. *J. Bacteriol.* **53,** 509–518.
Endemann, W., Ramson, A. und Egerer, A. (1962): Die Anfälligkeit verschiedener Sorten und Zuchtstämme von Nicotiana tabacum L. gegenüber Peronospora tabacina Adam. *Ber. Inst. Tabakforsch., Dresden,* **9,** 5–49.
Engler, R. and Schramm, G. (1960): Die Bildung der infektiösen Ribonucleinsäure während der Vermehrung des Tabakmosaikvirus. *Z. Naturforsch.* **15b,** 38–45.
English, H. and Davis, J. R. (1960): The source of inoculum for bacterial canker and blast of stone fruit trees. *Phytopathology* (Abstr.) **50,** 634.
Ercolani, G. L. (1968): Schizomycetes. In: *Manuale di Patologia Vegetale* III. (Ed.) G. Goidanich. Edizione Agricole, Bologna.

Eriksson, J. (1894): Über die Spezialisierung des Parasitismus bei den Getreiderostpilzen. *Ber. Dtsch. Bot. Ges.* **12,** 292–331.

Erwin, D. C., Zentmyer, G. A., Galindo, J. and Niederhauser, J. S. (1963): Variation in the genus Phytophthora. *Ann. Rev. Phytopath.* **1,** 375–396.

Esau, K. (1948): Some anatomical aspects of plant virus disease problems. II. *Bot. Rev.* **14,** 413–449.

Fantechi, F. and Lorito, N. (1965): Observations on resistance to Peronospora tabacina of several resistant hybrids and Italian cultivars. *Tobacco,* (Roma) **713,** 241–258.

Farkas, G. L., Király, Z. and Solymosy, F. (1960): Role of oxidative metabolism in the localization of plant viruses. *Virology* **12,** 408–412.

Farkas, G. L., and Solymosy, F. (1962): Activation of hydrogen and electron transport systems in a virus-induced local lesion host. *Biochem. J.* **84,** 113.

Farkas, G. L. and Solymosy, F. (1965): Host metabolism and symptom production in virus-infected plants. *Phytopath. Z.* **53,** 85–93.

Fehrmann, H. and Dimond, A. E. (1967): Peroxidase activity and Phytophthora resistance in different organs of the potato plant. *Phytopathology* **57,** 69–72.

Feinberg, J. G. (1957): Identification, discrimination and qualification in Ouchterlony gel plates. *Int. Arch. Allergy* **11,** 129–152.

Fennell, D. I. (1960): Conservation of fungous cultures. *Bot. Rev.* **26,** 79–141.

Fenwick, H. S. and Guthrie, J. W. (1969): An improved Pseudomonas phaseolicola pathogenicity test. *Phytopathology* (Abstr.) **59,** 11.

Fischer, G. W. and Holton, C. S. (1957): *Biology and Control of the Smut Fungi.* The Ronald Press Co., New York.

Flor, H. H. (1956): The complementary genetic systems in flax and flax rust. *Adv. Genetics* **8,** 29–54.

Ford, D. H. and Rawlins, T. E. (1956): Improved cytochemical methods for differentiating Cronartium ribicola from Cronartium occidentale on Ribes. *Phytopathology* **46,** 667–668.

Fraenkel-Conrat, H. (1966): Preparation and testing of tobacco mosaic virus-RNA. In: *Procedures in Nucleic Acid Research.* (Ed.) G. L. Cantoni and D. R. Davies. Harper and Row, Publ., New York and London. Pp. 480–487.

Fraenkel-Conrat, H., and Singer, B. (1962): The absence of phosphorylated chain ends in tobacco mosaic virus ribonucleic acid. *Biochemistry* **1,** 120–128.

Fraenkel-Conrat, H., Singer, B. and Tsugita, A. (1961): Purification of viral RNA by means of bentonite. *Virology* **14,** 54–58.

Fraenkel-Conrat, H., Singer, B. and Veldee, S. (1958): The mechanism of plant-virus infection. *Biochim. Biophys. Acta* **29,** 639–640.

Fraenkel-Conrat, H., Staehelin, M. and Crawford, L. V. (1959): Tobacco mosaic virus reconstitution using inactivated nucleic acid. *Proc. Soc. Exp. Biol. Med.* **102,** 118–121.

Fraenkel-Conrat, H. and Williams, R. C. (1955): Reconstitution of active tobacco mosaic virus from its inactive protein and nucleic acid components. *Proc. Nat. Acad. Sci. U. S.* **41,** 690–698.

Fraenkel-Conrat, H. and Williams, R. C. (1957): Infectivity of viral nucleic acid. *Biochem. Biophys. Acta* **25,** 87–96.

Francki, R. I. B., Boardman, N. K. and Wildman, S. G. (1965): Protein synthesis by cell free extracts from tobacco leaves. I. Amino acid incorporating activity of chloroplasts in relation to their structure. *Biochemistry* **4,** 865–872.

Frandsen, N. O. (1958): Resistenzzüchtung gegen pilzliche und bakterielle Krankheiten der Kartoffel. In: *Handbuch der Pflanzenzüchtung.* (Ed.) H. Kappert and W. Rudorf. 2. Aufl. Paul Parey, Berlin und Hamburg. Vol. 3. Pp. 71–97.

Franklin, R. E., Caspar, D. C. D. and Klug, A. (1959): The structure of viruses as determined by X-ray diffraction. In: *Plant Pathology. Problems and Progress* 1908–1958. (Ed.) C. S. Holton, et al. University of Wisconsin Press, Madison, Pp. 447–461.

Frey, K. J. (Ed.) (1966): *Plant Breeding. A Symposium Held at Iowa State University.* Iowa State Univ. Press, Ames, Iowa.

Frisch-Niggemeyer, W. and Steere, R. L. (1961): Chemical composition of partially purified alfalfa mosaic virus. *Virology* **14,** 83–87.

Fuchs, Eva (1965): Untersuchungen über die physiologische Spezialisierung des Weizengelbrostes (Puccinia striiformis West. f. sp. tritici Erikss. et Henn. in den Jahren 1959–1964 und über das Anfälligkeitsverhalten einiger Weizensorten. *Nachr. blatt Dtsch. Pfl. schutzdienstes.* **17,** 161–176.

Fuchs, Eva (1967): Vorläufige Mitteilung über das Auftreten einer neuen und gefährlichen Weizengelbrostrasse. *Nachr. blatt Dtsch. Pfl. schutzdienstes* **19,** 77–78.

Fukushi, T. (1933): Transmission of the virus through the eggs of an insect vector. *Proc. Acad. Imp. Japan* **9,** 457–460.

Fukushi, T. (1940): Further studies on the dwarf disease of rice plant. *J. Fac. Agr. Hokkaido University* **45,** 83–154.

Fulton, R. W. (1952): Mechanical transmission and properties of rose mosaic virus. *Phytopathology* **42,** 413–416.

Galindo, A. and Gallegly, M. E. (1960): The nature of sexuality in Phytophthora infestans. *Phytopathology* **50,** 123–128.

Gallegly, M. E. (1962): Greenhouse tests for multigenic resistance to Phytophthora infestans. *Phytopathology* **52,** 1218.

Gallegly, M. E. (1968): Genetics of pathogenicity of Phytophthora infestans. *Ann. Rev. Phytopath.* **6,** 375–396.

Gallegly, M. E. and Niederhauser, J. S. (1959): Genetic controls of host-parasite interactions in the Phytophthora late blight disease. In: *Plant Pathology. Problems and Progress 1908–1958.* (Ed.) C. S. Holton, et al. University of Wisconsin Press, Madison. Pp. 168–182.

Garber, E. D. (1956): A nutrition-inhibition hypothesis of pathogenicity. *Amer. Naturalist* **90,** 183–194.

Garber, E. D. (1961a): Wildfire disease of tobacco. *J. Bacteriol.* **81,** 974–978.

Garber, E. D. (1961): Nutritional aspects of the host-parasite relationship. *Rec. Adv. Botany* 1004–1007.

Garber, E. D. and Heggestad, H. E. (1958): Observations on the pathogenicity of biochemical mutants of Pseudomonas tabaci. *Phytopathology* **48,** 535–537.

Garber, E. D. and Shaeffer, S. G. (1957): Free histidine content of turnip varieties and their resistance to histidine requiring mutants of Erwinia aroideae. *J. Bacteriol.* **74,** 392–395.

Garber, R. H. and Houston, B. R. (1967): Nature of Verticillium wilt resistance in cotton. *Phytopathology* **57,** 885–888.

Garrett, C. M. E., Panagopoulos, C. G. and Crosse, J. E. (1966): Comparison of plant pathogenic pseudomonads from fruit trees. *J. Appl. Bacteriol.* **29,** 342–356.

Gaskin, T. A. and Schafer, J. F. (1962): Some histological and genetic relationships of resistance of wheat to loose smut. *Phytopathology* **52,** 602–607.

Gäumann, E. (1946): *Pflanzliche Infektionslehre.* Birkhäuser, Basel.

Gäumann, E. (1951): *Pflanzliche Infektionslehre.* 2. Augl. Birkhäuser, Basel.

Gendron, Y. and Kassanis, B. (1954): The importance of the host species in determining the action of virus inhibitors. *Ann. Appl. Biol.* **41,** 183–188.

Gicherman, G. and Loebenstein, G. (1968): Competitive inhibition by foreign nucleic acids and induced interference by yeast-RNA with the infection of tobacco mosaic virus. *Phytopathology* **58,** 405–409.

Gierer, A. and Mundry, K. W. (1958): Production of mutants of tobacco mosaic virus by chemical alteration of its ribonucleic acid, in vitro. *Nature* **182,** 1457–1458.

Gierer, A. and Schramm, G. (1956): Die Infektiosität der Nucleinsäure aus Tabakmosaikvirus. *Z. Naturforsch.* **11b** 138–142.

Ginoza, W. and Norman, A. (1957): Radiosensitive molecular weight of tobacco mosaic virus nucleic acid. *Nature* **179,** 520–521.

Girbardt, M. (1961): Licht- und elektronenmikroskopische Untersuchungen an Polystictus versicolor. II. Die Feinstruktur von Gründplasma und Mitochondrien. *Arch. Mikrobiol.* **39,** 351–359.

Gloyer, W. O. (1924): The effect of late planting on the bacterial blight of beans. *Phytopathology* **14,** 27.
Goldin, M. I. (1938): On the crystalline inclusion bodies of tobacco mosaic virus (In Russian). *Mikrobiologija* **7,** 353–359.
Goldin, M. I. (1954): *Virus Inclusion Bodies of Plant Cells.* Izd. Akad. Nauk U.S.S.R. Moskow.
Goldin, M. I. and Vostrova, N. G. (1959): Histochemical investigation of virus and non-virus inclusion bodies using Sakaqushi reaction. *Vopr. Virus.* Pp. 367–369.
Goode, M. J. (1958): Physiological specialization in Colletotrichum lagenarium. *Phytopathology* **48,** 79–83.
Gooding, G. V. and Hebert, T. T. (1967): A simple technique for purification of tobacco mosaic virus in large quantities. *Phytopathology* **57,** 1285.
Goodman, R. N. (1965): In vitro and in vivo interactions between components of mixed bacterial cultures from apple buds. *Phytopathology* **55,** 217–221.
Goodman, R. N. (1967): The protection of apple tissue against Erwinia amylovora infection by avirulent strains and three other bacterial species. *Phytopathology* **57,** 22–24.
Goodman, R. N. (Ed.) (1967): *Proceedings of the First Workshop on Phytobacteriology* (Second ed.). Univ. of Missouri, Columbia, Mo.
Goodman, R. N., Király, Z. and Zaitlin, M. (1967): *The Biochemistry and Physiology of Infectious Plant Disease.* Van Nostrand, Princeton, New Jersey.
Gordon, M. P., Singer, B. and Fraenkel-Conrat, H. (1960): The terminal phosphate groups of tobacco mosaic virus. *J. Biol. Chem.* **253,** 1014–1018.
Gordon, M. P. and Smith, C. (1960): Multiplication of tobacco mosaic virus in a normally insusceptible host. *J. Biol Chem.* **235,**
Gordon, M. P. and Smith, C. (1961): The infection of Rhoeo discolor by tobacco mosaic virus ribonucleic acid. *J. Biol. Chem.* **236,** 2762–2763.
Gordon, W. L. (1965): Pathogenic strains of Fusarium oxysporum. *Canad. J. Bot.* **43,** 1309–1318.
Goss, R. W. (1940): The relation of temperature to common and halo blight of beans. *Phytopathology* **30,** 258–264.
Goss, R. C. and Marr, J. L. (1963): Media sterilization with propylene oxide. *Iowa Acad. Sci.* **70,** 125–129.
Graham, D. C. (1964): Taxonomy of the soft rot coliform bacteria. *Ann. Rev. Phytopath.* **2,** 13–42.
Graham, D. C. and Dowson, W. Y. (1960): The coliform bacteria associated with potato black-leg and other soft rots. II. Biochemical characteristics of low-and high-temperature strains. *Ann. Appl. Biol.* **48,** 58–64.
Graham, K. M., Niederhauser, J. S. and Romero, S. (1959): Observations on races of Phytophthora infestans in Mexico during 1956–1957. *Amer. Potato J.* **36,** 196–203.
Graham, S. O. (1956): Germination responses of Ustilago tritici (Pers.) Rostr. teliospores in relation to lyophylization. I. Some factors affecting mortality before and after sublimation. *Res. Stud. State Coll. Washington* **24,** 307–317.
Gretchushnikov, A. I. (1939): Role of peroxidase in immunity against Phytophthora infestans de Bary. (In Russian.) *Izv. Acad. Sci. USSR* N. S. **25,** 683–687.
Groth, R. W. (1965): Puncture method for isolating bacterial blights of bean. *Phytopathology* **55,** 930–931.
Gupta, B. M. and Price, W. C. (1950): Production of plant virus inhibitors by fungi. *Phytopathology* **40,** 642–652.
Groves, A. B. (1965): The greenhouse production of apple and pear seedlings for experimental uses. *Proceedings of the First Workshop on Phytobacteriology*, University of Missouri, Columbia, Missouri. April. 12–16. 1965.
Haarring, F. (1930): Eine Infektionsmethode für Haferflugbrand (U. avenae Jens.) und ihre Anwendung zu Beiz- und Immunitätsversuchen in Laboratorium und Feld. *Bot. Arch.* **29,** 444–473.

Hadwiger, L. A. and Hall, C. V. (1961): The relation of pigmentation and free amino acid content with resistance to Colletotrichum lagenarium in watermelons. *Plant Dis. Reptr.* **45,** 373–374.
Halisky, P. M. (1965): Physiologic specialization and genetics of the smut fungi. III. *Botan. Rev.* **31,** 114–150.
Hamers-Casterman, C. and Jeener, R. (1957): An initial ribonuclease-sensitive phase in the multiplication of tobacco mosaic virus. *Virology* **3,** 197–206.
Hamlyn, B. M. G. (1953): Quantitative studies on the transmission of cabbage black ring spot virus by Myzus persicae (Sulz.). *Ann. Appl. Biol.* **40,** 393–402.
Hampton, R. E. and Fulton, R. W. (1961): The relation of polyphenol oxidase to instability in vitro of prune dwarf and sour cherry necrotic ringspot viruses. *Virology* **13,** 44–52.
Hansen, H. P. (1956): Correlations and interrelationships in viruses and in organisms. I. Classification and nomenclature of plant viruses. *Kgl. Vet. Landb. Arsskr.* Pp. 108–137.
Hansing, E. D. (1954): Physiologic races of Ustilago avenae and U. kolleri in Kansas. *Phytopathology* **44,** 491.
Harpaz, I. (1959): Needle transmission of a new maize virus. *Nature* **184,** 77–78.
Harrison, B. D. (1958): Ability of single aphids to transmit both avirulent and virulent strains of potato leaf roll virus. *Virology* **6,** 278–286.
Harrison, B. D. (1960): The biology of soil-borne plant viruses. *Adv. Virus Res.* **7,** 131–161.
Harrison, B. D. and Nixon, H. L. (1959): Some properties of infective preparation made by disrupting tobacco rattle virus with phenol. *J. Gen. Microbiol.* **21,** 591–599.
Hart, R. G. (1956): Morphological changes accompanying thermal denaturation of tobacco mosaic virus. *Biochim. Biophys. Acta* **20,** 388–389.
Harter, L. L. and Zaumeyer, W. J. (1941): Differentiation of physiologic races of Uromyces phaseoli typica on bean. *J. Agr. Res.* **62,** 717–731.
Hastie, A. C. (1962): Genetic recombination in the hop-wilt fungus Verticillium albo-atrum. *J. Gen. Microbiol.* **27,** 373–382.
Hastie, A. C. (1967): Mitotic recombination in conidiophores of Verticillium albo-atrum. *Nature* **214,** 249–252.
Hawker, L. E. (1965): Fine structure of fungi as revealed by electron microscopy. *Biol. Rev.* **40,** 52–92.
Hayashi, Y. (1962): Amino acid activation in tobacco leaves infected with tobacco mosaic virus. *Virology* **18,** 140–141.
Hayes, H. K., Immer, F. R. and Smith, D. C. (1955). *Methods of Plant Breeding.* McGraw-Hill Book Co., New York and London.
Hayword, A. C. (1964): Characteristics of Pseudomonas solanacearum. *J. Appl. Bacteriol.*
Hedges, F. (1926): Bacterial wilt of bean (Bacterium flaccumfaciens including comparisons with Bacterium phaseoli. *Phytopathology* **16,** 1–21.
Heinze, K. (1957): Das pflanzliche Virus im Überträger und seine Einbringung in die Pflanze. *Z. angew. Zool.* **44,** 187–227.
Heinze, K. (1959): *Phytopathogene Viren und ihre Überträger.* Duncker—Humbolt Verl., Berlin.
Henniger, H. (1959): Versuche zur Kultur verschiedener Rassen von Phytophthora infestans (Mont.) de Bary auf künstlichen Nährböden. *Phytopath. Z.* **34,** 285–306.
Henniger, H. (1965): Untersuchungen über Knollen- und Lagerfäulen der Kartoffel. I. Zur Methodik der Resistenzprüfung mit dem Erreger der bakteriellen Knollennassfäule (Pectobacterium carotovorum) var. atrosepticum (van Hall) Dowson. *Züchter* **35,** 174–180.
Henniger, H. and Bartel, W. (1963): Die Eignung des Peroxydaseaktivität-Testes zur Bestimmung der »relativen Phytophthora-Resistenz« (Feldresistenz) bei Kartoffeln. *Züchter* **33,** 86–91.
Hermansen, J. E. (1966): Use of detached leaves to test the viability of stored conidia of barley mildew (Erysiphe graminis DC.). *Kgl. Vet. Landb. Arsskr.* Pp. 61–67.
Hermansen, J. E. (1966): Identification of physiologic races by means of detached leaves. *Cereal Rust Conf.* 1964. Plant Breeding Inst., Cambridge. Pp. 90–91.

Hershey, A. D. (1946): Spontaneous mutations in bacterial viruses. *Cold Spring Harb. Symp. Quant. Biol.* **11**, 67.
Hershey, A. D. and Chase, M. (1952): Independent function of viral protein and nucleic acid in growth of bacteriophage. *J. Gen. Physiol.* **36**, 39–56.
Hewitt, W. B. (1956): Fanleaf virus of grapevines is soilborne. *Phytopathology* (Abstr.) **45**, 15.
Hidaka, Z. and Tomaru, K. (1960): Vacuum freeze-drying of tobacco leaf tissues infected with cucumber mosaic virus. *Virology* **12**, 8–13.
Hildebrand, D. C. and Schroth, M. N. (1963): Relation of arbutin-hydroquinone in pear blossoms to invasion by Erwinia amylovora. *Nature* **197**, 513.
Hildebrand, D. C. and Schroth, M. N. (1964): Arbutin-hydroquinone complex in pear as a factor in fire blight development. *Phytopathology* **54**, 640–645.
Hill, A. V. (1961): Dissemination of conidia of Peronospora tabacina Adam. *Austr. J. Biol Sci.* **14**, 208–222.
Hill, A. V. (1966): Effect of inoculum spore load, length of infection period, and leaf washing on occurrence of Peronospora tabacina Adam (blue mold) of tobacco. *Austr. J. Agric. Res.* **17**, 133–146.
Hille, M. (1965): Die Beurteilung von Kartoffelsorten hinsichtlich ihres Verhaltens gegenüber Synchytrium endobioticum (Schilb.) Perc., dem Erreger des Kartoffelkrebses. *Nachr. Blatt. Dtsch. Pfl. schutzdienstes* **17**, 137–142.
Hirai, T. (1958): Intracellular evidence of the inhibition of tobacco mosaic virus multiplication by antiviral substances. *Virology* **6**, 732–742.
Hirai, T. and Shimomura, T. (1960): The mode of action of some antibiotics in their inhibitory effect on tobacco mosaic virus multiplication. *Phytopath. Z.* **40**, 35–44.
Hirai, T. and Wildman, S. G. (1963): Cytological and cytochemical observations on the early stage of infection of tomato hair cells by tobacco mosaic virus. *Plant Cell Physiol.* **4**, 265–275.
Hiura, U. and Heta, H. (1955): Studies on the disease-resistance in barley. III. Further studies on the physiologic races of Erysiphe graminis hordei in Japan. *Ber. Ohara Inst. Landw. Biol. Okayama Univ.* **10**, 135–156.
Hodgson, W. A. (1961): Laboratory testing of the potato for partial resistance to Phytophthora infestans. *Amer. Potato J.* **38**, 259–264.
Hodgson, W. A. (1962): Studies on the nature of partial resistance in the potato to Phytophthora infestans. *Amer. Potato J.* **39**, 8–13.
Hoffmann, G. M. (1966): Heterokaryose und parasexuelle Vorgänge bei Fusarium oxysporum. *Naturwissenschaften* **53**, 45–46.
Hoffmann, J. A., Kendrick, E. L. and Meiners, J. P. (1962): Pathogenic races of Tilletia controversa in the Pacific Northwest. *Phytopathology* **52**, 1153–1157.
Hoffmann, J. A., Kendrick, E. L. and Metzger, R. J. (1967): A revised classification of pathogenic races of Tilletia controversa. *Phytopathology* **57**, 279–281.
Holley, R. W., Apgar, J., Everett, G. A., Madison, J. T., Marquissee Merrill, S. H., Penswick, J. R. and Zamir, A. (1965): Structure of ribonucleic acid. *Science* **147**, 1462–1465.
Holliday, R. (1961): Induced mitotic crossing-over in Ustilago maydis. *Genet. Res.*, Cambridge **2**, 231–238.
Hollings, M. and Lelliot, R. A. (1960): Preservation of some plant viruses by freezedrying. *Plant Pathol.* **9**, 63–66.
Holmes, F. O. (1929): Local lesions in tobacco mosaic virus. *Bot. Gaz.* **87**, 39–55.
Holmes, F. O. (1931): Local lesion of mosaic in Nicotiana tabacum L. *Contrib. Boyce Thompson Inst.* **3**, 163–172.
Holmes, F. O. (1948): Order Virales. In: *Bergey's Manual of Determinative Bacteriology* Pp. 1125–1296.
Holmes, F. O. (1954): Inheritance of resistance to viral disease in plants. *Adv. Virus Res.* **2**, 1–30.
Holowczak, J., Kuć, J. and Williams, E. B. (1962): Metabolism of DL- and L-phenylalanine

in Malus related to susceptibility and resistance to Venturia inaequalis. *Phytopathology* **52**, 699–703.
Holton, C. S. and Rodenhiser, H. A. (1948): Physiologic specialization in the oat smut fungi and its relation to breeding oats for smut resistance. *US Dept. Agr. Techn. Bull.* 952.
Holton, C. S., Hoffmann, J. A. and Duran, R. (1968): Variation in the smut fungi. *Ann. Rev. Phytopath.* **6**, 213–242.
Honecker, L. (1937): Die Bestimmung der physiologischen Rassen des Gerstenmehltaues (Erysiphe graminis/hordei Marchal). *Phytopath. Z.* **10**, 197–222.
Horváth, J. (1967): Separation and determination of viruses pathogenic to potatoes with special regard to potato virus Y. *Acta Phytopath. Hung.* **2**, 319–360.
Hrushovetz, S. B. (1957): Effect of amino acids on the virulence of Helminthosporium sativum to wheat seedlings. *Phytopathology* **47**, 261–264.
Husain, A. and Kelman, A. (1958): Relation of slime production to mechanism of wilting and pathogenicity of Pseudomonas solanacearum. *Phytopathology* **48**, 155–165.
Iren, S. (1965): Türkiye'de Tütüm mildyosü (Peronospora tabacina Adam) belirtileri, biyologisi ve mücadelesi. (Blue mold (Peronospora tabacina Adam) of tobacco in Turkey, its symptoms, biology and control.) *Bitki Koruma Bült.* **5**, 3–24.
Israel, H. W. and Ross, A. F. (1967): The fine structure of local lesions induced by tobacco mosaic virus in tobacco. *Virology* **33**, 272–286.
Izrailsky, V. P. (Ed.) (1960): *Bacterial Diseases of Plants.* (In Russian). Gosud. Izd. Selkh. Lit., Moscow.
Ivanovsky, D. (1902): *Mosaic Disease of Tobacco.* Warsaw.
Izard, C. et Schiltz, P. (1961): Quelques précisions concentrant la méthode de culture de Peronospora tabacina Adam, sur des disques découpés dans le limbre des feuilles et primiéres observations «in vivo». *Ann. Inst. Exp. du Tabac de Bergerac.* **3**, 1–10.
Jankowski, F. (1965): Techniques of obtaining and conserving Peronospora tabacina conidia: the preparation of inoculum for tobacco resistance studies using the cotyledon test during the interseasonal period. *Biul. Centr. Lab. Przem. Tyron.* **1–2**, 29–34.
Jeener, R. (1954): A preliminary study of the incorporation in growing turnip yellow mosaic virus and its related non-infective antigen of labelled amino acids. *Biochim. Biophys. Acta* **13**, 307–308.
Jeener, R. (1956): Le role de l'acide ribonucléique dans la multiplication des virus. *Proc. 3rd Int. Congr. Biochem. Brussels.* Pp. 343–344.
Jensen, D. D. (1959): A plant virus lethal to its insect vector. *Virology* **8**, 164–175.
Jockusch, H. (1966): The role of host genes, temperature and polyphenoloxidase in the necrotization of TMV infected tobacco tissue. *Phytopath. Z.* **55**, 185–192.
Johnson, J. (1927): The classification of plant viruses. *Wisc. Agr. Exp. Sta. Res. Bull.* 76.
Johnson, J. (1951): Virus particles in various plant species and tissues. *Phytopathology* **41**, 78–93.
Johnson, J. (1927): The classification of plant viruses. *Wisc. Agr. Exp. Sta. Res. Bull.* 76.
Johnson, J. and Hoggan, I. A. (1935): A descriptive key for plant viruses. *Phytopathology* **25**, 328–343.
Johnson, L. F., Curl, E. A., Bond, L. H. and Fribourg, H. A. (1959): *Methods for Studying Soil Microflora-Plant Disease Relationships.* Burgess Publ. Co., Minneapolis.
Johnson, L. F. and Manka, K. (1961): A modification of Warcup's soil-plate method for isolating soil fungi. *Soil Sci.* **92**, 79–84.
Johnston, C. O. (1956): Unified number for races of Puccinia triticina. *Robigo* **1**, 2.
Johnston, C. O. and Browder, L. E. (1966): Seventh revision of the international register of physiologic races of Puccinia recondita f. sp. tritici. *Plant Dis. Reptr.* **50**, 756–760.
Kameraz, A. Ya. (1964): Breeding of potato for resistance to phytophthorosis and viruses (In Russian). *Sborn. Trud. Vses. Inst. Rast.* **2**, 13–19.
Kammermann, N. (1951): Undersökningar rörande potatisbladmöglet Phytophthora infestans (Mont.) de Bary. II. Sambandet mellan potatisbladsaftens proxidasa-aktivitet och Phytophthora resistens. *Statens Wáxtskyddsanstalt Mitteilung* Nr. 58. Stockholm.

Kaper, J. M. and Steere, R. L. (1959a): Infectivity of tobacco ringspot virus nucleic acid preparations. *Virology* **7**, 127–139.
Kaper, J. M. and Steere, R. L. (1959b): Isolation and preliminary studies of soluble protein and infectious nucleic acid from turnip yellow mosaic virus. *Virology* **8**, 527–530.
Kappert, H. und Rudorf, W. (1955, continuously): *Handbuch der Pflanzenzüchtung.* Paul Parey, Berlin.
Karamisheva, U. I. (1965): Resistance of apple trees to scab in the north-west zone (In Russian). *Sb. Trud, Aspir. molod. nauk. Sorrud. vsei Inst. Rast.* **6**, 279–282.
Kassanis, B. (1939): Intranuclear inclusions in virus-infected plants. *Ann. Appl. Biol.* **26**, 705–709.
Kassanis, B. (1952): Some effects of high temperature on the susceptibility of plants to infection with viruses. *Ann. Appl. Biol.* **39**, 358–369.
Kassanis, B. (1960): Comparison of the early stages of infection by intact and phenol-disrupted tobacco necrosis virus. *Virology* **10**, 353–369.
Kassanis, B. (1961): The transmission of potato aucuba mosaic virus by aphids from plants also infected by potato viruses A or Y. *Virology* **13**, 93–97.
Katznelson, H. and Sutton, M. D. (1951): A rapid phage plaque count method for detection of bacteria as applied to the demonstration of internally borne bacterial infections of seed. *J. Bacteriol.* **61**, 689–701.
Katznelson, H., Sutton, M. D. and Bailey, S. T. (1954): The use of bacteriophage of Xanthomonas phaseoli in detecting infection in beans with observations of its growth and morphology. *Canad. J. Microbiol.* **1**, 22–29.
Kavanagh, T. (1961): Inoculating barley seedlings with Ustilago nuda and wheat seedlings with U. tritici. *Phytopathology* **51**, 175–177.
Kedar (Kammermann), N. (1959): The peroxidase test as a tool in the selection of potato varieties resistant to late blight. *Amer. Potato J.* **36**, 315–324.
Keitt, G. W. and Palmiter, D. H. (1938): Heterothallism and variability in Venturia inaequalis. *Amer. J. Bot.* **25**, 338–345.
Kelman, A. (1953): The bacterial wilt caused by Pseudomonas solanacearum. A literature review and bibliography. *North Carolina Agr. Exp. Sta. Tech. Bull.* 99. Pp. 1–149.
Kelman, A. (1954): The relationship of pathogenicity in Pseudomonas solanacearum to colony appearance on a tetrazolium chloride medium. *Phytopathology* **44**, 693–695.
Kelman, A. (1956): Factors influencing viability and variation in cultures of Pseudomonas solanacearum. *Phytopathology* **46**, 16–17.
Kelman, A. and Person, L. H. (1961): Strains of Pseudomonas solnacearum differing in pathogenecity to tobacco and peanut. *Phytopathology* **51**, 158–161.
Kendrick, E. L. (1961): Race groups of Tilletia caries and Tilletia foetida for varietal-resistance testing. *Phytopathology* **51**, 537–540.
Kendrick, E. L. (1965): The reaction of varieties and hybrid selections of winter wheat to pathogenic races of Tilletia caries and T. foetida. *Plant Dis. Reptr.* **49**, 843–846.
Kendrick, E. L. and Hoffmann, J. A. (1963): Reactions of wheat varieties and selections to pathogenic races of Tilletia controversa. *Plant Dis. Reptr.* **47**, 736–738.
Kennedy, B. W. and Crosse, J. E. (1966): Inoculation procedure for comparing reaction of soybeans to bacterial blight. *Plant Dis. Reptr.* **50**, 560–565.
Kikuchi, M. and Yamaguchi, A. (1960): Polyphenol oxidase activity of Nicotiana glutinosa leaves infected with tobacco mosaic virus. *Nature* **187**, 1048–1049.
Király, Z. (1968): *A növényi betegségellenállóság élettana. (Physiology of Plant Disease Resistance)* Akadémiai Kiadó, Budapest.
Király, Z., El Hammady, M. and Pozsár, B. I. (1968): Susceptibility to tobacco mosaic virus in relation to RNA and protein synthesis in tobacco and bean plants. *Phytopath. Z.* **63**, 47–63.
Király, Z. and Klement, Z. (1968): Inhibition of plant virus infections by saprophytic bacteria and foreign proteins. Paper presented at *Nat. Conf. Gen. and Appl. Microbiol., Bucharest.*
Király, Z. and Lelley, J. (1956): Contribution to the hypersensitive reaction of wheat to loose smut (Ustilago tritici [Pers.] Rostr.) infection. *Phytopath. Z.* **26**, 143–146.

Kirby, A. H. M., and Frick, E. L. (1963): Greenhouse evaluation of chemicals for control of powdery mildews. I. A. method suitable for apple and barley. *Ann. Appl. Biol.* **51**, 51–60.

Klarman, W. L. and Craig, J. (1960): Sterilization of agar media with propylene oxide. *Phytopathology* **50**, 868.

Klein, D. T. and Klein, R. M. (1953): Transmittance of tumor inducing ability to avirulent crown-gall and related bacteria. *J. Bacteriol.* **66**, 220–228.

Klein, D. T. and Klein, R. M. (1956): Quantitative aspect of transformation of virulence in Agrobacterium tumefaciens. *J. Bacteriol.* **72**, 308–313.

Klement, Z. (1963): Rapid detection of the pathogenicity of phytopathogenic pseudomonads. *Nature* **199**, 299–300.

Klement, Z. (1965): Method of obtaining fluid from the intercellular spaces of foliage and the fluid's merit as substance for phytobacterial pathogens. *Phytopathology* **55**, 1033–1034.

Klement, Z. (1968): Pathogenicity factors in regards to relationships of phytopathogenic bacteria. *Phytopathology* **58**, 1218–1221.

Klement, Z., Farkas, G. L. and Lovrekovich, L. (1964): Hypersensitive reaction induced by phytopathogenic bacteria in the tobacco leaf. *Phytopathology* **54**, 474–477.

Klement, Z. and Goodman, R. N. (1966): Hypersensitive reaction induced in apple shoots by an avirulent form of Erwinia amylovora. *Acta Phytopath. Hung.* **1**, 177–178.

Klement, Z. and Goodman, R. N. (1967): The role of the living bacterial cell and induction time in the hypersensitive reaction of tobacco plants. *Phytopathology* **57**, 322–323.

Klement, Z. and Goodman, R. N. (1967): The hypersensitive reaction to infection by bacterial plant pathogens. *Ann. Rev. Phytopath.* **5**, 17–44.

Klement, Z., Király, Z. and Pozsár, B. I. (1966): Suppression of virus multiplication and local lesion production in tobacco following inoculation with a saprophytic bacterium. *Acta Phytopath. Hung.* **1**, 11–18.

Klement, Z. and Lovrekovich, L. (1960): Comparative study of Pseudomonas species affecting hemp and the mulberry tree. *Acta Microbiol. Hung.* **7**, 113–119.

Klement, Z. and Lovrekovich, L. (1961): Identification of phytopathogenic bacteria by means of phages. *Conf. Sci. Problems Plant Prot.* Vol. 1. *Phytopathology* (Ed.) Z. Király, Budapest, 1960. Pp. 187–192.

Klinkowski, M. (1958): *Pflanzliche Virologie.* Bd. I. u. II. Akademie Verlag, Berlin.

Klinkowski, M. and Schmiedeknecht, M. (1960): Der falsche Mehltau des Tabaks, Peronospora tabacina Adam, eine für Deutschland bisher unbekannte Tabakkrankheit. *Nachr. blatt Dtsch. Pfl. schutzdienst.* **14**, 61–74.

Knight, C. A. (1963): Chemistry of viruses. *Protoplasmatologia*, Bd. IV. Springer Verl. Wien.

Knight, C. A. and Woody, B. R. (1958): Phosphorus content of tobacco mosaic virus. *Arch. Biochem. Biophys.* **78**, 460–467.

Koch, W. J. (1956): Studies on the motile cells of Chytridis. I. *Amer. J. Bot.* **43**, 811–819.

Koch, W. J. (1958): Studies on the motile cells of Chytridis. II. *Amer. J. Bot.* **45**, 59–72.

Köhler, E. (1955): Beitrag zum Prämunitätsproblem. Studien am Tabakmosaik. *Phytopath. Z.* **23**, 323–327.

Köhler, E. (1956): Über die Ausbreitung von Mosaikviren in der Tabakpflanze. I. Das Verhalten der Kartoffelviren X und Y *Phytopath. Z.* **26**, 147–160.

Köhler, H. und Klinkowski, M. (1954): *Viruskrankheiten.* Paul Parey, Berlin and Hamburg.

Kreitlow, K. W. (1961): The pipe cleaner, an inexpensive tool for inoculating plants with viruses. *Phytopathology* **51**, 808–809.

Kröber, H. (1965): Über die Lebensdauer der Konidien von Peronospora tabacina Adam. *Phytopath. Z.* **54**, 328–334.

Kröber, H. and Massfeller, D. (1962): Untersuchungen über das Resistenzerhalten bei Tabak gegen Peronospora tabacina Adam. *Nachr. blatt Dtsch. Pfl. schutzdienst.* **14**, 82–85.

Kröber, H. and Weinmann, W. (1964): Der Nachweis von Blauschimmelinfektionen durch Oosporen von Peronospora tabacina. Adam. *Phytopath. Z.* **51**, 79–84.

Kubo, S. (1966): Chromatographic studies of RNA synthesis in tobacco leaf tissues infected with tobacco mosaic virus. *Virology* **28**, 229–235.

Kubo, S., Tomaru, K., Nitta, T., Shiroya, T. and Hidaka, Z. (1965): Chromatography of tobacco mosaic virus, its constituents, and nucleic acids extracted from infected tobacco leaf tissues. *Virology* **26**, 406–412.
Kunkel, L. O. (1940): Genetics of viruses pathogenic to plants. *Publ. Amer. Assoc. Adv. Sci.* **12**, 22–27.
Lammering, J. (1965): Six pathogenic races of Plasmodiophora brassicae Wor. in New Zealand. *New Zealand J. Agr. Res.* **8**, 156–164.
Lange-de la Camp, Maria (1936): Gewinnung und Kultur der Haplonten von Ustilago tritici. *Phytopath. Z.* **9**, 455–477.
Lapwood, D. H. (1961a): Laboratory assessments of the susceptibility of potato haulm to blight (Phytophthora infestans [Mont.] de Bary). *Europ. Potato J.* **4**, 117.
Lapwood, D. H. (1961b): Potato haulm resistance to Phytophthora infestans. I. Field assessment of resistance. *Ann. Appl. Biol.* **49**, 140–151.
Lapwood, D. H. (1963): Potato haulm resistance to Phytophthora infestans. IV. Laboratory and field estimates compared, and further field analyses. *Ann. Appl. Biol.* **51**, 17–28.
Lauffer, M. A., Trkula, D. and Buzzell, A. (1956): Mechanism of inactivation of tobacco mosaic virus by X-rays. *Nature* **177**, 890.
Lea, H. W. (1961): A review of two commercial flue-cured blue mould resistant tobacco varieties. *Extrait de Bull. Inf. Coresta* 1961. Pp. 21–27.
Leach, J. G., Lilly, V. G., Wilson, H. A. and Purvis, M. R. (1957): Bacterial polysaccharides: The nature and function of the exudate produced by Xanthomonas phaseoli. *Phytopathology* **47**, 113–120.
Lederberg, J. and Lederberg, E. M. (1952): Replica plating and indirect selection of bacterial mutants. *J. Bacteriol.* **63**, 399–406.
Lee, C. L. and Black, L. M. (1955): Anatomical studies of Trifolium incarnaturm infected by wound-tumor virus. *Amer. J. Bot.* **42**, 160–168.
Lelliott, R. A. (1956): Slow wilt of carnations caused by a species of Erwinia. *Plant Path.* **5**, 19–23.
Lelliott, R. A. (1967): Taxonomy and determination of phytopathogenic Pseudomonadales. *2nd Intern. Conf. Phytopathogenic Bacteria*, Oeiras, Portugal.
Lelliott, R. A., Billing, E. and Hayward, A. C. (1966): A determinative scheme for the fluorescent plant pathogenic pseudomonads. *J. Appl. Bacteriol.* **29**, 470–489.
Limasset, P. (1956): Les virus se multipliant dans l'organisme de leurs insectes vecteurs et les «gros virus» des plantes. *Rev. Path. Gén. Phys. Clin.* **674**, 211–234.
Lincoln, R. E. and Gowen, J. W. (1942): Mutations of Phytomonas stewarti by X-ray irradiation. *Genetics* **27**, 441–462.
Lippincott, J. A. (1961): Properties of infectious ribonucleic acid preparations obtained from tobacco mosaic virus by a heat method. *Virology* **13**, 348–362.
Lippincott, J. A. and Commoner, B. (1956): Reactivation of tobacco mosaic virus infectivity in mixtures of virus protein and nucleic acid. *Biochim. Biophys. Acta* **19**, 198–199.
Litvinov, M. A. (1964): The principal trends and methods of taxonomic studies on the causal agents of Verticillium wilt of plants. (In Russian) *Trans. of the Allunion Symp. on Cotton Wilt Control Tashkent*, Izdat. «Uzbekiztan», 1964. Pp. 17–18.
Littau, V. C. and Maramorosch, K. (1960): A study of the cytological effects of aster yellows virus on its insect vector. *Virology* **10**, 483–500.
Littrell, R. H. and Epps, W. M. (1965): Standardization of a procedure for artificial inoculation of cucumbers with Colletotrichum lagenarium. *Plant Dis. Reptr.* **49**, 649–653.
Loebenstein, G. (1960): Native protein, an immunizing agent against virus infection. *Nature* **185**, 122.
Loebenstein, G. (1962): Inducing partial protection in the host plant with native virus protein. *Virology* **17**, 574–581.
Loebenstein, G. and Lovrekovich, L. (1966): Interference with tobacco mosaic virus local lesion formation in tobacco by injecting heat-killed cells of Pseudomonas syringae. *Virology* **30**, 587–591.

Loebenstein, G., Rabina, S. and van Praagh, T. (1965): Induced interference phenomena in virus infections. In: *Viruses of Plants.* (Ed.) A. B. R. Beemster and J. Dijkstra. Pp. 151–157.

Loebenstein, G., Rabina, S. and van Praagh, T. (1968): Sensitivity of induced localized acquired resistance to actinomycin D. *Virology* **34,** 264–268.

Loegering, W. Y., McKinney, H. H., Harmon, D. L. and Clark, W. A. (1961): A long term experiment for preservation of urediospores of Puccinia graminis tritici in liquid nitrogen. *Plant Dis. Reptr.* **45,** 384–385.

Lovrekovich, L. (1959): Egyszerű szerológiai mikromódszer a burgonya X-virus kimutatására. (A simple serological micromethod for detection of potato X virus). *Növénytermelés* **8,** 73–76.

Lovrekovich, L. and Farkas, G. L. (1965): Induced protection against wildfrie disease in tobacco leaves treated with heat-killed bacteria. *Nature* **205,** 823–824.

Lovrekovich, L. and Klement, Z. (1961): Species-specific antigens of Pseudomonas tabaci. *Acta Microbiol. Hung.* **8,** 303–310.

Lovrekovich, L. and Klement, Z. (1963): A practical method to demonstrate the bacterial infection of bean seeds. *Acta Agronomica Hung.* **12,** 83–88.

Lovrekovich, L. and Klement, Z. (1965): Serological and bacteriophage sensitivity studies on Xanthomonas vesicatoria strains isolated from tomato and pepper. *Phytopath. Z.* **52,** 222–228.

Lovrekovich, L., Klement, Z. and Dowson, W. J. (1963): Serological investigation of Pseudomonas syringae and Pseudomonas morsprunorum strains. *Phytopath. Z.* **47,** 19–24.

Lovrekovich, L., Lovrekovich, H. and Stahmann, M. A. (1967): Inhibition of phenol oxidation by Erwinia carotovora in potato tuber tissue and its significance in disease resistance. *Phytopathology* **57,** 737–742.

Luke, H. H. (1964): Identification and distribution of oat smuts of the Southeastern United States. *Phytopathology* **54,** 792–794.

Luke, H. H., Pfahler, P. L. and Hadden, S. J. (1961): Oat smut races of the South Atlantic States. *Phytopathology* **51,** 490–491.

Lyr, H. (1957): Ein neues Peroxydase-Bestimmungsverfahren. *Biochem. Z.* **329,** 91–96.

Macer, R. C. F. and Doling, D. A. (1966): Identification and possible origin of a race of Puccinia striiformis isolated in 1966. *Nature* **212,** 643.

MacMillan, P. C. and Brandt, W. H. (1966): Possible role of a peroxidative system in melanin synthesis in Verticillium. *Antonie van Leeuwenhoek* **32,** 202–211.

MacNeill, B. H. and Barron, G. L. (1966): Avirulence in prototrophs of Penicillium expansum. *Canad. J. Bot.* **44,** 355–358.

MacNeill, B. H. and Sabanayagam, J. V. (1968): The induction of dodine tolerance in Fusarium oxysporum f. melonis: A technique applicable to the study of the bionomics of soil-borne fungal pathogens. *Canad. J. Microbiol.* **14,** 1262–1263.

Magdoff, B. S. (1960): Sub-units in southern bean mosaic virus. *Nature* **185,** 673–674.

Malca, I., Erwin, D. C., Moje, W. and Jones, B. (1966): Effect of pH and carbon and nitrogen sources on the growth of Verticillium albo-atrum. *Phytopathology* **56,** 401–406.

Malec, K. (1964): Z badai nad powstawanien nowych baradzeij wirulentnych biotypow gryzba Synchytrium endobioticum (Schilb.) Perc. (Investigations on the formation of new more virulent biotypes of Synchytrium endobioticum (Schilb.) Perc.) *Hodowla Rosl. Aklim. Nasienn.* **8,** 665–675.

Malec, K. (1965): Metodyka badania rakoodpornosci materialow hodowlanych Ziewniaka. (Method for investigating resistance of potato breeding material to wart.) *Biul. Inst. Hodowl. Alchim. Rosl.* **5–6,** 143–148.

Mandel, M. (1967): DNA base composition and taxonomy of Erwinia. Intern. *2nd, Conf. Phytopathogenic Bacteria,* Oeiras, Portugal.

Mandeles, S. (1967): Base sequence of the 5' linked terminus of tobacco mosaic virus ribonucleic acid. *J. Biol. Chem.* **242,** 3103–3107.

Manil, P. (1947): Action négative sur le virus de la Mosaique du Tabac, de la pénicilline et du prontosil. *Ann. Inst. Pasteur* **73,** 294.

Maramorosch, K. (1952): New evidence for multiplication of aster yellows virus in its vector. *Phytopathology* **42,** 14.
Maramorosch, K. (1955): Multiplication of plant viruses in insect vectors. *Adv. Virus Res.* **3,** 221–249.
Maramorosch, K. (1958): Viruses that infect and multiply in both plants and insects. *Trans. New York Acad. Sci. Ser.* II. **20,** 383–393.
Maramorosch, K. (1960): Leafhopper-transmitted plant viruses. *Protoplasma* **52,** 457–466.
Markham, R. (1959): The biochemistry of plant viruses. In: *The Viruses.* Vol. 2. Pp. 33–125. Academic Press, New York and London.
Markham, R. (1963): Plant virus nucleic acids. *Progr. Nucl. Acid Res.* **2,** 61–81.
Markham, R. and Smith, K. M. (1949): Studies on the virus of turnip yellow mosaic. *Parasitology* **39,** 330–342.
Markham, R. and Smith, J. D. (1952): The structure of ribonucleic acid. 2. The smaller products of ribonuclease digestion. *Biochem. J.* **52,** 558–571.
Martin, G. W. (1955): Are fungi plants? *Mycologia* **47,** 779–792.
Martin, G. W. (1960): The systematic position of the Myxomycetes. *Mycologia* **52,** 119–129.
Martinson, C. A. and Englander, L. (1967): Use of growth rates at high temperatures to confirm identifications of Verticillium albo-atrum and Verticillium dahliae. *Phytopathology* (Abstr.) **57,** 821.
Matsushita, K. (1965): Tobacco mosaic virus in chloroplast and cytoplasm of infected tobacco leaf. *Plant and Cell. Physiol.* **6,** 1–6.
Matthews, R. E. F. (1951): Effect of some substituted purines on the development of plant virus infection. *Nature* **167,** 892.
Matthews, R. E. F. (1953): Incorporation of 8-azaguanine into nucleic acid of tobacco mosaic virus. *Nature* **171,** 1065.
Matthews R. E. F. (1956): Thiouracil in tobacco mosaic virus. *Biochim. Biophys. Acta* **19,** 559.
Matthews, R. E. F. (1957): *Plant Virus Serology.* Univ. Press Cambridge.
Matthews, R. E. F. (1959): Virus inactivation in vitro and in vivo. In: *Plant Pathology. An Advanced Treatise.* (Ed.) J. G. Horsfall and A. E. Dimond. Academic Press, New York and London.
Matthews, R. E. F. (1960): Properties of nucleoprotein fractions isolated from turnip yellow mosaic preparations. *Virology* **12,** 521–539.
Matthews, R. E. F., Bolton, E. T. and Thompson, H. R. (1963): Kinetics of labelling of turnip yellow mosaic virus with P^{32} and S^{35}. *Virology* **19,** 179–189.
Mazzone, H. M., Incardona, N. L. and Kaesberg, P. (1962): Biochemical and biophysical properties of squash mosaic virus and related macromolecules. *Biochim. Biophys. Acta* **55,** 164–175.
McKinney, H. H. (1944b): Genera of the plant viruses. *J. Washington Acad. Sci.* **34,** 139–154.
Medina, A. C. and Grogan, R. G. (1961): Seed transmission of bean mosaic virus. *Phytopathology* **51,** 452–456.
Mehta, B. K., Frederiksen, R. A., Collier, J. and Futrell, M. C. (1967): Evaluation of physiologic specialization in Sphacelotheca reiliana. *Phytopathology* **57,** 925–928.
Meiners, J. P. (1959): Methods of infecting wheat with the dwarf bunt fungus. *Phytopathology* **49,** 4–8.
Menzites, J. D. and Griebel, G. E. (1967): Survival and saprophytic growth of Verticillium dahliae in uncropped soil. *Phytopathology* **57,** 703–709.
Merrett, M. J. (1962a): The effect of uncoupling agents on the multiplication of tomato-aucuba mosaic virus in tobacco leaves. *Physiol. Plant.* **15,** 200–200.
Merrett, M. J. (1962b): Oxidase activity of tissues systemically infected by tobacco mosaic virus. *Physiol. Plant.* **15,** 465–472.
Michail, S. H. and Carr, A. J. H. (1966): Use of culture filtrates as a rapid technique for screening lucerne for resistance to Verticillium albo-atrum. *Trans. Brit. Mycol. Soc.* **49,** 133–138.
Middendorf, M. (1958): Untersuchungen über Methoden zur Infektion mit Maisbrand

(Ustilago zeae (Beckm.) Unger) und ihre Abhängigkeit von Alter, Temperatur und Sorte. *Züchter* **28**, 80–94.
Miličić, D. and Udjbinać, Z. (1960): Die Eiweisskristalle von Capsicum annuum sind Viruskörper. *Protoplasma* **52**, 446–456.
Milinkó, I. (1956): Vizsgálatok és védekezési kísérletek a paradicsom vírusbetegségeinek leküzdésére. (Investigations and control measures of virus diseases of tomato) *Növénytermelés* **5**, 165–176.
Milinkó, I. (1958a): Dohánymozaik-vírus ellenállóságra nemesítés kilátásai paradicsomnál. (Prospects of breeding for TMV resistance of tomato.) *Kísérl. Közlem.* Pp. 85–93.
Milinkó, I. (1958b): Hozzászólás Mészöly Gyula: Rezisztenciára-nemesítés feladatai és módszerei c. előadásához. (Comments on the lecture by Gyula Mészöly on Problems and methods of breeding for disease resistance) *MTA Agrártud. Oszt. Közl.* **XIV**, Pp. 211–213.
Milinkó, I. (1961): Vizsgálatok és védekezési kísérletek a paradicsomot fertőző dohánymozaikvírus leküzdésére. (Investigations and control measures for tobacco mosaic virus on tomato). Thesis Pp. 1–194.
Milinkó, I. (1962): Paradicsom rezisztenciára-nemesítési munkánk növénykórtani eredménye. (Plant pathological results of breeding tomato for disease resistance). *Duna Tisza Közi Mezőg. Kísérl. Int. Évkönyve, Kecskemét.* Pp. 19–34.
Miller, C. E. (1958): Morphology and cytology of the zoosporangia and cystosori of Sorosphaera veronicae. Jr. E. Mitchell *Sci. Soc.* **74**, 49–64.
Miller, G. L. and Stanley, W. M. (1941): Derivatives of tobacco mosaic virus. I. Acetyl and phenylureido virus. *J. Biol. Chem.* **141**, 905–920.
Miller, H. N. and McFadden, L. A. (1961): A bacterial disease of Philodendron. *Phytopathology* **51**, 826–831.
Milo, G. E. and Santilli, V. (1967): Changes in the ascorbate concentration of Pinto bean leaves accompanying the formation .of TMV-induced local lesions. *Virology* **31**, 197–206.
Minsov, I. N. and Peruanskii, Yu. V (1965): On a method of submerged cultivation of Piricularia oryzae and Trichometasphaeria turcica. (In Russian). *Vest. Sel'-khoz Nauki, Alma-Ata* **8**, 20–24.
Mirzabekyan, R. O. (1965): Culture of isolated tissues from warty outgrowth of potato. (In Russian). *Trudy Inst. Genet.* **35**, 83–87.
Miura, K. I., Kimura, I. and Suzuki, N. (1966): Double-stranded ribonucleic acid from rice dwarf virus. *Virology* **28**, 571–579.
Miyamoto, Y. (1959a): The nature of soil transmission in soil-borne plant viruses. *Virology* **7**, 250–251.
Miyamoto, Y. (1959b): Studies on soil-borne cereal mosaics. VI. Further studies on the nature of soil transmission in soil-borne cereal mosaic viruses. Ann. Phytopath. Soc. Japan 24, 207–212.
Moericke, V. (1961): Virusartige Körper in Speicheldrüsen von kartoffelblattrollinfektiösen Myzus persicae (Sulz.). *Z. Pflanzenkrankh. u. Pflanzenschutz* **68**, 581–587.
Moore, M. B. (1936): A method of inoculating wheat and barley with loose smuts. *Phytopathology* **26** 397–400.
Moore, R. T. (,1963): Fine structure of Mycota. 10. Thallus formation in Puccinia podophylli aecia. *Mycologia* **55**, 644–652.
Moore, R. T. (1965): The ultrastructure of fungal cells. In: *The Fungi.* (Ed.) G. C. Ainsworth and A. S. Sussman. Academic Press New York.
Moore, R. T. and McAlear, J. H. (1960): Fine structure of Mycota. 2. Demonstration of the haustoria of Lichens. *Mycologia* **52**, 805–807.
Moore, R. T. and McAlear, J. H. (1961): Fine structure of Mycota. 5. Lomasomes previously uncharacterized hyphal structures. *Mycologia* **53**, 194–200.
Moore, R. T. and McAlear, J. H. (1962): Fine structure of Mycota. 7. Observations on septa of Ascomycetes and Basidiomycetes. *Amer. J. Bot.* **49**, 86–94.
Morton, D. J. (1961): Trypan blue and boiling lactophenol for staining and clearing barley tissues infected with Ustilago nuda. *Phytopathology* **51**, 27–29.

Moseman, J. G. (1956): Physiological races of Erysiphe graminis f. sp. hordei in North America. *Phytopathology* **46**, 318–322.
Moseman, J. G. (1964): Inheritance of pathogenicity in Erysiphe graminis f. sp. hordei in relation to breeding for resistance to powdery mildew. *Crop. Sci.* **4**, 61–66.
Moseman, J. G. (1966): Genetics of powdery mildews. *Ann. Rev. Phytopath.* **4**, 269–290.
Moseman, J. G. and Greeley, L. W. (1966): Reaction of one barley plant to several cultures of Erysiphe graminis f. sp. hordei. *Phytopathology* **56**, 1428–1429.
Moseman, J. G. and Powers, H. R., Jr. (1957): Function and longevity of cleistothecia of Erysiphe graminis f. sp. hordei. *Phytopathology* **47**, 53–56.
Moseman, J. G. and Reid, D. A. (1958): Reaction of spring and winter barley to cultures of four races and culture P35 of Ustilago nuda. *Plant Dis. Reptr.* **42**, 614–617.
Moseman, J. G., Wiberg, A. and Greeley, L. W. (1965): Genes conditioning pathogenicity in Erysiphe graminis f. sp. hordei on 23 barley varieties. *Phytopath. Z.* **52**, 209–221.
Mumford, D. L. and Rasmusson, D. C. (1963): Resistance of barley to Ustilago nuda after embryo infection. *Phytopathology* **53**, 125–128.
Mundry, K. W. (1959): The effect of nitrous acid on tobacco mosaic virus: Mutation, not selection. *Virology* **9**, 722–726.
Mundry, K. W. and Gierer, A. (1958): Die Erzeugung von Mutation des Tabakmosaikvirus durch chemische Behandlung seiner Nucleinsäure in vitro. *Z. Vererbungslehre* **89**, 614–630.
Murti, V. D., Venkatareddy, T. C., Kotesvararao, D. and Govindarao, P. (1963): Nuinidhan: A blast resistant rice from Andhra. *Andhra Agric. J.* **10**, 168–172.
Müller, K. O. (1928): Über die Züchtung krautfäuleresistenter Kartoffelsorten. *Z. Pflanzenzücht.* **13**, 143–156.
Müller, K. O. (1933): Über die Biotypen von Phytophthora infestans und ihre Verbreitung in Deutschland. *Nachrblatt Dtsch. Pfl. schutzdienst.* **13**, 91.
Müller, K. O. (1956): Einige einfache Versuche zum Nachweis von Phytoalexinen. *Phytopath. Z.* **27** 237–254.
Müller, R. (1957): Untersuchungen über die Biologie und Bekämpfung des Apfelmehltaues Podosphaera leucotricha (Ell. et Ev.) Salm. *Dissert.* Hohenheim.
Nelson, R. R. (1956): Transmission of factors for uredospore color in Puccinia graminis var. tritici by means of nuclear exchange between hyphae. *Phytopathology* **46**, 538–540.
Nelson, G. A. and Olsen, O. A. (1967): Staining reactions of resting sporangia of Synchytrium endobioticum with a tetrazolium compound. *Phytopathology* **57**, 965–968.
Neu, H. C. and Heppel, L. A. (1964): Nucleotide sequence analysis of polyribonucleotides by means of periodate oxidation followed by cleavage with an amine. *J. Biol. Chem.* **239**, 2927–2934.
Newell, J. (1954): Milk spray cured tomato mosaic. *Grower* **41**, 1409.
Newmark, P. and Frazer, D. (1956): Composition of an abnormal protein present in tobacco plant inefected with tobacco mosaic virus. *J. Amer. Chem. Soc.* **78**, 1588–1590.
Newton, M. and Cherewick, W. J. (1947): Erysiphe graminis in Canada. *Canad. J. Res.*, Sect. C. **25**, 73–93.
Niederhauser, J. S. (1962): Evaluation of multigenic „field resistance" of the potato to Phytophthora infestans in 10 years of trials at Toluca Mexico. *Phytopathology* (Abstr.) **52**, 746.
Niederhauser, J. S. (1968): Resistance to Phytophthora infestans in Mexico. *First Int. Congr. Pl. Pathology, London*, Abstr. of Papers p. 138.
Niederhauser, J. S. and Mills, W. R. (1953): Resistance of Solanum species to Phytophthora infestans in Mexico. *Phytopathology* **43**, 456–457.
Niemann, E. (1956): Taxonomie und Keimungsphysiologie der Tilletia-Arten von Getreide und Wildgrasern. *Phytopath. Z.* **28**, 113–166.
Niemann, G. J. (1964): Investigations on the chemical control of powdery mildew. *Dissert.* Univ. of Utrecht. Pp. 1–128.
Nixon, H. L. and Gibbs, A. J. (1960): Electron microscope observations on the structure of turnip yellow mosaic virus. *J. Molec. Biol.* **2**, 197–200.

Noble, M. and Graham, D. C. (1956): A selective medium for the isolation of coliform soft rot bacteria from plant tissue. *Nature* **178,** 1479–1480.
Novacky, A. (1964): Studium sneti zakrpatenej (Tilletia controversa Kühn). Kandidátska dizertacná práca. Bratislava. (Studies on dwarf bunt, Tilletia controversa Kühn. Dissertation for Candidate of Science. Bratislava.)
Nover, I. (1957): Sechsjährige Beobachtungen über die physiologische Spezialisierung des echten Mehltaus (Erysiphe graminis DC.) von Weizen und Gerste in Deutschland. *Phytopath. Z.* **31,** 85–107.
Nover, I. (1963): Probleme der Resistenzphysiologie. Zur Frage der biologischen Spezialisierung des Getreidemehltaus. *Sitz. ber. Deutsch. Akad. Landw. wiss. Berlin* **12,** 23–30.
Noviello, C. and Korf, R. P. (1961): A simple technique for investigating stromatal formation in the Sclerotiniaceae. *Mycologia* **53,** 237–243.
O'Brien, L. M., Morton, D. J., Manning, W. J. and Scheetz, R. W. (1967): Serological differences between apparently typical pepper and tomato isolates of Xanthomonas vesicatoria. *Nature* **215,** 532–533.
Ohms, R. E. and Bever, W. M. (1955): Types of seedling reaction of Kawvale and Wabash winter wheat to three physiologic races of Ustilago tritici. *Phytopathology* **45,** 513–516.
Okabe, N. and Goto, M. (1955): Studies on Pseudomonas solanacearum. X. Genetic change of the bacterial strains induced by the temperate phage T-c 200. *J. Fach. Agric. Shizuoka Univ.* Pp. 57–62.
Okabe, N. and Goto, M. (1961): Studies on Pseudomonas solanacearum. XI. Pathotypes in Japan. *Shizuoka Univ. Fac. Agric. Rept.* **11,** 25–42.
Okabe, N. and Goto, M. (1963): Bacteriophages of plant pathogens. *Ann. Rev. Phytopath.* **1,** 397–418.
Oort, A. J. P. (1939): Inoculation experiments with loose smuts of wheat and barley (Ustilago tritici and U. nuda). *Phytopathology* **29,** 717–728.
Oort, A. J. P. (1944): Onderzoekingen over Stuifbrand. II. Overgevoeligheid van Tarwe voor Stuifbrand (Ustilago tritici). *Meded. Inst. Phytopath. Wageningen* **103,** 1–106.
Ossiannilsson, F. (1961): Negative evidence against passage of whole plant virus particles through gut wall in aphid vectors. *Nature* **190,** 1222.
Ouchterlony, Ö. (1958): Diffusion-in-gel methods for immunological analysis. *Progr. Allergy* **5,** 1–78.
Padmanabhan, S. Y. (1965): Studies on forecasting outbreaks of blast disease of rice. I. Influence of meteorological factors on blast incidence at Cuttar. *Proc. Indian Acad. Sci.* Sect. B. **62,** 117–129.
Pady, S. M. (1959): A continuous spore sampler. *Phytopathology* **49,** 757–760.
Panagopoulos, C. G. and Crosse, J. E. (1964): Blossom blight and related symptoms caused by Pseudomonas syringae van Hall on pear trees. *Rept. East Malling Res. Sta.* 1963. Pp. 119–122.
Papavizas, G. C. (1967): Evaluation of various media and antimicrobial agents for isolation of Fusarium from soil. *Phytopathology* **57,** 848–852.
Parish, C. L., Zaitlin, M. and Siegel, A. (1965): A study of necrotic lesion formation by tobacco mosaic virus. *Virology* **26,** 413–418.
Patil, S. S. and Dimond, A. E. (1967a): Effect of sugars and sugar alcohols on production of polyglacturonase by Fusarium oxysporum f. sp. lycopersici. *Phytopathology* (Abstr.) **57,** 825.
Patil, S. S. and Dimond, A. E. (1967b): Induction and repression of polygalacturonase synthesis in Verticillium albo-atrum. *Phytopathology* (Abstr.) **57,** 825.
Patil, S. S. and Dimond, A. E. (1967c): Inhibition of Verticillium polygalacturonase by oxidation products of polyphenols. *Phytopathology* **57,** 492–496.
Paton, A. M. (1958): Pectin-decomposing strains of Pseudomonas. *Nature* **181,** 61–62.
Paul, H. L. (1961): Die Gestalt und Struktur pflanzenpathogener Viren. *Zentralbl. Bakteriol. Parasitknd.* **114,** 694–717.

Pawlik, A. (1961): Zur Frage der Überwinterung von Peronospora tabacina Adam. Beobachtungen über Oosporenkeimung. *Z. Pflanzenkrankh.* **58,** 193–197.
Pawlik, A., Schmid, K., Spran, F. und Krauss, E. (1963): Resistenzprüfung des Tabaks im Keimblattstadium gegen Peronospora tabacina Adam. Erfahrungen mit dem Schiltz-Izard-Test. *Z. Pflanzenkrankh.* **70,** 332–339.
Pénzes, A. (1962): Az élővilág új rendszere különös tekintettel az élet keletkezésének körülményeire. (A new system of the world living beings with special regard to the circumstances of the origin of life.) *Természettud. Közl.* **6,** 160–162.
Person, C. (1959): Gene-for-gene relationships in host parasite systems. *Canad. J. Bot.* **37,** 1101–1130.
Person, C., Samborski, D. J. and Forsyth, F. R. (1957): Effect of benzimidazole on detached wheat leaves. *Nature* **180,** 1294–1295.
Person, C., Samborski, D. J. and Rohringer, R. (1962): The gene-for-gene concept. *Nature* **194,** 561–562.
Person, L. H. (1961): A method of maintaining viability and ability to sporulate in isolates of Ascochyta. *Phytopathology* **51,** 797–798.
Peyton, G. A. and Bowen, C. C. (1963): The host-parasite interface of Peronospora manshurica on Glycine max. *Amer. J. Bot.* **50,** 787–797.
Philipps, D. V., Leben, C. and Allison, C. C. (1967): A mechanism for the reduction of Fusarium wilt by a Cephalosporium species. *Phytopathology* **57,** 916–919.
Pirie, N. W. (1959): The properties of fragments of tobacco mosaic virus. *Proc. IV. Int. Congr. Biochem.* Pergamon Press, Oxford. Pp. 45–57.
Podhradszky, J. (1961): Provokatív vizsgálati módszerek rizsfajták Piricularia-rezisztenciájának elbírálására. (Provocative methods for the evaluation of resistance of rice varieties to Piricularia infection). *Növénytermelés* **10,** 67–76.
Podhradszky, J. and Südi, J. (1957): Rizsfajtáki „bruzone rezisztenciájának" szabadföldi elbírálása. (Evaluation of resistance of rice varieties to Piricularia in field trials). *Növénytermelés* **6,** 239–248.
Poehlman, J. M. (1945): A simple method of noculating barley with loose smut. *Phytopathology* **35,** 640–644.
Pontecorvo, G. (1946): Genetic systems basedon heterokaryosis. *Cold Spring Harbor Symp. Quant. Biol.* **11,** 193–201.
Pontecorvo, G. (1956): The parasexual cycle in fungi. *Ann. Rev. Microbiol.* **10,** 393–400.
Pontecorvo, G., Roper, J. A. and Forbes, E. (1953): Genetic recombination without sexual reproduction in Aspergillus niger. *J. Gen. Microbiol.* **8,** 198–210.
Popp, W. (1955): A comparative study of spore germination of Ustilago tritici and U. nuda *Phytopathology* **45,** 585–590.
Popp, W. (1958): A new procedure for embryo examination for mycelium of smut fungi. *Phytopathology* **48,** 19–22.
Porter, C. A. and Weinstein, L. H. (1957): Biochemical changes induced by thiouracil in cucumber mosaic virus-infected and non-infected tobacco plants. *Contr. Boyce Thompson Inst.* **19,** 87–106.
Powers, H. R., Jr., and Moseman, J. G. (1957): Pathogenic variability within cleistothecia of Erysiphe graminis. *Phytopathology* **47,** 136–138.
Preece, T. F. (1959): A staining method for the study of apple scab infections. *Plant Pathology* **8,** 127–129.
Presypkin, V. F. and Arkhipova, F. D. (1964): Pro metodiku vydilennya monosporovykh kultur i ras Peronospora tabacina Adam. (On a method of isolating monospore cultures and races of Peronospora tabacina Adam.) *Visn. sil'-hos Nauki* **7,** 50–53.
Price, W. C. (1935): Acquired immunity from cucumber mosaic in Zinnia. *Phytopathology* **25,** 776–789.
Protzenko, A. E. (1960): Morphology of phytopathogenic viruses. In: *Virus Diseases of Agricultural Crops and Their Control.* (Ed.) K. S. Sukhov. (In Russian) Izd. Min Selkh. Hoz. Moscow. Pp. 25–40.

Protzenko, A. E. and Legunkova, R. M. (1960): *Electron microscopy of phytopathogenic viruses*. (In Russian) Izd. Akad. Nauk USSR. Pp. 1–95.
Purdy, H. A. (1929): Immunologic reactions with tobacco mosaic virus. *J. Exp. Med.* **49,** 919–935.
Quanjer, H. M. (1931): The methods of classification of plant viruses and an attempt to classify and name potato viruses. *Phytopathology* **21,** 577–613.
Rai, P. V. and Strobel, G. A. (1967): Phytotoxins of Corynebacterium. *Phytopathology* **57,** 1008.
Ralph, R. K. and Clark, M. F. (1966): Intracellular location of double-stranded plant viral ribonucleic acid. *Biochim. Biophys. Acta* **119,** 29–36.
Ralph, R. K., Matthews, R. E. F., Matsui, A. I. and Mandel, H. G. (1965): Isolation and properties of double-stranded viral RNA from virus-infected plants. *J. Mol. Biol.* **11,** 202–212.
Rands, R. D. and Brotherton, W. (1925): Bean varietal tests for disease resistance. *J. Agr. Res.* **31,** 101–154.
Raper, K. B. and Fennell, D. I. (1965): *The Genus Aspergillus*. Williams and Wilkins Co., Baltimore.
Rappaport, I. and Siegel, A. (1955): Inacivation of tobacco mosaic virus by rabbit antiserum. *J. Immunol.* **74,** 106–116.
Rappaport, I. and Wildman, S. G. 1957t): A kinetic study of local lesion growth on Nicotiana resulting from tobacco mosaic (rus infection. *Virology* **4,** 265–274.
Rawlins, T. E. (1957): A modificvition of Bald's strain for viruses and for cell inclusions associated with virus infections.a*Phytopathology* **47,** 307.
Rawlins, T. E. and Tompkins, C. M. (1934): The use of carborundum as an abrasive in plant-virus inoculations. *Phytopa thology* **24,** 1147.
Rawlins, T. E., Weierich, A. J. and Schlegel, D. E. (1956): A histochemical study of certain plant viruses by means of the Sakaguchi reaction for arginine. *Virology* **2,** 308–311.
Reddi, K. K. (1963a): Studies on the formation of tobacco mosaic virus ribonucleic acid. II. Degradation of host ribonucleic acid following infection. *Proc. Natl. Acad. Sci.* **50,** 75–81.
Reddi, K. K. (1963b): Studies on the formation of tobacco mosaic virus ribonucleic acid. III. Utilization of ribonucleosides of host ribonucleic acid. *Proc. Natl. Acad. Sci.* **50,** 419–425.
Reddi, K. K. (1964): Studies on the formation of tobacco mosaic virus ribonucleic acid. V. Presence of tobacco mosaic virus in the nucleus of the host cell. *Proc. Natl. Acad. Sci.* **52,** 397–401.
Reddi, K. K. (1966): Studies on the formation of tobacco mosaic virus ribonucleic acid. VII. Fate of tobacco mosaic virus after entering the host cell. *Proc. Natl. Acad. Sci.* **55,** 593–598.
Reed, G. M. and Faris, J. A. (1924): Influence of environmental factors on the infection of sorghums and oats by smuts. II. *Amer. J. Bot.* **11,** 579–599.
Reichmann, M. E. (1960): Degradation of potato virus X. *J. Biol. Chem.* **235,** 2959–2963.
Reichmann, M. E. and Stace-Smith, R. (1959): Preparation of infectious ribonucleic acid from potato virus X by means of guanidine denaturation. *Virology* **9,** 710–712.
Reid, M. S. and Matthews, R. E. F. (1966): On the origin of the mosaic induced by turnip yellow mosaic virus. *Virology* **28,** 563–570.
Rider, N. E., Cruickshank, I. A. M. and Bradley, E. F. (1961): Environment and sporulation in phytopathogenic fungi. III. Peronospora tabacina Adam.: Field environment, sporulation and forecasting. *Austr. J. Agric. Res.* **12,** 1119–1125.
Riker, A. J., Berbee, J. G. and Smalley, E. B. (1959): Effects of crown gall and hairy root on the growth of apple trees. *Phytopathology* **49,** 88–90.
Ritter, F. J. and Meyer, G. M. (1962): Preparative thin-layer chromatography as an alternative for column chromatography. *Nature* **193,** 941–942.
Roberts, D. A. (1964): Local-lesion assay of plant viruses. In: *Plant Virology*. (Ed.) M. K. Corbett and H. D. Sisler. Univ. of Florida Press, Gainesville, USA. Pp. 194–210.

Rodenhiser, H. A. and Holton, C. S. (1937): Physiologic races of Tilletia tritici and T. levis. *J. Agric. Res.* **55,** 483–496.
Roemer, Th. and Bartholly, R. (1933): Die Aggressivität verschiedener Steinbrandherkünfte und ihre Veränderung durch die Wirtssorte. *Phytopath. Z.* **6,** 469–506.
Roemer, Th., Fuchs, W. H. and Isenbeck, K. (1938): *Die Züchtung resistenter Rassen der Kulturpflanzen.* Paul Parey, Berlin.
Rogerson, C. T. (1958): Kansas aeromycology. I. Comparison of media. *Trans. Kansas Acad. Sci.* **61,** 155–161.
Roland, G. (1949): Etude de la transmission du virus X (Solanum virus 1, Orton) a la descendance végétative chez la Pomme de terre. *Parasitica* **5,** 105–109.
Roland, G. (1957): La thermothérapie des viruses végétales. *Meded. Landbouwhogschool en Opzoekinssta. Gent* **22,** 553–560.
Rosen, H. R. (1936): The influence of dry air on the longevity of the fire-blight pathogen. *Phytopathology* **26,** 439–449.
Rosen, H. R. (1938): Life span and morphology of the fire-blight bacteria as influenced by relative humidity temperature and nutrition. *J. Abric. Res.* **56,** 239–258.
Rosenstiel, K. (1929): Untersuchungen über die Widerstandsfähigkeit von Haferarten und sorten gegen Haferflugbrand (U. avenae Pers.) und ihre Vererbung. *Phytopath. Z.* **1,** 317–360.
Ross, A. F. (1961): Localized acquired resistance to plant virus infection in hypersensitive hosts. *Virology* **14,** 329–339.
Ross, A. F. (1965): Systemic effects of local lesion formation. In: *Viruses of Plants.* (Ed.) A. B. R. Beemster and J. Dijkstra. Pp. 127–150.
Ross, A. F. (1950): Local lesion formation and virus production following simultaneous inoculation with potato viruses X and Y. *Phytopathology* **40,** 24.
Ross, R. G. and Hamlin, S. A. (1965): Influence of nutrients on perithecial production of Venturia inaequalis (Cke.) Wint. *Canad. J. Bot.* **43,** 959–965.
Rowell, J. B. and DeVay, J. E. (1953): Factors affecting the partial vacuum inoculation of seedling corn with Ustilago zeae. *Phytopathology* **43,** 654–658.
Röttger, B. (1965): Ribonucleic acids of healthy and tobacco mosaic virus infected tobacco leaves. *Biochim. Biophys. Acta* **95,** 525–531.
Rushizky, G. W. and Knight, C. A. (1959): Ribonuclease-sensitive infectious units from tomato bushy stunt virus. *Virology* **8,** 448–455.
Rushizky, G. W. and Knight, C. A. (1960a): A mapping procedure for nucleotides and oligonucleotides. *Biochem. Biophys. Res. Commun.* **2,** 66–70.
Rushizky, G. W. and Knight, C. A. (1960b): An oligonucleotide mapping procedure and its use in the study of tobacco mosaic virus nucleic acid. *Virology* **11,** 236–249.
Rushizky, G. W., Knight, C. A., Roberts, W. K. and Dekker, C. A. (1962): Studies on the mechanism of action of micrococcal nuclease. II. Degradation of ribonucleic acid from tobacco mosaic virus. *Biochim. Biophys. Acta* **55,** 674–682.
Rushizky, G. W., Knight, C. A. and Sober, H. A. (1961): Studies on the preferential specificity of pancreatic ribonuclease as deduced from partial digests. *J. Biol. Chem.* **236,** 2732–2737.
Rushizky, G. W. and Sober, H. A. (1962): Characterization of the major compounds in ribonuclease T_1 digests of ribonucleic acid. I. Mono-, di and trinucleotides. *J. Biol. Chem.* **237,** 834–840.
Ryzhkov, V. L. (1952): Systematics of viruses in the recent literature. (In Russian) *Mikrobiologija* **21,** 458–476.
Ryzhkov, V. L. (1957):: Metabolites and antimetabolites in investigating the multiplication of tobacco mosaic virus. *Izv. Akad. Nauk USSR.* Ser. Biol. Pp. 41–54.
Ryzhkov, V. L. (1959): Biochemische Bedingungen der Vermehrung der Viren. *Proc. 4th Intern. Congr. Biochem.* Pergamon Press, London. Pp. 31–44.
Sakai, R. and Tomiyama, K. (1964): Relation between physiological factors related to phenols and varietal resistance of potato plant to late-blight. I. Field observation. *Ann. Phytopath. Soc. Japan* **29,** 33–38.

Sänger, H. L. and Knight, C. A. (1963): Action of actinomycin D on RNA synthesis in healthy and virus-infected tobacco leaves. *Biochim. Biophys. Commun.* **13,** 455–461.

Sansome, E. (1963a): Meiosis in the oogonium and antheridium of Pythium debaryanum Hesse. *Nature* **191,** 827–828.

Sansome, E. (1963b): Meiosis in Pythium debaryanum Hesse and its significance in the life-history of the Biflagellatae. *Trans. Brit. Mycol. Soc.* **46,** 63–72.

Sato, K. and Egami, F. (1957): Studies on ribonucleases in Takadiastase 1. *J. Biochem.* (Tokyo) **44,** 753–767.

Schick, R. (1932): Über das Verhalten von Solanum demissum, Solanum tuberosum und ihren Bastarden gegenüber verschiedenen Herkünften von Phytophthora infestans. *Züchter* **4,** 233–237.

Schick, R. and Hopfe, A. (1962): Die Züchtung der Kartoffel. In: *Die Kartoffel.* Herausg. R. Schick und M. Klinkowski. VEB Deuttch. Landw. Verlag Berlin. Vol. 2. Pp. 1461–1583.

Schick, R. and Schick, E. (1959): Die Differenzierung der verschiedenen Rassen der Phytophthora infestans auf Sämlingen von S. demissum (Lindl.) und S. stoloniferum (Schlecht. et Bouche). *Züchter* **29,** 220–225.

Schiltz, P. and Izard, C. (1962): Susceptibilité cotylédonnaire et résistance á Peronospora tabacina Adam. *Acad. Agric. de France.* Pp. 561–564.

Schlegel, D. E. (1960): Highly infectious phenol extracts from tobacco leaves infected with cucumber mosaic virus. *Virology* **11,** 329–338.

Schmelzer, K. (1956a): Beiträge zur Kenntnis der Übertragbarkeit von Viren durch Cuscuta-Arten. *Phytopath. Z.* **28,** 1–56.

Schmelzer, K. (1956b): Beiträge zur Kenntnis der Virushemmstoffe in Cuscuta-Arten. *Zentrbl. Bakteriol. Parasitenkunde* Abt. II. **109,** 482–515.

Schmelzer, K. (1957): Die Passage durch Stellaria media in ihrer Bedeutung für die mechanische Übertragung von Viren an Nelken. *Phytopath. Z.* **28,** 457–460.

Schmidt, H. B. (1960): Die Übertragung pflanzlicher Viren durch Insekten. *Monatsber. Deut. Akad. Wiss. Berlin.* **2,** 214–223.

Schmitthenner, A. F. (1962): Isolation of Pythium from soil particles. *Phytopathology* **52,** 1133–1138.

Schmutterer, H. (1960): Zur Kenntnis der Beissinsekten als Überträger pflanzlicher Viren I.–II. *Z. Angew. Entomol.* **47,** 416–439.

Schnathorst, W. C. (1966): Unaltered specificity in several xanthomonads after repeated passage through Phaseolus vulgaris. *Phytopathology* **56,** 58–60.

Schramm, G. (1954): *Die Biochemie der Viren.* Springer Verl. Berlin – Göttingen – Heidelberg.

Schramm, G. and Engler, R. (1958): The latent period after infection with tobacco mosaic virus and virus nucleic acid. *Nature* **181,** 916–917.

Schramm, G. and Müller, H. (1940): Zur Chemie des Tabakmosaikvirus. Über die Einwirkung von Keton und Phenylisocyanat auf das Virusprotein. *Z. Physiol. Chem.* **266,** 43–55.

Schramm, G. and Röttger, B. (1959): Untersuchungen über das Tabakmosaikvirus mit fluoreszierenden Antikörpern. *Z. Naturforsch.* **14b,** 510–515.

Schroth, M. N. and Hildebrand, D. C. (1965): Some differences in the arbutin-hydroquinone complex between fireblight-resistant and susceptible Pyrus spp. *Phytopathology* **55,** 1075.

Schuhmann, G. (1966): Untersuchungen über die physiologische Spezialisierung von Tilletia caries (DC.) Tul. und Tilletia controversa Kühn in Deutschland und das Resistenzverhalten von Weizensorten. Habil. schrift. *Mitt. Biol. Bundesanst. Land u. Forstwirt.* Heft 117.

Schuster, M. L. (1955): A method for testing resistance of beans to bacterial blights. *Phytopathology* **45,** 519–520.

Schuster, H. and Schramm, G. (1958): Bestimmung der biologisch wirksamen Einheit in der Ribonucleinsäure des Tabakmosaikvirus auf chemischen Wege. *Z. Naturforsch.* **13b,** 697–704.

Schwinghammer, E. A. (1959): Induced mutations of pathogens for the study of host-parasite interactions. In: *Plant Pathology. Problems and Progress 1908–1958.* (Ed.) C. S. Holton et al. Univ. of Wisconsin Press, Madison. Pp. 192–201.

Schwochau, M. E. and Hadwiger, L. A. (1968): Stimulation of pisatın production in Pisum sativum by actinomycin D and other compounds. *Arch. Biochem. Biophys.* **126,** 731–733.
Seaman, W. L., Walker, J. C. and Larson, R. H. (1963): A new race of Plasmodiophora brassicae affecting Badger Shipper cabbage. *Pytopathalogy* **53,** 1426–1429.
Sela, I. and Appelbaum, S. W. (1962): Occurrence of antiviral factor in virus-infected plants. *Virology* **17,** 543–548.
Shaffer, W. W. (1967): Flagella stain method. *Proceedings af the First Workshop on Phytobacteriology.* (Second Edition): 32–33.
Shainoff, J. R. and Lauffer, M. A. (1957): An application of ion exchange chromatography to the identification of virus activity with characteristic particles. *Virology* **4,** 418–434.
Shalla, T. A. (1959): Relations of tobacco mosaic virus and barley stripe mosaic virus to their host cells as revealed by ultrathin tissue-sectioning for the electron microscope. *Virology* **7,** 193–219.
Shay, J. R. and Hough, F. L. (1952): Evaluation of apple scab resistance in selections of Malus. *Amer. J. Bot.* **39,** 288–297.
Shay, J. R. and Williams, E. B. (1956): Identification of three physiologic races of Venturia inaequalis. *Phytopathology* **46,** 190–193.
Scheffield, F. M. L. (1936): The role of plasmodesms in the translocation of virus. *Ann. Appl. Biol.* **23,** 506–508.
Sheffield, F. M. L. (1941): The cytoplasmic and nuclear inclusions associated with severe etch virus. *J. Royal Microscop. Soc.* **61,** 30–45.
Shepherd, C. J. and Mandıyk, M. (1962): Auto-inhibitors of germination and sporulation in Peronospora tabacina Adam. *Trans. Erit. Mycol. Scc.* **45,** 233–244.
Shepherd, C. J. and Mandryk, M. (1963): Germination of conidia of Peronospora tabacina Adam. *Austr. J. Biol. Sci.* **16,** 77–87.
Shepherd, C. J., Stuart, F. and Mandryk, M. (1963): Isolation and maintenance of single spore lines of Peronospora tabacina Adam. *Nature* **197,** 515.
Shepherd, C. J. and Tosic, L. (1966): The role of riboflavin and inhibitors in conidial germination in Peronospora tabacina Adam. Austr. *J. Biol. Sci.* **19,** 335–337.
Shigematsu, A., Mizusawa, Y. and Hirai, T. (1966): Incorporation of two radioactive amino acids into the nucleic acid fractions in tobacco leaves infected with tobacco mosaic virus. *Virology* **28,** 331–338.
Shipp, W. and Haselkorn, R. (1964): Double-stranded RNA from tobacco leaves infected with TMV. *Proc. Natl. Acad. Sci.* U. S. **52,** 401–408.
Siebs, E. (1958): Ergebnisse zu Problemen des Mehltaus und der Mehltauresistenz des Apfels. I. Mehltau. *Phytopath. Z.* **34,** 86–102.
Siegel, A., Ginoza, W. and Wildman, S. G. (1957): The early events of infection with tobacco mosaic virus nucleic acid. *Virology* **3.** 554–559
Siegel, A., Wildman, S. G. and Ginosa, W. (1956): Sensitivity to ultra-violet light of infectious tobacco mosaic virus nucleic acid. *Nature* **178,** 1117–1118.
Siegel, A., Zaitlin, M. and Sehgal, O. W. (1962): The isolation of defective tobacco mosaic virus strains. *Proc. Natl. Acad. Sci.* U. S. **48,** 1845–1851.
Sill, W. H., Jr. and Walker, J. C. (1952): A virus inhibitor in cucumber in relation to mosaic resistance. *Phytopathology* **42,** 349–352.
Simons, J. N. (1958): Titers of three nonpersistent aphid-borne viruses affecting pepper in South Florida. *Phytopathology* **48,** 265–268.
Simons, J. N. (1959): Variation in efficiency of aphid transmission of southern cucumber mosaic virus and potato virus Y in pepper. *Virology* **9,** 612–623.
Simmonds, P. M. (1946): Detection of the loose smut fungi in embryos of barley and wheat. *Sci. Agric.* **26,** 51–58.
Singer, B. and Fraenkel-Conrat, H. (1961): Effects of bentonite on infectivity and stability of TMV-RNA. *Virology* **14,** 59–65.
Singh, R. S. and Mitchell, J. E. (1961): A selective method for isolation and measuring the population of Phytium in soil. *Phytopathology* **51,** 440–444.

Skoog, H. A. (1952): Studies on a host-parasite relation of bean varieties resistant and susceptible to Pseudomonas phaseolicola and toxin production by the parasite. *Phytopathology* **42,** 475.

Skvortzoff, S. S. (1937): A simple method for detecting hyphae of loose smut in wheat grains. *Plant Prot.* **15,** 90–91. (Ref. *Rev. Appl. Mycology* **17,** 382. 1938.)

Slagle, C. W., Wolcyrz, S. and Price, W. C. (1952): Inhibition of plant virus infection by growth products of Neurospora. *Phytopathology* **42,** 240–245.

Slogteren, D. H. M., van (1955a): Serological analysis of some plant viruses with the gel-diffusion method. *Proc. Second Conf. Potato Virus Dis.* 1954. Pp. 45–50.

Slogteren, E., van (1955b): Serological diagnosis of plant virus diseases. *Ann. Appl. Biol.* **42,** 122–128.

Small, B. C. and Worley, J. F. (1956): Evaluation of 2,3,5-triphenyl tetrazolium chloride for obtaining pathogenic types from stock cultures of halo blight and common blight organisms. *Plant Dis. Reptr.* **40,** 628.

Smith, K. M. (1937): *A Textbook of Plant Virus Diseases.* Blakiston and Son, Inc., Philadelphia. 2nd ed. J. and A. Churchill Ltd. London, 1957.

Smith, K. M. (1951): *Recent Advances in the Study of Plant Viruses.* J. and A. Churchill Ltd. London.

Smith, K. M. (1958): Transmission of plant viruses by Arthropods. *Ann. Rev. Entomol.* **3,** 469–482.

Smith, S. H. and Schlegel, D. E. (1965): The incorporation of ribonucleic acid precursors in healthy and virus-infected plant cells. *Virology* **26,** 180–189.

Smith, W. K. (1958a): A survey of the production of pectic enzymes in plant pathogenic and other bacteria. *J. Gen. Microbiol.* **28,** 33–41.

Smith, W. K. (1958b): Chromatographic examination of the products of digestion of pectic materials by culture solutions of plant pathogenic and other bacteria. *J. Gen. Microbiol.* **18,** 42–47.

Smoot, J. J., Gough, F. J. and Gallegly, M. E. (1957): Oospore formation in Phytophthora infestans. *Phytopathology* (Abstr.) **47,** 33.

Smoot, J. J., Gough, F. J., Lamey, H. A., Eichenmuller, J. J. and Gallegly, M. E. (1958): Production and germination of oospores of Phytophthora infestans. *Phytopathology* **48,** 165–171.

Snyder, W. C. and Hansen, H. N. (1940): The species concept in Fusarium. *Amer. J. Bot.* **27,** 64–67.

Sokolova, I. I. (1964): The degree of resistance in rice varieties to piriculariosis. (In Russian) *Sborn. Trud. vses. Inst. Rasten.* **2,** 66–71.

Solymosy, F. (1960): Identification of the cucumber mosaic virus strain causing the so called „újhitűség" of red pepper. *Acta Agr. Hung.* **10,** 177–196.

Solymosy, F. and Farkas, G. L. (1962): Simultaneous activation of pentose phosphate shunt enzymes in a virus-infected local lesion host plant. *Nature* **195,** 835.

Solymosy, F. and Farkas, G. L. (1963): Metabolic characteristic at the enzymatic level of tobacco tissues exhibiting localized acquired resistance to viral infection. *Virology* **21,** 210–221.

Solymosy, F., Farkas, G. L. and Király, Z. (1959): Biochemical mechanism of lesion formation in virus infected plant tissues. *Nature* **184,** 706–707.

Solymosy, F., Tener, G. M. and Reichmann, M. E. (1965): Partial nucleotide sequences in turnip yellow mosaic virus RNA. *Virology* **27,** 409–417.

Sowell, G. (1960): Bacterial spot resistance of introduced peppers. *Plant Dis. Reptr.* **44,** 587–590.

Spencer, D. and Wildman, S. G. (1964): The incorporation of amino acids into protein by cell-free extracts from tobacco leaves. *Biochemistry.* **3,** 954–959.

Staehelin, M. (1960): Chemical modifications of virus infectivity. Reactions of tobacco mosaic virus and its nucleic acid. *Experientia* **16,** 1–11.

Staehelin, M. (1961a): Studies on nucleotide sequences in ribonucleic acids. I. Separation of oligonucleotides on DEAE-cellulose. *Biochim. Biophys. Acta* **49,** 11–19.

Staehelin, M. (1961b): Studies on nucleotide sequences in ribonucleic acids. II. Spectroscopic properties of oligoribonucleotides. *Biochim. Biophys. Acta* **49,** 20–26.
Staehelin, M. (1961c): Studies on nucleotide sequences in ribonucleic acids. III. Amounts of oligonucleotides in pancreatic digests. *Biochim. Biophys. Acta* **49,** 27–35.
Stakman, E. C. and Christensen, J. J. (1927): Heterothallism in Ustilago zeae. *Phytopathology* **17,** 824–834.
Stakman, E. C. and Christensen, J. J. (1953): Problems of variability in fungi. In: *Plant Diseases. The Yearbook of Agriculture.* U. S. Dept. Agriculture. Pp. 35–62.
Stakman, E. C. and Harrar, J. G. (1957): *Principles of Plant Pathology.* The Ronald Press Co., New York.
Stakman, E. C., Levine, M. N. and Cotter, R. U. (1930): Origin of physiologic forms of Puccinia graminis through hybridization and mutation. *Ann. Sci. Agric. Bot.* **10,** 707–720.
Stakman, E. C., Levine, M. N. and Loegering, W. Q. (1944): Identification of physiologic races of Puccinia graminis tritici. *U. S. Dept. Agric. Bur. Entom. Plant Quarant.* E. p. 617.
Stakman, E. C. and Piemeisel, F. J. (1917): Biologic forms of Puccinia graminis on cereals and grasses. *J. Agric. Res.* **10,** 429–495.
Stakman, E. C., Stewart, D. M. and Loegering, W. Q. (1962): Identification of physiologic races of Puccinia graminis var. tritici. *U. S. Dept. Agric. Res. Serv.* E617.
Stall, R. E. and Cook, A. A. (1966): Multiplication of Xanthomonas vesicatoria and lesion development in resistant and susceptible pepper. *Phytopathology* **56,** 1152–1154.
Stapp, C. (1942): Der Pflanzenkrebs und sein Erreger Pseuodmonas tumefaciens. XI. Zytologische Untersuchungen des bakteriellen Erregers. *Z. Parasitenkunde* Abt. II. 105–1
Stapp, C. (1957): Zellkerne und Kopulationsvorgänge bei Bakterien. *Umschau* **57,** 206–208.
Stapp, C. (1958): *Pflanzenpathogene Bakterien.* Paul Parey, Berlin und Hamburg.
Stapp, C. (1961): *Bacterial Plant Pathogens.* Oxford Univ. Press. London.
Stapp, C. and Bortels, (1931) M. Der Pflanzenkrebs und sein Erreger Pseudomonas tumefaciens. I. Mitt. Konstitution und Tumorbildung der Wirtspflanze. *Z. Parasitenk.* **3,** 654–663.
Stapp, C. and Marcus, D. (1944): Beiträge zur weiteren Vereinfachung der serologischen Virusdiagnose. *Zentr.bl. Bakt. Parasitenkunde* Abt. II. **106,** 465–471.
Starr, M. P. (1947): The causal agent of bacterial root and stem disease of guayule. *Phytopathology* **37,** 291–300.
Starr, M. P. (1959): Bacteria as plant pathogens. *Ann. Rev. Microbiol.* **13,** 211–238.
Steere, R. L. (1959): The purification of plant viruses. *Adv. Virus Res.* **6,** 1–73.
Steere, R. L. (1952): Virus increment curves obtained from counts of particles in clarified plant juice. *Amer. J. Bot.* 39, 211–220.
Steere, R. L. (1956): Purification and properties of tobacco ringspot virus. *Phytopathology* **46,** 60–69.
Stewart, D. M., Radulescu, E. and Negulescu, Fl. (1967): Reactions of adult wheats to stripe, leaf, and stem rusts in Rumania in 1965 and reaction to stem rust of seedlings of 6 winter wheats from Rumania to cultures in the United States in 1966. *Plant Dis. Reptr.* **51,** 362–366.
Steel, K. J. (1961): The oxygen reaction as a taxonomic tool. *J. gen. Microbiol.* **25,** 297–306.
Stoddard, E. M. (1947): The X-disease of peach and its chemotherapy. *Conn. Agric. Exp. Sta. Bull.* 506.
Stoddart, J. L. and Carr, A. J. H. (1966): Properties of wilt-toxins produced by Verticillium albo-atrum Reinke et Berth. *Ann. Appl. Biol.* **58,** 81–92.
Stolp, H. (1961): Neue Erkenntnisse über phytopathogene Bakterien und die von ihnen verursachten Krankheiten. I. Verwandschaftsbeziehungen zwischen phytopathogenen Pseudomonas »Arten« und saprophytischen Fluoreszenten auf der Grundlage von Phagenreaktionen. *Phytopath. Z.* **42,** 197–262.
Stolp, H., Starr, M. P. and Baigent, N. L. (1965): Problems in speciation of phytopathogenic pseudomonads and xanthomonads. *Ann. Rev. Phytopath.* **3,** 231–264.
Stoughton, R. H. (1930): *Ann. Appl. Biol.* **17,** 162. Cited by Dowson, W. J.: *Manual of Bacterial Plant Diseases.* Adam and Charles Black, London, 1949. Pp. 59–60.

Strandberg, J. O. and Williams, P. H. (1967): Inheritance of clubroot resistance in Chinese cabbage. *Phytopathology* **57**, 330.
Strobel, G. A. (1967): Purification and properties of a phytotoxic polysaccharide produced by Corynebacterium sepedonicum. *Plant Physiol.* **42**, 1433–1441.
Strobel, G. A. and Hess, W. M. (1968): Biological activity of a phytotoxic glycopeptide produced by Corynebacterium sepedonicum. *Plant Physiol.* **43**, 1673–1688.
Strobel, G. A. and Rai, P. V. (1968): A rapid serodiagnostic test for potato ring rot. *Plant Dis. Reptr.* **52**, 502–504.
Südi, J. and Podhradszky, J. (1958): A short-term glasshouse test for estimating Piricularia resistance of different rice varieties. *Acta Microbiol. Hung.* **5**, 139–145.
Südi, J. and Podhradszky, J. (1959): Rizsfajták Piricularia-rezisztenciájának gyors elbírálása üvegházi vizsgálattal. (Rapid test for determining resistance to Piricularia disease in greenhouse) *Kísérl. Közl.* 52/A, 3–16.
Sugiyama, T. and Fraenkel-Conrat, H. (1961): Identification of 5-linked adenosine as end-group of TMV-RNA. *Proc. Natl. Acad. Sci.* U. S. **47**, 1393–1397.
Sugiyama, T. and Fraenkel-Conrat, H. (1963): The end-groups of tobacco mosaic virus. RNA. II. Nature of the 3-linked chain end in TMV and of both ends in four strains. *Biochemistry* **2**, 332–334.
Sukhov, K. S. (1956): *Viruses.* (In Russian) Izd. Akad. Nauk USSR. Moscow.
Sukhov, K. S. and Razvyazkina, G. M. (1955): *Biology of Viruses and Virus Diseases of Plants* (In Russian). Sovietskaja Nauka. Moscow.
Sylvester, E. S. (1956): Beet yellows virus transmission by the green peach aphid. *J. Econ. Entomol.* **49**, 789–800.
Symons, R. H., Reese, M. W., Short, M. N. and Markham, R. (1963): Relationships between the ribonucleic acid and protein of some plant viruses. *J. Mol. Biol.* **6**, 1–15.
Szirmai, J. (1941): A fűszerpaprika leromlását megindító, újhitűségnek nevezett vírusbetegségről. (On a new paprika disease caused by a virus) *Növényeg. Évkönyv* **1**, 110–113.
Szirmai, J. (1950): Kísérletek egészséges vetőmag nyerésére vírusbeteg növényekről. (Experiments on the production of healthy seeds from virus infected plants.) *Ann. Inst. Prot. Plant. Hung.* **5**, 215–227.
Takahashi, W. N. (1948): The inhibition of virus increase by malachite green. *Science* **107**, 226.
Takahashi, W. N. (1951): A groundglass homogenizer for preparing virus samples from plant tissue. *Phytopathology* **41**, 481–482.
Takahashi, W. N. (1955): Anomalous proteins associated with three strains of tobacco mosaic virus. *Virology* **1**, 393–396.
Takahashi, W. N. (1959): The role of anomalous noninfectious protein in virus synthesis. *Virology* **9**, 437–445.
Takahashi, W. N. and Ishii, M. (1952): An abnormal protein associated with tobacco mosaic virus infection. *Nature* **169**, 419.
Takahashi, W. N. and Ishii, M. (1953): A macromolecular protein associated with tobacco mosaic virus infection: its isolation and properties. *Amer. J. Bot.* **40**, 85–90.
Takahashi, W. N., Karler, A. and Knight, C. A. (1958): Electrochromatographic studies on infectious nucleic acid from tobacco mosaic virus. *Virology* **6**, 637–648.
Talbot, P. H. B. (1954): Micromorphology of the lower Hymenomycetes. *Bothalia* **6**, 249–299.
Talboys, P. W. (1960): A culture-medium aiding the identification of Verticillium albo-atrum and V. dahliae. *Plant Pathol.* **9**, 57–58.
Tapke, V. E. (1935): An effective and easily applied method of inoculating seed barley with covered smut. *Phytopathology* **25**, 1038–1039.
Tapke, V. E. (1937): A method of inoculating seed barley with black loose smut for use in studies on physiologic races. *Phytopathology* **27**, 115–116.
Tapke, V. E. (1945): New physiologic races of Ustilago hordei. *Phytopathology* **35**, 970–976.
Tapke, V. E. (1951): New physiologic races of Ustilago nigra from the United States and Israel. *Phytopathology* **41**, 139–141.

Tapke, V. E. (1955): Physiologic races in Ustilago nuda and techniques for their study. *Phytopathology* **45**, 73–78.
Tatum, E. L. (1946): Induced biochemical mutations in bacteria. *Cold Spring Harbor Symp. Quant. Biol.* **11**, 278–284.
Tatum, E. L. and Lederberg, J. (1946): Gene recombination in Escherichia coli. *Nature* **158**, 558–560.
Tavel, P. (1959): The p.rification of tobacco mosaic virus by chromatography. *Arch. Biochem. Biophys.* **85**, 491–498.
Taylor, R. H. (1962): The location of tobacco mosaic virus in tobacco seed. *Austr. J. Exp. Agric. Anim. Husb.* **2**, 86–89.
Taylor, R. H., Grogan, R. G. and Kimble, K. A. (1961): Transmission of tobacco mosaic virus in tomato seed. *Phytopathology* **51**, 837–842.
Teakle, D. S. (1960): Assotiation of Olpidium brassicae and tobacco necrosis virus. *Nature* **188**, 431–432.
Teakle, D. S. and Sylvester, E. S. (1962): Infection of plants by the feeding of aphids on leaves sprayed with tobacco mosaic virus. *Virology* **16**, 363–365.
Ternovskii, M. F., Dashkecva, K. N. and Popuskai, I. S. (1965): Results of testing species and varieties of tobacco for resistance to false mildew. *Rast. Mold.* **4**, 15–28.
Thomas, W. D. and Baker, R. R. (1949): Chemical inactivation of the carnation mosaic virus in vivo. *J. Colo. Wyo. Acad. Sci.* **4**, 51.
Thompson, H. C. and Gardemann, J. W. (1962): An improved method for sterilization of agar media with propylene oxide. *Phytopathology* **52**, 167–168.
Thren, R. (1937): Gewinnung und Kultur von monokaryotischem und dikaryotischem Myzel. Ein Beitrag zur Physiologie und Genetik des Gerstenflugbrandes (Ustilago nuda [Jens.] Kellerm. et Sw.). *Z. Bot.* **31**, 337–391.
Thresh, J. M. (1956): Some effects of tannic acid and of leaf extracts which contain tannins on the infectivity of tobacco mosaic and tobacco necrosis viruses. *Ann. Appl. Biol.* **44**, 608–618.
Thung, T. H. (1931): Smetstof en plantencel bij enkele virusziekten van de tabaksplant. *Handel. Ned. Ind. Natuurwetenschap 6th Congr.* Pp. 450–463.
Thurston, H. D. (1957): The culture of Phytophthora infestans. *Phytopathology* **47**, 186.
Thurston, H. D., Wilde, P. and Sudia, T. W. (1959): Growth of six races of Phytophthora infestans on artificial media. *Phytopathology* **49**, 450–451.
Tinline, R. D. and MacNeill, B. H. (1969): Parasexuality in plant pathogenic fungi. *Ann. Rev. Phytopath.* **7**, 147—170.
Tisdale, W. H. and Johnston, C. O. (1926): A study of smut resistance in corn seedling grown in the greenhouse. *J. Agric. Res.* **32**, 649–668.
Tomlinson, R. V. and Tener, G. M. (1963a): The effect of urea, formamide and glycols on the secondary binding forces in the ion-exchange chromatography of polynucleotides on DEAE-cellulose. *Biochemistry* **2**, 697–702.
Tomlinson, R. V. and Tener, G. M. (1963b): A proposed general procedure for isolating end-groups of nucleic acids. *Biochemistry* **2**, 703–706.
Torgeson, D. C. and Lindberg, C. G. (1961): A greenhouse method for evaluation of chemicals to control apple powdery mildew. *Contr. Boyce Thompson Inst.* **21**, 33–34.
Toussoun, T. A. and Nelson, P. E. (1968): *A Pictorial Guide to the Identification of Fusarium Species According to the Taxonomic System of Snyder and Hansen.* The Pennsylvania State Univ. Press. Univ. Park and London.
Townsley, P. M. (1959a): The electrophoretic analysis of plant virus preparations. *Canad. J. Biochem. Physiol.* **37**, 119–126.
Townsley, P. M. (1959b): Zone electrophoresis of tobacco mosaic virus. *Canad. J. Biochem. Physiol.* **37**, 1025–1031.
Tseng, T. C., Lee, Y. S. and Wu, L. C. (1965): Sporulation by physiologic races of Piricularia oryzae Cav. *Bot. Bull. Acad. Sinn. Taipei.* N. S. **6**, 182–188.
Tsugita, A., Gish, D. T., Yound, J., Fraenkel-Conrat, H., Knight, C. A. and Stanley, W. M.

(1960): The complete amino acid sequence of the protein of tobacco mosaic virus. *Proc. Nat. Acad. Sci. U. S.* **46,** 1463–1469.

Tuboly, L. (1966): The appearance of a new biotype of Peronospora tabacina Adam in Hungary. *Extr. Bull. Coresta* 1966. Pp. 27–34.

Turner, E. M. C. (1961): An enzymic basis for pathogenic specificity in Ophiobolus graminis. *J. Exp. Bot.* **12,** 169–175.

Ubrizsy, G. and Vörös, J. (1966): A new conception in the review on the phylogeny and system of fungi. *Acta Bot. Hung.* **12,** 199–220.

Ubrizsy, G. (1965) (Ed.): *Növénykórtan.* Második kiadás. (Plant Pathology, 2nd edition). Akadémiai Kiadó, Budapest.

Ulrychová—Zelinková, M. (1960): The effect of some pyrimidine derivatives on the multiplication of tobacco mosaic virus. *Biol. Plant.* **3,** 240–243.

Umaerus, V. (1959): The relationship between peroxidase activity in potato leaves and resistance to Phytophthora infestans. *Amer. Potato J.* **36,** 124–131.

Van Arsdel, E. P. (1965): Micrometeorology and plant disease epidemiology. *Phytopathology* **55,** 945–950.

Van Rysselberge, C. and Jeener, R. (1957): Plant virus synthesis and the abnormal protein constituents of infected leaves. *Biochim. Biophys. Acta* **23,** 18–23.

Vakili, N. G. and Caldwell, R. M. (1957): Recombination of spore color and pathogenicity between uredial clones of Puccinia recondita f. sp. triciti. *Phytopatgology* **47,** 536.

Vavilov, N. I. (1935): *Theoretical bases of plant breeding* (In Russian) Moscow.

Vavilov, N. I. (1951): *The Origin, Variation, Immunity and Breeding of Cultivated Plants.* Transl. by K. Starr Chester. Chronica Botanica Col. Waltham, Mass.

Vavilov, N. I. (1957): World supply of cereal and bean cultivars and their use in plant breeding. 1. Results of agroecological investigations on the main field crops. Moscow—Leningrad.

Volkin, E. and Cohn, W. E. (1953): On the structure of ribonucleic acids. II. The products of ribonuclease action. *J. Biol. Chem.* **205,** 767–782.

Waitz, L. and Schwartz, W. (1956): Untersuchungen über die von Pseudomonas phaseolicola (Burk.) hervorgerufene Fettfleckenkrankheit der Bohne. II. Untersuchungen der Pathogenese. *Phytopath. Z.* **26,** 297–312.

Walker, J. C. (1941): Disease resistance in the vegetable crops. *Bot. Rev.* **7,** 458–506.

Walker, J. C. (1950): *Plant Pathology.* McGraw-Hill Book Co., Inc. New York, Toronto, London. (3rd. ed. in 1969.)

Walker, J. C. (1952): *Diseases of Vegetable Crops.* McGraw-Hill Book Co., Inc. New York, Toronto, London.

Walker, J. C. (1953): Disease resistance in the vegetable crops. II. *Bot. Rev.* **19,** 606–643.

Walker, J. C. (1965): Disease resistance in the vegetable crops. III. *Bot. Rev.* **31,** 331–380.

Walker, J. C. (1966): The role of pest resistance in new varieties. In: *Plant Breeding.* A Symposium Held at Iowa State Univ. (Ed.) Kenneth J. Frey, Iowa State Univ. Press, Ames, Iowa. Pp. 219–242.

Walker, J. C., Link, K. P. and Angell, H. R. (1929): Chemical aspects of disease resistance in the onion. *Proc. Nat. Acad. Sci. U. S.* **15,** 845–850.

Warcup, J. H. (1950): The soil plate method for isolation of fungi from soil. *Nature* **166,** 117.

Wark, D. C., Hill, A. V., Mandryk, M. and Cruickshank, I. A. M. (1960): Differentiation in Peronospora tabacina Adam. *Nature* **187,** 710–711.

Watson, A. G. (1967): The movement of Plasmodiophora brassicae in soil. *Phytopathology* (Abstr.) **57,** 835.

Watson, I. A. (1957): Further studies on the production of new races of Puccinia graminis var. tritici on wheat seedlings. *Phytopathology* **47,** 507–509.

Watson, I. A. and Luig, N. H. (1961): The classification of Puccinia graminis var. tritici in relation to breeding resistant varieties. *Proc. Linnean Soc.* N. S. Wales **88,** 235–258.

Watson, M. A. (1938): Further studies on the relationship between Hyosciamus virus 3 and the aphis Myzus persicae (Sulz.) with special reference to the effects of fasting. *Proc. Royal Soc. Ser.* B. **125,** 144–170.

Watson, M. A. (1960a): The ways in which plant viruses are transmitted by vectors. *Reptr. 7th Comm. Entom. Conf.*
Watson, M. A. (1960b): Evidence for interaction or genetic recombination between potato virus Y and C in infected plants. *Virology* **1**, 211–232.
Watson, M. A. and Nixon, H. L. (1953): Studies on the feeding of Myzus persicae (Sulz.) on radioactive plants. *Ann. Appl. Biol.* **40**, 537–545.
Watson, M. A. and Roberts, F. M. (1939): A comparative study of the transmission of Hyosciamus virus 3, potato virus Y and cucumber virus 1 by the vectors Myzus persicae (Sulz.), M. circumflexus (Buckton), and Macrosiphum gei (Koch). *Proc. Royal Soc. Ser.* B. **127**, 543–576.
Watson, M. A. and Sinha, R. C. (1959): Studies on the transmission of European wheat striate mosaic virus by Delphacodes pellucida Fabricius. *Virology* **8**, 139–163.
Wetter, C. und Brandes, J. (1956): Untersuchungen über das Kartoffel-S-Virus. *Phytopath. Z.* **26**, 81–92.
Whittaker, R. H. (1959): On the board classification of organisms. *Quart. Rev. Biol.* **34**, 210–226.
Wilhelm, S., Sagen, J. E. and Tietz, H. (1967): Some Verticillium wilt-resistant cottons obtained by integrated laboratory and field tests. *Phytopathology* (Abstr.) **57**, 836.
Wilcoxson, R. D., Tuite, J. F. and Tucker, S. (1957): Anastomosis in Puccinia graminis. *Phytopathology* **47**, 537.
Williams, E. B., Dayton, D. F. and Shay, J. R. (1966): Allelic genes in Malus for resistance to Venturia inaequalis. *Proc. Ann. Soc. Hort. Sci.* **88**, 52–56.
Williams, P. G., Scott, K. J., Kuhl, J. L. (1966): Vegetative growth of Puccinia graminis f. sp. triciti in vitro. *Phytopathology* **56**, 1418–1419.
Williams, P. G., Scott, K. J., Kuhl, J. L. and Maclean, D. J. (1967): Sporulation and pathogenicity of Puccinia graminis f. sp. tritici grown on an artificial medium. *Phytopathology* **57**, 326–328.
Williams, P. H. (1966): A system for the determination of races of Plasmodiophora brassicae that infect cabbage and rutabaga. *Phytopathology* **56**, 624–626.
Williams, P. H. and Keen, N. T. (1967): Relation of cell permeability alterations to water congestion in cucumber angular leaf spot. *Phytopathology* **57**, 1378–1385.
Williams, P. H. and Walker, J. C. (1963): Races of clubroot in North America. *Plant Dis. Reptr.* **47**, 608–611.
Williams, P. H. and Yukawa, Y. B. (1967): Ultrastructural studies on the host-parasite relations of Plasmodiophora brassicae. *Phytopathology* **57**, 682–687.
Willison, R. S., Tremaine, J. H. and Weintraub, M. (1961): Serological and physical properties of some stone-fruit viruses: Non-virus particles associated with infection. *Canad. J. Bot.* **39**, 1447–1452.
Wilson, E. E. (1936): Symptomatic and etiologic relations of the canker and the blossom blast of Pyrus and the bacterial canker of Prunus. *Hilgardia* **10**, 259–298.
Wilson, E. M. (1958): Rust-TMV crossprotection and necrotic-ring reaction in bean. *Phytopathology* **48**, 228–231.
Wilson, H. A., Lilly, W. G. and Leach, J. G. (1965): Bacterial polysaccharides. IV. Longevity of Xanthomonas phaseoli and Serratia marcescens in bacterial exudates. *Phytopathology* **55**, 1135–1138.
Winstead, N. N., Goode, M. J. and Barham, W. S. (1959): Resistance in watermelon to Colletotrichum lagenarium race 1, 2, and 3. *Plant Dis. Reptr.* **43**, 1–9.
Wittmann, H. G. (1960): Comparison of the tryptic peptides of chemically induced and spontaneous mutants of tobacco mosaic virus. *Virology* **12**, 609–612.
Wittmann, H. G. (1961): Ansätze zur Entschlüsselung des genetischen Codes. *Naturwissenschaften* **48**, 729–734.
Wolfe, M. S. (1963): Resistance of wheat to the powdery mildew disease caused by Erysiphe graminis DC. f. sp. tritici and the influence of the host on the physiologic specialization and growth of the fungus. *Ph.D. Thesis. Univ. of Cambridge.*

Wolfe, M. S. (1965): Physiologic specialization of Erysiphe graminis f. sp. tritici in the United Kingdom. *Brit. Mycol. Soc. Trans.* **48,** 315–326.
Wolfe, M. S. (1967): Physiologic specialization of Erysiphe graminis f. sp. tritici in the United Kingdom, 1964–65. *Trans. Br. Mycol. Soc.* **50,** 631–640.
Wolfe, M. S. and Macer, R. C. F. (1966): The use of kinetin in the detached leaf culture of Puccinia striiformis. *Cereal Rust Conf.* 1964. Plant Breeding Inst., Cambridge. Pp. 58–63.
Wood, H. N. (1967): Plant carcinogenesis. *Progr. Exp. Tumor Res.* **9,** 286–311.
Woodbury, W., Macko, V. and Stahmann, M. A. (1967): A collector for fungal spores. *Phytopathology* **57,** 455.
Woods, M. W. (1940): Reversible inhibition of tobacco-mosaic virus in living cells with 0.0002 molar sodium cyanide. *Science* **91,** 295–296.
Wormald, H. (1932): Bacterial diseases of stone-fruit trees in Britain. IV. The organism causing bacterial canker of plum trees. *Trans. Brit. Mycol. Soc.* **17,** 157.
Wright, D. S. C. (1966): Verticillium wilt of tobacco. (A note.) *New Zealand J. Agric. Res.* **9,** 448–451.
Wu, J. H. and Rappaport, I. (1961): An analysis of the interference between two strains of tobacco mosaic virus on Phaseolus vulgaris L. *Virology* **14,** 259–263.
Yamazaki, H. and Kaesberg, P. (1961): Biophysical and biochemical properties of wild cucumber mosaic virus and of two related virus-like particles. *Biochim. Biophys. Acta* **51,** 9–18.
Yamaguchi, A. and Hirai, T. (1959): Effect of local infection with tobacco mosaic virus on respiration in leaves of Nicotiana glutinosa. *Phytopathology* **49,** 447–449.
Yarwood, C. E. (1951a): Fungicides for powdery mildews. *Proc. IInd Int. Congr. Crop Protection, London,* 1949. P. 500.
Yarwood, C. E. (1951b): Associations of rust and virus infections. *Science* **114,** 127–128.
Yarwood, C. E. (1952): The phosphate effect in plant virus inoculations. *Phytopathology* **42,** 137–143.
Yarwood, C. E. (1953): Quick virus inoculation by rubbing with fresh leaf discs. *Plant Dis. Reptr.* **37,** 501–502.
Yarwood, C. E. (1955): Mechanical transmission of an apple mosaic virus. *Hilgardia* **23,** 613–628.
Yarwood, C. E. (1957a): Mechanical transmission of plant viruses. *Adv. Virus Res.* **4,** 243–274.
Yarwood, C. E. (1957b): Powdery mildews. *Bot. Rev.* **23,** 235–301.
Yarwood, C. E. (1960): Localized acquired resistance to tobacco mosaic virus. *Phytopathology* **50,** 741–744.
Yarwood, C. E. and Sylvester, E. S. (1959): The half-life concept of longevity of plant pathogens. *Plant Dis. Reptr.* **43,** 125–128.
Yasuda, Y. and Hirai, T. (1964): Incorporation of H^3-uracil into tobacco leaf epidermis infected with tobacco mosaic virus. *Exp. Cell Res.* **34,** 210–212.
Youden, W. J. and Beale, H. P. (1943): A statistical study of the local lesion method for estimation tobacco mosaic virus. *Contrib. Boyce Thompson Inst.* **6,** 437–454.
Young, H. C. and Browder, L. E. (1965): The North American 1965 set of supplemental differential wheat varieties for identification of races of Puccinia recondita tritici. *Plant Dis. Reptr.* **49,** 308–311.
Yu, Ch. T. and Zamecnik, P. C. (1960): A hydrolytic procedure for ribonucleotides and its possible application to the sequential degradation of RNA. *Biochim. Biophys. Acta* **45,** 148–154.
Zade, A. (1928): Massinfektion mit Haferflugbrand nach einem neuen Verfahren. *Z. Pfl. bau Pfl. schutz* **5,** 43.
Zadoks, J. C. (1961): Yellow rust on wheat, studies on epidemiology and physiologic specialization. *Tijdschr. Plantenziekten* **67,** 69–256.
Zaitlin, M. and Boardman, N. K. (1958): The association of tobacco mosaic virus with plastids. I. Isolation of virus from the chloroplast fraction of diseased-leaf homogenates. *Virology* **6,** 743–757.

Zakopal, J. and Spitzová, B. (1964): Nová ponorovaci laboratorni metoda zkouseni hybrid a krizencn bramboru na vzdornost k rakovine (Synchytrium endobioticum [Schilb.] Perc.). (A new laboratory submersion method for testing the resistance of potato varieties and hybrides to Synchytrium endobioticum [Schilb.] Perc.). *Ved. Pr. Vyzk. Ust. Rostl. Vyroby* v *Praze-Ruzyni* **7,** 165–184.

Zaracovitis, C. (1964): Factors in testing fungicides against powdery mildews. I. A slide spore-germination method of evaluating protectant fungicides. *Ann. Appl. Biol.* **54,** 361–374.

Zatykó, J. and Garay, A. (1959): A hervadási toxinok és a kajszi gutaütésének kérdése. (Wilt toxins and the problem of apoplexy of apricot). *Kísérl. Közl.* **52/C,** 39–52.

Zaumeyer, W. J. and Thomas, H. R. (1957): A monographic study of bean diseases and methods for their control. *U. S. Dept. Agric. Techn. Bull. No.* 868.

Zech, H. (1954): Morphologische und cytochemische Beobachtungen an Tabakmosaikvirus-infizierten Protoplasten von Nicotiana tabacum. *Exp. Cell Res.* **6,** 560–562.

Zech, H. and Vogt-Köhne, L. (1955): Ultraviolettmikrospektrographische Untersuchungen am Tabakmosaikvirus in situ. *Naturwissenschaften* **42,** 337–339.

Zinder, N. D. and Lederberg, J. (1952): Genetic exchange in Salmonella. *J. Bact.* **64,** 679–699.

AUTHOR INDEX

A

Adair, C. R., 436, 437
Agnihothrudu, V., 272
Akai, S., 426
Alexopoulos, C. J., 324
Al-Sohaily, I. A., 419
Anderer, F. A., 29, 70, 98
Anderson, H. W., 351, 353
Angell, H. R., 305
Apgar, J., 31
Ark, P. A., 295
Arkhipova, F. D., 344
Armstrong, G. M., 316
Armstrong, J. K., 316
Arx, J. A., 421
Asai, G. N., 434
Astier-Manifacier, S., 70
Atkins, I. M., 409
Atkins, J. G., 436, 437
Averre, C. W., 198
Avery, O. T., 298
Ayers, G. W., 326

B

Babayan, A. A., 348
Babos, P., 70, 73
Badami, R. S., 59
Baigent, N. L., 125
Bailey, S. T., 191
Bailey, W. T., 294
Baker, R. R., 94
Bald, J. G., 67–69, 86
Barham, W. S., 426
Barron, G. L., 295, 301
Barrus, M. F., 291
Barsegyan, S. G., 348
Bartel, W., 339, 340
Bartels, R., 52
Bartholly, R., 405
Batts, C. C. W., 411
Bawden, F. C., 24, 27, 30, 44, 53, 54, 56, 58, 77, 86, 92, 93, 96
Baylis, R. J., 406
Beadle, G. W., 295
Beale, H. P., 75
Benada, J., 357
Benda, G. T. A., 64
Bennett, C. W., 43, 61, 64, 82, 96, 97, 101
Beraha, L., 54
Berbee, J. G., 230
Bercks, R., 52
Berend, St., 429
Best, R. J., 294
Bever, W. M., 410, 412, 413
Billeter, M. A., 68
Billing, E., 125
Bils, R. F., 35
Black, C. M., 58
Black, L. M., 22, 59, 71
Black, W., 295, 334
Blattny, C., 54, 331
Blendon, D. C., 145
Blumer, S., 97
Boardman, N. K., 73, 74
Bollard, E. G., 72
Bolton, E. T., 72
Bond, L. H., 272, 284, 285
Borlaug, N. E., 313–315
Bortels, M., 231
Bos, L., 41
Bozarth, R. R., 308
Boynansky, V., 332
Bowen, C. C., 262, 305
Böning, K., 93
Bradley, R. H. E., 56, 57, 59
Brakke, M. K., 22
Brandes, J., 27, 28, 86, 102
Brandt, W. H., 428, 429
Braun, A. C., 120, 198, 199
Brcak, J., 59
Bridgmon, G. H., 300

Bromfield, K. R., 344
Brotherton, W., 227
Browder, L. E., 392, 393
Browning, J. A., 313, 314
Brückner, F., 357, 365
Buddenhagen, J., 197, 229
Burdon, R. H., 68
Burill, T. J., 117
Burkholder, W. H., 119, 216, 223
Busch, L. V., 425, 432
Buxton, E. W., 297, 300, 439

C

Cadman, C. H., 19, 37, 62, 73
Caldwell, J., 45, 49, 65
Caldwell, R. M., 300, 301, 357
Canova, A., 48
Capoor, S. P., 82
Carpenter, C. C., 286
Carr, A. J. H., 429, 431
Caspar, D. C. D., 29
Cech, M., 54
Chalcroft, J., 74
Chang, W. L., 437
Chapman, G. B., 260
Chapman, R. A., 209
Chase, M., 36
Cherewick, W. J., 358, 365
Chessin, M., 64
Chester, K. S., 78, 102
Christensen, J. J., 296, 297, 418
Cholodny, N. G., 272
Chu, O. M. Y., 434
Clark, M. F., 69, 74
Clark, W. A., 372
Clements, F. E., 265
Cohn, W. E., 32
Collier, J., 419, 420
Commoner, B., 37, 72, 94
Curtis, K. M., 329
Curl, E. A., 272, 284, 285
Cook, A. A., 195, 214
Cook, M. T., 85
Cooke, W. M. B., 272
Copeland, H. F., 265
Corey, R. R., 298
Cornuet, P., 70
Cotter, R. U., 295
Couch, J. N., 323
Craig, J., 282
Craigie, J. H., 368
Crispin, A., 399
Crook, P., 286
Crosse, J. E., 195, 217–220
Crowley, N. C., 44, 48, 91
Cruickshank, I. A. M., 309, 341, 343, 347

D

Dashkecva, K. N., 348
Davis, D., 439
Davis, J. R., 219
Davison, A. D., 397-399
Day, P. R., 400
Dayton, D. F., 354
Dean, A. C. R., 297
Dekker, J. 365
Delbrück, M., 183, 294
De Ley, J., 125
Deppe, K., 340, 341
Devaux, A. L., 430
DeVay, J. E., 296, 410, 418
D'Herelle, F., 183
Dickson, J. G., 432, 439, 442
Diener, T. O., 37, 65, 69, 87
Diercks, R., 93
Dimond, A. E., 209, 223, 339, 429, 440
Doling, D. A., 394, 396
Dongo, S., 399
Donk, M. A., 257
Dowson, W. J., 117, 140, 175, 222, 224, 231
Drandarevski, Ch., 278, 357
Duran, R., 400
Dutta, S. K., 425, 426
Dye, D. W., 213
Dyson, J. G., 64

E

Egami, F., 31
Egerer, A., 346, 347
Ehrlich, H. G., 261, 262
Ehrlich, M. A., 261, 262
Eichenmuller, J. J., 334
El Hammady, M., 90
Ellingboe, A. H., 301
Elliott, E. W., 323
Elrod, R. P., 120
Endemann, W., 347
Englander, L., 431
Engler, R., 66, 69
English, H., 219
Epps, W. M., 422, 423
Ercolani, G. L., 117
Eriksson, J., 291
Erwin, D. C., 340, 428
Esau, K, 85
Everett, G. A., 31

F

Fantechi, F., 345, 348
Farkas, G. L., 87–89, 194
Fehrmann, H. 339
Feinberg, J. G., 178
Fennell, D. I., 274, 275
Fenwick, H. S., 217
Fischer, G. W., 406
Forbes, E., 300
Ford, D. H., 278
Forsyth, F. R., 396
Fraenkel-Conrat, H., 25, 29, 35, 37, 38, 40, 66, 70, 98, 295
Frandsen, N. O., 340
Francki, R. I. B., 74
Franklin, R. E., 29, 35
Frazer, D., 71
Fredericksen, R. A., 419, 420
Freeman, G. G., 44
Frey, K. J., 313
Fribourg, H. A., 272, 284, 285
Frick, E. L., 362–364
Frisch-Niggemeyer, W., 30
Fuchs, Eva, 394
Fuchs, W. H., 317, 333, 340, 350, 355, 367, 416
Fukushi, T., 58
Fulton, R. W., 53, 54
Futrell, M. C. 419, 420

G

Gabriel, G. E., 428
Galindo, A., 334
Galindo, J., 340
Gallegly, M. E., 333, 334, 338, 340
Ganong, R. J., 57
Garay, A., 429
Garber, E. D., 193, 194
Garber, R. H., 429, 431
Gardemann, J. W., 282
Garrett, C. M. E., 219, 220
Gaskin, T. A., 410
Gäumann, E., 103, 307
Gendron, V., 44
Gibbs, A. J., 35
Gicherman, G., 92
Gierer, A., 36–38, 65, 98, 295
Ginosa, W., 66, 67
Girbardt, M., 261
Gish, D. T., 29, 70, 98
Gloyer, W. O., 215
Godberg, H. S., 145
Goldin, M. I., 86
Goode, M. J., 424, 426
Gooding, G. V., 25
Goodman, R. N., 117, 195, 197, 213, 306, 309, 396
Gordon, M. P., 34, 66
Gordon, W. L., 441
Goss, R. C., 282
Goss, R. W., 215
Goto, K., 436, 437
Goto, M., 189, 229, 299
Gough, F. J., 334
Govindarao, P., 437
Gowen, J. W., 295
Graham, D. C., 125, 222
Graham, K. M., 334
Greeley, L. W., 356, 365
Gretchushnikov, A. I., 339
Groves, A. B., 364
Grogan, R. G., 48
Groth, R. W., 216
Gupta, B. M., 44
Guthrie, J. W., 217

H

Haarring, F., 416
Hadden, S. J., 416
Hadwiger, L. A., 309, 426
Hagedorn, D. J., 41
Halisky, P. M., 400
Hall, C. E., 35
Hall, C. V., 425, 426
Hamers-Casterman, C., 65
Hamlin, S. A., 351
Hamlyn, B. M. G., 56
Hampton, R. E., 54
Hansen, H. N., 441
Hansen, H. P., 102
Hansing, E. D., 416, 417
Hanson, J. B., 91
Harder, A., 97
Harmon, D. L., 372
Harpaz, I., 51, 52
Harrar, J. G., 317, 374, 400
Harrison, B. D., 37, 58, 62, 63, 65
Harter, L. L., 398
Haselkorn, R., 68
Hastie, A. C., 301, 428
Hawker, L. E., 265
Hayes, H. K., 317
Hayward, A. C., 125, 229
Hebert, T. T., 25
Hedges, F., 119, 208, 226
Heggestad, H. E., 194

Heinze, K., 56, 57, 59
Henniger, H., 224, 286, 337, 339, 340
Heppel, L. A., 34
Hermansen, J. E., 357, 396
Hershey, A. D., 36, 294
Hess, W. M., 198
Heta, H., 360
Hewitt, W. B., 63
Heyne, E. G., 425, 426
Hidaka, Z., 70
Hildebrand, D. C., 194
Hill, A. V., 341, 343, 347
Hille, M., 331
Hinshelwood, C. N., 297
Hirai, T., 67–69, 71, 87, 94, 95
Hiura, U., 360
Hodgson, W. A., 338
Hoffmann, G. A., 407, 408
Hoffmann, G. M., 439
Hoffmann, J. A., 400
Hoggan, I. A., 102
Holley, R. W., 31
Holliday, R., 301
Holmes, F. O., 50, 74, 87, 93, 101, 102
Holton, C. S., 400, 403, 406, 416, 417
Honecker, L., 357
Hopfe, A., 340
Horváth, J., 111, 112
Hough, F. L., 352
Houston, B. R., 429, 431
Hrushowetz, S. B., 300
Husain, A., 198

I

Immer, F. R., 317
Incardona, N. L., 73
Iren, S., 347
Isenbeck, K., 317, 333, 340, 350, 355, 367, 416
Israel, H. W., 90
Ishii, M., 71
Ivanovsky, D., 85
Izard, C., 345
Izrailsky, V. P., 117

J

Jankowski, F. 344
Jeater, A., 411
Jeener, R., 36, 65, 71, 72
Jensen, D. D., 58, 59
Jockusch, H., 88
Johnson, J., 27, 101, 102
Johnson, L. F., 272, 284, 285
Johnston, C. O., 387, 418
Jones, B., 428
Jones, M. W., 434

K

Kaesberg, P., 35, 73
Kameraz, A. Ya., 332
Kammermann, N., 339
Kaper, J. M., 37
Kappert, H., 317
Karamisheva, U. I., 353
Kassanis, B., 37, 39, 44, 59, 65, 66, 73, 86, 93, 96
Katznelson, H., 191
Kavanagh, T., 410
Keclzek, N. A., 348
Kedar (Kammermann), N., 339
Keen, N. T., 157, 158
Keitt, G. W., 350
Kelman, A., 197, 198, 228, 229, 274
Kendrick, E. L., 403–405, 407, 408
Kennedy, B. W., 195
Kikuchi, M., 88
Kimble, K. A., 48
Kimura, I., 71
Király, Z., 87, 88, 90, 92, 195, 306, 309, 396, 410
Kirby, A. H. M., 362–364
Klarman, W. L., 282
Kleczkowski, A., 54
Klein, D. T., 298, 299
Klein, R. M., 298, 299
Klement, Z., 92, 120, 121, 126, 153, 155, 157, 159, 173, 175, 177, 179, 184–186, 189, 193, 195, 197, 205, 213
Klens, P. F., 286
Klinkowski, M., 56, 341
Klug, A., 29
Knight, C. A., 29–31, 37, 65, 68, 70, 98, 99
Koch, W, J., 265
Korf, R. P., 284
Kotesvararao, D., 437
Kozaka, T., 436, 437
Köhler, E., 82, 96
Köhler, H., 56
Krauss, E., 345
Kreitlow, K. W., 50
Kröber, H., 341, 342, 347
Kubo, S., 70
Kuć, J., 354
Kuhl, J. L., 270
Kunkel, L. O., 295

L

Lamey, H., 334
Lammering, J., 326
Lange-de la Camp, Maria, 409
Lapwood, D. H., 338, 339
Larson, R. H., 326
Lea, H. W., 347
Leach, J. G., 119
Lederberg, E. M., 297
Lederberg, J., 294, 297, 299
Lee, Y. S., 434
Legunkova, R. M., 26
Lehoczky, J., 233
Lelley, J., 410
Lelliott, R. A., 125, 223
Levine, M. N., 295, 303
Li, H. W., 434
Lilly, V. G., 119
Limasset, P., 59
Lincoln, R. E., 295
Lindberg, C. G., 364
Link, K. P., 305
Lippincott, J. A., 37, 38
Littau, V. C., 58
Littrell, R. H., 422, 423
Litvinov, M. A., 430
Loebenstein, G., 90–92
Loegering, W. Q., 302, 303, 372, 373, 375, 385
Lorito, N., 345, 348
Lovas, B., 121, 184–186
Lovrekovich, L., 78, 92, 120, 173, 175, 177, 179, 189, 194, 213, 222, 396
Luig, N. H., 385, 386
Luke, H. H., 416
Luria, S. E., 183
Lyr, H., 340

M

Macer, R. C. F., 394, 396
Macko, V., 288, 370
Maclean, D. J., 270
MacLeod, C. M., 298
MacMillan, P. C., 429
MacNeill, B. H., 295, 300, 301, 440
Madison, J. T., 31
Malec, K., 331
Malca, I., 428
Mandel, M., 69, 125
Mandeles, S., 35
Mandryk, M., 341, 343, 345–347
Manka, K., 272
Mankin, C. J., 419
Manil, P., 95
Manning, W. J., 213
Maramorosch, K., 58, 59
Marcus, D., 77
Markham, R., 24, 30–32, 71, 72
Marquissee Merrill, S. H., 31
Marr, J. L., 282
Martin, G. W., 239, 265, 323
Martinson, C. A., 431
Massfeller, D., 347
Mastenbroek, C., 334
Matsui, A. I., 69
Matsushita, K., 73
Matthews, R. E. F., 69, 72, 74, 80, 81, 94, 95
Mazzone, H. M., 73
McAlear, J. H., 260–262
McCarty, M., 298
McFadden, L. A., 223
McKinney, H. H., 102, 372
Medina, A. C., 48
Mehta, B. K., 419, 420
Meiners, J. P., 406, 407
Menzites, J. D., 428
Mercer, F. L., 94
Merrett, M. J., 89
Metzger, R. J., 407
Meyer, G. M., 288
Michail, S. H., 431
Middendorf, M., 418
Milinkó, I., 48, 94, 95
Miller, C. E., 324
Miller, G. L., 36
Miller, H. N., 223
Mills, W. R., 334
Milo, G. E., 89
Minsov, I. N., 434
Mirzabekyan, R. O., 330
Mitchell, J. E., 286
Miura, K. I., 71
Miyamoto, Y., 62
Mizusawa, Y., 71
Moje, W., 428
Moore, M. B., 409
Moore, R. T., 259–262
Morton, D. J., 213, 411, 412
Moseman, J. G., 354, 356–358, 365, 412
Mumford, D. L., 410
Mundry, K. W., 37, 98, 295
Murti, V. D., 437
Müller, K. O., 36, 309, 334
Müller, R., 364

N

Naylor, A. W., 64
Negulescu, Fl., 386
Nehari, V., 72
Nelson, G. A., 330
Nelson, P. E., 438
Nelson, R. R., 300
Neu, H. C., 34
Newell, J., 95
Newmark, P., 71
Newton, M., 358, 365
Niederhauser, J. S., 334, 336, 340
Niemann, E., 406
Niemann, G. J., 365
Nitta, T., 70
Nixon, H. L., 27, 35, 37, 56, 65
Noble, M., 224
Novacky, A., 407
Nover, A., 407
Nover, I., 358, 359, 365
Noviello, C., 284

O

O'Brien, L. M., 213
Ochoa, R. O., 68
Ohms, R. E., 410, 412
Okabe, N., 189, 229
Olsen, O. A., 330
Oort, A. J. P., 365, 409, 410
Ouchterloni, Ö., 176, 178

P

Padmanabhan, S. Y., 433
Pady, S. M., 288
Palmiter, D. H., 350
Panagopoulos, C. G., 219
Parish, C. L. 88
Patil, S. S., 429, 440
Paton, A. M., 224
Paul, H. L., 26
Pawlik, A., 342, 345
Penswick, J. R., 31
Pénzes, A., 265
Perrin, D. R., 309
Person, C., 312, 396
Person, L. H., 229, 274
Peruanskii, Yu., V., 434
Peterson, L. C., 334
Peyton, G. A., 262, 305
Pfahler, P. L., 416
Piemeisel, F. J., 291
Pirie, N. W., 24, 30, 37, 93
Podhradszky, J., 434, 435, 437
Poehlman, J. M., 409
Pontecorvo, G., 300
Popp, W., 409, 411
Popuskai, I. S., 348
Porter, C. A., 94
Powers, H. R., Jr., 354, 356, 358
Pozdena, J., 54
Pozsár, B. I., 90, 92, 195
Praagh, T., 90, 91
Presypkin, V. F., 344
Price, W. C., 44
Protzenko, A. E., 26
Purvis, M. R., 119

Q

Quanjer, H. M., 102
Quantz, L., 41

R

Rabina, S., 90, 91
Radulescu, E., 386
Rai, P. V., 180, 181, 198
Ralph, R. K., 69, 74
Ramson, A., 347
Rands, R. D., 227
Raper, K. B., 274
Rappaport, I., 82, 97
Rasmusson, D. C., 410
Rawlins, T. E., 54, 57, 86, 278
Razvyazkina, G. M., 58
Reddi, K. K., 67, 70, 71, 73
Reese, M. W., 32
Reichmann, M. E., 32, 35, 37, 38, 65
Reid, D. A., 412
Reid, M. S., 74
Riker, A. J., 230
Ritter, F. J., 288
Robert, A. L., 436, 437
Roberts, D. A., 75
Roberts, F. M., 55
Rodenhiser, H. A., 403, 416, 417
Roemer, Th., 317, 333, 340, 350, 355, 367, 405, 416
Rogerson, C. T., 286
Rohringer, R., 312
Roland, G., 93
Romero, S., 334
Roper, J. A., 300
Rorie, F. G., 434
Rosen, H. R., 119

Ross, A. F., 89–91, 96, 308
Ross, R. G., 351
Rowell, J. B., 410, 418
Röttger, B., 70, 73
Rudorf, W., 317
Rushizky, G. W., 31, 37, 65
Ryzhkov, V. L., 94, 101, 102

S

Sabanayagam, J. V., 440
Sackston, W. E., 430
Sagen, J. E., 432
Sakai, R., 339
Samborski, D. J., 312, 396
Sansome, E., 342
Santilli, V., 89
Sato, K., 31
Sänger, H. L., 68
Schaeffer, W. W., 145, 194
Schafer, J. F., 410
Scheetz, R. W., 213
Schick, E., 335
Schick, R., 334, 335, 340
Schiltz, P., 345
Schlegel, D. E., 40, 57, 68
Schmelzer, K., 53, 61
Schmid, K., 345
Schmidt, H. B., 59
Schmiedeknecht, M., 341
Schmitt, C. G., 344
Schmitthenner, A. F., 286
Schmutterer, H., 59
Schnathorst, W. C., 193
Schramm, G., 29, 36, 38, 65, 66, 69, 70, 73, 98, 102, 295, 296
Schroth, M. N., 194
Schuhmann, G., 406
Schuster, H., 296
Schuster, M. L., 217
Schwartz, W., 215
Schwochau, M. E., 309
Scott, K. J., 270
Seaman, W. L., 326
Seehofer, F., 341
Sehgal, O. W., 19, 37, 73
Semeniuk, G., 419
Sequeira, L., 229
Shalla, T. A., 82
Shanmuganathan, N., 220
Shay, J. R., 352–354
Shear, C. L., 265
Sheffield, F. M. L., 82, 86
Shepherd, C. J., 341, 345, 346
Shigunatsu, A., 71
Shimomura, T., 95
Shipp, W., 68
Shiroya, T., 70
Short, M. N., 32
Siebs, E., 364
Siegel, A., 19, 37, 66, 67, 73, 88
Sill, W. H., 55
Simmonds, P. M., 410
Simons, J. N., 56, 59
Singer, B., 35, 37, 40, 66
Singh, R. S., 286
Sinha, R. C., 58
Skoog, H. A., 215
Skvortzoff, S. S. 410
Slagle, C. W., 44
Small, B. C., 217
Smalley, E. B., 230
Smith, C., 66
Smith, D. C., 317
Smith, K. M., 24, 31, 59, 72, 83 101
Smith, S. H., 68
Smith, W. K., 222
Smooth, J. J., 334
Snyder, W. C., 441
Sober, H. A., 31
Sokolova, I. I., 433, 437
Solymosy, F., 32, 87–89
Sowell, G., 214
Spencer, D., 73
Spitzová, B., 330
Spran, F., 345
Stace-Smith, R., 37, 38, 65
Staehelin, M., 32
Stahmann, M. A., 222, 288, 370
Stakman, E. C., 291, 295, 296, 302, 303, 317, 320, 373–375, 385, 400, 418
Stalder, L., 97
Stall, R. E., 195, 214
Stanley, W. M., 29, 36, 48, 70
Stapp, C., 77, 117, 120, 170, 224, 231, 232
Starr, M. P., 125, 170, 193, 197, 213, 223, 224, 298
Steel, K. J., 152
Steere, R. L., 24, 30, 37
Stewart, D. M., 302, 373, 375, 385, 386
Stoddard, E. M., 94, 209
Stoddart, J. L., 429
Stolp, H., 125, 189, 213
Strandberg, J. O., 327
Strobel, G. A., 180, 181, 198
Stuart, F., 345, 346

Sugiyama, T., 35
Sukhov, K. S., 58
Sutton, M. D., 191
Suzuki, N., 71
Südi, J., 434, 435, 437
Schwinghammer, E. A., 295
Sylvester, E. S., 49, 55, 57, 58
Symons, R., 32
Szirmai, J., 48, 53

T

Takahashi, W. N., 37, 49, 71, 72, 94
Talbot, P. H. B., 257
Talboys, P. W., 430
Tapke, V. E., 412–414
Tatum, E. L., 294, 295
Taylor, R. H., 48
Teakle, D. S., 57, 63
Tener, G. M., 32, 33
Terasawa, H., 426
Ternovskii, M. F., 348
Thomas, H. R., 214
Thomas, W. D., 94
Thompson, H. C., 282
Thompson, H. R., 72
Thornberry, H. H., 54
Thren, R., 409
Thresh, J. M., 54
Thung, T. H., 96
Thurston, H. D., 337
Tietz, H., 432
Tinline, R. D., 300
Tisdale, W. H., 418
Tomaru, K., 70
Tomiyama, K., 339
Tomlinson, R. V., 32, 33
Tompkins, C. M., 54
Torgeson, D. C., 364
Tosic, L., 346
Tousson, T. A., 438
Townsley, P. M., 23
Tremaine, J. H., 73
Tseng, T. C., 434
Tsugita, A., 29, 37, 70, 98
Tuboly, L., 341, 347
Tucker, S., 300
Tuite, J. F., 300
Turner, E. M. C., 305
Twort, F. W., 183

U

Ubrizsy, G., 239, 262, 323, 331
Uhling, H., 29, 70, 98
Ulrychová-Zelinková, M., 94
Umaerus, V., 339

V

Vakili, N. G., 300, 301
Van Rysselberge, C., 71
Van Slogteren, D. H. M., 77
Vartanyan, V. S., 348
Varzondek, M., 54
Vaughan, E. K., 397–399
Vavilov, N. I., 317
Veldee, S., 66
Velden, F. J. J. A., 421
Venkatareddy, T. C., 437
Vogt-Köhne, L., 67, 69
Volkin, E., 32
Vostrova, N. G., 86
Vörös, J., 239, 265, 323

W

Waitz, L., 215
Walker, J. C., 55, 305, 316, 317, 326, 332, 425, 432, 440, 442
Wang, M. K., 437
Warcup, J. H., 272
Wark, D. C., 347
Warner, C., 68
Watson, A. G., 323
Watson, I. A., 301, 385, 386
Watson, M. A., 55, 56, 58, 59, 294
Weaver, M. L., 37, 65
Weber, E., 29, 70, 98
Weierich, A. J., 57
Weinmann, W., 342
Weinstein, L. H., 94
Weintraub, M., 73
Weismann, C., 68
Weltzien, H. C., 278, 357
Wetter, C., 28, 102
Whittaker, R. H., 265
Wiberg, A., 365
Wilcoxson, R. D., 300
Wildman, S. G., 66, 67, 69, 73, 74, 82
Wilhelm, S., 432
Williams, E. B., 353, 354
Williams, P. G., 270
Williams, P. H., 157, 158, 324–327
Williams, R. C., 37, 38, 295
Willison, R. S., 73
Wilson, E. M., 97
Wilson, H. A., 119, 219
Winstead, N. N., 426

Wittmann, H. G., 37, 99
Wolcyrz, S., 44
Wolfe, M. S., 357, 361, 362, 365, 396
Wood, H. N., 199
Woodbury, W., 288, 370
Woods, M. W., 94
Woody, B. R., 30
Worley, J. F., 217
Wormald, H., 219
Wright, D. S. C., 431
Wu, J. H., 97
Wu, L. C., 434

Y

Yamaguchi, A. 87, 88
Yamazaki, H., 35, 73
Yang, S. C., 437
Yarwood, C. E., 49–51, 53, 55, 89, 97, 364, 365
Yasuda, Y., 67, 68
Yasumori, H., 426
Youden, W. J., 75
Yound, J., 29, 70, 98
Young, H. C., 392, 393
Yu, Ch. T., 34
Yukawa, Y. B., 324

Z

Zade, A., 416
Zadoks, J. C., 394
Zaitlin, M., 19, 37, 73, 88, 306, 309, 396
Zakopal, J., 330
Zamecnik, P. C., 34
Zamir, A., 31
Zaracovitis, C., 364
Zatykó, J., 429
Zaumeyer, W. J., 214, 398
Zech, H., 67, 69, 86
Zentmyer, G. A., 340
Zinder, N. D., 299

SUBJECT INDEX

A

Abnormal protein 71
Abrasive 50, 54
Absidia sp. 251
Absorption maximum 81
— spectrum 81
Abutilon-mosaic virus 53
Abutilon striatum var. *thompsonii* 53
Acceptor host 51, 55, 93
Aceratogallia sanguinolenta 59
Acervulus 247
Acidic fuchsin 278
Acquired immunity 308, 309
Acquisition feeding 55, 57, 58, 60
— period 56
Actidione 95
Actinomycetales 125
Actinomycetes 260, 286
cultures 284
Actinomycetin 95
Actinomycin 90, 95
Actinomycin-D 68
Active growth of soilborne pathogens 305
Adaptation 296–298
Adenine 29
Adenosine 29, 35
Adenylic acid 29
Adsorption test 178
Adult-plant resistance 317, 386, 396
Aecial primordium 366
Aeciospore 260, 366–368
Aerial fungal spore, determination of 288
Aerobic bacteria 124
Affinity 307
Agar-agar 138
Agar-gel diffusion method 77, 78, 176, 177, 181
Agaricales 253
Agar layer plate 188, 191
Ageing *in vitro* 43
Agglutination 78, 171, 174
— titre 171, 173, 175
Agrobacterium 120, 123, 125, 167
— *radiobacter* 232
— *rhizogenes* 232
— *rubi* 232
— *tumefaciens* 129, 199, 230–233, 298, 299
artificial inoculation 233
determination of 231
isolation of 232
medium for 231, 232
star formation 231
symptoms caused by 230
virulence of 232
Airborne fungal pathogens 305
Air sampling equipment 288
Albugo candida 316, 342
Aleurispore 247
Alfalfa mosaic virus 26, 30, 55
Alkaline degradation of RNA 33, 35
— phosphomonoesterase 33
Alternaria sp. 269
Amann solution 276
Amaranthus caudatus 83, 108
Amino acids 28, 49, 94
— — sequence 98
Ammonia production 149
Ammonium sulphate 54
Amputative hypersensitivity 64
Anaphylaxis 80
Anastomoses 300
Aneuploidy 301
Angular leaf spot of
cucumber 127, 129, 134, 157, 158
hemp 131
Anisometric viruses 26
Antheridium 249, 250, 252
Anthracnose of cucurbits 421–427
acervulus 421, 422
amino acids and susceptibility to 426
artificial inoculation 422, 423
biochemical mutants 425

Anthracnose of cucurbits
biochemical specialization 424, 425
control of 426, 427
differential hosts for races 424, 425
inhibitors of 426
life cycle of the fungus 422, 423
physiologic races 424, 425
resistance 425, 426
symptoms 422
taxonomic position 421
Antibiotics in media 139
Antibody 76, 79, 80, 170, 171
Antigen 76, 79, 80, 170, 171
— comparison 178
Antigenic mosaic 171
Antiserum 76–78, 80, 81
Antiviral factor 91, 92
Aphids 53, 55–57, 59, 60
Aphis gossypii 59
Apical necrosis 83
Aplanospore 248
Apothecium 254, 255
Apple-juice agar 284
Apple mildew 362–365
aetiology 362
artificial inoculation 352–364
control 364
fungicides against 364
inoculation tower 363
life cycle 362
physiologic races 364
primary infection phase 362
secondary infection phase 362
slide-spore germination method 364
Apple mosaic virus 53
Apple scab 349–354
artificial inoculation 352, 353
control 354
cultivars resistant to 353, 354
epidemic spread 351
races 353
rating scale for symptoms 352
symptoms 351, 352
Appressorium 239, 305
Approach graft 47
Arabis mosaic virus 62
Arthrospore 242
Artificial inoculation 154, 155, 224, 318
Ascocarp 253
Ascochyta sp. 269, 274
Ascogenous hypha 252, 258
Ascogonium 249, 252
Ascoloculares 255, 266, 268, 349
Ascomycetes 249, 252–255, 258, 260, 261, 265, 267, 268, 270, 349, 422
Ascorbic acid 53, 54, 88, 89
Ascospore 252
Ascostroma 254, 255
Ascus 248, 252, 256
Asexual fungi 300
— reproduction of 242
— spores 242, 243
Aspergillus 252, 273, 285, 287, 300
glaucus 285
umbrosus 252
Aspermy virus 83
Aspirator 61
Aster yellows virus 57, 93
Atomizer 157, 159
Atypical necrosis 126, 153, 156
Aucuba strain of TMV 83
Aureomycin 95
Autoclave 282
Autoradiography 68
Autotrophic organisms 124, 264
Auxin 64, 305
Auxotrophs 295
Avirulence 295
Avirulent bacteria 195
Axeny 307–309
8-Azaguanine 95
Azide 89

B

Backcross 315, 316
Bacteria
biochemical characteristics 146
morphological characteristics 143
nutrient requirements 138, 193
pH in relation to 139
reproduction of 120, 122
resistance to 193
selection of pathogenic isolate 142
Bacterial
action on carbon compounds 147
blight of
Chrysanthemum 223
pepper and tomato 211–214
artificial inoculation of 214
isolation of the pathogen 213
resistance to 214
symptoms of 211
blossom blight 217–220
artificial inoculation 220
isolation of the pathogen 219
seasonal cycle 218
symptoms 218

Bacterial
canker 132
of stone fruits 217–220
artificial inoculation 220
isolation of the pathogen 219
seasonal cycle 218
symptoms 218
capsule 119
cell morphology 143
— number 155–157, 161, 196
— — in the living plant 201
— structure 119
— viability 119
colony morphology 143, 146
disease types 123
exudate 119, 127
growth curve 122
serology 170
slime 119, 127
spot diseases 211
of paprika 211–213
of tomato 211–213
toxins 127, 180, 181, 197, 198
viruses 183
wilt of
beans 225–227
artificial inoculation of 226
identification of the pathogen 226
isolation of the pathogen 226
resistance to 227
symptoms 225
cucumber 316
Solanaceae 228, 229
biotypes 229
isolation of the pathogen 228
identification of the pathogen 228
symptoms 228
virulence of the pathogen 229
Bacteriology 115
Bacteriophage 36, 183, 299
adsorption of 183, 184
characteristics of 183
in plant pathology 189
isolation of 189
lysis of 186
nucleic acid of 186
plaques 187
protein coat of 186
reproduction of 183
synthesis of 185
titre 188
Bacterium lactis aerogenes 297
Bait methods for isolation of soil-fungi 272
Barley mildew 354–360
aetiology 354
artificial inoculation 355, 356
biology of the fungus 354
control 360
cultures in test tube 356
detached leaf cultures 356, 357
differential sets for races 358–360
direct microscopic observation of 357
infection types 357
life cycle 355
mature plant resistance 360
physiologic races 357–360
pure-line cultures 356
resistant cultivars 360
yellow mosaic virus 62
Barriers to infection 305
Basic nutrient solutions for bacteria 147
Basidiomycetes 252–255, 257–261, 265, 267, 268, 366, 399
Basidiospore 242, 252, 366, 367
Basidium 248, 252, 253, 257, 366–368
— mother cell 257
Bean mosaic virus 48
— rust 397–399
artificial inoculation 397, 398
differential cultivars 398, 399
physiologic races 398, 399
rust grading scale 399
— yellow mosaic virus 48
Bentonite 37, 40
Benzidine-ascorbic acid test for peroxidase activity 340
Benzimidazole 356, 357, 361, 396
Benzyladenine 90
Berberis vulgaris 366, 368
Binomial 263
— nomenclature of plant viruses 101
Biochemical characteristics of bacteria 146
Biological activity of plant viruses 36
Biotype 293, 296, 300, 302, 307, 313
detection of 302
identification of 302
Bitunicate ascus 254
Blackleg of potato 221
Black loose smut 318, 413, 414
cultivar reactions 414
differential set 413, 414
physiologic races 414
Blend of
isogenic lines 314
susceptible and resistant plants 314
Blepharoplast 248

Blue mould of tobacco 340–348
artificial inoculation 343–347
breeding for resistance 347
control of 348
cotyledon test for resistance 345
leaf disc method for resistance 346
overwintering 342
quarantine regulations 344
resistant cultivars 347, 348
seed transmission 346
symptoms 343
Boiled antigen 171
Boletus sp. 254
Botrytis sp. 269
Bottom component of plant viruses 72
Brassica sp. 112
— *chinensis* 113
— *napobrassica* 326
— *oleracea* var. *capitata* 326
— *pekinensis* 327
— *rapa* 327
Breeding for disease resistance 289, 312, 313, 316, 317
Bromphenol blue 37
Bromthymol blue 147
Brush-extraction of plant viruses 50, 51
Brusone 433–437
Budding 47, 242
Buffer capacity of plant 194
Bunt of wheat 318, 400, 401, 403–406
artificial inoculation 403
differential set 405
L races 403–405
physiologic races 403–405
race groups 404, 405
spore germination test 406
teliospores of 402, 403
testing of fungicides to 406
T races 403–405
Bürker chamber 279

C

Caffeic acid 429
Calcium carbonate agar 167
Capsicum annuum 112
Capsule of bacteria 119
Carborundum 50, 51, 54
Carnation mosaic virus 53, 94
Carpenteles 268
Caryophyllaceae 53
Catechol method of peroxidase test 340
Celite 54
Cell division of fungi 260, 265
Cellulose cell wall 265
Cellulolytic enzymes 305
Cell wall degradation 305
of bacteria 119
Centrifugation of bacterial cells 202, 203
Cercospora sp. 269
Cercosporella sp. 318
Chelating agents 53
Chemical resistance 308
Chemotherapy 93, 94, 103
Chenopodium amaranticolor 106
— *hybridum* 108
— *quinoa* 108
Cherry rosette virus 62
— yellow virus 73
Chiastobasidium 257
Chinese cabbage 69, 70, 74
Chitin cell wall 265
Chlamidospore 246, 247
Chloramphenicol 90, 91
Chlorogenic acid 429
Chlorophyta 265
Chloroplast 69, 73, 74, 78, 79, 90
Chlorosis 83, 375
Chlorotic halo 198
Chylotrophic organisms 264
Chytridiales 249, 266, 267, 328
Cilia of bacteria 265
Citovirin 95
Citrulline 426
Clamp connection 257–259, 261
Class 264
Classification of
phytopathogenic bacteria 125, 126
viruses 102
Claviceps sp. 268
Cleome spinosa 113
Cleistothecium 353–355
Clone 292
Clover mosaic virus 55
— yellow mosaic virus 68
— wound tumour virus 51
Clubroot 318, 323–327
control of 327
disease index 326
resistant cultivars 327
symptoms 325
Coenocytic hypha 267
— mycelium 239, 260
Co-evolution of host and parasite 312
Collection of spores 288, 370
Collections of phytopathogenic bacteria 168
Collective order 264
Colletotrichum sp. 269
— *lagenarium* 277, 421–427

Colletotrichum
— *lindemuthianum* 287
— *orbiculare* 421–427
acervulus 421, 422
amino acids and susceptibility to 426
artificial inoculation 422, 423
biochemical mutants 425
biological specialization 424, 425
control of 426, 427
differential host 424, 425
inhibitors of 426
life cycle 422, 423
physiologic races 424, 425
resistance 425, 426
symptoms caused by 422
taxonomic position 421
Colonies of bacteria 122, 123
Colonization 305
Columella 248
Common antigens 172
Compatible host-parasite relation 154, 307
Complement fixation 79, 80
Congenial bacteria 193, 195
Conidiophore 243, 244
Conidium 242
germination 244
Constriction 242
Continuous cultures 162
— spore sampler 288
Control of bacterial disease 209
Coremium 247
Corn-meal agar 283
Corn smut 313, 314, 417–419
artificial inoculation 418
drop method of inoculation 418
field resistance 418
hypodermic injection 418
monosporidial lines 418
physiologic races 419
resistance 313, 314, 419
seedling inoculation 418
solopathogenic lines 418
Corynebacteriaceae 125
Corynebacterium sp. 120, 123, 125, 145
— *flaccumfaciens* 117, 163, 179, 184–186, 206, 214, 216, 220, 225–227
artificial inoculation test 227
identification of 226
inoculation with 226
isolation of 226
resistance to 227
symptoms caused by 225
— *insidiosum* 181, 198
— *michiganenrse* 129, 132, 181, 198, 295
— *sepedonicum* 180–182, 198
Cotton blue 277, 411
Covered smut 318
of barley 414, 415
artificial inoculation 414
differential cultivars 415
physiologic races 414, 415
m-Cresol 38
Cronartium sp. 268
— *occidentale* 278
— *ribicola* 278
Crop loss 320
Crosier hook 259
Crossing over 301
Cross-pollinated plants 313
Cross protection test 96, 97, 99, 100
Crown gall 129, 132, 230–233
artificial inoculation 233
isolation of the pathogen 232
symptoms 230
the pathogen 231
Cruciform cell division 324
Crystallization of viruses 24
C-terminal amino acids 29
Cucumber mosaic virus 35, 53, 55, 56, 59, 60, 83, 93, 95, 97, 108
Cucurbitaceae 426
Cultivar 291–293, 307, 314
of cross-pollinated plants 292
variability 314
Culture filtrates 202
Culturing of rust fungi 270
Cuscuta campestris 61
— *subinclusa* 61
Cyanide 89
Cyclohexylamine 33
Cyclone separator 370
Cysteine 53, 88, 89
Cytidine 29, 35
Cytidylic acid 29
Cytochrome 89
Cytokinins 90, 305, 396
Cytoplasm 66, 67, 69, 73, 74, 90
of bacteria 119
Cytoplasmic membrane of bacteria 119
— sterility 315
Cytosine 29, 295
Czapek's solution 285

D

Damping off 439, 440
Datura innoxia 112
— *stramonium* 108, 112
DEAE-cellulose chromatography 32

Deamination of nucleotides 296
Death phase of bacterial growth 122, 123
Deep-freezing of viruses 21, 22
Defence reaction 153, 154, 193, 195, 197, 200, 307–309, 320
Dehydrogenases 94
of *E. carotovora* 222
Demissum resistance to late blight 337, 339
Densitometer 162, 163
Density gradient centrifugation 22, 70, 72
– of bacterial suspension 161, 162
Detached leaf culture method
powdery mildews 356, 361
rusts 396
Detergents 38
Determination of
host range of bacteria 153, 156
inoculum cell number 161
living cell number 161
phage number 188
phytobacteria 142, 143
Detection of pathogens by phages 191
Destructiveness of disease 320
Deuteromycetes 268, 300, 421, 427, 433, 438
Dianthus caryophyllus 53
Dichotomously branching conidiophores 243
Didymella sp. 268
Die-back 123, 127, 129, 160, 161
of shoots 217–220
artificial inoculation 220
isolation of the pathogen 219
seasonal cycle 218
symptoms 218
Differential centrifugation 25
Differential sets 303, 304
– staining 137
Dikaryophase 248, 252, 253, 267
of rusts 366, 367
Dilution end point in virus serology 42
– loop 165, 188
– method for isolation of fungi 271
– plate 166
– row 164, 165, 187, 188
Diplococcus pneumoniae 298
Diplodia sp. 269
Dioecious fungi 259
Discomycetes 254, 255, 266, 268
Disease frequency 320
– resistance 193, 320
determination of 320
Disinfection of leaf surface 135
Disposition prophylaxis 103
Distortions 83
DNA (deoxyribonucleic acid) 66, 67, 90, 298, 299
DNA-dependent RNA synthesis 67, 68
DNase (deoxyribonuclease) 68
Dodder latent mosaic virus 43
transmission 61
Dodine-resistant *Fusarium* 440
Dolipore 261, 267
Donor host 50–53
Doralis fabae 59
Double cross hybrid 314
Double-stranded RNA 68, 69
Dowex-1 32
Downy mildew 270
of tobacco 340–348
Drop-diffusate technique for phytoalexin 309, 310
Drying of agar plates 139
Durham's tube 147, 148
Dwarf bunt of wheat 406–408
artificial inoculation 406, 407
differential set 407, 408
physiologic races 407, 408
resistance testing 406
teliospores of 407

E

Ectoparasitic fungi 240
Effective temperature 351
Electrochromatography 37
Electron microscopy 26, 82
– transport 89
Electrophoresis 23, 31
Elsinoe sp. 268
Enations 83
Encapsulation 305
End group determination of RNA 32
Endoconidia 243
3′-end of RNA 33, 35
5′-end of RNA 30
Endomycetales 253, 266, 268
Endoparasitic fungi 240
Endoplasmic reticulum 90
in fungus cells 261, 265
Enterobacteriaceae 125
Enzymes 66, 67, 70
causing soft rot 200
Enzymic digestion of RNA 31
Embryo 48
Epibasidium 257
Epidemics 314
artificially induced 207

Epidemiology 288
Epidermis 51, 56
Equilibrium centrifugation 22
Ergot of rye 319
Erwinia 123, 125, 127, 167, 222, 231
– *amylovora* 117, 119, 124, 127, 129, 194, 206
– *aroideae* 223
– *atroseptica* 223
– *carotovora* 117, 129, 200, 218, 221–223
dehydrogenase 222
differentiation of 222
inoculation with 224
isolation of 223
medium for 224
symptoms 221
– *chrysanthemi* 223
Erysiphales 268
Erysiphe sp. 268
– *betae* 364
– *graminis* 244
f. sp. *hordei* 354–360
aetiology 354, 355
artificial inoculation 355
biology 354, 355
control 360
cultures in test tube 356
detached leaf culture 356
differential cultivars 358–360
direct microscopic observation 357
infection types 357
life cycle 355
mature plant resistance 360
physiologic races 357–360
pure-line cultures 356
resistant cultivars 360
f. sp. *tritici* 361–362
– *horridula* 364
– *Martii* 364
– *polyphaga* 97
Escherichia coli 99, 297
Estimation of crop loss 320
Euascomycetidae 268, 349
Eubacteriales 125
Eucarpic reproduction 241, 242
Eumycetes 265
Eurotiales 266, 268
Eurotium sp. 268
Exobasidiomycetes 268
Exoconidia 243
Exonucleases 33
Eyespot 318

F

Facultative anaerobes 124
– parasitic fungi 270
– saprophytic fungi 270
– synonym 263
Factor-serum 179
Family 264
Fangsortiment 385
Ferricytochrome 89
Fertility restorer 315
Field assessment of rust resistance 371
– collection of rust 302, 375
– races 385, 386, 394
– resistance 317
to late blight 338–340
Filtration of bacterial cells 202, 203
– of viruses 26
– for sterilization 282
Fireblight of apple 127
Flagellation of bacteria 120, 143
Flagellum 242
fine structure of 265
– types 248
Flower inoculation 319
Fluid from intercellular spaces 204
– in the intercellular spaces 193
Foot rot 318
Forecasting 288
Formalin 58
Forma specialis (f. sp.) 291, 293
Fragmentation 242
Free-cell formation 242
Freeze drying of fungi 275
Fruiting bodies of fungi 253, 287
Fructification 287, 320
Fuchsin (acidic) 411
Fungal spore collector 288
Fungi 239
Fungi Imperfecti 268, 301, 421, 427
Fungitoxic substances 309
Fungus diseases (typical) 321
Fusaric acid 440
Fusarium oxysporum 300, 438–442
artificial inoculation 440, 441
Dodine-resistant strains 440
formae speciales 441, 442
isolation of 440, 441
control 442
life cycle 438, 439
polygalacturonase 440
population-estimation of 440
sporodochium of 438
races 441

Fusarium oxysporum
resistance to 316, 442
symptoms caused by 439, 440
toxins 439, 440
taxonomic position 437
— — f. sp. *batatas* 441
— — f. sp. *callistephi* 439, 441
— — f. sp. *conglutinans* 316, 441
— — f. sp. *cucumerinum* 441
— — f. sp. *dianthi* 441
— — f. sp. *lini* 441
— — f. sp. *lupini* 441
— — f. sp. *lycopersici* 316, 439–441
— — f. sp. *melonis* 440
— — f. sp. *niveum* 440, 441
— — f. sp. *pisi* 297, 298, 301, 316, 439, 441
— — f. sp. *tracheiphilum* 441
— — f. sp. *tuberosi* 439, 441
— — f. sp. *vasinfectum* 316, 441, 442
Fusarium wilt 437–442
artificial inoculation 440, 441
control of 442
Dodine-resistant strains 440
formae speciales 441, 442
life cycle of the fungus 438
polygalacturonase 440
population estimation 440
races 441
resistance 316, 442
symptoms 439, 440
toxins 439, 440
Fusicladium dendriticum 350
Fusion of bacteria 294

G

Gamate 249
Gametangium 249, 329
Gasteromycetes 253, 257, 268
Gelatin 138
— liquefaction 149
Gemma 247
Gene-for-gene theory 312
Genetic code 69
— information 65, 98
— polymorphism 312
— recombination 294, 368, 301
— variability 313
Genus 264
Geotrichum sp. 242
Germ tube 239, 242
Glass filter 202, 203, 282
— rod for streaking bacteria 166
— spatula in virus transmission 50
γ-Globulin 76
Gloeosporium 269
— *orbiculare* 421
Glomerella 268
— *cingulata* 412
Glucose 94
Glucosides (fluorescent) 305
Glutamine 426
Glutathione 88
Glycine 94
Gomphrena globosa 110
Gossypium barbadense 432
— *hirsutum* 432
Grading scale for bean rust 399
Graft transmission of viruses 47, 53
Gram stain 145
Grapevine fanleaf virus 63
Green-fluorescent pigments 123, 146
Growth substances 305
Guanidine 38
Guanine 29
Guanosine 29, 35
Guanylic acid 29
Gummoses 85

H

Haemacytometer 279
H-agglutination 171
Hair root 132
Half-leaf method 76
Half-life time of viruses 43, 44
Half series of differential set 302, 373, 375
Halo blight of beans 132, 214–217
differentiation of the four bean pathogens 216
symptoms 215
virulence of pathogens 217
Hanging-drop method 144
— preparation 276
Haploid cells 48
— recombination 301
Haustorium 239, 261, 262
Head smut 419, 420
artificial inoculation 419
differential cultivars 419, 420
monosporidial lines 419
physiologic races 419, 420
resistance 420
of corn 318
Heat denaturation of viruses 38
— killed bacteria 195
— therapy of virus diseases 93, 103
Heller type nutrient medium 330

Helminthosporium 269
— *sativum* 300
— *turcicum* 244
Hemiascomycetes 253
Hemiascomycetidae 268
Hemicellulose cell wall 265
Hemiptera 55
Hemolysis 79, 80
Hemolytic amboceptor 80
Hemp seed as a medium 283
Hermaphroditic fungi 259
Heterobasidiomycetes 266
Heterobasidiomycetidae 268
Heterobasidium 257
Heterogamy 249
Heterogeneity in cultivars 313
Heterogeneous sex cells 248
Heterokaryosis 259, 300, 301
Heterologous antiserum 77
Heterothallic nature of barley mildew 354
Heterothallism 259, 334
Heterothallism of rusts 366, 368
Heterotic yield effect 316
Heterotrophic organisms 124, 264
High pressure spraying 157, 159, 202
Histidine 94
Histopathology 85, 86
Holocarpia 328
Holocarpic reproduction 242
Holozoic organisms 264
Homobasidiomycetes 266
Homobasidiomycetidae 268
Homobasidium 257
Homogeneous sex cells 248
Homogeneity of virus preparations 24
Homogenizer 49
Homokaryotic organisms 300
Homologous antiserum 77
Homothallism 259
Honecker-races of barley mildew 357, 358
— *spontaneum nigrum* 358
Hordeum vulgare var. *distichon* 358
— — var. *hexastichon* 358
Host-parasite relation 307, 154, 155
gene-for-gene theory of 312
Host range of bacteria 126, 153
of viruses 62
Host plant
intraspecific subdivision of 293
proteins 49
Host specificity 193
H_2S production 150
Humidity 157, 158
Hybridization 68
Hybrid vigour 315
Hydroquinone 194
8-Hydroxyquinoline 38
Hymenium 253, 254
Hymenomycetes 257, 268
Hyosciamus mosaic virus 96
Hyperergic defence reaction 64, 308
Hypertrophy 197
caused by bacteria 123, 160, 161, 197
Hypersensitive reaction 64, 87, 88, 90, 91, 196, 197, 200, 308, 320, 371, 372, 386
induced by bacteria 126, 142, 153–156, 193, 195, 197, 200, 214, 217, 220
to loose smut 410
Hypersensitivity 332
Hypha 239
Hyphal growth 305
Hyphomycetales 269
Hyphomycetes 427
Hypobasidium 257
Hypodermic injection method 370, 371
Hypodermic needle 160

I

Identification of phytopathogenic bacteria 135, 142, 189, 190
Immune reaction 308, 309
— serum 171
— state 308
Immunity 308
Immunization 76, 77
of rabbits 172
Imperfect fungi 300
Inactivation of viruses 41
Incidence of attack 320, 371, 372
Inclusions in virus-infected plants 68, 85, 87, 94
Incompatible host 154, 157, 195, 200
— host-parasite relation 307
Incubation period 57
— resistance 320
Indicator bacterium 188, 190
— plants 41
Indoleacetic acid 64
Indole production 150
Induced defence reaction 308, 309
— immune reaction 308, 309
— resistance 193, 194, 308, 309
Infection
artificial 318
— feeding 55–57, 60
— modes and sites of 130
— process of fungi 305
— prophylaxis 103

Infection structures 305
— types 320
— of stem rust 373–384
Infectious sterility of hops 54
Infectivity of viruses 36, 37, 40, 75
Infiltration by vacuum 202
— methods 156, 202
Inhibitors of viruses 41, 44–46, 52, 53, 61, 74, 96
Injection 318
— infiltration method 142, 144, 159, 160, 202
— into the intercellular spaces 142, 144, 159, 160, 202
— with rust spores 370
— with smut spores 409, 418, 419
Ijuries and phytoalexins 309
Inoculating, loop 140
— needle 208
Inoculation 318, 319
in the field 208
in the seedling-bed 208
— methods 126, 155, 202
— tests 153, 161, 202
Inoculum 155–157, 161
buildup of 320
Insect transmission of viruses 55
Insecticides 49
Interaction between viruses 95
Interference 96, 183
Interferon-like substances 91, 92
Intermediate host 366, 368
Internal seed transmission 48
Intracellular inclusions 86
Intranuclear inclusions 86
Intraspecific subdivision
of host 293
of pathogen 293
Iodine fumes for staining of fungi 278
Ion exchange chromatography 23, 31
Ipomeamarone 309
Irradiation 41
Isocoumarin 309
Isoelectric point 22
Isogamy 249, 251
Isolation
of bacteriophages 189, 190
of fungi 270–272
of plaques 191
of bacteria 135, 140, 143
from leaves 136
from seeds 136
from vascular bundles 136
from woody tissues 136
of physiologic races 303
of soil fungi 285–286

K

Katznelson's method for detection of bacteria 192
Karyogamy 248, 252, 267
of rusts 367, 368
Kinetin 90, 330, 396
King's medium 146
Klement's injection infiltration method 142, 144, 159, 160
Koch's pot 281
Koch's postulate 135

L

Labile viruses 22, 51
Laboratory methods
in bacteriology 137
in mycology 276–280
Lactophenol 276, 277, 412
Lag phase of bacterial growth 122, 123
Late blight 332–340
aetiology 333
artificial inoculation 337
assessment of field resistance 338–340
demissum genes 335
— resistance 335–337, 339
differential set for races 335
field resistance to 336, 338–340
hypersensitivity to 337
major genes 335
minor gene resistance 336
peroxidase test for resistance 339, 340
resistance to 337–340
resistance genes 334
R genes 334–338
world testing ground of resistance 334
Latency 55–58, 64
Latent phages 299
— phase of bacterial growth 122, 123
Latin-square design 75
Leaf abscission 64
Leafhoppers 51, 55, 57–60
transmitted viruses 58
Leaf rot of *Phylodendron* 223
Leaf rust of wheat 368, 369, 387–393
differential set 387
physiologic races 387–393
UN races 387
supplemental differential set 392, 393
Leaf spot (bacterial) 123, 127, 132, 136
Leguminosae 112
Leonian's solution agar 287
Leptosphaeria 256, 268

Less-mutant 295
Less-strain 295
Liburnia striatella 58
Light scattering 26
Litmus milk 151
solution 152
Local epidemic 319
— lesions 45, 46, 54, 66, 75, 83, 87–93
— — test 74
Localized acquired resistance 89–92
Locules (in the stroma) 254
Loculoascomycetes 254
Log(arithmic) phase of bacterial growth 122, 123
Lomasome 259–261, 265
Loose smut 278, 319, 408–413
artificial inoculation 409, 410
detection of mycelium in embryo 410, 411
detection of mycelium in seedlings 412
differential sets 412, 413
embryo test 412
flower inoculation 409
haploid cultures 409
hypersensitive reaction 410
hypodermic needle for inoculation 409
physiologic races 409, 412, 413
promycelium of 409
resistance to 409, 410, 413
seedling inoculation 410
seedling test 412
stains for mycelium 411
teliospore germination 409
vacuum method for inoculation 409
Lophotrichous flagellation 120, 121, 123
Low pressure spraying 156
L races of wheat bunt 403–405
Lugol solution 146, 150
Lupine stem as a medium 283
Lycium halinifolium 112
Lyophilization 167, 372
of fungi 275
Lysine 34
Lysogenic bacteria 299

M

Macroconidia 243
Mahonia spp. 366
Maintenance,
of bacterial cultures 167
of fungi 273–265
MAK column chromatography 70
Malachite green 94
Malt-Czapek's agar 287
Malt-extract agar 284
Malus spp. 352
— *astrosanguinea* 353
— *baccata* 353
— *floribunda* 353
— *kaido* 353
— *prunifolia* 353
— *prunifolia-microcarpa* 353
— *pumila* var. *niedzwetzkyana* 353
— *sargenti* 353
— *sieboldii* 353
— *sylvestris* 353
Mapping procedure 31
Mating types
of *Phytophthora* 333
of rusts 366
Mature plant resistance 317, 360, 386
McFarland's scale 162
Mechanical transmission of viruses 49–51
Media
for fungi 281–287
pH of 138, 139
preparation of 138
sterilization of 139
Medulla 241
Melampsora 268
— *lini* 312
Melanconiales 269, 421
Meristematic tissues 49, 91
Merosporangium 248
Merthiolate 77
Merulius lacrymans 253, 258
Messenger RNA 66
Metabasidium 257
Methionine 94
metabolism of plants 198
Microautoradiography 67
Micrococcal nuclease 31
Microconidia 243
Microscopic examination of diseased plants 136, 137
— investigation methods of fungi 276–280
Microsomes 73
Microsphaera 268
Mildews 320
races of 304
Milk, as inhibitor of viruses 95
Mitochondria 69
of fungi 265
Mitomycin C 95
Mitotic crossovers 301, 428
Moist chamber 271, 319
Molecular sieving 22
Mollisia sp. 260

Momordica balsamina 54
Monilia 269
Moniliaceae 439
Moniliales 269, 427, 433, 438
Monoecious fungi 259
Monogenic inheritance of resistance 316
Monospore culture 292
Monosporidial lines of smuts 401
Monotrichous flagellation 120, 121, 123
Morchella sp. 254
Morphological resistance 308
Morphology,
of bacterial cell 143
of bacterial colony 146
of fungi 239
Mosaic mottling 83
— of rose 53
— symptoms 56
Motility of bacteria 120, 121, 123, 137, 144
Moulds 287
Mounting medium for fungi 276
M strain of TMV 31
Mucor 251, 268, 284, 287
Multigenic inheritance of resistance 317
Multiline cultivars 313, 314
— hybrid cultivars 314, 315
Multinucleated spore cell 292
Multiplication of viruses 65
Mutation 295, 296, 298
induced 295
in pathogenicity 295
of viruses 37, 98
Müller's drop-diffusate technique 309, 310
Myceliales 269
Mycelia sterilia 269
Mycelium 239
Mycobacterium avium 126
— *bovis* 126
— *tuberculosis* 126
Mycogone rosae 246
Mycological analysis of air 286
Mycology 235, 237
Mycosphaerella sp. 254, 268
Myxomycetes 265–267, 323
Myxomycetes-Plasmodiophorales line 265, 267
Myzus persicae 59, 60, 111

N

Naming of fungi 263
Naphtalene-acetic acid 64
Naramycin 95
Native preparation of fungi 276
Natural media 283
Necrogenic hypersensitivity 64
Necrosis 83, 375
Necrotic ring spot virus 73
Necrotization 87, 88
Necrotizing toxins 197
Nectria sp. 127, 268
Needle transmission
in brush 160
of viruses 52
Nematodes 63
Nessler reagent 149
Neurospora sitophila 95
Nicotiana debneyi 347, 348
— *exigua* 347, 348
— *glutinosa* 57, 64, 65, 74, 82, 83, 93, 106–108, 112, 334
— *goodspedii* 347, 348
— *ingulba* 348
— *knightiana* 347
— *megalosiphon* 347
— *plumbaginifolia* 347
— *suaveolens* 347
— *sylvestris* 106
— *tabacum* 88, 90, 91, 106, 112, 347
— *velutina* 347
Nicotine 54
— sulphate 54
Nitrate broth 148
— reduction 148
Nitrite mutants of viruses 98, 295, 296
Nitrous acid 98, 295
Noformycin 95
Nomenclature
bacterial 125, 126
botanical 263
viral 101
Non persistant viruses 55–57
Normal serum 77, 78
Normergic defence reaction 64, 308
— (normosensitive) reaction 308
Nonspecific acquired immunity 194
N-terminal amino acid 29
Nuclear-cap in zoospores 248
— DNA 68
— substance of bacteria 119
Nucleic acids (viral) 82, 94, 98, 295
Nucleolus 67–69
Nucleolytic enzymes 31
Nucleosides 29
Nucleotide composition of RNA 67
— sequence of RNA 30
Nucleotides 29, 49, 98
Nucleus 66–69, 73, 74, 86

Nutrient broth 138, 139
— requirements of pathogens 138
Nutritional mutants 194
— theory of resistance 193

O

O-agglutination 171
Oat dwarf disease 58
Oatmeal-agar 284
Obligate parasites 239, 270
— synonym 263
Ocular micrometer 278
Oidium 242
Oidium begoniae 364
Oil for preservation of fungi 274
Oligonucleotides 32
Olpidium 249, 267
— *brassicae* 63, 106
Onion yellow dwarf virus 56
Oogamy 250, 253
Oogonium 249, 250
Oomycetes 249, 250, 266, 267, 340
Oospore 250, 333
Open pollinated cultivars 314
Ophiobolus sp. 268, 318
Optical density 81
Optimal growth temperature for bacteria 124
— proportion method 81
Orchinol 309
Ordo 264
Organic acids 94
Ostiole 247, 254
Ouchterlony method in serology 176, 178, 181
Overdosage of inoculum 318
Overwintering of fungi 242
Oxidase reaction 152

P

Paper chromatography 31
Parasexuality 259, 300, 301, 428, 433, 434, 439
Parasitism 312
Paratrophs 124
Parenchyma 51, 56, 81, 82, 85
Parenthesome 261, 267
Partial sequences in RNA 30
Pathogen
  intraspecific subdivision of 293
Pathogenesis 305
Pathogenic variations 298
Pathogenicity 303, 312
Pathogenicity
  bacterial 126, 153
  stable 316
  test 142
Pea enation mosaic virus 83
Peach yellows 93
Pectate medium 224
Pectin-decomposing enzymes 127
Pectobacterium carotovorum 221, 222
  var. *aroideae* 222
— — var. *atroseptica* 222
— — var. *chrysaothemi* 222
Pectolytic enzymes 305
Penetration
  of bacteria 153
  of fungi into hosts 305
  of viruses into hosts 65
Penicillin 95
Penicillium 273, 285, 287, 300
— *expansum* 295
Pentachloronitrobenzene 63
Peptide bond 28
Peptone-glucose agar 285
Pepton water 147, 149
Peridium 254
Periodate oxidation of RNA 33
Periphysis 255
Perithecium 254, 255
Peritrichous flagellation 120, 123
Peronosporaceae 332
Peronosporales 344
Peronospora 267, 270
— *manchurica* 262
— *tabacina* 340–348
    artificial inoculation 343–347
    biotypes 347
    breeding for resistance 347
    control of 348
    cotyledon test for resistance 345
    germination of conidia 341
    inoculation material 344
    leaf disc method for resistance 346
    life cycle 340–342
    monospore cultures 344, 345
    oospore formation 341, 342
    overwintering 342
    seed transmission 346
    symptoms caused by 343
    taxonomic position 340
Peroxidase test for late blight resistance 339, 340
Persistent viruses 55, 57, 58
Petunia sp. 83
Phage DNA 299

Phage sensitivity test 192
Phaseollin 309
Phenol 36, 38, 39, 40, 77
Phenoloxidase 429
Phialide 243
Phloem 51, 57, 61, 64, 82, 85
pH of media 124, 138, 139
Phoma sp. 269
Phomales 269
Phosphate effect in virus inoculation 55
Phosphomonoesterase 33
Photoelectric measurements 162
Photosynthesis of infected plant 305
Phragmidium sp. 268
Phragmobasidium 257
Phycomycetes 249, 261, 265, 267, 323 328, 332, 340
Phycomycetes-Ascomycetes-Basidiomycetes line 261, 266, 267
Phylogeny of fungi 264–269
Phylum 264
Physalis floridana 60, 111, 112
Physiological sodium chloride 171
Physiologic races 291–293, 316, 320
Phytoalexin
  detection of 309
  drop-diffusate technique 309, 310
Phytopathogenic bacteria
  classification of 125, 126
  identification of 135
  isolation of 135
  symptoms caused by 127, 129, 130, 132, 153
Phytophthora 267, 272, 287, 342
— *infestans* 223, 295, 332, 340
    artificial inoculation 337
    assessment of field resistance 338
    compatibility types 334
    cultivation of 337
    demissum genes 335
    resistance 335, 336
    differential set 335
    heterothallism 334
    major genes 335
    mating types of 333, 334
    minor genes 335
    oospore formation 333
    peroxidase test for field resistance 339, 340
    physiologic races 334–336
    resistance genes to 334–338
    R genes 334–338
Phytotoxic glycopeptide 180, 198
— polysaccharides 198
Pierces' disease virus 82, 85
Pigments of bacterial colony 123, 124
Pinto beans 55, 89–91
Pipe cleaner 50
Piricularia oryzae 269, 433, 437
  artificial inoculation 434, 435
  differential sets 436
  life cycle 433, 434
  physiologic races 436
  resistance 435, 437
  symptoms caused by 434
  taxonomic position 433
  types of symptoms 435, 436
Pisatin 309
Pisum sp. 112
Planogamete 249, 329
Plaque 187
— count technique 187, 188
— types 191
Plasmatic defence reaction 64
Plasmodesmata 82, 239
Plasmodiophora brassicae 267, 318, 323–327
  artificial inoculation 325
  biological specialization 326
  differential cultivars 327
  disease index 326
  life cycle 323
  races 326, 327
  taxonomic position of 323
Plasmodiophorales 266
Plasmodiophoromycetes 323
Plasmodium 324, 325
Plasmogamy 248, 252, 267,
— of rusts 366, 367
Plasmopara sp. 250
Plate-count technique 161, 162, 165, 201, 279
— dilution technique 271
Plectenchyma 240
Plectomycetes 253
Pleospora 268
— *herbarum* 256
Pleosporales 350
Plexi-glass plate 166
PM_1 strain of TMV 19, 73
PM_2 strain of TMV 73
Podosphaera 268
— *leuchotricha* 362–365
    aetiology 362
    artificial inoculation 362–364
    control 364
    fungicides against 364
    inoculation tower 363
    life cycle 362
    physiologic races 364

— *leuchotricha*
primary infection phase 362
secondary infection phase 362
slide spore-germination 364
Pollen 48
Polyphenol 87–89, 305, 309
Polyphenoloxidase 53, 54, 87–89, 91, 429
Polyvirulent phage 189, 192
Population 269, 291, 292
of host 312, 313
of parasite 312, 313
Pores in septa 239
Powdered potato tuber medium 286
Powdery media proparations 138
— mildew 242, 270, 278, 318
apple 362–365
barley 354–360
wheat 361, 362
Potato, aucuba mosaic virus 59
— black ring virus 62
— dextrose-agar 138, 284
— disc method for evaluation of soft rot 224
— leaf roll virus 57, 60, 83, 93, 110, 111
- plug 283
— streak disease 111
— virus A 59
— — S 28
— — X 23, 30, 35, 52, 57, 79, 82, 96, 109
— — Y 52, 56, 59, 82, 93, 96, 111
wart 328–332
cultivars resistant to 332
symptoms of 330
— witches' broom 93
— yellow dwarf virus 59
Preacquisition fasting 56–58
Precipitation 172, 176, 177, 181
— band 178
— end point method 80
— reaction 77
— titre 172
Preformed resistance 193, 307
Premunity 96, 97, 193, 194, 308, 309
Preservation of bacterial cultures 167
— of fungi in soil 274
Pressure and temperature
correlation between 282
Pretreatment of plants for inoculation 156
195
Pricking of stems and shoots 160
Primary factors of pathogenesis 197, 200
— symptoms 83
Priority 263
Proliferation of tissue 230
Promycelium 257, 366, 368
Propagation of fungi 242
Propylene oxide for sterilization 282
Prosenchyma 240
Prosorus 329
Protein (viral) 28, 66, 70, 71, 98
contamination 36, 37
synthesis 48, 66, 89, 94
Protophyta 125
Protosterigma 257
Prototroph 295
Prune-juice agar 283
Prunus virus A 53
— — B 53
— — E 53
— — G 53
Pseudomonadaceae 125, 232
Pseudomonadales 125
Pseudomonas spp. 123–125, 147, 155, 167, 175, 176, 189, 195, 228
basal medium for, 147
— *cannabina* 131
— *fluorescens* 195
— *lachrymans* 121, 127, 129, 134, 157, 158
— *marginata* 134
— *mors-prunorum* 117, 127, 206, 217–220
as compared to *P. syringae* 219, 220
inoculaton with 220
isolation of 219
symptoms caused by 218
— *oryzicola* 122
— *panici* 131
— *phaseolicola* 117, 132, 133–135, 155 179, 214–217
biotypes of 217
differentiation 216
inoculation methods 217
symptoms caused by 214
virulence 217
— *pisi* 206
— *savastonii* 232
— *solanacearum* 117, 197, 198, 228, 229, 299
artificial inoculation 229
biotypes 229
isolation and identification 228
symptoms caused by 228
tetrazolium medium for 229
virulence 229
— *syringae* 92, 122, 127, 195, 206, 217–220
as compared to *P. mors-prunorum*, 219
inoculation with 220
isolation of 219
symptoms caused by 218
— *tabaci* 127, 154, 155, 194, 198
— *tonelliana* 232

Pseudoparaphysis 255
Pseudoparenchyma 240
Pseudoperithecium 254
Pseudomycelium 242
Puccinia 268
Pucciniaceae 291
Puccinia graminis 262, 291, 300, 301, 303, 366
— — f. sp. *tritici* 368–388
adult-plant resistance 371, 386
artificial inoculation 368
collection of spores 370
cultivar reactions 373–384
culturing *in vitro* 270
culturing on detached leaves 396, 397
differential cultivars 302, 303, 373, 375
field races 385, 386
hypersensitive reaction 371, 386
infection types 373–384
injection with spores 370
life cycle 367, 368
local epidemics 371
lyophilization 372
mature plant resistance 386
physiologic races 372–386
preservation of inoculum 372
reaction classes 373, 374
resistance 371
rust nurseries 371, 386
rust trial projects 385, 386
scale for susceptibility 371
seedling resistance 386
— *polysora* 313
— *recondita* 300, 301, 368, 369, 387–393
— — f. sp. *tritici* 387, 393
detached leaf culture 396
differential set 387
physiologic races 387–392
supplemental differential set 392, 393
UN races 387
— *sorghi*, 313
— *striiformis* f. sp. *tritici* 387, 394, 395
detached leaf culture 396
European races 394
field races 394
physiological races 394, 395
reduced differential set 394
Pure cultures
bacterial 140
fungal 271
— line 291
cultivars 313, 314
culture of powdery mildew 356
— stand 314
Purification of viruses 21
Purine bases 29
Pycnidiospore 247
Pycnidium 245, 247
Pycnium 366, 367
Pyrenochaeta terrestris 316
Pyrenomycetes 253–255, 266, 268
Pyrenophora 268
Pyrimidine bases 29
Pyrimidine-derivatives 94
Pythium sp. 127, 267, 286, 287, 342

Q

Quinones 88, 89

R

Races (physiologic) 296, 300, 313, 315
detection of 302
identification of 302, 303
Rapid serodiagnostic test
for bean diseases 179
for potato ring rot 180
Raspberry ringspot mosaic virus 62
Rate zonal centrifugation 22
Reaction types 303, 304
Reannealing 68
Receptive hypha 249, 366
Recombination 298–300, 368
sexual 294
Recombinant viruses 294
Reconstitution of viruses 37
Red pepper 48, 53, 64, 83, 108
Reducing agents 53
Re-infection 308
Replica plate technique 297
Replicative form of viruses 68, 69, 71
Reproductive organs of fungi 241, 242
Resistance 307–309, 312
breeding for 312, 313
if pathogens are stable 316
methods of determining 320
monogenic inheritance of 316
multigenic inheritance of 317
to antibiotics 297
to viral infection 64, 87, 89–91, 93
Resistant host plant 195
Respiration of infected plant 305
Respiratory processes 94
Resting sporangium 329
Rhizobiaceae 125, 232
Rhizobium japonicum 232
— *leguminosarum* 232
— *lupini* 232

Rhizobium meliloti 232
phaseoli 232
trifolii 232
Rhizoctonia sp. 269
Rhizoid 239
of *Rhizopus stolonifer*, 240
Rhizomorph 240, 242
Rhizoplast 248
Rhizopus 268
— *stolonifer*, 246, 247
Rhodophyta 265
Rhaeo discolor 66
Rhitisma sp. 268
Ribonuclease 31, 40, 65, 67, 68
— T_1, 31
Ribonucleic acid 19, 29, 30, 36–38, 65, 67–72, 89, 90, 92, 95
synthesis of 68, 69, 95
Ribosomal RNA 70
Ribosomes 27, 66, 70, 71, 74, 90
bacterial 119
fungal 265
Rice blast 433–437
artificial inoculation 434–436
differential series 436
life cycle of the fungus 433, 434
physiologic races 436
resistance 435, 437
symptoms 434
taxonomic position 433
types of symptoms 435, 436
— dwarf virus 71
— grain, as a medium 283
— stunt virus 58
Ring rot of potato 180
Ringspot 83
RNA-directed RNA synthesis 68
Rod-shaped viruses 27
Root exudates 305
Rosette formation of potato wart 329
Rossi-Cholodny "contact-slide" method 272
Rotting enzymes 197
Rough dwarf disease of maize 51, 52
Rust 257, 270, 295, 312, 314, 316, 318, 320, 366–399
buildup 314
diseases 366
nursery 371, 386
races 303, 304
resistance 313, 315
susceptibility in field 371
trial projects 386
Rye seed extract 286

S

Saccharomyces 268
Saltation 300
Salting-out 22, 24
Sand for preservation of fungi 274
Saprophytic bacteria 193, 195
Saprolegnia 283
Scab (apple) 349–354
— spot of tomato 127, 133
Scale for susceptibility to rust 371
Schiff's staining 277
Schizomycetes 119, 125
Sclerotinia 127, 268
— *fructicola* 309
— *sclerotiorum* 241
Sclerotiniaceae 284
Sclerotium 240, 242, 269
Seasonal cycle of the bacterial blossom blight 218
Secondary factors of pathogenesis 197, 200
— .symptoms, 83
Secretory bodies 305
— process 305
Sedimentation 26
Seedborne pathogens 277
— inoculation 318
— transmission of viruses 48
— treatment 209
Seedling resistance 316, 386
Segregation of nuclei 300
Seitz fritted-glass filter 282
Selection 298
— effect 313
— value 312
Self-pollinated plants 313
Self-sterile monoecic fungi 259
Semi-persistent viruses 55, 58
Semi-synthetic media 287
Senescence 90
Sensitive host plant 195
Separation of colonies 140
Sephadex for molecular sieving 22, 23
Septobasidium 257
Septoria sp. 269
Septum 239
— formation 260
— pore 260, 261, 267
Serological reaction 77, 80
Serology 170
detection of bacterial infection by 179
Serum 79
— dosage 174
— titre 173, 175

Sexual compatibility of fungi 259
— fungi 300
— recombination 294
— reproduction of bacteria 120
— — of fungi 248, 265
Sexuality 294
Shot holes on peach leaf 131
Sida micrantha 53
Single cell culture method 292
— — spore 292
— cross hybrid 315
— pustules of spores 292
— spore culture method 292
— — — of powdery mildew 356
Slide agglutination 174, 179, 180
Slime substance of bacteria 119, 127
Slow wilt of carnation 223
Smut 257, 295, 313, 318, 399–420
races of 304
diseases 399–420
aetiology, 399
artificial inoculation, 401
biology 399
dikaryophase, 399–400
diploid phase, 400
flower infection 401
haploid lines 401
haplophase 399
life cyle of smuts 400
local infection 402
monosporidial lines 401
physiologic races 401
seed infection 401
spore germination 401
Smuts of oats 415–417
artificial inoculation 416
cultivar reactions 417
differential sets 416, 417
life cycle 416
physiologic races 416, 417
resting mycelium 415, 416
vacuum method of inoculation 416
Snake venom phosphodiesterase 33, 35
Sodium cyanide 94
— propionate for elimination of bacteria 286
— pectate medium 224
— sulphite 53
Soft rot (bacterial) 123, 129, 156, 160, 161
— — causing enzymes 200
— — of potato tuber 221
inoculation 224
isolation of the pathogen 223
medium for the pathogen 224
resistance in tuber 222
Soft rot of patato tuber
symptoms 221
Soilborne pathogens 305
Soil fungi
isolation of 271, 272, 285, 286
— transmission of viruses 62
Solanaceae 330, 433
Solanum andigenum 336
— *demissum* 112, 332, 334, 335, 336, 338
— *longipedicellatum* 332
— *nudiflorum* 86
— *phureja* 336
— *stoloniferum* 335, 336
— *tuberosum* 334–336, 338
Soluble antigen 71
— RNA 70
Sorosporium 268
Southern bacterial wilt 228
— bean mosaic virus 24, 35, 48, 112
— cucumber mosaic virus 58
Source of infection 207
Sour cherry necrotic ringspot virus 54
Specialization
of host 293
of pathogen 291, 293
Species 264, 291–293, 307
Specific infectivity of viruses 24, 37
— phage 189
Spectrophotometry 81
Spermatium 249, 366
Spermogonium 366
Sphacelotheca reiliana 318, 401, 419, 420
— — var. *reiliana* 419
— — var. *zeae* 419
artificial inoculation 419
differential cultivars 419, 420
monosporidial lines 419
physiologic races 419, 420
resistance 420
Sphaerotheca 268
Sphaeropsidales 269
Spherical viruses 27
Spiculum 257
Spilocea pomi 350
Spleen phosphodiesterase 33
Split blade symptom 52
Spontaneous immune reaction 308, 309
Sporangiospore 247, 248
Sporangiolum 248
Sporangium 246, 247
— sorus, 328, 329
Spore
bacterial 119
fungal 242

Spore buidup, 314
— collector 288
— germination 305
— formatiɔn (fungi) 287
— trap 288, 314
Sporidium 242, 366, 367
Sporodochium 247
Spot diseases (bacterial) 132, 134, 156, 161
Spraying leaves with bacteria 156
Spreader rows of rust-susceptible cultivars 371
Spread of pathogenic bacteria 206, 207
Squash mosaic virus 73
Stable viruses 51
Staining
of bacteria in tissue 137
of cellulose cell wall 137
of flagella 143, 144
of lignified tissue 137
Star formation of *Agrobacterium* 120
Strach hydrolysis 150
— iodine lesions 84, 85
Starr's method for isolation of bacteria 224
Stationary phase of bacterial growth 122, 123
Stellaria media 53
Stem necrosis 83
Stem rust
races 302, 303
of wheat 368–388
adult-plant resistance 386
artificial inoculation 368
collection of spores 370
cultivar reactions 373–384
culturing *in vitro* 270
— in detached leaves 396–397
differential cultivars 302, 303, 373, 375
field races 385, 386
hypersensitive reaction 371, 386
infection types 373–384
injection with spore suspension 370, 371
life cycle 367, 368
local epidemic 371
lyophilization 372
mature plant resistance 386
physiologic races 372, 386
preservation of inoculum 372
reaction classes 373, 374
resistance to 371, 386
rust-nurseries 371, 386
— trial project 385, 386
— scale for susceptibility to 371
— seedling resistance 386
— supplemental differentials 385
Stepwise degradation of RNA 30, 32
Sterigma 243, 252, 257
Sterile water for preservation of fungi 274
Sterilization 137–139, 148, 281, 282
media 281
thermolabile substances 282
Stichobasidium 257
Stimuli for spore germination 305
Stipe 255
Stock culture
degeneration of 274
Stolbur virus 57
Stoughton's method of differential staining 137
Strains of viruses 32, 96, 97
Streak method for isolation of fungi 271
Streaming steam for sterilization 281
Streptomyces spp. 125, 260
Streptomycetaceae 125
Streptomycin 95
— resistant indicator bacterium 190
— spraying 209, 210
Stripe rust of wheat 394–396
detached leaf culture 396
European races 394
field races 394
physiological races 394, 395
reduced differential set 394
Striped spot of millet 132
Stroma 241, 242, 254
Structure of viruses 21, 28
Subclass 264
Subdivision of races 385
Subraces 385
Subfamily 264
Subfibrils of flagellum 248
Subicle (subiculum) 247
Suborder 264
Subphylum 264
Substructure of viruses 35
Sugar-beet curley top virus 51, 58, 61, 85, 97
— mosaic virus 59, 85
— yellows virus 26, 58, 83
Super-infection 308
Supplemental differentia cultivars for rust races 385, 392, 393
Surface antigens of bacteria 120
— seed transmission of viruses, 48
Survival capacity 316
Suscept 50, 51
Susceptibility 307, 320
to viral infection 46, 55, 64, 91–93, 95
Swarm cells 248
— spores 242, 329
Symbiotic phage carriers 299

Symptomatology 83
Symptoms caused by bacteria 127, 129, 130, 132, 161
Synchytrium endobioticum 267, 328–332
  artificial inoculation 330, 331
  biological specialization 331
  life cycle 328
  physiologic races 331
  resistance to 332
  taxonomic position 328
Syncytium 300
Synnema 247
Synonym 263
Synthetic acid medium 285
— cultivars 314
— media 285
— nutrient solution 147, 286
Systematic mycology 263
— acquired resistance 91, 92

T

Takátsy's dilution loop 165, 188
Take-all disease 318
Tannins 54
Taphrina sp., 268
Taphrinales 253, 266, 268
Taxonomic categories 264
Taxonomy of viruses 101
Teleutospore 257
Teliospore 268, 366–368
Telosporae 257
Temperature and pressure, correlation between 282
— requirements of bacteria 124
Template 67
Terramycin 95, 209
Tetragonia expansa 108
Tetramethyl-thiuram-disulphide 63
Tetrazolium medium 229
Thallus 242
Thermal death point of bacteria 124
Thermolabile antigens 175
Thermostabile antigens 175
Therapy of virus diseases 103
Thermal inactivation coefficient 42, 43
Thielaviopsis basicola 243, 245
Thiouracil 94
Tilletia caries 318, 400–406
  artificial inoculation 403
  differential set 405
  L races 403–405
  physiologic races 403–405
Tilletia caries
  race groups of 404, 405
  spore germination test 406
  testing of fungicides to 406
  teliospores of 402
  T races 403–405
— *controversa* 318, 400, 401, 406–408
  artificial inoculation 406, 407
  differential set 407, 408
  physiologic races 407, 408
  resistance testing 406
  teliospore of 407
— *foetida* 318, 400, 401–403
  artificial inoculation 403
  differential set 405
  L races 403–405
  physiologic races 403, 405
  race groups of 404, 405
  spore germination test 406
  teliospore of 403
  testing of fungicides to 406
  T races 403–405
Tirotrycin 95
Tinsel flagellum 248, 265
Tissue proliferation 127, 136
Titre of phage suspension 188
TMV (tobacco mosaic virus) 295
Tobacco etch virus 86, 96
Tobacco, mosaic virus 23–25, 29, 30, 35, 36, 39, 40, 48, 52, 55, 57, 62, 64–66, 69–74, 82–87, 89–95, 97, 98, 101, 104, 107, 295
— — — protein 73
— — — ribonucleic acid 31, 34, 35, 37, 39, 40, 65–68, 70, 71, 94, 98
— necrosis virus 24, 30, 55, 62, 63, 66, 73, 106
— rattle virus 19, 73
— ringspot virus 24, 30, 48, 64, 83, 95
— rosette disease 85
Tolerance 308
  to virus diseases 64
Tomato, aspermy virus 107, 108
— aucuba virus 86
— black ring virus 62
— bushy stunt virus 24, 35, 83, 92
— ringspot virus 91
— spotted wilt virus 51, 83
Top component of viruses 72
Toxin 127, 180, 181, 306
  bacterial 127, 180, 181
  extracted from plant tissue 182
  purification 181
Traces of wheat bunt 403–405
Trachea 61

Tracheomycosis 439
Transfer RNA 31, 66
Transformation 298
Translocation of viruses 65, 81
Transmission of viruses 47
Tremellales 257, 268
Trichothecium roseum 95
Trinomials 263
Tripan blue 278, 411, 412
Triphenyl tetrazolium chloride (TTC) 330
Triplet code 70, 99
Triticum aestivum 372, 413
— *carthlicum* 386
— *compactum* 372
— *dicoccum* 373, 386
— *durum* 372, 373, 386
— *monococcum* 373
— *timopheevi* 386, 413
Trychogyne 249
Trytophane broth 150
Tube agglutination 174, 175, 176
Tuberculariaceae 438
Tulip mosaic virus 83
Turbidimetric measurement 162
Turnip crinkle virus 32
— mosaic virus 91
— yellow mosaic virus 24, 30–32, 69, 70, 72, 74, 113
Typical fungus diseases 321

U

Ultracentrifugation 22
Ultrasonic treatment 41
Ultrastructure of fungi 259
Ultraviolet, irradiation 57, 66
— light 65
— microscopy 67
Uncongenial bacteria 193, 195, 196
Uncinula sp. 268
Unified number of races of leaf rust 387–392
UN races of leaf rust 387–392
Uracil 29, 94, 295
Urea 32
Uredial generation 368
Uredinales 253, 257, 268
Urediospores 366, 367, 368
Uridine 29, 35
Uridylic acid 29
Uromyces phaseoli 97, 397–399
  artificial inoculation 397–398
  differential set 398–399
  physiologica races 398–399
  rust grading scale 399
Ustilaginales 253, 257, 268
Ustilago avenae 400–402, 415–417
  artificial inoculation 416
  cultivar reactions 417
  differential set 417
  life cycle 416
  physiologic races 416–417
  resting mycelium 415–416
  vacuum method of inoculation 416
— *hordei* 318, 400, 401, 414–415
  artificial inoculation 414
  differential set 415
  physiologic races 414, 415
— *kolleri* 318, 400–402, 415–417
  artificial inoculation 416
  cultivar reactions 417
  differential set 417
  physiologic races 416, 417
— *nigra* 318, 400, 401, 413–414
  cultivar reactions 414
  differential set 413–414
  physiologic races 414
— *nuda* 278, 400–402, 408–413
  artificial inoculation 409, 410
  detection of mycelium in embryo 410, 411
    in seedlings 412
  differential sets 412, 413
  embryo test 412
  flower inoculation 409
  haploid cultures 409
  hypersensitive reaction 410
  hypodermic needle for inoculation 409
  physiologic races 409, 412–413
  promycelium of 409
  resistance 409, 410, 413
  seedling inoculation 410, 412
  stain for mycelium 411
  teliospore germination 409
  vacuum method for inoculation 409
— *tritici* 400, 401, 408–413
  field resistance 411
  physiologic races 412, 413
  teliospore germination 409
— *zeae* 301, 399–402, 417–419
  artificial inoculation 418
  drop method of inoculation 418
  field resistance 419
  hypodermic injection 418
  monosporidial lines 418
  physiologic races 419
  resistance 313, 314, 419
  seedling inoculation 418
  solopathogenic lines 418

V

Vacuum collector for spores 370
Variability 293
genetic 313
in cross-pollinated cultivars 313
in self-pollinated cultivars 313, 314
of pathogens 291
Variation 302
Varietas 293
Variety (varietas) 293, 307
Vascular diseases (bacterial) 123, 128, 132, 136, 197, 208
— wilt 440
Vector 55, 57, 58
— specificity 59
Vegetative mycelium 239
Vein banding 83
— clearing 83, 104
Venturia inaequalis 268, 349–354
artificial inoculation 352
ascus of 349
control of 354
differential hosts for races 353
life cycle 350, 351
preparation of inoculum 352
races of 353
rating scale for symptoms 352
resistant cultivars 353–354
symptoms 351, 352
taxonomic position 349, 350
Venturiaceae 350
Verticillate conidiophores 243
Verticillium albo-atrum 301, 318, 427–433
artificial inoculation 431, 432
conidial formation 428
conidiophores 427
control of 432
differentiation of *Verticillium* species. 430
life cycle 427, 428
melanin production 429
mitotic crossover 428
parasexual process 428
peroxidase 429
phenoloxidase 429
polygalacturonase 429
root dipping method 432
root infection 431
saprophytic survival 428
seedling toxin method 431
selective medium for differentiation of 430
symptoms 428, 429
taxonomic position 427
temperature requirements 431
Vericillium albo-atrum
toxins of 428, 429
utilization of carbon sources 428
— *dahliae* 427–433
— *nigrescens* 428, 430, 431
— *nubilum* 430
— *tricorpus* 428, 430
— wilt 427–433
Vesicle 243
Viability of bacteria 167
Vicia faba 68, 112
Vigna sinensis 53, 55, 108
Viral protein 66
— RNA 66
Virology 17
Virulence 296, 303, 307, 312, 313, 316
Virulent bacteria 195
Virus aggregates 75
— concentration 52, 74
— multiplication 95
— yields 21
Voges-Proskauer reaction 152

W

Warcup's soil plate method 272
Wart of potato 319
Washing of bacterial cells 202, 204
Water agar 287
Wheat mildew 361–362
— mosaic virus 62
— yellow mosaic virus 62
Whiplash flagellum 248, 265, 328
Wild cucumber mosaic virus 32, 73
Wildfire disease 127, 194, 198, 199, 207, 209
Wilt diseases 123, 127, 128, 132, 136, 160, 161, 180, 197, 318, 427, 440
— toxin 180
Wire loop 140
Wood destroying fungi 284
Wound tumour virus 35, 71, 85
Wounding 88, 89

X

Xanthomonas 120, 123–125, 131, 167, 176, 189
— *carotae* 121
— *juglandis* 295
— *malvacearum* 117, 133, 206
— *phaseoli* 117, 132, 135, 179, 214, 216, 298
— — var. *fuscans* 135, 167, 179, 187, 214, 216

Xanthomonas
— *pruni* 131, 134
— *stewartii* 295
— *vesicatoria* 117, 127, 133, 151, 167, 173, 177, 211–214, 216
artificial inoculation 214
biotypes 213
isolation of 213
resistance to 214
symptoms 211
X-bodies 86
X-disease 94
X-protein 71, 72
X-ray diffraction 35
Xylem 82, 85
Xyphinema index 63

Y

Yeast extract-Czapek agar 287
Yellow type symptoms 57
Yellows 94

Z

Zinc chloride 94
— sulphate 94
Zinnia elegans 95, 108
Zone of apposition 305
Zoospore 248
Zoosporangium 248, 324
Zygogamy 251, 253
Zygomycetes 251, 253, 266, 268
Zygospore 251